Lorenzo Magnani, Walter Carnielli, and Claudio Pizzi (Eds.)

Model-Based Reasoning in Science and Technology

T0140473

Studies in Computational Intelligence, Volume 314

Editor-in-Chief

Prof. Janusz Kacprzyk
Systems Research Institute
Polish Academy of Sciences
ul. Newelska 6
01-447 Warsaw
Poland
E-mail: kacprzyk@ibspan.waw.pl

Lorenzo Magnani, Walter Carnielli, and
Claudio Pizzi (Eds.)

Model-Based Reasoning in Science and Technology

Abduction, Logic, and Computational Discovery

 Springer

Lorenzo Magnani
Dipartimento di Filosofia
Università di Pavia
Piazza Botta 6
27100 Pavia
Italy
E-mail: lmagnani@unipv.it

Walter Carnielli
Centre for Logic
Epistemology and the History of Science
CLE State University of Campinas
UNICAMP P.O. Box 6133
13083-970 Campinas -SP
Brazil
E-mail: carniell@cle.unicamp.br

Claudio Pizzi
Dipartimento di filosofia e
scienze Sociali
Università di Siena
Via Roma 47
53100 Siena
Italy
E-mail: pizzic@msn.com

ISBN 978-3-642-26467-2

ISBN 978-3-642-15223-8 (eBook)

DOI 10.1007/978-3-642-15223-8

Studies in Computational Intelligence

ISSN 1860-949X

Typeset & *Cover Design:* Scientific Publishing Services Pvt. Ltd., Chennai, India.

Printed on acid-free paper

9 8 7 6 5 4 3 2 1

springer.com

The psychologists undertake to locate various mental powers in the brain; and above all consider it as quite certain that the faculty of language resides in a certain lobe; but I believe it comes decidedly nearer the truth (though not really true) that language resides in the tongue. In my opinion, it is much more true that the thoughts of a living writer are in any printed copy of his book than that they are in his brain.

Charles Sanders Peirce

Preface

This volume is a collection of selected papers that were presented at the international conference *Model-Based Reasoning in Science and Technology. Abduction, Logic, and Computational Discovery* (MBR09_BRAZIL), held at the University of Campinas, Campinas, Brazil in December 2009.

A previous volume, *Model-Based Reasoning in Scientific Discovery*, edited by L. Magnani, N.J. Nersessian, and P. Thagard (Kluwer Academic/Plenum Publishers, New York, 1999; Chinese edition, China Science and Technology Press, Beijing, 2000), was based on the papers presented at the first "model-based reasoning" international conference, held at the University of Pavia, Pavia, Italy in December 1998. Other two volumes were based on the papers presented at the second "model-based reasoning" international conference, held at the same place in May 2001: *Model-Based Reasoning. Scientific Discovery, Technological Innovation, Values*, edited by L. Magnani and N.J. Nersessian (Kluwer Academic/Plenum Publishers, New York, 2002) and *Logical and Computational Aspects of Model-Based Reasoning*, edited by L. Magnani, N.J. Nersessian, and C. Pizzi (Kluwer Academic, Dordrecht, 2002). Another volume *Model-Based Reasoning in Science and Engineering*, edited by L. Magnani (College Publications, London, 2006), was based on the papers presented at the third "model-based reasoning" international conference, held at the same place in December 2004. Finally, volume *Model-Based Reasoning in Science and Medicine*, edited by L. Magnani and L. Ping (Springer, Heidelberg/Berlin 2006), was based on the papers presented at the fourth "model-based" reasoning conference, held at Sun Yat-sen University, Guangzhou, P. R. China.

The presentations given at the Campinas conference explored how scientific thinking uses models and explanatory reasoning to produce creative changes in theories and concepts. Some speakers addressed the problem of model-based reasoning in technology, and stressed the issue of science and technological innovation. The study of diagnostic, visual, spatial, analogical, and temporal reasoning has demonstrated that there are many ways of performing intelligent and creative reasoning that cannot be described with the help only of traditional notions of reasoning such as classical logic. Understanding the contribution of modeling practices to discovery and conceptual change in science requires expanding scientific reasoning to include complex forms of creative reasoning that are not always successful and can lead to incorrect solutions. The study of these heuristic ways of reasoning is situated at the crossroads of philosophy, artificial intelligence, cognitive psychology, and logic; that is, at the heart of cognitive science. There are several key ingredients common to the various forms of model-based reasoning. The term "model" comprises both internal and external representations. The models are intended as interpretations of target physical systems, processes, phenomena, or situations. The models are retrieved or constructed on the basis of potentially satisfying salient constraints of the target domain. Moreover, in the modeling process, various forms of abstraction are used. Evaluation and adaptation take place in light of structural, causal, and/or functional constraints. Model simulation can be used to produce new states and enable evaluation of behaviors and other factors.

The various contributions of the book are written by interdisciplinary researchers who are active in the area of modeling reasoning and creative reasoning in logic, science and technology: the most recent results and achievements about the topics above are illustrated in detail in the papers.

The editors express their appreciation to the members of the Scientific Committee for their suggestions and assistance: Atocha Aliseda, Instituto de Investigaciones Filosoficas, Universidad Nacional Autonoma de Mexico (UNAM) – Silvana Borutti, Department of Philosophy, University of Pavia, ITALY – Eduardo Bustos, Department of Logic, History and Philosophy of Science at UNED (Spanish Open University), Madrid, SPAIN – Marcelo Esteban Coniglio, Department of Philosophy, CLE and Institute of Philosophy and Human Sciences, State University of Campinas, BRAZIL – Itala D'Ottaviano, CLE and Department of Philosophy, Institute of Philosophy and Human Sciences, State University of Campinas, BRAZIL – Roberto Cordeschi, Department of Philosophy, "La Sapienza" University of Rome, ITALY – Roberto Feltrero, Department of Logic, History and Philosophy of Science at UNED (Spanish Open University), Madrid, SPAIN – Marcello Frixione, Department of Communication Sciences, University of Salerno, ITALY – Michel Ghins, Institut Supérieur de Philosophie, Université de Louvain, BELGIUM - David Gooding, Department of Psychology, University of Bath, UK – Mike E. Gorman, Technology, Culture & Communications, SEAS University of Virginia, USA – Marcello Guarini, Department of Philosophy, University of Windsor, CANADA – Ricardo Gudwin, Department of Computer Engineering and Industrial Automation, the School of Electrical Engineering and Computer Science, State University of Campinas, BRAZIL – Viorel Guliciuc, Stefan cel Mare University, Suceava, ROMANIA – Michael Leyton, Psychology Department, and DIMACS Center for Discrete Mathematics, & Theoretical Computer Science, Rutgers University, USA – Mamede Lima-Marques, Department of Information Science and

Documentation, University of Brasília, BRAZIL – Angelo Loula, Department of Exact Sciences, State University of Feira de Santana, BRAZIL – Shang-min Luan, Institute of Software, The Chinese Academy of Sciences, Beijing, P.R. CHINA – Cezar Augusto Mortari, Federal University of Santa Catarina, UFSC Florianópolis, Brazil – Alison Pease, Edinburgh University, Edinburgh, UK – João Queiroz, Department of Computer Engineering and Industrial Automation (DCA) School of Electrical Engineering and Computer Science, State University of Campinas, BRAZIL – Lucia Santaella, Center of Research in Digital Media, São Paulo Catholic University (PUCSP), BRAZIL – Gerhard Schurz, Institute for Philosophy, Heinrich-Heine University, GERMANY – Cameron Shelley, Department of Philosophy, University of Waterloo, Waterloo, CANADA - Colin Schmidt, Institut d'Informatique Claude Chappe, University of Le Mans, FRANCE – Frank Thomas Sautter, Department of Philosophy, University of Santa Maria, BRAZIL – Paul Thagard, Director of the Cognitive Science Program, and University Research Chair at the University of Waterloo, CANADA – Barbara Tversky, Department of Psychology, Stanford University, USA – Jon Williamson, Philosophy, School of European Culture and Languages, University of Kent, UK – Riccardo Viale, LaSCo-Laboratory of Cognitive and Complexity Sciences, Fondazione Rosselli, Torino, ITALY – John Woods, Department of Philosophy, University of British Columbia, CANADA – Woosuk Park, Humanities and Social Sciences, KAIST, Guseong-dong, Yuseong-gu Daejeon, SOUTH KOREA, and to the members of the local scientific committees: Itala D'Ottaviano, Alexandre Costa Leite, and Juliana Bueno-Soler (State University of Campinas UNICAMP), Emanuele Bardone, Tommaso Bertolotti, and Pino Capuano (University of Pavia).

Special thanks to Emanuele Bardone, Riccardo Dossena, and Tommaso Bertolotti for their contribution in the preparation of this volume. The conference MBR09_BRAZIL, and thus indirectly this book, was made possible through the generous financial support of MIUR (Italian Ministry of the University), University of Pavia, Fondazione CARIPLO, FAPESP- São Paulo Research Foundation, the Centre for Logic, Epistemology and the History of Science CLE of the State University of Campinas UNICAMP and the Brazilian Logic Society. Their support is gratefully acknowledged. The preparation of the volume would not have been possible without the contribution of resources and facilities of the Computational Philosophy Laboratory and of the Department of Philosophy, University of Pavia.

Several papers concerning model-based reasoning deriving from the previous conferences MBR98 and MBR01 can be found in Special Issues of Journals: in *Philosophica*: Abduction and Scientific Discovery, 61(1), 1998, and Analogy and Mental Modeling in Scientific Discovery, 61(2) 1998; in *Foundations of Science*: Model-Based Reasoning in Science: Learning and Discovery, 5(2) 2000, all edited by L. Magnani, N.J. Nersessian, and P. Thagard; in *Foundations of Science*: Abductive Reasoning in Science, 9, 2004, and Model-Based Reasoning: Visual, Analogical, Simulative, 10, 2005; in *Mind and Society*: Scientific Discovery: Model-Based Reasoning, 5(3), 2002, and Commonsense and Scientific Reasoning, 4(2), 2001, all edited by L. Magnani and N.J. Nersessian. Finally, other related philosophical, epistemological, and cognitive oriented papers deriving from the presentations given at the conference MBR04 have been published in a Special Issue of *Logic Journal of the IGPL*: Abduction, Practical Reasoning, and Creative Inferences in Science 14(1) (2006) and have been published in two Special Issues of *Foundations of Science*: Tracking Irrational Sets: Science, Technology, Ethics, and Model-Based Reasoning in Science and Engineering, all edited by L. Magnani.

Other more technical logical papers presented at (MBR09_BRAZIL) will be published in a special issue of the *Logic Journal of the IGPL*, edited by L. Magnani, W. Carnielli, and C. Pizzi.

Finally, the present book also includes a brief paper that two of the editors, Walter Carnielli and Lorenzo Magnani, have devoted to the 65th birthday of Claudio Pizzi.

Pavia, Italy
June 2010   Lorenzo Magnani
                                          University of Pavia, Pavia, Italy and
                            Sun Yat-sen University, Guangzhou, P.R. China

                                                               Walter Carnielli
                                    University of Campinas, Campinas, Brazil

                                                                  Claudio Pizzi
                                         University of Siena, Siena, Italy

Years of Reasoning
In Honor of the 65th Birthday of Claudio Pizzi

Walter Carnielli and Lorenzo Magnani

Abstract. This paper is devoted to Claudio Pizzi on the occasion of his 65th birthday. Looking at the essential bibliography reported below, Pizzi's reserved character is manifest in the few co-authors he had in his publications. However, it is hard to find among philosophers and logicians a colleague so unanimously recognized as gentlemanly in his attitude towards science and academic life. The blend of rigor and philosophical scope that characterizes Pizzi's work is, at the same time, analytic philosophy and philosophical logic as its best, as it is demonstrated in his intellectual career. Claudio Pizzi is also a fine cooker, and one of the best *connoisseurs* of Brazilian music we know.

1 Four Decades of Conditionals and Rational Inference

Claudio Pizzi obtained his degree in Philosophy In March 1969 at the State University of Milan, with a dissertation about subjunctive and counterfactual conditionals (titled *I Condizionali Congiuntivi: Aspetti Epistemologici e Problemi di Formalizzazione*).

This subject was considered quite abstruse in Italy in those years, and it is relevant to have in sight that the best accepted and most elaborated counterfactual theory of causation, represented by the work of David Lewis, would only see the light in 1973. The epistemological panorama was still dom-

Walter Carnielli
GTAL/CLE and Department of Philosophy–IFCH, State University of Campinas, Brazil
e-mail: `walter.carnielli@cle.unicamp.br`

Lorenzo Magnani
Department of Philosophy, University of Pavia, Pavia, Italy and Sun Yat-sen University, Guangzhou, P.R. China
e-mail: `lmagnani@unipv.it`

inated by logical positivism, and the only logic which in Italy was considered a reverent object of study was mathematical logic. Modal logic was almost completely ignored or considered an oddity, to be avoided even more than many-valued logic.

The attitude towards modal logic in Italy began to change after the publication of the Italian translation of Hughes and Cresswell's *Introduction to Modal Logic* (Il Saggiatore, Milan, 1973). Pizzi, as he says, had the luck to be charged with the translation and the editing of the book (for which he also wrote an introduction); the work was a good opportunity for him to get well acquainted with the basic elements of possible-worlds semantics.

The volume [3] was, at that time, an updated collection of papers on the topic of counterfactual reasoning and physical modalities. In preparing the long introduction for this book Pizzi developed an impression that the "consequentialist" view about conditionals (Chisholm-Goodman-Reichenbach) and the attempts to give it a formalization (especially in Angell and McCall's connexive logic) had been too hastily dismissed. [2] proposes to recover the basic idea of connexive logic by: (a) treating it as an extension of standard logic, and (b) introducing a distinction between an analytical and synthetical (context-dependent) variant of it *via* a simple axiomatization of Åqvist's circumstantial operator $*$. What is called *Boethius' Thesis* in the synthetical variant had the form $(*A \rightarrow B) \supset \neg(*A \rightarrow \neg B)$. Unfortunately, the system introduced in the paper could be accepted only as a limit case since the trivializing equivalence $p \equiv *p$ may be proved. What this family of systems needed was a decision procedures, and Pizzi tried to provide it in [8] and [9]. With respect to the 1977 paper, the analytical fragment of the basic system of [8], named **CI.0**, has the important difference of excluding the Factor Law $(A \wedge B) \supset (A \wedge C \rightarrow B \wedge C)$, which may be accepted only in a weakened variant. For this reason the logics belonging to this family cannot be qualified as connexive anymore, and need a different classification: Pizzi called them *logics of consequential implication*. The key idea of the decision procedure is very simple: every formula of form $A \rightarrow B$ is translated into a strict implication $A \dashv B$ conjoined with the assertion that A and B have the same modal status, i.e., the same position in Aristotle's square of modalities; the resulting translation is then tested with the tableaux methods used in standard modal logic. It turns out then that **CI.0** is definitionally equivalent to the modal system **KT**, and weaker consequential systems are equivalent to weaker systems of modal logics.

The paper [17] is centered on the idea that consequential logics grasp in a special sense the relevance of A to B; in fact, in order of $A \rightarrow B$ being true, A and B have actually something in common, namely their modal status. Implication relations weaker than $\rightarrow$ may be represented as "truncations" of analytical consequential implication preserving basic properties such as Aristotle's Thesis (namely $\neg(A \rightarrow \neg A)$). The interrelations between such different kinds of consequential implication may be visualized into three-dimensional pictures named "Aristotle's cubes" (see [23]). According to the

Italian historian Mauro Nasti de Vincentis consequential implication grasps some basic intuitions about conditionals that can be found in Aristotle and Chrysippus.

In two papers written in collaboration with T. Williamson [12, 18] two pathological extensions of systems of consequential implication are analyzed: the former by means of the Strong Boethius' Thesis $(A \rightarrow B) \rightarrow \neg(A \rightarrow \neg B)$ and the latter by means of the "Conditional Excluded Middle" $(A \rightarrow B) \vee (A \rightarrow \neg B)$. The first paper introduces a general schema of one-one translation between consequential-implication logics and normal modal logics, and it may be considered as providing the most advanced theoretical framework for any research on consequential implication.

Analytical consequential implication has been fundamentally studied at the propositional level only. A first attempt of studying it at first order level appears in [25], where the possibility of translating consequential implication directly in terms of quantifiers is explored for the first time.

Inquiries about consequential implication have mainly been concerned with analytical consequential implication. What about "synthetical" or context-dependent consequential conditionals? This leads to new proposals on two different directions. The first is a criticism of classical conditional logic, in which room is given to logical theses lacking a consequential nexus, such as for instance $(A \wedge B) \supset (A \Box\!\rightarrow B)$. Such logics may be interpreted, it was Pizzi's idea, as "holistic" conditional logics and it is argued that one may devise, in the extensions of the basic Lewis' system **CI**, at least "three grades of holistic involvement" (see [6]).

The second line of inquiry lends to the idea of working out an intuitive semantics allowing a definition of a consequential nexus on the basis of a context of factual presuppositions. Such a connection is identified in the result of performing "the best choice" between incompatible alternative consequents. This feature is seen (according to [20]) as a common feature of inductive, counterfactual and abductive inference .

The logics based on this choice are special cases of what Pizzi proposes to call *rational inference*, where it is understood that the basic aim in performing the choice of the best conclusion is the one of preserving maximal information. As a matter of fact, this idea was already at the center of a book published in Italian [4], and he has been encouraged to develop it by a positive review of the book written by Newton da Costa.

Moving from the mentioned considerations Pizzi attempted to stress the analogy between abductive and counterfactual reasoning. It was relevant in this connection, in his own words, the cooperation with Lorenzo Magnani who, he says, "had the merit of introducing in Italy the topic of abductive and model-based reasoning". Being engaged with Magnani in the joint organization of the six international conferences on Model-Based Reasoning – cf. for example [16] – (MBR98, MBR01, MBR04 in Pavia, Italy, MBR06_CHINA, in Guangzhou, P. R. China, and the last one MBR09_BRAZIL, held in Campinas, Brazil, in December 2009), he has been stimulated in going deeper into the analysis of causal and abductive reasoning.

2 Tense Logic, Causality and Abduction

In 1974 Pizzi edited an anthology about tense logic [1], in which the translations are preceded by a lengthy introduction. This collection was the first, and perhaps the unique, published on this subject. But Pizzi has also original contributions to the field of tense logic: in [23] a completeness proof of the logic K_t via tableaux has been published, quite different from a proof of the same theorem presented by Rescher and Urquhart.

A talk delivered in Campinas (published in [10]) discusses the problem of the existence of time without change. The problem is treated along the lines already developed by Prior, but carried out in the framework of second order tense logic, with the aim of distinguishing between "real changes" and "Cambridge changes", and by making use of determinate-determinable distinction. The question of determination among predicates has been treated in several papers Pizzi wrote in Italian, mainly with the aim of giving notice of an interesting and forgotten theory sketched by W.E. Johnson in his "Logic" (1921).

Starting from 1988, a certain number of papers have been devoted by Pizzi to the question of the iteration of conditionals, whose treatment is generally neglected even in classical conditional logic. The interest for this topic was due to Pizzi's conviction that the problem of causal redundancy (overdetermination and preemption) can be solved by making use of conditionals with iterated antecedents. Lewis' counterfactual theory of causality does not offer a satisfactory treatment to the problem of causal redundancy; furthermore, many logicians seem to ignore that the *conditio sine qua non* theory of causation is part of the European juridical tradition, and that in this realm the question has been discussed at length.

The idea of using nested conditionals to define "concurring" (i.e.redundant) causes is quite original. It is the core of a book [11] written in Italian for a publishing house specialized in juridical studies. A more formal treatment was anticipated in [5]. The idea, presented by the occasion of talks delivered in Lund and Uppsala in 1988, appears to be shared by the Swedish logician P. Gärdenfors (see "An Epistemic Analysis of Explanation and causal Beliefs", *Topoi* 9 (1990) 109–124 (p.122)).

The interest for abduction has been a natural corollary of the reflections about rational inference on one side, and causality on the other. Beginning from year 2007, beyond teaching at the Faculty of Letters in Siena, Pizzi was also charged with courses on "Logic of Proof" at the Faculty of Law of the Second State University of Milan. The contact with the new environment furthered his interest for abduction. On the one hand, the question of conditional iteration has been tackled also with reference to abductive reasoning (see [22]). On the other hand, the problem of applying the logic of abduction to concrete problems actually discussed by jurists also stimulated reflections: a recently published collection of essays [25] is devoted to this topic.

3   From Contingency to Multimodalities

The idea of defining necessity in terms of contingency received a negative answer from Cresswell for systems weaker than **KT**. However, Pizzi showed that the problem is solvable in linguistic extensions of contingential logic: to begin with, in a logic with propositional quantifiers (see [14]), but more economically in contingential logics whose language contains an axiomatized propositional constant [21]. The idea of using such systems to yield multi-modal logics has been proposed in a recent conference in Lisbon.

From the perspective of his interest in general modal logic, Pizzi not only devoted his energy to the Italian translation of Hughes and Cresswell's first handbook, but also promoted the Italian translation of "A companion to Modal Logic" for which he also wrote an introduction (CLUEB, Bologna, 1990). In October 1991 he was invited by the Centre for Logic, Epistemology and the History of Science (CLE) of the State University of Campinas to give a course in modal logic, which was then a new subject for many Brazilian logicians. It was only the beginning of a long cooperation, during which he also became member of the direction of CLE. The most recent fruit of it was a book written in Italian with W. Carnielli in 2001 [15], recently revised and published in English [24]. This is the first book in which the subject of multimodalities is proposed also at the level of a teaching tool, and which intends to provide a philosophically- and historically-based introduction to modal logic without casting out neither rigor nor the mathematical aspects.

Knowing how much Brazilian music won Pizzi's heart, an artistic inclination he certainly inherited from his father, the painter and musician Walter Pizzi-Bonafous (1908-1987), and from his grandfather Ercole Pizzi, a *belcanto*'s maestro, we cannot refrain from quoting the Brazilian idol João Gilberto. According to [7], João Gilberto is the only non-Italian (perhaps the only person) ever to turn an Italian song, the famous *Estate*, into a worldwide jazz success. He will certainly appreciate the comparison: for Claudio, many splendid summers!

Estate... che splendidi tramonti dipingeva! [1]

References

1. Pizzi, C. (ed.) *La logica del tempo*. Bollati Boringhieri, Torino (1974)
2. Pizzi, C.: Boethius' Thesis and conditional logic. *Journal of Philosophical Logic* **6**, 283–302 (1977)
3. Pizzi, C. (ed.) *Leggi di natura, modalità, ipotesi*. Feltrinelli, Milan (1979)

---

[1]"Estate", by Bruno Martino and Bruno Brighetti, voice of João Gilberto.

4. Pizzi, C.: *Una teoria consequenzialista dei condizionali*. Bibliopolis, Naples (1984)

5. Pizzi, C.: Counterfactuals and the complexity of causal notions. *Topoi* **9**, 147–156 (1990)

6. Pizzi, C.: Stalnaker-Lewis conditionals: Three grades of holistic involvement. *Logique et Analyse* **33**(131–132), 311–329 (1990)

7. Castro, R.: *Chega de Saudade: a História e as Histórias da Bossa Nova*. Companhia das Letras, São Paulo (1990)

8. Pizzi, C.: Decision procedures for logics of consequential implication. *Notre Dame Journal of Formal Logic* **32**, 618–636 (1991)

9. Pizzi, C.: Consequential implication: A correction. *Notre Dame Journal of Formal Logic* **34**, 621–624 (1993)

10. Pizzi, C.: Time without change: A logical analysis. In: F. Évora (ed.) *Espaço e Tempo*, pp. 133–156. CLE, Campinas (1995)

11. Pizzi, C.: *Eventi e cause. Una prospettiva condizionalista*. Giuffrè, Milan (1997)

12. Pizzi, C., Williamson, T.: Strong Boethius' thesis and consequential implication. *Journal of Philosophical Logic* **26**(5), 569–588 (1997)

13. Pizzi, C.: Iterated conditionals and causal imputation. In: P. McNamara, H. Praakken (eds.) *Norms, Logics and Information Systems*, pp. 147–161. IOS Press, Amsterdam (1999)

14. Pizzi, C.: Contingency logics and propositional quantification. *Manuscrito* **XXII**(2), 283–303 (1999)

15. Carnielli, W., Pizzi, C.: *Modalità e multimodalità*. Franco Angeli, Milan (2001)

16. Magnani, L., Nersessian, N.J., Pizzi, C. (eds.) *Logical and Computational Aspects of Model-Based Reasoning*. Kluwer Academic Publishers, Dordrecht (2002)

17. Pizzi, C.: Aristotle's thesis between paraconsistency and modalization. *Journal of Applied Logic* **3**, 119–131 (2004)

18. Pizzi, C., Williamson, T.: Conditional excluded middle in systems of consequential implication *Journal of Philosophical Logic* **34**(4), 333–362 (2005)

19. Pizzi, C.: A logic of contingency with a propositional constant. In: E. Ballo, M. Franchella (eds.) *Logic and Philosophy in Italy*, pp. 141–154. Polimetrica, Milan (2006)

20. Pizzi, C.: Gestalt effects in abductive and counterfactual inference. *Logic Journal of the IGPL* **14**(1), 257–270 (2006)

21. Pizzi, C.: Necessity and relative contingency. *Studia Logica* **85**, 305–410 (2007)

22. Pizzi, C.: Abductive inference and iterated conditionals. In: L. Magnani, P. Li (eds.) *Model-Based Reasoning in Science, Technology, and Medicine*, pp. 365-381. Series Studies in Computational intelligence. Vol. 64, Springer, Heidelberg/Berlin (2007)

23. Pizzi, C.: Aristotle's cubes and consequential implication. *Logica Universalis* **2**, 143–153 (2008)

24. Carnielli, W., Pizzi, C.: *Modalities and Multimodalities*. Springer (2008)

25. Pizzi, C.: The Problem of existential import in first-order consequential logics. In: W. Carnielli, M.E. Coniglio, I.M. Loffredo D'Ottaviano (eds.) *The Many sides of Logic*, pp. 133–150. College Publications, London (2009)

Contents

An Episodic Memory Implementation for a Virtual
Creature 393

Elisa Calhau de Castro, Ricardo Ribeiro Gudwin

Abduction and Meaning in Evolutionary Soundscapes 407

Mariana Shellard, Luis Felipe Oliveira, Jose E. Fornari,
Jonatas Manzolli

Part I
Abduction, Problem Solving, and Practical Reasoning

Virtuous Distortion
Abstraction and Idealization in Model-Based Science

John Woods and Alirio Rosales

> There are the obvious idealizations of physics – infinite potentials, zero time correlations, perfectly rigid rods, and frictionless planes. But it would be a mistake to think entirely in terms of idealization, of properties which we conceive as limiting cases, to which we can approach closer and closer in reality. For some properties are not even approached in reality.
>
> Nancy Cartwright

Abstract. The use of models in the construction of scientific theories is as wide-spread as it is philosophically interesting (and, one might say, vexing). Neither in philosophical analysis nor scientific practice do we find a univocal concept of model; but there is an established usage in which a model is constituted, at least in part, by the theorist's idealizations and abstractions. Idealizations are expressed by statements known to be false. Abstractions are achieved by suppressing what is known to be true. Idealizations, we might say, over-represent empirical phenomena, whereas abstractions underrepresent them. Accordingly, we might think of idealizations and abstractions as one another's duals. In saying what is false and failing to say what is true, idealization and abstraction introduce distortions into scientific theories. Even so, the received and deeply entrenched view of scientists and philosophers is that these distortions are both necessary and virtuous. A good many people who hold this view see the good of models as merely instrumental, in a sense intended to contrast with "cognitive". Others, however, take the stronger and more philosophically challenging position that the good done by these aspects of scientific modeling is cognitive in nature. Roughly speaking, something has instrumental value when it helps produce a result that "works". Something has cognitive value when it helps produce knowledge. Accordingly, a short way of making the cognitive virtue claim is as follows: Saying what's false and suppressing what is true is, for wide ranges of cases,

John Woods and Alirio Rosales
University of British Columbia Department of Philosophy, Vancouver, Canada
e-mail: {jhwoods,arosales}@interchange.ubc.ca

L. Magnani et al. (Eds.): Model-Based Reasoning in Science & Technology, SCI 314, pp. 3–30.
springerlink.com © Springer-Verlag Berlin Heidelberg 2010

indispensable to the production of scientific knowledge. Given the sheer volume of traffic in the modeling literature, focused discussions of what makes these distortions facilitators of scientific knowledge attracts comparatively slight analytical attention by philosophers of science and philosophically-minded scientists. This is perhaps less true of the distortions effected by abstraction than those constituted by idealization. Still, in relation to the scale of use of the models methodology, these discussions aren't remotely as widespread and, when even they do occur, are not particularly "thick". The principal purpose of this paper is to thicken the analysis of the cognitive virtuosity of falsehood-telling and truth-suppression. The analysis will emphasize the influence of these factors on scientific understanding.

1 Models

1.1 *Distortion*

The use of models in the construction of scientific theories is as widespread as it is philosophically interesting (and, one might say, vexing)[1]. In neither philosophical nor scientific practice do we find a univocal concept of model[2]. But there is one established usage to which we want to direct our particular attention in this paper, in which a model is constituted by the theorist's idealizations and abstractions. Idealizations are expressed by statements known to be false. Abstractions are achieved by suppressing what is known to be true. Idealizations over-represent empirical phenomena. Abstractions under-represent them. We might think of idealizations and abstractions as one another's duals. Either way, they are purposeful distortions of phenomena on the ground[3].

---

[1] See, for example, the online issue of *Synthese*, 172, 2 (2009).

[2] For the notion of model-based science see, for example, Peter Godfrey-Smith "The strategy of model based science", *Biology and Philosophy*, 21 (2006), 725–740. For model-based reasoning see, for example, Lorenzo Magnani, Nancy Nersession and Paul Thagard, editors, *Model-Based Reasoning in Scientific Discovery*, Dordrecht/New York: Kluwer/Plenum, 1999, and Lorenzo Magnani, editor, *Model Based Reasoning in Science and Engineering*, London: College Publications, 2006.

[3] An early discussion of idealization as distortion is Ernan McMullin's "Galilean idealization", *Studies in the History and Philosophy of Science*, 16, (1985), 247–273. A recent discussion is Michael Weisberg's "Three kinds of idealization", *Journal of Philosophy*, (2007), 639, 659. Weisberg characterizes idealization as "the intentional introduction of distortion into scientific theories" (p. 639)", and remarks that "idealization should be seen as an activity that involves distorting theories or models, not simply a property of the theory-world relationship" (p. 640). The link between false models and true theories is also a central preoccupation of William Wimsatt's *Re-Engineering Philosophy for Limited Beings: Piecewise Approximations to Reality*. Cambridge: Harvard University Press, 2007.

Biologists investigate the action of natural selection in populations by representing them as infinitely large. With stochastic effects safely ignored, changes in gene frequencies can be tracked by concentrating solely on the action of selection[4]. Physicists investigate mechanical movement in frictionless planes, by means of which they are able to investigate the simplest kind of movement: rectilinear motion[5]. Neoclassical economists work out the laws of supply and demand and downwards inelasticity in models where utilities are infinitely divisible and all motivational factors are suppressed, save for the optimization of subjective expected utility. Likewise, rational decision theorists elucidate the processes of decision-making in models in which, agent-motivation is the maximization of self-interest and nothing more and decisional agents are logically omniscient[6]. In the first kind of case, theorists impose something that is false. In the second, they omit things that are true. In the third and fourth, they do both.

There is a widespread and deeply entrenched view among scientists and philosophers that the distortions effected by idealization and abstraction are both necessary and virtuous. Many people who hold this view see the value of models as merely instrumental. Others take the stronger and more philosophically challenging position that the good done by these aspects of scientific modeling has a genuinely cognitive character. The distinction between instrumental and cognitive (or alethic) value is less an algorithmically exact one than an attractive expository convenience[7]. Roughly speaking, a theory has instrumental value when it works predictively or helps control events in new or better ways, without necessarily generating new knowledge. A theory has cognitive value when, whatever the character of its other virtues, it

---

[4] A good foundational discussion of mathematical models of natural selection can be found in Sean H. Rice, *Evolutionary Theory: Mathematical and Conceptual Foundations* Suderland: MA, Sinauer Press, 2004.

[5] See Roberto Torretti, *The Philosophy of Physics* New York: Cambridge University Press, 1999.

[6] Such idealizations are usually supposed to have *normative* significance; that is, they are descriptively wrong but normatively authoritative. We ourselves have reservations about this idea, which we lack the space to develop here. Interested readers might wish to consult Dov Gabbay and John Woods, "Normative models of rational agency: The disutility of some approaches", *Logic Journal of the IGPL*, 11 (2003), 597–613.

[7] On the instrumentalist side, recent writers include Ronald Laymon, "Idealiation, explanation and confirmation", *Proceedings of the Biennial Meeting of the Philosophy of Science Association*, Vol. 1, (1980), pp. 336–350 and Leslie Nowack, "The idealizational approach to science: A survey", in J. Brezinski and Leslie Nowack, editors, *Idealization III: Idealization and Truth*, Amsterdam: Rodopi 1992, pp. 9–63. On the cognitive side, see for example Stephan Hartmann "Idealization in quantum field theory", in Niall Shanks, editor, (1998), Robert Batterman, "Idealization and modelling", *Synthese*, 169 (2009), 427–446, Michael Weisberg, "Three kinds of idealization", *Journal of Philosophy*, 104 (2008), 639–659, and Michael Strevens, *Depth*, Cambridge: Harvard University Press, 2008.

enlarges our knowledge of its analytical targets. (For example, particle physics has a good predictive record, but it also brings us a knowledge of how particles behave.) So, then, a short way of making the present claim is this: For wide ranges of cases, saying what's false and suppressing what is true is indispensable to the *cognitive* good of science.

Notwithstanding the sheer volume of traffic in the modeling literature, focussed discussion of what makes these distortions facilitators of cognitive success attracts comparatively slight analytical attention by philosophers of science and philosophically-minded scientists. This is perhaps less true of the distortions effected by abstraction than those constituted by idealization. Still, in relation to the scale of use of the models methodology, these discussions aren't remotely as widespread and, even when they do occur, are not particularly "thick". The principal purpose of this paper is to thicken our understanding of the cognitive virtuosity of falsehood-promotion and truth-suppression.

The distinction we've drawn between idealization and abstraction is amply present in the contemporary literature, although not always with the clarity it both deserves and admits of. In McMullin's 1985 paper, "Galilean Idealization"[8], the distortions effected by idealization are simplifications facilitate "at least a partial understanding" of the objects modeled. At the risk of confusion, McMullin specifies a further form of distortion which he also identifies with idealization, but which corresponds to abstraction in our sense; this involves screening out factors from a complex situation with a view to achieving a better command of the simplified residue (p. 264). In 1989, Cartwright anticipates our own distinction[9]. Idealizations change properties of objects, perfecting them in certain ways, not all of which even approximate to reality. Abstraction on the other hand, is not a matter of "changing any particular features or properties, but rather of subtracting, not only the concrete circumstances but even the material in which the cause in embedded and all that follows from that" (p. 187)[10].

George Gale[11] draws this same contrast. Abstraction is a process "in which some features are chosen to be represented, and some rejected for representation [. . .]" (p. 167). In abstraction, "there is a loss of features when one goes

---

[8] *Studies in the History and Philosophy of Science* 16 (1985), 247–273.

[9] *Nature's Capacities and Their Measurement*, New York: Cambridge University Press, 1989.

[10] In "Abstraction via generic modelling in concept formation in science", in Martin R. Jones and Nancy Cartwright, editors, 2005, Nancy Nersessian resists Cartwright's distinction, and finds it more "salient" to speak of *various* abstractive processes, one of which is idealization in Cartwright's sense, and another is abstraction in Cartwright's sense. A third is what Nersessian calls "generic modeling" (p. 137).

[11] "Idealization and cosmology: A case study", in Neill Shanks, editor, *Idealization IX: Idealization in Contemporary Physics*, Amsterdam/New York, Rodopi, 1998, pp. 165–182.

from the subject to its model" (p. 168). Idealization is a "smoothing out" of the properties represented in the model, in which "the subject feature is 'perfected', with no net loss of features" (p. 168).

Anjan Chakravartty[12] like-mindedly proposes that abstraction is a process in which only some of the potentially many relevant factors or parameters present in reality are built-in to a model concerned with a particular class of phenomena" (p. 327). Abstraction is a matter of ignoring "other parameters that are potentially relevant to the phenomena at issue", whereas the "hall-mark of idealization is that model elements are constructed in such a way as to differ from the systems we take to be their subject matter, not merely by excluding certain parameters, but by enjoying assumptions that could never obtain" (p. 328).

More recently, Martin R. Jones[13] also makes the present distinction, and takes idealization to require the "assertion of a falsehood" and abstraction the "omission of a truth" (p. 175); and Peter Godfrey-Smith (2009)[14] sees idealization as "[t]reating things and having features they clearly do not have", and abstraction as "[l]eaving things out, while still giving a literally true description" (2009, p.47).

In their present characterizations, we are introduced to idealization and abstraction in their "pure" forms. Idealization in its pure form is the addition of something false without the suppression of anything true; and abstraction in its pure form is the suppression of something true without the addition of anything false. It bears repeating, however, that in actual practice idealization and abstraction often not only *co-occur*, but do so in intimately linked ways. As Gale points out, "while this difference between the two processes is clear in definition, it oftentimes collapses in practice" (p. 168). Consider, again, the frictionless plane in which surfaces are idealized as perfectly smooth and collateral forces (friction, component forces, and so on) are suppressed. The same applies to population genetics, in which populations are idealized as infinitely large and factors such as random fluctuations are suppressed. These cases help elucidate the "in-tandem" character of the idealization/abstraction duality. Abstraction on a domain of enquiry D is sometimes a precondition of idealization over it. What is more, a theory's domain may *itself* be an idealization of or abstraction from under-conceptualized "raw data". Patrick Suppes has an attractive way of making this point. He reminds us that there are cases in which the data for a scientific theory must themselves be

[12] "The semantic or model-theoretic view of theories and scientific realism", *Synthese* 127, (2001), 325–345.

[13] "Idealization and abstraction: A framework", in *Idealization XII: Idealization and Abstraction in the Sciences*, Martin R. Jones and Nancy Cartwright, editors, Amsterdam/New York: Rodopi, 2005, pp. 173–217.

[14] "Abstractions, idealizations, and evolutionary biology", in A. Barberousse, M. Morange, and T. Pradeu, editors, *Mapping the Future of Biology: Evolving Concepts and Theories, Boston Studies in the Philosophy of Science*, volume 266, Amsterdam: Springer, 2009, pp. 47–56.

modelled. In those cases, then, what the theory supplies is a model of the models of those data[15].

Notwithstanding ancient roots, contemporary excitement about models and their distortions may with convenience be dated from 1983, which saw the publication of Nancy Cartwright's *How the Laws of Physics Lie*[16]. Seen her way, some idealizations are of "properties that we conceive of as limiting cases, to which we can approach closer and closer in reality" (p. 153). On the other hand, "some properties are not even approached in reality" (p. 153). This is an important distinction. Some idealizations are real properties conceived of in the limit. Others are not real at all. Let us call the first group "within-limit" (WL) idealizations and the second "beyond-limit" (BL) idealizations. Then statements about WL-idealizations may be thought of as approximately true. Statements about BL-idealizations are unqualifiedly false, hence not in the least true. Both categories give rise to interesting philosophical questions, but BL-idealizations produce the greater challenge for the thesis that idealizations are cognitively virtuous. For how can the advancement of utter falsehoods ever rise to the status of knowledge?

1.2 Some Motivating Considerations

As we proceed, we want to be guided as much as possible by the reflectively intelligent practice of model-based science itself. This is where we will find our motivating data – the pre-theoretical intuitions which we want our account to preserve and elucidate. Accordingly, we take it as given that

- Science strives to attain a knowledge of things.
- At its best, science succeeds in this task. In other words, instrumentalism in science is an unnecessary refuge.
- Theories that advance our understanding of target phenomena possess cognitive significance.
- Models of successful science are distortive.
- Distortion is not incompatible with the attainment of scientific knowledge.
- With these things said, a simple methodological protocol falls out: *First, pay attention to the data. Second, do not abandon them without a fight.*

1.3 Idealization

In this section, our main target is the pure form of BL-idealization, the form of idealization to which there is no approximation in reality. As we said, examples abound: stars as perfect spheres, massless material points, perfect

---

[15] Patrick Suppes, "Models of data", in E. Nagel, P. Suppes and A. Tarski, editors *Logic, Methodology and philosophy of Science: Proceedings of the International 1960 Congress*, pp. 252–261, Stanford: Stanford University Press, 1962.

[16] Oxford: Oxford University Press.

fluids, infinite populations, perfectly isolated systems, perfectly smooth surfaces, and so on. They bear no closeness whatever to anything in nature and, in relation to a theory's target phenomena, statements about them are plain false; that is, false without an iota of truthlikeness. They are, as we might say, "extra-domain" with respect to those parts of the natural order that the theory in question seeks a knowledge of.

For a significant range of cases at a certain level of generality, a scientific theory is a triple $\langle D, T, M \rangle$. D is the *domain* of inquiry, and T the inquiry's *target* with respect to D. Consider a case – typical of theoretical science – in which T is the specification of a mathematically formulated set M of lawlike connections. M we may take as a *model* of the theory in question. We say that a property is *extra-domain* iff it is essential to the theory of D but not present there, and that a property is *intra-domain* iff it is present in D. Similarly, a sentence is extra-domain iff it is ineliminably true in a model of the theory of D and yet false in D itself.

A BL-idealization over D introduces a set of properties $Q_1, \ldots, Q_n$ which cannot be instantiated in D but only in M. The Q_i are true only of the objects in M, and assertions to the contrary are unqualifiedly false in D. Crucially, the lawlikeness of connections in M demand that the Q_i occur essentially. BL-idealizations (both as properties and as attributions of those properties) are extra-domain.

We are, by the way, not unaware of the self-referential character of our present task. We are trying for a philosophical account of the cognitive good of saying what is false and suppressing what is true. Like any number of philosophical theories, there is nothing to prevent the account we develop here from appropriating the devices of idealization and/or abstraction. In other words, we want to leave it open that a good account of these features of model-based reasoning might itself be an instance of model-based reasoning. No doubt there will be readers whose methodological scrupulosity will resist modes of treatment of given target concepts which themselves instantiate those very concepts. Perhaps if we were seeking for simple lexical definitions of "idealization" and "abstraction", a charge of circularity might be sustained. But since lexical biconditionals will bear little of the theoretical load here, we need not share – or welcome – our hypothetical objector's methodological finickiness. The modeling of modeling may in *some* sense be circular but, if so, the circle would be virtuous, in the manner of Goodman.

There is an obvious question about an idealization's Q_i: *Of what conceivable cognitive good are they?* It is an unruly question. Since the dynamic equations of natural selection contain the Q_i ineliminably, they are false "on the ground". They are false, that is to say, of the actual populations in D. What is more, no population in D even begins to approximate in size the populations of the model M. So there is no prospect of the truths about M's populations being even truth*like* about D's populations.

All the same, the dynamic equations of M manage to do something quite extraordinary. They don't tell us what populations on the ground are like, but they do play a role in telling us what *natural selection* on the ground is like[17].

This what-likeness relation is crucial to our account. Here is why. There is in natural speech a well-entrenched use of "like" in which the what-like relation is irreflexive. In this usage, nothing can be like itself, not even exactly like itself. True, in formal treatments of likeness, reflexivity is often stipulated as a limiting case. But this just reinforces the point at hand. For if perfect self-likeness is only an ideal to which the likenessness of nature come more or less close, then the likenesses of nature – which are what natural speech is designed to capture – are irreflexive.

This has an interesting bearing on the role of idealizations in science. Given the irreflexivity of "like", the respects in which x is *like* y are respects in which x is *not*. Accordingly,

FALSITY AND LIKENESS: *There are ways of saying what's false of x which reflect what y is like. (It is not ruled out that sometimes $y = x$)*[18].

Idealizing a thing is saying something false of it. Not all falsehoods are scientifically fruitful, needless to say. But when they are, they are *likeness-exposing*.

On reflection, perhaps this is no uncommon thing. There appear to be lots of cases in which a false description is genuinely illuminating; think here of the immense communicational value of metaphor. The point for our purposes is that knowing what things are like is a cognitive good, made so by the link between appreciating what things are like and having an *understanding* of them. Accordingly, getting population size wrong on purpose, is also a means of achieving an understanding of natural selection on the ground. So we have it, as promised, that:

AID TO UNDERSTANDING: *For significant ranges of cases, theoretical distortions of the sort examined here are indispensable to a scientific understanding of target phenomena*[19].

From which, also as promised:

COGNITIVE VIRTUOSITY: *For significant ranges of cases, theoretical distortions of the sort examined here are cognitively (and not just instrumentally) virtuous.*

---

[17] Let us take some care with the "on the ground" metaphor. In its use here, what happens on the ground is what happens in a theory's domain D, which, as we saw, may often itself be a model of the raw data.

[18] Consider, for example Newton's idealization of massive bodies as point-particles. The idealization is false on the ground, but it effects an understanding of physical motion as it actually occurs in nature for velocities that are low compared to the velocity of light.

[19] See here Batterman (2009): "... continuum idealizations are explanatorily ineliminable and ... a full understanding of certain physical phenomena cannot be obtained through completely detailed, non-idealized representations." (p. 1)

This is interesting. It tells us something about the cognitive character of model-based science. It tells us that getting things wrong is often an unavoidable way of getting things right.

1.4 Side-Bar: Classificatory Confusion

The likeness thesis and its link to understanding put a certain pressure on how idealizations are to be described. Consider again the difference between BL (beyond-limit) and WL (within-limit) idealizations. Let Q be an idealized property of the BL kind (e.g. the infinite size of populations). Let us say that Q's *natural image* is the size Q' of an actual population at its physically biggest. In that case, the Q of Q-populations does *not* in any significant way reveal what Q'-populations are like. Finite largeness is not like infinite largeness except in rather trivial ways. (For example, both are larger than smallness.) It is different with WL-idealizations. The athletic magnificence of an Achilles is only approximated to by the prowess of any actual athlete. A thinking agent who is smarter than any human actually is could be, but who is not infinitely smart, exhibits a smartness to which the smartnesses of nature clearly come more or less close. But all of them – the smartness of the super-smart agent and the various smartnesses of lesser beings bear the same closeness-relation to infinite smartness. They are all infinity and equally far from it. WL-idealizations, like finitely great smartness, have approximating natural images. BL-idealizations like infinite smartness have no natural images. There is nothing in nature that approximates to them. So it bears repeating that BL-idealizations don't reveal what their natural images are like; but when properly wrought they *can* show what some other natural property is like (again, natural selection in finite populations).

1.5 Abstraction

It is now time to say something about abstraction. In its pure form, abstraction tells the truth incompletely and selectively. Abstraction serves as a kind of filter or – to vary the figure – partition. It partitions the truths at hand into those that matter in a certain way and those that don't matter in that way. Thus what the filter captures is true, and what it leaves behind is also true. But these are truths with a difference. The former somehow count and the latter do not.

There exists in the literature no firmly established and precise unpacking of this notion of mattering. Certainly we know nothing in the literature that counts as a thick account of it. Distinctions have been made on the basis of several criteria. Among the more prominent we find *relevance* vs *irrelevance, simplicity* vs *complexity, manageability* vs *unmanageability* and *tractability* vs *intractability* (McMullin 1985, Cartwright 1989, Weisberg 2008, Strevens

$2008)^{20}$. To this we would add a further distinction which we'll make some use of in sections 9 and following. It is the distinction between *information* and *noise*.

As before, D is a domain of enquiry, and T is the enquiry's target with respect to D. Consider a case in which T is a mathematically formulated set M of lawlike connections. Here, too, M can be thought of a model. An abstraction AB is a *filter* on D which extracts real properties $P_1, \ldots, P_n$ featuring ineliminably in the lawlike connections worked up in M. Let us say that the elements $P_1, \ldots, P_n$ implicated in a lawlike connection are its *parameters* $P_1, \ldots, P_n$, and that what an abstraction from D leaves behind is its *residue* R. Both the abstracted elements P_i and the residual elements R_i are intra-domain. They are all properties occurring in D.

Abstractions come in three different forms, depending on how the residual elements behave in relation to M. Although frequently confused in the literature, it is ruinous not to respect their differences. They are:

- *Faute de mieux (FDM) abstractions.*
- *Essentiality (E) abstractions.*
- *Simplicity (S) abstractions.*

1.5.1   Faute de Mieux-Abstractions

We begin with the faute de mieux kind. There are classes of cases of which the following is true:

D = some set of natural phenomena including some event E.

T = to attain a knowledge of E's causes.

M = the set of conditions causally necessary and sufficient for E.

AB = an abstraction on D.

P_i = abstracted properties from D occurring essentially in the lawlike connections of M.

R_i = residual properties left behind – occurring in D but not in M and co-occurring in D with E.

Think here of isolated systems, or motion in the absence of external forces.

It is a point of considerable interest that for such cases, the residual R_i are such that if admitted to M, the desired lawlike connections break down. In other words, the causal law for E is extra-domain. So we will say that

TARGET-BUSTING: *Residues are target-busters.*

Accordingly, it is a distinctive feature of FDM-abstraction that what is true in the model is false on the ground. For example, if M contains conditions

---

[20] For more, see the essays in Shanks (1998), and Jones and Cartwright (2005).

causally necessary sufficient for E, they may be causally necessary for E in D, even though causally insufficient (which at least is something). Hence, they are faute de mieux.

We might note in passing that FDM models are plausible candidates for the role of partial structures and the support they lend to partial truths, in the manner of Newton da Costa and Steven French[21]. Note, too, that FDM-abstraction resembles WL-idealization (via partial or approximate truth).

A good question is why we have faute de mieux models in the first place. If we could find necessary and sufficient conditions for E in D itself, we wouldn't need M. We resort to M because it will give us what D itself won't give us. And it might do so in ways that also give us *some* of what we were looking for in D.

1.5.2 Essentiality-Abstractions

There are models aplenty in which lawlike connections are fashioned with causal minima, and are not subsequently broken when residual factors are added to the mix. Such models employ essentiality-abstraction. E-abstractions differ from FDM-abstractions in a quite particular way. They turn on a peculiar interplay of necessary and sufficient conditions. E-abstractions are compatible with the necessary and sufficient conditions on E affirmed in the model. They are also causally positive for E. But they are not causally necessary for E. Perhaps the most intuitive way of capturing what's peculiar to E-abstraction is to represent its residual elements as causally redundant. They are, so to speak, *causal surfeits*. Accordingly, the function of E-abstraction is to model a causally redundant world *irredundantly*. What's causally essential to E are its irredundant causes, and the inessential factors are the causally positive but redundant ones. E-abstraction captures something of Weisberg's "minimalist idealization", by means of which models capture "core causal factors which give rise to a phenomenon" (2007, p. 642).

It is easy to see the duality between FDM- and E-abstraction. In the FDM-cases, residual elements are target-busters. When added to M lawlike connections break down. In the E-cases, the addition of residual elements to M leave lawlike connections intact.

1.5.3 Simplicity-Abstractions

S-abstractions come in two varieties, both of which reduce the complexity of information. In one variation, their role is largely pedagogical. They filter complex information into forms that can be managed by the comparatively untutored recipient. (In a variation of this variation, they serve as a handy kind of executive summary.) The second principal function of the S-abstraction covers two further kinds of case, one in which the original

---

[21] *Science and Partial Truth: A Unitary Approach to Models and Scientific Reasoning.* Amsterdam: Springer, 2005.

material, while comprehensible to the theorist, is costly to a fault to process. The other is one in which the complexity of a fully detailed account of a target phenomenon is incomprehensible, even to experts. Incomprehensibility, in turn, attaches to yet another pair of cases. In some situations, unfiltered explanations occasion informational overload, hence are too much for the theorist "to get his mind around". In others, the informational surfeit is too much for the theory's mathematics to "get itself around". In the one case, we have a generalized inability to understand[22]. In the other, we have a more focused failure of mathematical intelligibility[23].

We see something of our S-abstractions in Strevens' recent book in which he distinguishes between *canonical* versus *textbook* causal explanations. Canonical explanations are fully detailed accounts, which contain no idealizations, and omit nothing of causal salience to the targeted explanandum. Canonical explanations are not "better" than textbook explanations. In Strevens' words, "an idealization does not assert, as it appears to, that some nonfactual factor is irrelevant to the explanandum; rather, it asserts that some actual factor is irrelevant" (Strevens, 2008, p. 26). Strevens' idealizations are our abstractions – a useful reminder of our earlier point about the absence of a fully settled usage of these expressions. They are filters of complex material whose particular purpose is to leave behind a residue of explanatorily irrelevant, unsalient, *non-difference-makers*. It is important to emphasize that what these residual elements fail to make a difference *to* is the theorist's intended causal *explanation*. It is not Strevens' contention that these factors are *causally* irrelevant; indeed the opposite is true. They are fully paid-up participants in the causal nexus on the ground. But they can be ignored without damage to the phenomenon's causal explanation.

This idea of explanatory unnecessariness, that is, of explanatory non-difference-making, covers two kinds of situation, both catered for here. In the one, non-difference-making factors are unneeded for the *construction* of the causal connections embedded in a successful explanation. In the second, the non-difference-making factors aren't necessary for the *comprehension* of the causal connections embedded in a successful explanation (and may be

---

[22] As mentioned earlier, the connection between models and understanding has been emphasized by Batterman. He cites a number of cases in which detailed models block the understanding. See Batterman (2009) and the further references cited there. Also of interest is Batterman's "A modern (= victorian?) attitude towards scientific understanding", *The Monist*, 83 (2000) 228–257, in which he draws a contrast between fully or highly detailed forms of explanation and understanding and their simplified forms which he himself favors. Writes Batterman: "In particular, the means by which physicists and applied mathematicians often arrive at understanding and physical insight involve a principled ignoring of the detailed mechanisms that form the essential core of the causal/mechanical views about understanding." (p. 233).

[23] On guarding against the loss of mathematical tractability, see Weisberg (2007), p. 640.

inimical to their comprehensibility.) In terms of the account developed here, Strevens' "idealizations" that leave behind non-difference-makers in this first sense are *essentiality-abstractions* in our sense. His "idealizations" that leave behind non-difference-makers in this second sense are *simplicity-abstractions* in our sense.

S-abstractions also feature in Robert Batterman's "On the explanatory role of mathematics in empirical science", *The British Journal for the Philosophy of Science* (Batterman, forthcoming)[24], is the distinction between "traditional" or Galilean idealizations (or abstractions, in our sense) and "non-traditional" or non-de-idealizable ones. Traditional idealizations are indisplaceable through de-idealization; they contribute to explanation and understanding, "to the extent that they can be eliminated through further work that fills in the details ignored or distorted in the idealized models" (p. 16). Non-traditional idealizations are not de-idealizable and are the more interesting type for Batterman, and for us. They play an essential ineliminable role in explanatory contexts (p. 17), by allowing the modeler to capture salient, robust features of target phenomena, which would lose their salience in a more detailed model. As Batterman says, "adding more details counts as explanatory noise – noise that often obscures or completely hides the features of interest" (p. 17). As we will see, in due course, this is a point of crucial importance for our own account.

It is useful to point out that S-abstractions have residues that are target-busters, but with a difference. In the case of FDL-abstraction, its residual elements upon admittance to the model falsify the causal connections that obtain there. In the case of simplicity-abstraction, residual factors are compatible with laws of M and are part of the causal nexus for E in D. The mischief they do is not the loss of a causal law, but rather, in the one class of cases, the *formulability* of the law and, in the second, the *comprehensibility* of the law once formulated.

In relation to a model M formulating lawlike connections with respect to a D-element E, our abstraction trichotomy pivots on the quite different causal relations borne by residual factors to M or to E or to both. These can be summed up as follows:

- With *FDM-abstraction*, the residual elements R_i override the lawlike connections that obtain in M and are sufficient conditions of their falsity in D.
- With *E-abstraction*, the R_i are compatible with the laws of M, are causally positive for E but not causally necessary for it.
- With *S-abstraction*, the R_i are compatible with the laws of M and necessary for E, but their recognition by M would render the desired account either unformulable or incomprehensible.

We note in passing the co-occurrability of S-abstraction with each of the other two. In the case of E-abstraction, it is perfectly possible that the

---

[24] Available online at doi: 10.1093/bjps/axp018.

elements they pick out as essential are also the very elements that make the
abstraction a simplifying one, hence one that gives us a graspable understanding of E's actual causal situation. Similarly for the case of FDM-abstraction.
There are cases in which the construal of conditions which, as a matter of
fact, are at best causally necessary for E as causally necessary and sufficient
achieve a simplification that effectuates a partial understanding of E's actual
causal situation[25].

When we introduced this trichotomy, we warned of the confusion caused by
not respecting relevant differences. A case in point is the oft-repeated claim
that the good of abstraction is a matter of its exposure of what's essential and
its simplification of what obtains on the ground. As we see, there are fair and
load-bearing interpretations on which these claims are clearly inequivalent.

There is something rather intriguing about simplicity-abstraction. There
are lots of cases in which there is more to E's causal set-up than can be
contained in any intelligible account of it. Yet grasping an *abstract* description
of this same set-up suffices for an understanding of it. The residual elements
of S-abstraction sometimes function as background conditions. Calling them
this calls to mind the old mediaeval distinction between *ordo cognescendi*
(the realm of knowledge) and *ordo essendi* (the realm of being).

With this distinction at hand, an interesting part of the story of simplicity-
abstraction can be retold as follows: Since they are "background," the elements R_i left behind by S-abstraction are causally *distal* in *ordo cognescendi*
and causally *proximate* in *ordo essendi*. They are causally distal in *ordo cognescendi* because, even when we have the causal understanding of something
furnished by an abstract description of it, we *know in a non-individuating
way* that there is more to the causal picture than is captured by the abstraction. They are causally proximate in *ordo essendi* because they are causally
active on the ground. This is rather important. If right, our knowledge of E's

---

[25] At the Model-Based Reasoning Conference in Campinas in 2009, Jaakko Hintikka proposed to us that no account of model-based science can pretend to
completeness in the absence of an adequate treatment of the boundary conditions of differential equations in mathematical models. We agree with this entirely, and hope to make some progress with it in future work. The neglect of
boundary conditions by philosophers of science is a regrettable omission. A significant exception is Mark Wilson's "Law along the frontier: Differential equations
and their boundary conditions", PSA 1990, 2 (1991), 565–575 and *Wandering
Significance: An Essay on Conceptual Behavior*, New York: Oxford University
Press, 2006. Concerning the expression "boundary conditions", Wilson writes
"...standard philosophy of science texts encourage a rather misty and misleading understanding of this term. The faulty picture arises, as stereotypes often do,
from inattention. Indeed, the standard philosophical texts say virtually nothing
about boundary conditions – they are scarcely mentioned before they are packed
off in an undifferentiated crate labeled "initial and boundary conditions" (usually
pronounced as one word). The salient fact about "initialandboundaryconditions"
is that, whatever else they might be, they are not laws and can be safely ignored."
(1991, p. 565)

causes is a distortion of E's actual causal set-up. Why? Because the cognitive state one is when one has a causal understanding of E is a *model* of the actual causal situation that E is in. And, in so saying, the apparent innocuousness of the heuristic side of simplicity-abstraction is called into question. For how can it be right to say that one has a causal knowledge of E when one's knowledge is a distortion of what's really there – when *ordo cognescendi* is out of step with *ordo essendi*? Doesn't having a knowledge of something require the smooth alignment of knowing and being?

The answer is No. Showing why is the business of Part II, immediately to follow.

2 Cognition

2.1 *Cognitive Systems*

Our question now is: *Why don't the misalignments between ordo cognescendi and ordo essendi wrought by S-abstraction discredit their use in model-based science?*

The short answer is "Because this is the way of knowledge *quite generally*". To see why, consider first some basic assumptions about the natural phenomenon of knowing things. In calling them to mind, we won't be following the standard philosophical practice of "unpacking" the concept of knowledge. We propose instead to say something fairly simple and general about the admittedly complex business of how knowledge acquisition actually works in the lives of beings like us. In so doing, we won't be checking the *bona fides* of these assumptions against their fit with this, that or the other pre-conceived philosophical account of knowledge.

Accordingly, we take it without further ado that the human individual is a *cognitive system*, that he and his ilk make their way through life *by knowing things* – by knowing what to believe and knowing what to do. We take it that the cognitive states in which a human individual finds himself is a *state of nature* and that, whatever the details, is the state which an agent is in when it is true of him that he has knowledge of something, a state that arises from the interplay of the agent's *cognitive devices* and the *causal surround* in which they are activated. Cognition and its modes of production are, therefore, a kind of *ecosystem* in which cognitive targets are roughly proportional in ambition to the availability of resources sufficient for their attainment in more or less efficient ways and with a non-trivial frequency[26].

---

[26] See H.A. Simon, *Models of Man*, New York: Wiley, 1957, Gerd Gigerenzer, *Adaptive Thinking: Rationality in the Real World*, New York: Oxford University Press, 2000, Dov M. Gabbay and John Woods, *Agenda Relevance: A Study in Formal Pragmatics*, Amsterdam: North-Holland, 2003, and Dov M. Gabbay and John Woods, *The Reach of Abduction: Insight and Trial*, Amsterdam: North-Holland, 2005.

Given the sundry distractions the human animal is heir to, it can only be said that, while perfection cannot be claimed, beings like us have done rather well in the cognitive arena. We survive and we prosper, and occasionally great civilizations arise. There are exceptions of course. There are cases galore in which we are mistaken in what we take for knowledge. Accordingly, there are two abundances that set out the basic contours of our cognitive lives.

COGNITIVE ABUNDANCE: *Beings like us have knowledge, lots of it.*

ERROR ABUNDANCE: *Beings like us make errors, lots of them.*

One of the striking things about the human individual is that error abundance doesn't constrain cognitive abundance in ways that threaten survival and prosperity. From which we may conclude:

ENOUGH ALREADY: *Human beings are right enough, enough of the time about enough of the right things to survive, prosper and occasionally build great civilizations.*

How could this be? Why wouldn't our being wrong about things on such a scale be enough to spoil our lives, if not simply wipe us out? The answer in part lies in the efficiency of our feedback mechanisms and the concomitant wherewithal for error-correction.

Here is a last given. We take it that knowledge is *valuable*, that having it is indispensable to our survival and our well-being. If we were conspicuously less good at knowing things than we historically have been and presently are, life would be unpleasant or nonexistent. We all know Hobbes' tag-line – nasty, brutish and short.

There are legions of problems that arise from the relationship between a state of knowledge and what it is knowledge of. A great many of these fall under the umbrella of what Ralph Barton Perry called the "egocentric predicament", which is a pessimistic reaction to the problem of the "veridicality" of direct experience. In one way or another, all these problems recognize the distortive circumstances in which cognitive states are formed, and virtually all of standard epistemology is a kind of answer to such questions. But this is not our focus here. Whatever might be the case – realism, idealism, Kantian transcendentalism or whatever else – we draw our line in the sand with the four assumptions currently at hand: cognitive abundance, error abundance, enough-already and the value-of-knowledge. There is one epistemological commonplace to which these assumptions give short shrift. It is scepticism. Scepticism is discouraged by the spirit of the cognitive abundance thesis and it is contradicted by the letter of the value-of-knowledge thesis when linked to the enough-already thesis. Beyond these modest musings, we will have as little to say of mainstream epistemology as we can get away with.

States of knowledge require information and states of information require causal provocation and causal support. In the interests of expository economy, we will focus on knowledge-states that can be considered both conscious and representational, but not before quickly adding that neither of these traits is a necessary condition on cognition[27].

2.2 Knowledge as Distortion

There are two general features of cognition on which we now want to focus:

- *A cognitive state is an* abstraction *of an information state.*
- *Irrespective of its abstractive character, cognitive states are* distortions *of reality.*

Beginning with distortion, you see a robin red-breast in the tree at the bottom of the garden. On the standard readings, the robin's breast *isn't* red. It is *seen* as red. This is Locke's problem of "secondary qualities", properties that are revealed in perception but are not part of nature's inventory. Some philosophers extend the point to "primary qualities", such as extension and magnitude. In a recent and extreme version, the denizens of the outer world are nought but "classes of quadruples of numbers according to an arbitrarily adopted system of coordinates"[28]. Whether a secondary quality problem or also a primary quality problem, the veridicality issue is old-hat epistemology. It is known – or at least widely taken to be known – that the ways in which things are seen are distortions of the causal stimuli of the perceptual states in question. Whatever the features of those perceptually provocative conditions on the ground, *they* aren't red; and in some versions they aren't breasted, and they aren't robinesque.

Locke and a great many other philosophers (e.g. Hobbes and Condorcet) was of the view that the non-existence of colors precludes our knowing that the robin is red-breasted. So let's ask: When you see the robin in the tree at the bottom of the garden, you are in a certain perceptual state. Is this a state that has cognitive value? Is it a state in virtue of which you know what's at the bottom of the garden? Is it a state thanks to which you know what color its breast is? If it is, then how can its cognitive virtue be denied? And if it is, we have significant ranges of cases in which knowledge of something is a distortion of its causally necessitating conditions.

Here is another case. The science of ornithology invests heavily in recording the distributions and conditions of change that affect the color of birds. Non-trivial correlations are drawn between variations in color and the presence or absence of other properties of significance. On the standard reading,

---

[27] There is abundant literature on what Gabbay and Woods call cognition "down below", that is, below the level of conscious awareness, some cases of which also fall short of the threshold of representational effect. See *Agenda Relevance: A Study in Formal Pragmatics*, Amsterdam: North-Holland, 2005, pp. 60 to 68 for a brief discussion and key citations.

[28] W.V. Quine, *Theories and Things*, Cambridge, MA: Harvard University Press, 1981; p. 17.

ornithology is riddled with falsehood. Given the standard closure-conditions for conjunction, ornithology itself is false. So there is no knowledge of birds that can be got from ornithology. Aside from an occasional philosophical zealot, is there anyone in the wide world who believes such a thing?

Not everyone will like this position. How, it will be demanded, is it possible to know that the breast is red if there aren't any colors? This is the wrong question. If you know that the robin's breast is red, it isn't true that nothing is red. What is true (as best we can yet tell) is that nothing red is *causally implicated* in knowing by looking that the robin's breast is red. There are two things to say about this, both rather important. First, it is an interesting claim. And, second, even if false, it is not false enough, so to speak, to warrant outright dismissal. But it is at odds with a good many of the standard philosophical moves. This is because the standard philosophical moves assume that irrealism is true for colors and that irrealism precludes knowledge. It is precisely this that the present example rightly bids us to question. Everything of a scientific nature that is known of such things supports the distortion-as-a-cognitive-necessity thesis, and with it that the standard philosophical responses to irrealism – *eliminativism, scepticism, reductionism, idealism, instrumentalism,* and so on – are overreactions.

What this gets us to see is something quite important about the extra-domainality of features that are necessary for the advancement of D-knowledge. When we look at the bottom of the garden and conditions are right, we know by looking that the robin's breast is red. The standard view is that since nothing in nature is red, the breast isn't red and we can't know otherwise. This is perverse. What is true is that the redness of the breast is extra-domain to the causal order of robins. Relative to that extra-domainality, robins themselves are causally intra-domain. Knowing by looking that the robin's breast is red is a distortion of the causal conditions necessary and sufficient for that knowledge to exist.

2.3 Information Suppression

We see the individual cognitive agent as a processor of information on the basis of which, among other things, he thinks and acts. Researchers interested in the behavior of human information-processors tend to suppose that thinking and deliberate action are modes of consciousness. Studies in information suggest a different view[29]. Consciousness has a narrow bandwidth. It

---

[29] We are not unaware of the challenges that attend philosophical appropriations of the mathematical concept of information. However, for our purposes here we need not engage these issues. A good recent account is Fred Dretske, "Epistemology and information", in Pieter Adriaans and Johan van Benthem, editors, *Philosophy of Information*, pages 29–48; a volume of the *Handbook of the Philosophy of Science*, edited by Dov M. Gabbay, Paul Thagard and John Woods, Amsterdam: North-Holland, 2008.

processes information very slowly. The rate of processing from the five senses combined – the sensorium, as the mediaevals used to say – is in the neighborhood of 11 million bits per second. For any of those seconds, something fewer than 40 bits make their way into consciousness. Consciousness therefore is highly entropic, a thermodynamically costly state for a human system to be in. At any given time, there is an extraordinary quantity of information processed by the human system, which consciousness cannot gain access to. The bandwidth of language is even narrower than the bandwidth of sensation. A great deal of what we in some sense know – most in fact – we aren't able to tell one another. Our sociolinguistic intercourse is a series of exchanges whose bandwidth is 16 bits per second[30].

In pre- or subconscious states, human systems are awash in information. Consciousness serves as an aggressive suppressor of information, preserving radically small percentages of amounts available pre-consciously. To the extent that some of an individual's thinking and decision-making is subconscious, it is necessary to postulate devices that avoid the noise, indeed the collapse, of information overload. Even at the conscious level, it is apparent that various constraints are at work to inhibit or prevent informational surfeit. The conscious human thinker and actor cannot have, and could not handle if he did have, information that significantly exceeded the limitations we have been discussing here[31].

Here, then, is the basic picture. Knowledge is the fruit of information-processing. But it is also an information suppressor. There is a basic reason for this. Consider the particular case of conscious representational knowledge. If what is suppressed by our cognitive processes were admitted to consciousness and placed in the relevant representational state, it would overload awareness and crash the representation. From the point of view of cognitive attainment, what is suppressed would otherwise be experienced as *noise*[32]. Noise, like

---

[30] M. Zimmerman, "The nervous system and the context of information theory", in R.E. Schmidt and G. Thews, editors, *Human Physiology*, pp. 166–175, Berlin: Springer Verlag, 2nd edition, 1989.

[31] Consciousness is a controversial matter in contemporary cognitive science. It is widely accepted that information carries negative entropy. Against this is the claim that the concept of information is used in ways that confuse the technical and common sense meanings of that word, and that talk of information's negative entropy overlooks the fact that the systems to which thermodynamic principles apply with greatest sure-footedness are *closed*, and that human agents are not. The complaint against the over-liberal use of the concept of information, in which even physics is an information system, is that it makes it impossible to explain the distinction between energy-to-energy transductions and energy-to-information transformations. See here, Stephen Wolfram, "Computer software in science and mathematics", *Scientific American*, 251:188, September 1984. Difficult as these issues are, we say again that they needn't be settled definitively for the concept of information to perform its intended role here.

[32] See, again, the noise-producing propensity of detail, noted in Batterman (forthcoming), p. 17.

bad money, drives out the good. It overloads perception and memory, and it hobbles conception. It blows the mind.

If we had larger capacities, there would be less of this suppression of noise. For beings with brains "the size of a blimp"[33], what is noise for us would be serviceable information. But for us it is too hot to handle, and we must make do with states that filter out the noise that would otherwise overwhelm them. This supports the abstraction thesis: *A cognitive state is an abstraction from an information state.* There is another way of saying the same thing: *A cognitive state is a model of an informational environment.*

2.4   Connection to Models

In what we have said so far, perhaps the connection with models is not wholly apparent. As with knowledge, a scientific model is a certain kind of information state. There are cases galore in which if abstractive-residues were admitted to them, they would lose their informational purchase. Information contained in the residue would operate as noise in the model. So the model excludes it. It is much the same way with cognition in the general case. A piece of knowledge is an information state meeting certain conditions. It is a state that filters manageable from unmanageable information. In the interesting cases, excluded information is the residue of abstractions; we might think of these as *informational surrounds.* Informational surrounds have some of the interesting properties we have been tracking here. If admissible, their elements $R_1, \ldots, R_n$ would make the knowing agent better informed about the thing at hand. But if actually admitted they would crash the host state. Informational surrounds are good information which the knowing agent cannot make use of – cannot get his head around. So when they are in good working order, his cognitive devices keep at bay the R_i of these informational surrounds.

Informational surrounds bear a certain likeness to background knowledge. Perhaps their most interesting point of similarity is that, while inaccessible to consciousness, informational surrounds and background information are causally implicated in the attainment of those informational states which, by wide agreement, qualify as states of knowledge, including the conscious ones. Thus when, looking into the garden, you come to know that there is a robin there, much of the information contained in your visual field is screened out. But it is screened out in a quite particular way. On the one hand, it is not part of the informational state that constitutes knowing the robin is there, but may well be causally necessary all the same to the creation of that state. One of the problems that launched us into these reflections arises from the fact that often a residue (or causal field) if admitted to a model that excludes it would blow the model, and hence is a defeater of the model's lawlike pronouncements on the ground.

---

[33] Stephen Stich, *The Fragmentation of Reason*, Cambridge, MA: MIT Press, 1990.

We wanted to know what was the cognitive good of models that exclude falsifying considerations in this manner. What we now propose is that scientific models mimic the structures and processes of cognition quite generally, and that a fully worked up model-based scientific theory would capture with some precision the constructive impossibility of knowing a thing through an awareness of most of what's true of it. With perception again as our guide, knowing of the world what you do when you see the robin in the tree is, in comparative terms, knowing hardly anything that's true of it. Such knowledge – a conscious awareness of the disclosures of your five senses – beckons paradoxically. It supplies you with ample knowledge on practically no information. It is not that the screened-out information is suppressed because of its uninformativeness about what you now perceive. It fairly brims with information about those objects. It is excluded because you can't process it, can't take it in. It is excluded because, if admitted, it would erase the little that you already do in fact know of it.

In these respects, the abstractions of model-based science mimic the abstractive character of knowledge *in general*. If, as we ourselves suppose, the abstractive character of perceptual states doesn't deny them cognitive value, why would the abstractive character of model-based reasoning deny it cognitive value? After all, shouldn't it be that what's sauce for the goose is sauce for the gander?

3 Artefactualism

4 Truth-Making

In the case of perceptual knowledge we have it thus. We see the redness of robins. We *know* that robins are red-breasted. It is *true* that robins are red-breasted. Since knowing is achieved by being in states which fulfill the relevant conditions, it is natural to ask what role redness has in producing the state that meets the conditions for knowing that the robin is red-breasted. The answer is none. Although it is true that the breast is red, although we know that it is, it is not true that redness makes any appearance in the causal arrangements necessary for that knowledge.

From the point of view of models, we can now say that

PERCEPTIONS AS MODELS: *In certain respects, a field of vision is a model of its causal surround, both of which are states of nature.*

In contrast, consider once more the case of population biology. We have long since settled it that infinite populations, which don't occur in nature, are necessary for a knowledge of natural selection, which does occur in nature. Since infinite populations don't occur in nature, they have no occurrence in the causal mix without which natural selection on the ground could not occur. Whatever the precise nature of the tie between infinite populations and the knowledge attained of natural selection, it is something different from their occurrence in the underlying causal surround.

Although the red-breastness of robins is extra-domain with regard to its causal surround, both the red breastness and the causal surround occur in nature. Similarly, although our knowledge of the red-breastness is extra-domain with respect to our knowledge of what causes the knowledge of red-breastness, both pieces of knowledge are true of the natural world. We know in each case *where* these things are true.

The infinite populations of population biology are different. Although infinite populations are extra-domain with respect to natural selection, natural selection occurs in nature but infinite populations do not. Similarly, although our knowledge of infinite populations is extra-domain to our knowledge of the processes of natural selection, what we know of natural selection is true of the natural world, whereas infinite populations are not true of it. We know in the one case *where* it is true, and in the other also know that it is *not* true where the former is. Accordingly, if we wish to retain the commonplace that only the true is known, there is a question that now presses:

TRUTH PLACES: *Where, if anywhere, is it true that populations are infinite? What, so to speak, is the truth-place of "Populations are infinite"?*

This is a philosophical question. It arises, quite properly, from the claims we have been making on behalf of the cognitive virtuosity of distortion. We said in section 1 that in this paper we would give short shrift to the philosophical questions provoked by what we would come to say of distortion; that beyond a short sketch of possible answers, we would be content to offer readers a promissory note, to be redeemed in future work[34]. Here, then, is a short sketch of an answer to the home-truth question for BL-idealizations. It is a version of *artefactualism* known as "stipulationism".

In its most general form, artefactualism is the doctrine that some things are made rather than found and that their main modes of production are either the linguistic acts or the cooperative behavior of human agents. The two main varieties of antifactualism are *conventionalism* and *stipulationism*. Conventionalism is the view that some facts are constituted by the cooperative behavior of agents, and that sentences reporting these facts are also made true by this behavior. Think here of the fact that in North America and Europe the right side of the road for motorists to drive on is the right-hand side. Stipulationism is the view that some facts are constituted and some statements about them made true by the linguistic acts of agents. Think here of the fact of legal guilt, constituted by a jury's assertion of it, or the fact that Parliament is now in session, made so by the Sovereign's proclamation. Stipulation is the form of artefactualism that matters most for our purposes here. So we won't trouble ourselves further with conventionalism.

---

[34] For example, in "Model-based realism", to appear.

5 Stipulationism

Stipulationism, then, is the view that some objects and some states of affairs are made rather than found, and that involved in their making is their being *thought up*. Stipulation likewise provides that objects and states of affairs are thought up in ways that make sentences about them true. Made objects make true the sentences that faithfully announce their presence and faithfully report their goings on. They are made, and made true, by fiat. *Fiat* is Latin for "Let it be made". "Let it be made" is a sentence in the hortatory subjunctive mood. This tells us something important about making. It tells us that artefactualism's objects and truths are, once they are thought up, uttered into being; they are, as we have said, the results of their makers' linguistic acts.

There are critical differences for the artefactualist to take note of. One's knowledge by looking that robins are red-breasted depends on our seeing the redness of the relevant breasts, and is the result of causal provocations from which redness is missing. It is true that robins are red-breasted because it is known that they are. A scientist's models are different. Their fruitfulness or otherwise for scientific enquiry has nothing directly to do with whatever may be the causal conditions of the theorist's thinking them up. It is true in nature, but not in the relevant causal nexus, that robins are red-breasted, and it is not true in nature that a population is infinite. If parity with the robin-example is to be got, we must find a sense of "true" and a place in which it is true that populations are infinite. If we succeed in this, the place in which the infinite population statement holds will be extra-domain with respect to the place in which the natural selection statements are intra-domain.

Where is this somewhere? According to our artefactualism, it is in the *model* of the theorist's theory. What makes it true there? From everything that is known of the natural history of scientific (and mathematical) enquiry, the answer is "By *fiat*". Consider the cases: *let* there be an x such that ...; *we put it* that φ; *we define* a function f over Σ; and so on. Artefactualists say that these sentences are made true by conditions other than designatory links to elements of the natural order. Stipulation answers the question: "How then *are* they made true?" The answer is that they are made true by the theorist's stipulation. Stipulation has a long history in modern philosophy, and has played long since a dominant role in mathematics[35]. Stipulations are a good deal more widely embodied in enquiry than are idealizations and

---

[35] For Kant, stipulation is synthesis, which is the making up of new concepts. For Russell, stipulation is mathematical definition, which makes things true without the generation of the concepts in virtue of which this is so. See Immanuel Kant, *Inquiry Concerning the Distinctness of Natural Philosophy and Morality*, Indianapolis: Bobbs-Merrill, 1974; first published in German in 1764, and *Logic*, Indianapolis: Bobbs-Merrill, 1974; first published in German in 1800; and Bertrand Russell, *Principles of Mathematics*, London: Allen and Unwin, 1937, p. 27; first published in 1903.

abstractions. When a mathematician stipulates a function f over a set Σ that meets certain conditions, he needn't think that f is either an abstraction or an idealization in the scientist's senses of these terms. We ourselves are drawn to stipulation for two particular reasons. One is that is *there*. That is, it is an established part of scientific – especially mathematical – practice, and not an *ad hoc* device of our own contrivance. (In other words, stipulation is not the product of our own stipulation.) The other is that it helps with the task of finding a place in which extra-D claims essential to a scientific knowledge of D can turn out true. In short, it helps in finding a *truth-place* for them.

Stipulationism is an answer to the question of how something false in D can be indispensable to a knowledge of D's phenomena. It provides that it really is the case that sentences false in D are nevertheless true, in the absence of which certain things known of D couldn't have been known. The central question to which this gives rise is: *How is it possible for the theorist's stipulations – his sayso - to generate truths?*

6 Fictionalism

This too is a large question, for whose honest consideration we lack the space here[36]. Suffice it to note a burgeoning literature which explores an interesting and attractive suggestion. It is that the sense in which sentences false on the ground can be true in a theory's model resembles the sense in which sentences false on the ground can be true in a *novel* or *short story*. It is a controversial issue, in which various issues of greater and less importance remain unsettled. But, for good or ill, the idea is now making the rounds in mainstream philosophy of science that the facts reported by true theories indispensable to a knowledge of D that have no possibility of occurrence in D are *fictions*. Since fictional sentences are true in *their place* - that is, stories are the truth-places for them - it is said that this gives the desired result, namely, a non-trivial sense of irrealism in science which evades the drab consolation prizes of idealism, reductionism, instrumentalism, and constructive empiricism, to say nothing of the sheer capitulations of eliminativism and scepticism.

Concerning literary fictions, a standing position is not only that there is an obvious sense in which they fail realist conditions, but that they fail them on purpose. No one – not even a literary creator – thinks that the objects and

---

[36] Interested readers may wish to consult our "Unifying fictions", in *Fictions and Models: New Essays*, which also contains a foreword by Nancy Cartwright and essays by Robert Howell, Amie Thomasson, Mark Balaguer, Otávio Bueno, Mauricio Suarez, Roman Frigg, Jody Azzouni, Alexis Burgess and Giovanni Tuzet. See also John Woods and Jillian Isenberg, "Psychologizing the semantics of fiction", forthcoming in *Methodos*, and Woods' *Préface*, in *Fiction: Logiques, Langages, Mondes*, Brian Hill *et al.*, editors, forthcoming in 2010 in the series *Les Cahiers de Logiques et d'Epistemologie* from College Publications, London.

goings-on of fiction are real, or that their faithful reports are true of reality. On the other hand, it is just about as widely held that there is a place in which these objects and events do occur and of which the sentences that faithfully report them are true. One of the main tasks of a theory of literary fictions is to supply a theory of truth that accommodates this truth-place approach. It tries to specify conditions under which sentences false (or not true) in the real world are true in the story. It is not an easy task, and we are a long way from a settled theoretical consensus. Even so, there are points of wide agreement about particular matters. One is that *authors* make true the sentences of the text of the story. Another is that, except where borrowed from the world, the story's truths are *false* (or at least not true) on the ground. A third is that, subject to conditions it is difficult to specify, fictional truths *semantically engage* with non-fictional truths to produce *further truths*. (For example, "Sherlock Holmes had tea with Dr. Watson" which is true in the story and, "Having tea with is a symmetrical relation" which is true in the world together imply "Dr. Watson had tea with Holmes", which is true in the story.

Stipulationists assert a parity with the creative efforts of model-based science. Accordingly, *theorists* make sentences true in their models. These sentences are false in the world, or more generally, in the theory's domain of enquiry. Nevertheless, sentences true in the model also semantically engage with sentences true in the domain to produce further sentences that are true in the domain[37]. There is substantial agreement that a principal task for a theory of fiction is to produce a theory that takes proper account of the three facts noted in the paragraph just above. And there is growing agreement that a principal task for a stipulationist theory of models is that it take proper account of the similar-seeming trio of facts noted in the present paragraphs. If these tasks were to be successfully discharged, then we would have a principled account of truth-places for fiction. We would also have a principled account of truth-places for model-based science.

Fictionalism in science is the view that the success-conditions for a theory of truth-places for fiction and the success-conditions for a theory of truth-places for science reflect a notion of fictionality which both those theories instantiate. Currently in question is whether the common notion is the literary one adapted to scientific contexts[38], or whether the common concept is something more generic than that.

Of course, it hardly needs saying that not all stipulationists, that is, all who accept the above three claims – true in the model; false in the domain;

---

[37] But let us note in passing what appears to be a significant difference between the two cases. In the scientific case, fictional truths can mingle with real-world truths to generate new real-world truths. In literature, it is the other way around. Fictional truths semantically engage with real-world truths to generate further fictional truths.

[38] For the pro case, see e.g. Roman Frigg, "Fictions in science", in *Fictions and Models: New Essays*. And for the con case, see again our own "Unifying the fictional".

yet subject to semantic integration – are happy with a fictionalist account of them. This, too, is an open question in the relevant research communities.

A theory of truth-places carries certain suggestions, all of them alarming to philosophers with pre-conceived leanings. One is that "true" is at least a *two-place*, rather than unary, predicate. Another, relatedly, is that truth is subject to a *relativism* of place. A third, again relatedly, is that truth-places commit the theory of truth to a kind of *semantic pluralism*. A fourth is that a theory of truth-places doesn't rule the consequence, and is not intended to, that good scientific theories can flourish notwithstanding – indeed on account of – sentences false in the theory's domain, hence the negations of sentences true in it.

These are consequences that shock orthodox opinion. Clearly something has to give. Either truth-places must go, or orthodoxies must be relaxed. This sets up two options. The consequences must be suppressed, and with them the renegade theory which gave rise to them. The other is to regard the consequences as constituting new research options, new projects for enquiring minds. As of this writing, these are projects already well-underway. Concerning the binary character of the truth predicate, its relativity to places and the promptings it gives to pluralism, there is now a brisk trade in logical pluralism, and an equally brisk one in truth-in-fiction[39]. And concerning the inconsistencies engendered by the domain-falsity of model truths, there are prosperous and beckoning projects in paraconsistent logics, which are logics that effectively de-claw the classical menance of inconsistency[40].

---

[39] For the first, see for example, John Woods and Bryson Brown, editors, *Logical Consequence: Rival Approaches*, Oxford: Hermes Science, 2001, John Woods, *Paradox and Paraconsistency: Conflict Resolution in the Abstract Sciences*, Cambridge: Cambridge University Press, 2003, and JC Beall and Greg Restall, *Logical Pluralism*, New York: Oxford University Press, 2006. For the second (and for an extensive bibliography) see Woods, "Fictions and their logic", in Dale Jacquette, editor, *Philosophy of Logic*, pp. 835–900, a volume of the *Handbook of the Philosophy of Science*, edited by Dov M. Gabbay, Paul Thagard and John Woods, Amsterdam: North-Holland, 2005.

[40] See, for example, Newton da Costa, "On the theory of inconsistent formal systems", *Notre Dame Journal of Formal Logic*, XV (1974) 497–510; Dederik Batens, "Paraconsistent extensional propositional logic", *Logique et Analyse* 23 (1980), 195–234; Graham Priest, Richard Routley and Jean Norman, editors, *Paraconsistent Logic*, Munich: Philosophia Verlag, 1989; Dov M. Gabbay and Anthony Hunter, "Making inconsistency respectable", part 1, in P. Jourand and J. Keleman, editors, *Fundamentals of Artificial Intelligence Research*, 535 (1993), pages 19–32, Berlin: Springer Verlag, 1991; and part 2 in *LNCS* 747, pages 129–136. Berlin: Springer-Verlag, 1993; Newton da Costa and Otávio Bueno, "Consistency, paraconsistency and truth", *Ideas of Valores*, 100 (1996), 48–60, Woods, *Paradox and Paraconsistency*, 2003, Da Costa and French, *Science and Partial Truth*, 2005, Woods, "Dialectical considerations on the logic of contradiction I", *Logic Journal of the IGPL*, 13 (2005), 231–260, and Peter Schotch, Bryson Brown and Raymond Jennings editors, *On Preserving: Essays on Preservationism and Paraconsistent Logic*, Toronto: University of Toronto Press, 2009.

We said earlier that to the extent possible we want to preserve the data to which truth-places are a measured response. Orthodox capitulation marks an end to this fidelity; it sends the data packing. The research programmes just mentioned, and others underway or in prospect, offer the data a reprieve. In our view, this is the way to go. As we said, we are not prepared to surrender our data in the absence of a hard case to the contrary. But since the orthodoxies against which our data rub were not fashioned in reaction to them, giving them standing now would be to violate the methodological requirement to *pay them due attention*. So we will stick with truth-places until such time as arguments tailor-made against them force us to stand down.

6.1   Irrealism Without Tears

One of the morals suggested by the preceding pages is that, as standardly taken, the realist/irrealist divide is a piece of antique taxonomy that lags behind the relevant facts. Properly understood, realism-irrealism is not a *partition*. There are philosophical appreciations that fall between the two classical extremes and their shared presumption that winners take all. If the claims that we have advanced here can be made to stand, then there is a perfectly good sense of "irrealism" and a perfectly good sense of "true" according to which some irrealist theories give us a knowledge of D-phenomena by way of truths that hold not in D but only in the theorists' model M. And it is a sense of "irrealism" for which none of the classical options to realism is supportable, never mind necessary. If so it breathes new life into an old distinction. In its refreshed form, the distinction is relativized to truth-places. Color-recognizing theories are realist as regards the place in nature in which color-statements are true and irrealist as regards the place in nature in which the causes of color-perception exercise their powers. Idealized and abstractive science are realist with respect to the places they are true (namely in models) and are often irrealist with respect to which their target phenomena occur (namely in nature). The concept of truth-places accommodates the extra- and intra-domain distinction in the obvious way. And it helps us see what is surely present in actual scientific practice, namely, that since parts of theories can be true (and false) in different *places*, they can be both realist and irrealist – irrealist in relation to the places where their claims are false and realist in relation to the places where they are true. It is perfectly true that there are monomaniacal monists lurking about, according to whom valid science must recognize only one place for the true. There's no accounting for philosophical tastes it seems, but nothing in reflectively intelligent scientific practice lends any encouragement to this vaunting monoplism[41].

---

[41] For a detailed exploration of this theme, see again our "Model-based realism", to appear.

Pages ago, we said that to the extent possible we would try for a thick account of the cognitive virtuosity of model-based distortion without pre-supposing this or that pre-set philosophical position on knowledge. We bring this essay to a close with a further few remarks on this same theme. At a certain level of generality, most by far of the going positions in epistemology all cleave to a common notion about knowledge. They see knowledge as *discovery*. Thus to come to know that ϕ you discover ϕ's truth. The discovery paradigm cuts no slack to embedded practices in mathematics and theoretical science. These are the practices of stipulation, of making ϕ true rather than discovering it. Clinging to the discovery paradigm carries hefty philosophical costs for what in many ways are the paradigm cases of knowledge - indeed are the best ways of knowing things. An inflexible attachment to the discovery paradigm infests the mathematical and theoretical sciences with irrealism, and with it one or other of the set-piece reactions to it. We said at the beginning that we weren't going to stand still for this, that we weren't prepared to countenance the nonsense that our best ways of knowing things fail to produce knowledge. Any philosophical theory requiring us to concede that we don't have a knowledge of natural selection on the ground deserves outright rejection. But it is a rejection with consequences. One of them is that the discovery-paradigm *can't* be right for all of knowledge.

This is the right thing to reject. Although one way of knowing something is by discovering what's true of it, another is by making things true of it. This is knowledge, not by discovery, but by creation; and it calls to mind Quine's jape that the knowledge conveyed by theories is "free for the thinking up", and Eddington's that it is a "put-up job." What is so striking about created knowledge is not so much that it lies within our power to make, but that it integrates so robustly with discovered knowledge. This – the phenomenon of semantic engagement – achieves a deep grip. For there are cases in which discovered knowledge is impossible in the absence of created knowledge. This is the cognitive message of model-based science.

Acknowledgements. The first version of this paper was presented in June 2007 to the Canadian Society for the History and Philosophy of Science in Vancouver. For some helpful prodding, both *via voce* and epistolary, we extend our thanks to Anjan Chakravartty, Michael Weisberg, and most especially Michael Strevens. The present version was given as an invited lecture at the MBR09, a conference on model-based reasoning, meeting in December 2009 in Campinas. For welcome support and instructive criticisms we are indebted and grateful to Lorenzo Magnani and Walter Carnielli (the conference organizers), Jaakko Hintikka, Paul Thagard, Balakrishnan Chandrasedaran, Peter Bruza, and Cameron Shelley. For correspondence or conversation on the role of models in science, we also thank Patrick Suppes, Bas van Fraassen, Steven French, Margaret Morrison, Dov Gabbay, Otávio Bueno, Roman Frigg, Sally Otto, Michael Whitlock, Mauricio Suarez, Shahid Rahman, John Beatty, and Andrew Irvine.

Naturalizing Peirce's Semiotics: Ecological Psychology's Solution to the Problem of Creative Abduction

Alex Kirlik and Peter Storkerson

> It is difficult not to notice a curious unrest in the philosophic atmo-
> sphere of the time, a loosening of old landmarks, a softening of oppo-
> sitions, a mutual borrowing from one another on the part of systems
> anciently closed, and an interest in new suggestions, however vague, as
> if the one thing sure were the inadequacy of extant school-solutions.
> The dissatisfactions with these seems due for the most part to a feeling
> that they are too abstract and academic. Life is confused and super-
> abundant, and what the younger generation appears to crave is more
> of the temperament of life in its philosophy, even though it were at
> some cost of logical rigor and formal purity.
>
> William James (1904)

Abstract. The study of model-based reasoning (MBR) is one of the most
interesting recent developments at the intersection of psychology and the philosophy of science. Although a broad and eclectic area of inquiry, one central
axis by which MBR connects these disciplines is anchored at one end in theories of internal reasoning (in cognitive science), and at the other, in C.S.
Peirce's semiotics (in philosophy). In this paper, we attempt to show that
Peirce's semiotics actually has more natural affinity on the psychological side
with ecological psychology, as originated by James J. Gibson and especially
Egon Brunswik, than it does with non-interactionist approaches to cognitive
science. In particular, we highlight the strong ties we believe to exist between
the triarchic structure of semiotics as conceived by Peirce, and the similar
triarchic stucture of Brunswik's lens model of organismic achievement in irre-
ducibly uncertain ecologies. The lens model, considered as a theory of creative
abduction, provides a concrete instantiation of at least one, albeit limited,
interpretation of Peirce's semiotics, one that we believe could be quite fruitful
in future theoretical and empirical investigations of MBR in both science and
philosophy.

Alex Kirlik and Peter Storkerson
University of Illinois at Urbana-Champaign
e-mail: kirlik@illinois.edu

L. Magnani et al. (Eds.): Model-Based Reasoning in Science & Technology, SCI 314, pp. 31–50.
springerlink.com   © Springer-Verlag Berlin Heidelberg 2010

1 Introduction

As is often the case, psychologist-philosopher William James' observations of more than 100 years ago remain true today, especially so in the philosophy of science. And we do not think it surprising that it took the work of another psychologist-philosopher, Patrick Suppes [96], to provide perhaps the most convincing challenge of its era to the highly rigorous, formal, yet overly simplistic view that scientific theories expressed in a logical calculus are given meaning by the nature of their connections to empirical data in an unambiguous and logically direct way: "One of the besetting sins of philosophers of science is to overly simplify the structure of science" [96, p. 260]. The reasons we are not surprised to see an affinity between James and Suppes in this regard are twofold. First, and as noted by Suppes himself, "the branches of empirical science that have the least substantial theoretical developments often have the most sophisticated methods for evaluating evidence" [96, p. 260]. Second, the need for coherence in thought and action [78] continually nags at those of us who are practicing psychological or cognitive scientists to keep our speculations on the origins and nature of scientific knowledge grounded in our everyday experience of formulating hypotheses, designing experiments, and modeling and analyzing empirical data.

To admittedly oversimplify matters ourselves, the next significant developments in the philosophy of science that bear on the theme and thesis of this paper were made in the 1980s by authors such as van Frassen and Cartwright. In *The Scientific Image* (1980) van Frassen made tangible and influential strides toward advancing a "semantic", or intensively model-based view of theories and scientific activity. And in *How the Laws of Physics Lie* (1983), Cartwright provided a compelling argument that scientific laws cannot be read to be true of the world, but only of scientific models. These ideas are compelling to us as they are largely consistent with what we observe in both the practice and products of science, at least in our home disciplines.

Work by van Frassen, Cartwright and others has been persuasive in highlighting the centrality of model-based reasoning (MBR), both as a topic for psychological investigation and as a conceptual approach to the history and philosophy of science [65]. In our preparations for, attendance at, and reflections upon, the conference on which this paper is based, we became aware of the fact that many researchers in this community, and especially those coming from a philosophical perspective, are now giving serious attention to the ideas of C.S. Peirce, and in particular his framework of semiotics and concept of abduction, as central to moving MBR research forward. This is true at both the disciplinary level, within psychology and philosophy individually, and especially at the multi-disciplinary level, to continue to forge ever stronger linkages between psychology and the philosophy of science with the goal to better understand the nature of creative abduction.

Our intent to contribute toward achieving this goal in this paper draws upon our backgrounds in (A.K.) basic and applied cognitive science (the latter often called "human factors engineering" or "human-computer inter-action"), and (P.S.) theoretical and applied graphic design. As such, we are simultaneously engaged in understanding the principles underlying effective human learning and performance in interacting with the ecology, and also in creating ecologies to foster ever more effective and efficient interaction and communication.

Perhaps interestingly, we find that pursuing these activities, which might naively be described as theory and application, are not at all at odds, but are instead highly mutually reinforcing. Psychology is a woefully empirically under-constrained science, as its "data" are being continually manufactured in laboratories using one contrived task or another. As such, these data are largely artifactual rather than natural, and as such, provide a somewhat free floating empirical foundation, rather than one anchored in consensual agree-ments concerning direct observations of the physical world. In situations such as this, it would be perverse for psychological theorists to not warmly wel-come the additional source of empirical constraint provided by application. If, for example, information display designs created by appeal to a particu-lar psychological principle or theory result in predictably improved human learning and performance, it is more than likely that the principle or theory in question contains at least a grain of truth. As we create and then har-vest these grains, we strive to contribute toward scientific goals and practical purposes at one and the same time.

2 Peirce

Peirce (1839-1914) was a mathematician, a chemist, and a philosopher of ana-lytical bent. He was the founder of modern pragmatism: the view that things are what they do. In short a thing can only be known through its interactions with other things as those interactions disclose characteristics of that thing relative to the other things. Experience comes about through the interaction of an individual with the environment and others. Thus, all knowledge is ul-timately based on such concrete experience. Objects, like a tree or a dollar *as experienced* are cognitive objects, created in the mind through cognitive processes of interpretation, that is, semiosis. Peirce described semiosis as a triadic relationship of representamen, cognitive object and intepretant or sig-nification. The "representamen", also called a "sign vehicle", or "signal" is what comes into the eye or other sense organ. The cognitive object is what is perceived or apperceived. It is what the representamen is interpreted to be, for instance when a piece of paper is recognized as a dollar bill. The "inter-pretant", "significance" or "meaning" in the vernacular, is the content of the cognitive object, the notion of what being a dollar bill entails, that is, what money is, what can be bought with it, etc.

Peirce constructed a multi-faceted taxonomy of signs, starting with sym-bols, indexes and icons. Symbols are "arbitrary signs" in which the form of the sign is not related to its signification and the signification is assigned by convention. The dollar bill is a symbol. Its characteristics and markings are codes that have been assigned to connect the paper object as representamen to its cognitive object. Indexes are indicators. The angle of the sun and the shadows it projects can be used as clocks. A train can be used as a clock if one knows its schedule. These are natural signs. They reflect causal observations such as the movement of shadows. Icons function by having a similarity of resemblance or analogy. In a line chart, the line that rises as it goes to the right can be an iconic signification of "rising prices".

3 Abduction, Iconicity, Diagrammatic Thinking, and Model Building

Peirce was strongly anti-nominalist. In his view, knowledge is not based on concepts, but on interactions. For example, he discusses philosophers' accounts of the apprehension of self-evident truths as reflecting "the light of reason", nature or grace [79, pp. 12–13]. For Peirce, these accounts are theological rather than scientific. The history of science is equivocal, showing that "there is a natural light of reason, that is, that man's guesses at the course of nature are more often correct than could be otherwise accounted for, while the same facts actually prove that this light is extremely uncertain and deceptive, and consequently unfit to strengthen the principles of logic in any sensible degree" [79, p. 13].

How, then, does one get from observations to knowledge? Peirce's answer is in the chain of abduction, diagrammatic thinking, model building and empirical testing. The passage just quoted gives an account of abduction as a phenomenological sense of knowing or logically determining that something is or might be the case. Its track record is too good for abduction to be merely random guessing, yet it is probabilistic guessing nonetheless, and as we shall demonstrate in the remainder of this paper, it can be surprisingly adaptive or functional given that the conditions for learning through experience are met (e.g., the availability of timely and accurate feedback). Another aspect of abduction is the cognitive function of creating frameworks for interpreting and analyzing phenomena, which allows for the transition from solely intuitive to systematized knowledge: "An initial abduction makes a guess about how to formalize a given phenomenon, the deductive diagrammatic phase ... follows, and finally an inductive investigation concludes the picture, in which the diagrammatic result is compared to the actual empirical data" [94, p. 104].

Peirce's model of thinking is the diagram, which is a model of a phenomenon's structure/function; a diagram is "primarily an icon of relations" [81, p. 341]. It is an "operational account of similarity", [94, p. 90] which displays the structural relations of the object it diagrams in a way that enables one to see how the object functions. Looking at a diagram of a triangle, for example, involves using the physical diagram or figure drawn on paper as a sign that signifies first, the abstract triangle, with precisely straight lines of zero thickness and sharp vertices. From that cognitive object comes the next signification, of the abstract triangle as equivalent at any angle or scale, and finally, there is the abstract universal triangle as the combinations of lengths and angles that are possible in triangles [94, pp. 99–102]. These semiotic transformations of cognitive objects from the material figure on paper, to the abstract figure, to its potential permutations and structural interactions, mark the development of a mental model, which can be manipulated either mentally, or physically by drawing the diagram, folding and measuring, to reveal, in this case, the geometry of triangles. This is what prototyping in design and engineering does. Finally, Peirce states clearly that these diagrammatic models are not descriptions of things that exist in the universe, but models of what such things would be like were they to exist. Empirical research and experimental testing actually determines the relations of models to facts.

In short, Peirce's model is of the human being as a cognizing individual, in the world, whose knowledge has its origins in sensory experience (i.e., interaction with the world). One's knowledge is developed through a cognitive, semiotic system that is continually creating and computing models, building concepts and theories, and testing its models against reality. Peirce provides a philosophy of inquiry from which middle range theories and research methods can be developed, though he does not supply those theories and research tools.

4 Saussure

It would be convenient to end the discussion of semiotics here, but the general term has also been used to cover, conflate and confuse Peirce's semiotics with Fernand De Saussure's semiology. Semiology is also a theory of signs, but it is different from Peirce's semiotics. Semiology deals with symbols and language. It has a two-part theory of signification: the symbol or word and the meaning assigned to it. While Peirce explicitly bases his theory on the individual thinking human, Saussure's semiology is based on language as a collective set of meanings belonging to the culture, with individuals having little or no autonomy. By concentrating on language systems as primary constituents of cultural meanings, it enables language to be viewed as the primary source of meaning, rather than an expression of underlying cognitive function.

Semiology has its distinct uses and its notion of studying cultures through their languages has been enormously influential in the humanities, shaping twentieth century hermeneutics, anthropology, and cultural studies, enabling structuralism, post structuralism, deconstruction, and the "linguistic turn" in philosophy.

Further confusion has resulted from mixing semiotics and semiology. For example, Umberto Eco [23] described semiotics and semiology as a division of labor. In his "watergate" model, Eco described a system for regulating water flow in which a series of lights serve as arbitrary signs (symbols) indicating the flow and level of water according a code. He demonstrated that given such a code, it is possible to infer meanings, outside of those defined by the code. In this way, Eco wrapped the indexical sign, like the train that is used as a clock, around the coded arbitrary symbols. Eco used the same method in reverse order to describe recognizing a cat. Peirce's model is used in perceptual semiosis, seeing a shape as a cat, while the signification "cat" is a cultural/linguistic object, under the purview of semiology and the collective institution of language. This combining creates confusions and contradictions, as noted by Tomas Maldonado [67, pp. 119–123].

5   Semiotics and Semiology

Semiology was a key part of graphic design for much of the last century. It was quite useful as a tool in the design of the visual languages for specialized or multilingual contexts such as road signs and international gatherings. Peircian semiotics is used in technical communication and semiotic concepts are used in human factors to decompose and analyze interpretation in human judgment. Semiotics is also used in human-computer interaction, design of virtual environments and education [77]. Semiotics has great potential as a framework to unify quickly developing but scattered literatures in naturalistic thinking as they are relevant to design. The semiotic model of diagrammatic thinking is of great importance, making possible a comprehensive understanding not only of diagrams, but the principles behind visual and spatial thinking, and abstraction in general (cf. Vorms, this volume [101]). It demonstrates the profound importance of graphical communication in the human leap from experiences in the world to the ability to think about those experiences in abstract terms: to make order of what is and to imagine what could be.

6   Reception of Semiotic Theory in Graphical and Industrial Design

For historical reasons, graphical and industrial design programs are most often located in art schools or departments, and are influenced by them. Academic fine arts cultures are often both humanist and decidedly anti-science. Here is one educator's reaction not only to semiotics, but to theory in general:

> Semiotics is academic and abstract. I would venture that for many studio instructors, theory is simply beside the point. Better to discuss successful graphic design or the art canon with students and let them get to work. [14]

Professional graphic designers are often similarly inclined. Whatever influence semiotics and semiology have had in practice, they have often been viewed as problematical theories for many areas of design. Maldonado [67] criticized semiotics in a number of ways:

The attempt to make use of a semiotic set of ideas to describe communicative (and even aesthetic) phenomena in the fields of architecture, urbanistics, and "industrial design" have not yielded the results that many expected, for may reasons, but above all for the lack of maturity in the semiotic itself. [67, p. 119]

This "lack of maturity" was reflected in semiotics-semiology confusion and the differing interpretations of Peirce by later theorists, but, particularly, the problem of operationally applying semiotics:

The semiotics [or the semiology] of architecture still remains at the metaphorical level. It would seem that, up to now, all efforts have been directed exclusively toward a substitution of the terminology of another, and little more. [67, p. 123]

That semiotics has not been widely used in design does not mean that it is not appropriate to the task of informing design. In fact, Peirce's semiotics is disciplined and well defined, and design fields based in science or engineering programs, rather than the arts, have been able to put semiotic thinking to good use. They have done so not much by consciously drawing on Peirce's works *per se*, but instead by drawing upon scientific theory that, to us, can be viewed as a quite legitimate and powerful naturalization of semiotics. We now turn our attention to this theory.

7 Ecological Psychology: James J. Gibson and Egon Brunswik

Although it is possible to trace selected aspects of ecological psychology to its functionalist origins in William James or John Dewey (see Heft [45]), ecological psychology emerged in mature form in the middle of the 20th century in the pioneering work of Egon Brunswik [9, 11] and James J. Gibson [33, 34, 35] on the problem of perception. These two theorists rejected idealist and Gestalt theorizing of the day, which stood upon an empirical foundation of perceptual biases, illusions, and errors, cataloged by laboratory scientists who thought they were investigating the fundamental aspects of perception by presenting simple stimuli to subjects using bite bars and tachistoscopes. Brunswik and Gibson agreed on the primary importance of a psychology focused on *organism-environment relations* (rather than solely the organism, human, or brain alone), a focus that defines what it means to take an ecological approach to psychology. These theorists also agreed that psychology should strive first and foremost to understand functional achievement, rather than to be satisfied merely by accumulating a body of findings on the frailties of perception in illusions, biases, or other errors. For more elaborate discussions of the relationships between Gibsonian and Brunswikian theory and method, see Kirlik [57, 52], Vicente [48], and Araujo, Davids and Passos [3].

Brunswik once followed a student around the Berkeley campus, cataloging her judgments of object sizes and distances, and found a much more accurate perceptual system than would have been expected considering the laboratory findings of the day. Gibson's early studies of military aviators alerted him to the existence of more perceptual richness and dynamism in optical stimulation than was then ever made available to laboratory subjects. His realist perceptual theory was based on the assumption that perception evolved to access this molar information, and thus what was then being observed in the laboratory was how perception had to contort itself to perform the "fundamental" task of accessing briefly displayed stimuli.

Brunswik and Gibson disagreed, however, on at least a few points. Most notably (and perhaps most fundamentally), they disagreed on whether the organism-environment relationship was fully deterministic and thus "lawful" (Gibson) or instead largely probabilistic (Brunswik). Brunswik once characterized his position on this matter with the statement "God may not gamble, animals and humans do" [10, p. 236]. In contrast, Gibson noted his discomfort with this position in his (1957/2001) *Contemporary Psychology* review of Brunswik's (1956) book, the latter largely representing Brunswik's life's work. Characterizing Brunswik's position, Gibson [32] wrote:

> Perception is based on insufficient evidence but, surprisingly, it is generally correct. By rights, the animal should not have functional contact with the environment, and yet it does. In his struggle with this dilemma, Brunswik never took refuge in subjectivism or in the sterile theory of the private phenomenal world. He was too well aware that functional behavior demands veridical perception. Nor would he accept the Gestalt theory of a brain process which would spontaneously produce the correct object in phenomenal experience. Instead he was driven to consider the difficult position of supposing that both the perceptual process and the environment itself are probabilistic, that is to say, imperfectly lawful. This is not a comfortable theory. Brunswik himself could not rest comfortably in the lap of uncertainty. Nevertheless he disciplined himself to make a virtue of what he considered a necessity. (p. 245)

Despite Gibson's discomfort with Brunswik's probabilism, he concluded his review by stating "His [Brunswik's] work is an object lesson in theoretical integrity" [32, p. 246].

8   Brunswik's Probabilistic Functionalism

Brunswik's ([9, 11]; also see Hammond & Stewart [44]) ecological perspective on perception is reflected in his theory of "probabilistic functionalism". Brunswik's frame is pragmatic. The organism (a human or any other creature that acts in the world) seeks to act appropriately with the environment, in order to further its goals [97, p. 13]. This is the objective level at which the organism succeeds or fails–it stops at the cliff or falls off. To succeed and survive, it needs an internal model of its environment that functionally corresponds that environment in terms of the organism's ability to perceive and act. The organism's cognitive job is to use "proximal" sensory information it receives as indexes, signifying objects and events comprising the "distal" environment, to make that environment predictable. This is difficult in natural environments, because a distal cause in the environment can have any number of proximal sensory effects, and proximal effects (or stimuli) can have many distal causes.

The organism receives sensory information in different modes (sight, sound, touch) and from different organs (eyes, ears, skin). There is often redundancy between sensory inputs, and the organism integrates and weighs those various indicators in order to come up with a reliable picture or what is happening to what (seeing and hearing the hammer hit the nail). Put simply, by weighing many sensory signals, any of which can be in error, a very high degree of reliability is possible. People rely on their senses to perceive their environments, and their senses are generally highly reliable, and even more reliable when considered collectively rather than in isolation.

Brunswik' crystalized his approach in his "lens model" of perception, as shown in Figure 1. It models the functional correspondence between the environment and the organism's representation of the environment. The initial focal variable, which is the distal object, is available to the organism through a series of mediating sensory signals or signs, which Brunswik called "cues", along with spurious noise and errors. The organism's achievement of a "stable relationship" or functional correspondence of the terminal focal variable with respect to the environment, is effected through "vicarious functioning", in which the organism decides which signs or cues to pay attention to and what distal object, event, or property they signify. Most of the time, this happens spontaneously and below awareness. We human beings, for example, do not actually experience the proximal light on our retinas (the initial focal variable). We see the distal scene of objects around us (final focal variable), and we see them as the same objects ("stable relationship") under widely differing conditions of light, distance and angle. This is an achievement of perceptual interpretation, in which many different "cues" are weighed, so that we spontaneously see the clock on a distant church tower as bigger than the alarm clock on the night table next to us.

The lens model takes its name from the similarity between a graphical representation of the encounter between an organism and the environment and a convex lens [9, p. 20]. Note that the lens model was meant to describe both an organism's perceptual encounter with the environment as vicariously (opportunistically) mediated by a number of perceptual cues, as well as its overt, behavioral encounter with the environment as mediated by a number of actions, means-to-ends, or "habits". In the current context, however, we will mainly restrict the following discussion to solely perceptual encounters for simplicity.

Brunswik [9, 11] originally proposed measuring functional achievement (the top arc spanning, and connecting, the "terminal focus variable" or environmental criterion, and the observer's perceptual judgment) by linear association, or bivariate correlation. Achievement is therefore measured by

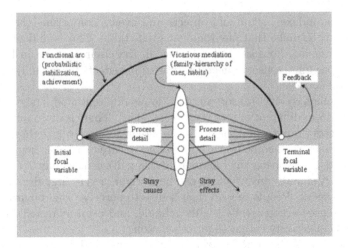

Fig. 1 The lens model (from Brunswik [9, p. 20]).

correlating the observer's judgments with the true values of the perceptual variable being judged. Hammond [43] adopted this measure and first introduced the use of multiple linear regression to create a multiple correlation measurement of judgment as a function of fallible or probabilistic cues.

9   The Lens Model Equation

Figure 2 depicts a modern version of the lens model with associated quantitative variables and measures labeled. Achievement in perceptual judgment is indicated by the correlation coefficient r_{YO}, where the subscripts represent the observer's judgment (Y) of an environmental "Object" of judgment (O). Perhaps the most important extension of Brunswik's original theory of probabilistic functioning was the development of the lens model equation [47, 99]. The lens model equation (LME) provides a mathematical representation of the lens model and partitions the overall correlation represented by the level of achievement (r_{YO}) into correlations related to the "ecological validities" or "trustworthiness" [9, 11] of the perceptual cues (that is, their statistical correlation the criterion - the "ev" values in Figure 2), the observer's cue utilizations (the "cu" values in Figure 2), the overall predictability of the environment, and the consistency with which an observer implements his or her perceptual judgment (cue-weighting) strategy.

At the basis of the LME are two parallel (typically linear regression) models, which represent the environmental and the observer sides of the lens model shown in Figure 2. The environmental model describes the overall correspondence between the cues (X_i's) and the object of judgment (O), and the observer model describes the overall correspondence between the cues (X_i's)

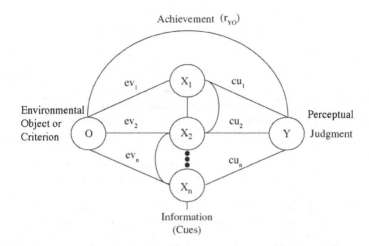

Fig. 2 The lens model with variables and measures identified (after Hammond [43]).

and the observer's judgments (Y). Based on these two models, the resulting decomposition of achievement is depicted conceptually in Equation 1.

$$(r_{YO}) = \text{Environmental Predictability} \times \text{Consistency} \times \text{Knowledge} + $$
$$+ \text{ Unmodeled Knowledge} \quad (1)$$

Environmental Predictability is the degree of correspondence between the perceptual cues and ecological criterion or object of perceptual judgment as reflected in the environmental model. *Consistency* is the degree of correspondence between the cues and observers' judgments as reflected in the observer model. *Knowledge* is the degree of correspondence between the outputs or predictions of the environmental and the observer models. This component is called *Knowledge* because it indicates the degree to which the observer correctly adjusts to the regularities of the ecology, or weights the cues adaptively (i.e., in accordance with their true ecological validities). *Unmodeled Knowledge* in Equation 1 is the degree of correspondence between the unpredictable portions of the environmental and observer models. Typically, but not always, this value is found to be marginal in human judgment [8]. For a meta-analysis of lens model applications in more than 250 tasks, see Karelaia & Hogarth [50].

The decomposition shown in Equation 1 is accomplished with multiple linear regression. Thus, *Environmental Predictability, Consistency, Knowledge*, and *Unmodeled Knowledge* are measured using multiple correlation statistics. The mathematical form of the LME is shown in Equation 2, and its components are summarized below in Table 1.

$$r_{YO} = R_{O.X} G R_{Y.X} + C\sqrt{1 - R_{Y.X}^2}\sqrt{1 - R_{O.X}^2} \quad (2)$$

Table 1 Components of the lens model equation.

Component	Name	Description
r_{YO}	Achievement	Correlation between perceptual judgments and the task criterion
$R_{O.X}$	Criterion (Environmental) Predictability	Correlation between perceptually available variables or cues and the ecological criterion.
G	Knowledge	Correlation between predictions of a cue/variable-criterion model of the environment and the cue-response model of the observer.
$R_{Y.X}$	Perceptual or Cognitive Control or Consistency	Correlation between perceptual cues/variables and perceptual judgments.
C	Unmodeled Knowledge or Ability	Correlation between the residuals of environmental and observer models: Adaptivity to ecological structure that is not captured in the scientists' models.

It is natural to identify lens model parameter G (which measures the degree to which the observer's cue utilization pattern corresponds to the actual ecological validities of the cues) as a measure of the current level of "education of attention" that the observer has achieved [34]. As discussed by Jacobs and Michaels [48], education of attention is Gibsonian terminology for a perceptual learning process in which observers "converge on more useful nonspecifying variables [cues] or even on variables that specify the to-be-perceived properties" [48, p. 131]. Because G reflects the current degree of exactly this form of adaptivity, it is technically more precise to identify positive changes in G as the education of attention, where any current, particular level of G reflects the product of the learning or education process at any point in time.

Brunswik [9, 11] believed that if anything is likely to be fundamental and universal about human perception, it is exactly this context-specific, adaptive and highly opportunistic mode of cue search and utilization. Brunswik thought that this contextual and opportunistic aspect of perception (and indeed of all behavior, including habits and the achievement of distal goals by proximal means or actions – see [98], that is, *vicarious functioning*, was so fundamental that he constructed his entire theory of probabilistic functionalism and methodology of representative design around it (in addition to Brunswik's original work see Hammond & Stewart [44]).

As noted by Goldstein [42], Brunswik borrowed the term vicarious functioning from Hunter [46]. Hunter noted that, whereas physiological functions normally carried out by one organ can rarely be carried out by another, a fundamental aspect of behavior studied by psychologists was that, when

parts of the body typically used to carry out some type of behavior are unavailable, other parts of the body are often vicariously (opportunistically,
adaptively) recruited to perform that behavior. Brunswik mainly used the
term vicarious functioning to highlight the context-specific, highly opportunistic inter-substitutability of cues used on various occasions to support
perception.

In essence, Brunswik's view was that if what was truly fundamental about
perception was locally adaptive opportunism, rather than globally-adaptive
laws, achievement in any particular ecological situation would have to be
acquired through a feedback-guided, abductive process of trial and error
learning. Much recent research conducted from a Brunswikian tradition (e.g.,
Gigerenzer & Goldstein [37]; Gigerenzer & Selten [40]; Gigerenzer, et al. [41];
Rothrock & Kirlik [86]) finds that, when accurate feedback is available, people do learn to gravitate to using the most useful cues available to them
in a particular ecology. As such, those working in a Brunswikian tradition
recommend a shift away from a search for "global cognitive principles", and
toward an "adaptive toolbox" [40] of largely "local solutions to local problems", (Gigerenzer & Goldstein, [37]; Gigerenzer & Kurz [38]).

Not surprisingly, then, research in this Brunswikian tradition takes a positive attitude toward passing epistemic judgment on these "local solutions" or
"heuristics". In the Brunswikian tradition, a person using a locally adaptive
(yet perhaps globally fallible) heuristic or solution is hardly viewed as being
less than fully rational. Nor is that person viewed as demonstrating the limitations of cognitively constrained "bounded rationality" [89]. Instead, such
a person is viewed as using the most elegant and robust solution to the ecological challenge at hand (see Todd [64]).

That is, think of this set of ecologically local solutions to ecologically local problems (i.e., cognitive competence or even expertise, which is always
restricted or local in some sense) as a product of evolution akin to the evolution of multiple species of plants and animals. Each has a niche, and none,
including Homo sapiens, can simultaneously outperform all of those species
more finely, albeit perhaps more narrowly, attuned to their particular niche
or sub-ecology. Just as evolution has not been able to produce a context-free
super-organism capable of eradicating all of the more ecologically narrowly
located species, it has similarly been impossible for evolution to create a
Laplacean super-mind [37] or super-observer as an engine for context-free,
global rationality, as was once envisioned in the early, heady days of AI-
inspired cognitivism, and in what was once called the "received view" in the
philosophy of science. An attractive alternative to such a super-mind (e.g., a
fully general logical calculus of rationality) is a menagerie of narrower and locally restricted, yet reliable and robust, heuristic solutions to local problems,
akin to an ecologically diverse and locally, rather than globally, adapted set
of species.

10   Integration

Gibson [34, 35] put forth a radical epistemology placing the lion's share of the burden for functional or adaptive behavior on the ecology and evolution. In his view, the world's lawful nature, and its almost limitless supply of information available in ambient energy arrays, combined with evolutionary time scales, equip organisms to seize on these resources to solve (or perhaps more appropriately, dissolve) the majority of their epistemic problems (e.g., Michaels and Carello [71]). In contrast, early, classical approaches to cognitive-representational cognitive science (e.g., Fodor [26]; Fodor & Pylyshyn [27]) instead placed nearly the entire burden for functional or adaptive behavior on largely context-free, internal, cognitive computation and ratiocination, assuming a nearly totally disordered, and thus minimally informative ecology (cf. Cutting [16]).

We now know that neither of these extreme positions is tenable. A hearty dialogue and debate between, and among, those espousing views at opposing poles (nature vs. nurture, genes vs. environment, etc.) may be an efficient path forward in science. But we also know that largely dichotomous arguments such as these hardly ever result in truth being found at one pole or the other, but instead, and often inconveniently for those involved, lying stubbornly and (apparently) inelegantly in a theoretical middle ground.

Due to exactly this *apparent* inelegance, Brunswik's [9, 11] sophisticated, Darwinian theory of probabilistic functionalism did not receive a warm reception when it was proposed. In Brunswik's day, the debate between dichotomous empiricist and rationalist poles in psychology, though having a long tradition in philosophy, was only beginning to be informed by experimental evidence. Why accept an apparently inelegant middle ground when much more apparently elegant, fully empiricist and fully rationalist, positions were then still empirically viable?

Today, they are not, and the empirical research reviewed here indicates a convergence to a quite elegant, Darwinian solution able to accommodate aspects of both Gibson's [34, 35] theory of direct perception and Brunswik's [9, 11] theory of probabilistic functionalism. For empirical research demonstrating the utility of drawing simultaneously on both Brunswikian and Gibsonian theory see Kirlik [55, 53, 54, 56] and Davids, Button, Araujo, Renshaw & Hristovski [17]. Recent Brunswikian research (e.g., Gigerenzer & Engel [36]; Gigerenzer & Goldstein [37]; Gigerenzer & Selten [40]; Gigerenzer et al. [41]) is extending this approach in creative directions.

Further evidence for a convergence in this direction can be found in alternative, yet possibly complementary research programs emerging with one foot planted firmly in ecological-evolutionary insights and themes (e.g. J. R. Anderson's rational analysis – see Anderson [1]) and the other planted firmly in the organism's cognitive contribution (e.g., J. R. Anderson's ACT-R – see Anderson, Bothell, Byrne, Douglass, Lebiere & Quin [2]). Rational analysis and ACT-R can be viewed as complementary to Brunswik's and Gigerenzer's programs in many respects, albeit grounded in Bayesian rather than correlative statistics, and taking different phenomena as their core explananda (for additional discussion, see Kirlik [54, 57]). Each of these studies illustrates the successful use of a scientific theory in which the organism's cognitive contribution was assumed to consist largely in the reflection of, and adaptive response to the ecological regularities present in some particular (local) region of the human ecology.

11 Application

In psychology, Hammond & Stewart [44] present a broad range of successful applications of probabilistic functionalism and the lens model in areas such as perceptual size constancy, interpersonal perception and interpersonal learning, social perception and moral judgment, medical judgment, vicarious functioning in teamwork, cognitive aging, emotional communication in musical performance, guidance counseling of adolescents, and evolutionary psychology. In human factors and human-computer interaction, Kirlik [54] presents a range of successful applications of probabilistic functionalism and the lens model in display design, fault diagnosis, searching for information on the Internet, collision avoidance in commercial aviation (both in air, and during surface taxi navigation), tactical judgment and decision making, the effects of time pressure on team performance, training, the design of decision aids, and achieving an effective coupling between people and cognitive automation.

12 Conclusion: Peirce Revisited

The large and rapidly growing body of work embracing and profitably using Brunswik's probabilistic functionalism, much of it design related, stands in sharp contrast to the way in which Peirce's semiotics has been (unprofitably) received in design disciplines based in the arts and humanities. We believe that this difference owes to the fact that Brunswik's triarchic, criterion-cue-judgment theory expressed in the lens model is tangible in a manner that Peirce's similar triarchic theory of semiotics is not. While semiotics is certainly more sweeping and encompassing than probabilistic functioning, this conceptual richness comes at a steep price.

We suggest that probabilistic functionalism, and more concretely, Brunswik's lens model, can legitimately be considered as a scientific naturalization of Peirce's semiotics. This theory and model captures how abductive knowledge is gained by an organism's ability to learn cue-criterion (Brunswik) or sign-referent (Peirce) relations over time, and in the presence of noise, through closed-loop interaction with the world that generates feedback revealing these relations. We would hope that this theory and model, viewed as a naturalization of semiotics, might similarly provide the philosophy of science with concrete resources to further its own inquiries into model-based reasoning and the nature of abduction.

Authors Note

Portions of this paper have been drawn, in amended form, from Storkerson, P.: Antinomies of semiotics in graphic design. *Visible Language* **44**, 6–39 (2010), Special Issue: Communication Design Failures, S. Poggenpohl, D. Winkler (eds.); and from Kirlik, A.: Brunswikian resources for event perception research. *Perception* **38**, 376–398 (2009).

References

1. Anderson, J.R.: The Adaptive Character of Thought. Lawrence Erlbaum Associates, Hillsdale (1990)
2. Anderson, J.R., Bothell, D., Byrne, M.D., Douglass, S., Lebiere, C., Quin, Y.: An integrated theory of the mind. Psychological Review 111, 1036–1060 (2004)
3. Araujo, D., Davids, K., Passos, P.: Ecological validity, representative design, and correspondence between experimental task constraints and behavioral setting: Comments on Rogers, Kadar, and Costall. Ecological Psychology 19(1), 69–78 (2007)
4. Barthes, R.: Elements of Semiology. Hill and Wang, New York (1964)
5. Bonsiepe, G.: Visual/verbal rhetoric. UIM 14/15/16, 23–40 (1965)
6. Bonsiepe, G.: A method for quantifying order in typographic design. The Journal of Typographic Research 2(3), 203–220 (1968)
7. Bonsiepe, G.: The uneasy relationship between design and design research. In: Mitchel, R. (ed.) Design Research Now, pp. 25–39. Springer, Berlin (2008)
8. Brehmer, B.: The psychology of linear judgment models. Acta Psychologica 87, 137–154 (1994)
9. Brunswik, E.: The Conceptual Framework of Psychology. University of Chicago Press, Chicago (1952)
10. Brunswik, E.: Representative design and probabilistic theory in a functional psychology. Psychological Review 62, 193–217 (1955)
11. Brunswik, E.: Perception and the Representative Design of Psychological Experiments. University of California Press, Berkeley (1956)
12. Candlin, F.: Practice-based doctorates and questions of academic legitimacy. International Journal of Art and Design Education 19(1), 96–101 (2000)
13. Cartwright, N.: How the Laws of Physics Lie. Oxford University Press, New York (1983)
14. Crisp, D.: Book review of visible signs: An introduction to semiotics and the fundamentals of creative design. Print 58(2), 34 (2004)
15. Crotty, M.: The Foundations of Social Research: Meaning and Perspective in the Research Process. Sage Publications, London (1998)
16. Cutting, J.E.: Two ecological perspectives: Gibson vs. Shaw and Turvey. The American Journal of Psychology 95(2), 199–222 (1982)
17. Davids, K., Button, C., Araujo, D., Renshaw, I., Hristovski, R.: Movement models from sports provide representative task constraints for studying adaptive behavior in human movement systems. Adaptive Behavior 14(1), 73–95 (2006)
18. Deely, J.: Basics of Semiotics. University of Indiana Press, Bloomington (1990)

19. Basel School of Design Visual Communication Institute: Welcome (2009), http://www.baselschoolofdesign.ch (Retrieved, June 23, 2009)
20. Domandi, M., Mead, M.: Anthropology and glyphs. Print 23(6), 50–53 (1969)
21. Dreyfus, H.: Symbol Sourcebook. Van Nostrand Reinhold, New York (1984)
22. Dreyfus, P.: Design education today: Turmoil and transition. Print 23(5) (1969)
23. Eco, U.: A Theory of Semiotics. Indiana University Press, Bloomington (1979)
24. Ehses, H.: Representing Macbeth: A case study in visual rhetoric. Design Issues 1(1), 53–63 (1984)
25. Flyvberg, B.: Making Social Science Matter: Why Social Inquiry Fails and how it Can Succeed Again. Cambridge University Press, Cambridge (2001)
26. Fodor, J.A.: Methodological solipsism considered as a research strategy in cognitive psychology. Behavioral and Brain Sciences 3, 63–110 (1980)
27. Fodor, J.A., Pylyshyn, Z.W.: How direct is visual perception? some reflections on Gibson's ecological approach. Cognition 9, 139–196 (1981)
28. Frascara, J.: User-Centered Graphic Design: Mass Communications and Social Change. Taylor and Francis, London (1997)
29. Frassen van, B.: The Scientific Image. Oxford University Press, New York (1980)
30. Frutiger, A.: Signs and Symbols: Their Design and Meaning. Studio Editions, Zurich (1978/1989)
31. Frutiger, A.: Type Sign Symbol. Editions ABC, Zurich (1980)
32. Gibson, J.: Survival in a world of probable objects. In: Hammond, K.R., Stewart, T. (eds.) The Essential Brunswik, pp. 244–246. Oxford University Press, Oxford (2001) (Reprinted)
33. Gibson, J.J.: The Perception of the Visual World. Houghton Mifflin, Boston (1950)
34. Gibson, J.J.: The Senses Considered as Perceptual Systems. Houghton Mifflin, Boston (1966)
35. Gibson, J.J.: The Ecological Approach to Visual Perception. Lawrence Erlbaum, Hillsdale (1986), Originally work published in 1979
36. Gigerenzer, G., Engel, C.: Heuristics and the Law. MIT Press, Cambridge (2006)
37. Gigerenzer, G., Goldstein, D.G.: Reasoning the fast and frugal way: Models of bounded rationality. Psychological Review 103, 650–669 (1996)
38. Gigerenzer, G., Kurz, E.M.: Vicarious functioning reconsidered: A fast and frugal lens model. In: Hammond, K.R., Stewart, T.R. (eds.) The Essential Brunswik, pp. 342–347. Oxford University Press, New York (2001)
39. Gigerenzer, G., Murray, D.: Cognition as Intuitive Statistics. Lawrence Erlbaum, Hillsdale (1987)
40. Gigerenzer, G., Selten, R.: Bounded Rationality: The Adaptive Toolbox. MIT Press, Cambridge (2001)
41. Gigerenzer, G., Todd, P.M.: The ABC Research Group: Simple Heuristics that Make Us Smart. Oxford University Press, New York (1986)
42. Goldstein, W.M.: Introduction to Brunswikian theory and method. In: Kirlik, A. (ed.) Adaptive Perspectives on Human-Technology Interaction, pp. 10–24. Oxford University Press, New York (2006)
43. Hammond, K.: Probabilistic functionalism and the clinical method. Psychological Review 62, 255–262 (1955)

44. Hammond, K., Stewart, T.R.: The Essential Brunswik. Oxford University Press, Oxford (2001)
45. Heft, H.: Ecological Psychology in Context: James Gibson, Roger Barker, and the Legacy of William James Radical Empiricism. Lawrence Erlbaum, Mahwah (2001)
46. Hunter, W.S.: The psychological study of behavior. Psychological Review 39, 1–24 (1932)
47. Hursch, C.J., Hammond, K.R., Hursch, J.L.: Some methodological considerations in multiple-cue probability studies. Psychological Bulletin 71, 42–60 (1964)
48. Jacobs, D.M., Michaels, C.F.: On the apparent paradox of learning and realism. Ecological Psychology 14, 127–139 (2002)
49. James, W.: A world of pure experience. Journal of Philosophy, Psychology, and Scientific Methods 1(20, 21) (1904)
50. Karelaia, N., Hogarth, R.M.: Determinants of linear judgment: A meta-analysis of lens studies. Psychological Bulletin 134(3), 404–426 (2008)
51. Kinross, R.: Semiotics and designing. Information Design Journal 4(3), 190–198 (1986)
52. Kirlik, A.: On Gibson's review of Brunswik. In: Hammond, K.R., Stewart, T.R. (eds.) The Essential Brunswik, pp. 238–242. Oxford University Press, New York (2001)
53. Kirlik, A.: Reiventing intelligence for an invented world. In: Sternberg, R.J., Preiss, D.D. (eds.) Intelligence and technology: The impact of tools on the nature and development of human abilities, pp. 105–134. Lawrence Erlbaum Associates, Mawhah (2005)
54. Kirlik, A.: Adaptive Perspectives on Human-Technology Interaction: Methods and models for cognitive engineering and human-computer interaction. Oxford University Press, New York (2006)
55. Kirlik, A.: Requirements for psychological models to support design: Toward ecological task analysis. In: Flach, J., Hancock, P., Caird, J., Vicente, K. (eds.) Global Perspectives on the Ecology of Human-Machine Systems, pp. 68–120. Lawrence Erlbaum Associates, Mahwah (2006)
56. Kirlik, A.: Ecological resources for modeling embedded cognition and interactive behavior. In: Gray, W. (ed.) Integrated Models of Cognitive Systems, pp. 194–210. Oxford University Press, New York (2007)
57. Kirlik, A.: Brunswikian resources for event perception research. Perception 38(3), 376–398 (2009)
58. Koffka, K.: Principles of Gestalt Psychology. Brace and World, Harcourt (1935)
59. Krampen, M.: Symbols in graphic communication. Design Quarterly 62, 1–31 (1965)
60. Kripopendorff, K.: The Semantic Turn: A New Foundation for Design. Taylor and Francis, London (2006)
61. Lakoff, G., Johnson, M.: Philosophy in the Flesh. Basic Books, New York (1999)
62. Lausen, M.: Design for Democracy: Ballot and Election Design. University of Chicago Press, Chicago (2007)
63. Lausen, M.: Designing change. Design matters. Univeristy of Illinois, Urbana Champaign (2009),
http://designmatters.art.uiuc.edu/Videos/marcia-lausen-2/
(Retrieved, June 24, 2009)

64. Todd, P.M.: Fast and frugal heuristics for environmentally bounded minds. In: Gigerenzer, G., Selten, G. (eds.) Bounded Rationality: The Adaptive Toolbox, pp. 51–70. MIT Press, Cambridge (2002)
65. Magnani, L., Nersessian, N.J., Thagard, P. (eds.): Model-Based Reasoning in Scientific Discovery. Kluwer Academic/Plenum Publishers, New York (1980)
66. Maier, M.: Basic Design Principles: The Foundation Program at the School of Design. Van Nostrand Reinhold Publishers, Basel (1977)
67. Maldonado, T.: Nature, Design and Revolution: Toward a Critical Ecology. Harper & Row, New York (1972)
68. Maldonado, T., Bonsiepe, G.: Science and design. ULM 10/11, 10–16 (1965)
69. Manning, A., Amare, N.: Visual-rhetoric ethics: Beyond accuracy and injury. Technical Communications 53(2), 195–211 (2006)
70. Marcus, A.: m-locos ui: A universal visible language for global mobile communication. In: Jacko, J.A. (ed.) HCI 2007. LNCS, vol. 4552, pp. 144–153. Springer, Heidelberg (2007)
71. Michaels, C.F., Carello, C.: Direct Perception. Prentice-Hall, Englewood Cliffs (1981)
72. Moles, A.: The legibility of the world: A project of graphic design. Design Issues 3(1), 43–53 (1986)
73. Morris, C.: Foundations of the Theory of Signs. University of Chicago Press, Chicago (1938)
74. Neurath, M.: Isotype. Instructional Science 3(2), 127–150 (1974)
75. Neurath, O.: BASIC by Isotype. Kegan Paul, Trench, Trubner & Co., New York (1937)
76. Ockerse, T., van Dijk, H.: Semiotics and graphic design education. Visible Language 13(4), 358–378 (1979)
77. de Oliveira, O., Baranauskas, M.: Semiotics as a basis for educational software design. British Journal of Educational Technology 31(2), 153–161 (2000)
78. Thagard, P.: Coherence in Thought and Action. MIT Press, Cambridge (2002)
79. Peirce, C.: Logic. In: Hartshorne, C., Weiss, P. (eds.) The Collected Papers of Charles Sanders Peirce, vol. I-II. Harvard University Press, Cambridge (1933)
80. Peirce, C.: Pragmatism and pragmaticism. In: Hartshorne, C., Weiss, P. (eds.) The Collected Papers of Charles Sanders Peirce, vol. V. Harvard University Press, Cambridge (1933)
81. Peirce, C.: The simplest mathematics. In: Hartshorne, C., Weiss, P. (eds.) The Collected Papers of Charles Sanders Peirce, vol. V. Harvard University Press, Cambridge (1933)
82. Polanyi, M.: The Tacit Dimension. Routledge & Kegan Paul, London (1966)
83. Ramachandran, S.: Purple numbers and sharp cheese (2003), http://www.bbc.co.uk/radio4/reith2003/lecture4.shtml, BBC Reith Lectures 2003: The Emerging Mind. BBC Radio 4. (Retrieved June 1, 2009)
84. Ramachandran, V., Hubbard, E.: Synesthesia: What does it tell us about the emergence of qualia, metaphor, abstract thought, and language? In: van Hemmen, J., Sejnowski, T. (eds.) 23 Problems in Systems Neuroscience, pp. 432–473. Oxford University Press, Oxford (2006)
85. Ramachandran, V., Hubbard, E., Butcher, P.: Synesthesia, cross-activation and the foundations of neuroepistemology. In: Robertson, L., Sagiv, N. (eds.) Synesthesia: Perspectives from cognitive neuroscience, pp. 147–190. Oxford University Press, Oxford (2004)

86. Rothrock, L., Kirlik, A.: Inferring rule-based strategies in dynamic judgment tasks. IEEE Transactions on Systems, Man, and Cybernetics: Part A: Systems and Humans 33(1), 58–72 (2003)
87. Rust, C.: Unstated contributions – how artistic inquiry can inform interdisciplinary research. International Journal of Design 1(3), 69–76 (2008)
88. Saussure, F.: Course in General Linguistics. In: Baskin (Tr), E., Bally, C., Reidlinger, A. (eds.) Philosophical Library (1966)
89. Simon, H.: Models of Man. Wiley and Sons, New York (1957)
90. Skaggs, S.: Introduction. Zed 4, 8–9 (1997)
91. Skaggs, S., Shank, G.: Codification, inference and specificity in visual communication design. Zed 4, 54–59 (1997)
92. Sless, D.: Design philosophy. communication research institute of australia (2009), http://www.communication.org.au/dsblog/?p=43 (Retrieved June 23, 2009)
93. STC: Stc ethical principles for commmunicators (1998),
    http://www.stc.org/about/policy_ethicalPrinciples.asp,
    Society for Technical Communication (Retrieved July 5, 2009)
94. Stjernfeldt, F.: Diagrammatology: An Investigation on the Borderlines of Phenomenology, Ontology, and Semiotics. Springer, Berlin (2007)
95. Storkerson, P.: Is disciplinary research possible in communication design? Design Research Quarterly 3(2), 1–8 (2008), http://www.drsq.org (Retrieved, July 15, 2009)
96. Suppes, P.: Models of data. In: Nagel, E., Suppes, P., Tarski, A. (eds.) Logic, Methodology, and Philosophy of Science: Proceedings of the 1960 International Congress, pp. 252–161 (1960)
97. Tolman, E.: Purposive Behavior in Man and Animals. University of California Press, Berkeley (1951)
98. Tolman, E.C., Brunswik, E.: The organism and the causal texture of the environment. Psychological Review 43, 43–77 (1935)
99. Tucker, L.R.: A suggested alternative formulation in the developments of Hursch, Hammond, and Hursch, and Hursch, Hammond & Todd. Psychological Review 71(6), 528–530 (1964)
100. Vicente, K.J.: Beyond the lens model and direct perception: Toward a broader ecological psychology. Ecological Psychology 15, 241–267 (2003)
101. Vorms, M.: The theoretician's gambits: Scientific representations, their formats and content. In: SCI, vol. 314. Springer, Heidelberg (2010)
102. Wertheimer, M.: Principles of perceptual organization. In: Wertheimer, M., Beardslee, M. (eds.) Readings in Perception, pp. 115–135. Van Nostrand, Princeton (1958)
103. Winkler, D.: Design practice and education: Moving beyond the Bauhaus model. In: Frascara, J. (ed.) User-centered Graphic Design: Mass Communications and Social Change, pp. 129–135. Taylor & Francis, London (1997)
104. Winn, W., Hoffman, H., Osberg, K.: Semiotics, cognitive theory and the design of objects, actions and interactions in virtual environments. Journal of Structure, Learning and Intelligent Systems 14(1), 29–49 (1999)

Smart Abducers as Violent Abducers
Hypothetical Cognition and "Military Intelligence"

Lorenzo Magnani

Abstract. I will describe the so-called coalition enforcement hypothesis, which sees humans as self-domesticated animals engaged in a continuous hypothetical activity of building cooperation through morality, incorporating punishing policies at the same time: morality and violence are seen as strictly intertwined with social and institutional aspects, implicit in the activity of cognitive niche construction. *Hypothetical thinking* (and so *abduction*) is in turn very often embedded in various linguistic kinds of the so-called fallacious reasoning. Indeed, in evolution, coalition enforcement works through the building of social cognitive niches seen as new ways of diverse human adaptation, where guessing hypotheses is central and where guessing hypotheses is occurring as it can, depending on the cognitive/moral options human beings adopt. I will also stress the moral and violent effect played by human natural languages, focusing on the analysis of the relationships between language, logic, fallacies, and abduction. This "military" nature of abductive hypothetical reasoning in linguistic communication (*military intelligence*) is intrinsically "moral" (protecting the group by obeying shared norms), and at the same time "violent" (for example, harming or mobbing others – members or not of the group – still to protecting the group itself). However, the "military" power can be considered also active at the level of *model-based cognition*: taking advantage of the naturalistic perspective on abductive "hypothesis generation" at the level of both instinctual behavior and representation-oriented behavior, where nonlinguistic features drive a "plastic" model-based cognitive role, cognition gains a fundamental semiotic, eco-physical, and "military" significance, which nicely furnishes further insight into a kind of "social epistemology".

Lorenzo Magnani
Department of Philosophy, University of Pavia, Pavia, Italy,
and Department of Philosophy, Sun Yat-sen University, Guangzhou, P.R. China
e-mail: lmagnani@unipv.it

L. Magnani et al. (Eds.): Model-Based Reasoning in Science & Technology, SCI 314, pp. 51–82.
springerlink.com

1 Multimodal Abduction

Hypothetical thinking (and so *abduction*) is very often embedded in various linguistic kinds of the so-called fallacious reasoning (which in turn constitutes a relevant part of the linguistic cognitive niches where human beings are embedded). As I will illustrate in the following section, coalition enforcement works through the building of social cognitive niches – so entering evolution by giving rise to a coevolutionary interplay between genes and culture – seen as new ways of diverse human adaptation, where guessing hypotheses is central and where guessing hypotheses is occurring as it can, depending on the cognitive/moral options human beings adopt. In the fourth section I will stress the moral and violent effect played by human natural languages, focusing on the analysis of the relationships between language, logic, fallacies, and abduction. This "military" nature of abductive hypothetical reasoning in linguistic communication (*military intelligence*) is intrinsically "moral" (protecting the group by obeying shared norms), and at the same time "violent" (for example, harming or mobbing others – members or not of the group – still to protecting the group itself).

Hence, to the aim of showing how "smart abducers" can become "violent abducers", thanks to the role of hypothetical reasoning in coalition enforcement and language, it if first of all necessary to illustrate the multimodal character of abductive cognition. This multimodality of abduction will also show that, however, the "military" power can be considered also active at the level of *model-based cognition*, and so not only in language.

As I have already stressed in my book [22, chapter 1], Peirce considers inferential any cognitive activity whatever, not only conscious abstract thought; he also includes perceptual knowledge and subconscious cognitive activity. For instance in subconscious mental activities visual representations play an immediate role (cf. [28]). Many commentators criticized this Peircean ambiguity in treating abduction at the same time as inference and perception. It is important to clarify this problem, because perception and imagery are kinds of that model-based cognition which we are exploiting to explain abduction: I contend that we can render consistent the two views [23], beyond Peirce, but perhaps also within the Peircean texts, partially taking advantage of the concept of *multimodal abduction*, which depicts hybrid aspects of abductive reasoning. Thagard [35, 41] observes, that abductive inference can be visual as well as verbal, and consequently acknowledges the sentential, model-based, and manipulative nature of abduction[1]. Moreover, both data and hypotheses can be visually represented:

> For example, when I see a scratch along the side of my car, I can generate the mental image of a grocery cart sliding into the car and producing the scratch. In this case both the target (the scratch) and the hypothesis (the collision) are visually represented. [...] It is an interesting question whether hypotheses can be represented using all sensory modalities. For vision the

---

[1] I introduced the concept of manipulative abduction in [23].

answer is obvious, as images and diagrams can clearly be used to represent events and structures that have causal effects.

Indeed hypotheses can be also represented using other sensory modalities:

> I may recoil because something I touch feels slimy, or jump because of a loud noise, or frown because of a rotten smell, or gag because something tastes too salty. Hence in explaining my own behavior my mental image of the full range of examples of sensory experiences may have causal significance. Applying such explanations of the behavior of others requires projecting onto them the possession of sensory experiences that I think are like the ones that I have in similar situations. [...] Empathy works the same way, when I explain people's behavior in a particular situation by inferring that they are having the same kind of emotional experience that I have in similar situations (cf. Thagard [41]).

Thagard illustrates the case in which a colleague with a recently rejected manuscript is frowning: other colleagues can empathize by remembering how annoyed they felt in the same circumstances, projecting a mental image onto the colleague that is a non-verbal representation able to explain the frown. Of course a verbal explanation can be added, but this just complements the empathetic one. It is in this sense that Thagard concludes that abduction can be fully multimodal, in that both data and hypotheses can have a full range of verbal and sensory representations.

Thagard also insists on the fact that Peirce noticed that abduction often begins with puzzlement, but philosophers rarely acknowledged the emotional character of this moment: so not only is emotion an abduction itself, like Peirce maintains, but it is at the starting point of most abductive processes, no less than at the end, when a kind of positive emotional satisfaction is experienced by humans.

2 Coalition Enforcement: Morality and Violence

2.1 Abduction, Respecting People as Things, and Animals as Moral Mediators

In this subsection I aim at introducing two key concepts that will be in the background of various argumentations contained in this paper: 1) the importance of abduction, as a way of guessing hypotheses, in moral reasoning and in describing the role of fallacies in moral and violent settings; 2) the philosophical usefulness of framing the problem of violence in a biological and evolutionary[2] perspective, where – so to say – the "animality of human animals" is a tool for avoiding traditional and robust philosophical illusions, prejudices, and ingenuities about violence.

---

[2] Not in the sense of sociobiology and evolutionary psychology, though.

My study on hypothetical reasoning in terms of abductive cognition [22] has demonstrated that the human activity of guessing hypotheses – that is, abductive cognition – touches the important subject of morality and moral reasoning, and therefore violence and punishment. In the framework of "distributed morality", a term coined in my [21], I have illustrated the role of abduction in moral decision, both in deliberate and unconscious cases, and its relationship with hard-wired and trained emotions[3]. Moreover, the result was that the abductive guessing of hypotheses is partly explicable as a more or less "logical" operation related to the "plastic" cognitive endowments of all organisms and partly as a biological phenomenon.

Furthermore, the analysis of the interplay between fallacies and abductive guessing I will present in section 4 will also illustrate how: 1) abduction and other kinds of hypothetical reasoning are involved in dialectic processes, which are at play in both everyday agent-based settings and rather scientific ones; 2) they are strictly linked to so-called smart heuristics and to the fact that very often less information gives rise to better performance; 3) heuristics linked to hypothetical reasoning like "following the crowd", or social imitation, are differently linked to fallacious aspects – which involve abductive steps – and they are often very effective. I will especially stress that these and other fallacies, are linked to what René Thom calls "military intelligence", which relates to the problem of the role of language in the so-called *coalition enforcement*, that is in the affirmation of morality and the related perpetration of violent punishment. It is in this sense that I will point out the importance of fallacies as "distributed military intelligence".

Indeed the aim of the following subsections is to clarify the idea of "coalition enforcement". This idea illustrates a whole theoretical background for interpreting the topics above concerning morality, violence, and hypothetical reasoning as well as my own position, which resorts to the hypothesis about the existence of a strict multifaceted link between morality and violence. The theme is further related to some of the issues I will illustrate in the section 3 of this paper, where I make profitable use of Thom's attention to the moral/violent role of what he calls "proto-moral conflicts": I will fully stress that, for example, the fundamental function of language can only be completely understood in the light of its intrinsic *moral* (and at the same time *violent*) purpose, which is basically rooted, as I have already said, in an activity of *military intelligence*.

2.2   Coalition Enforcement Hypothesis

The coalition enforcement hypothesis, put forward by Bingham [3, 4], aims at providing an explanation of the "human uniqueness" that is at the origin of human communication and language, in a strict relationship with the

---

[3] The problem of the role of emotions (and their relationships with reason) in moral decision from the point of view of neuroscience is treated in [25, 17, 24].

spectacular ecological dominance achieved by *H. Sapiens*, and of the role of cultural heritage. In this perspective, and due to the related constant moral and policing dimension of *Homo*'s coalition enforcement history (which has an approximately two-million-year evolutionary history), human beings can be fundamentally seen as *self-domesticated animals*. I think the main speculative value of this hypothesis consists in stressing the role of the more or less stable stages of cooperation *through morality* (and through the related unavoidable violence). Taking advantage of a probabilistic modeling of how individual learning – coupled with cultural transmission and a tendency to conform to the behavior of others – can lead do adaptive norms, Boyd [6] contend that it is possible for norms to guide people toward sensible behaviors that they would not choose if left to their own devices.

The hypothesis implicitly shows how both *morality* and *violence* are strictly intertwined in their social and institutional aspects, which are implicit in the activity of cognitive niche construction. In an evolutionary dimension, coalition enforcement is enacted through the construction of social *cognitive niches* as a new way of diverse human adaptation. Basically the hypothesis refers to cooperation between related and unrelated animals to produce significant mutual benefits that exceed costs and are potentially adaptive for the cooperators. Wilson *et al.* [42] aim at demonstrating the possibility that groups engage in coordinated and cooperative cognitive processes, thus exceeding by far the possibilities of individual thinking , by recurring to a hypothesized "group mind" whose role would be fundamental in social cognition and group adaptation. The formation of appropriate groups which behave according to explicit and implicit more or less flexible rules of various types (also moral rules, of course) can be reinterpreted – beyond Wilson's strict and puzzling "direct" Darwinian version – as the "social" constitution of a cognitive niche, that is a cognitive modification of the environment which confronts the coevolutionary problem of varying selective pressures in an adaptive or maladaptive way.

In hominids, cooperation in groups (which, contrary to the case of non-human animals, is largely independent from kinship) fundamentally derived from the need to detect, control, and punish social parasites, who for example did not share the meat they had hunted or partook of the food without joining the hunting party (cf. Boehm [5]) (also variously referred to as free riders, defectors, and cheaters). These social parasites were variously dealt with by killing or injuring them (and also by killing cooperators who refused to punish them) *from a distance* using projectile and clubbing weapons. In this case injuring and killing are cooperative and remote (and at the same time they are "cognitive" activities). According to the coalition enforcement hypothesis, the avoidance of proximal conflict reduces risks for the individuals (hence the importance of *remote killing*). Of course, cooperative morality that generates "violence" against unusually "violent" and aggressive free riders and parasites can be performed in other weaker ways, such as denying a future access to the resource, injuring a juvenile relative, gossiping to persecute dishonest communication and manipulative in-group behaviors or waging war against less cooperative groups, etc.[4]

In such a way group cooperation (for example for efficient collective hunting and meat sharing through control of free riders) has been able to evolve adaptively and to render parasitic strategies no longer systematically adaptive. Through cooperation and remote killing, individual costs of punishing are greatly reduced and so is individual aggressiveness and violence, perhaps because violence is morally "distributed" in a more sustainable way: "Consistent with this view, contemporary humans are unique among top predators in being relatively placid in dealing with unrelated conspecific nonmates under a wide variety of circumstances" (cf. Bingham [3, p. 140]). [I have to note, "contrarily to the common sense conviction", formed by the huge amount of violence human beings are still everyday faced with!]. Hence, it has to be said that humans, contrarily to non-human animals, exchange a fundamental and considerable amount of relatively reliable information with unrelated conspecifics [3, p. 144].

Here the role of docility is worth citing (which relates to the already recognized distressing human tendency to conform, displace responsibility, comply, and submit to authority of dominant individuals, emphasized by social psychology – cf. Dellarosa Cummins [8, p. 11]). According to Herbert Simon, humans are docile in the sense that their fitness is enhanced by "[...] the tendency to depend on suggestions, recommendations, persuasion, and information obtained through social channels as a major basis for choice" [32, p. 156]. In other words, humans support their limited decision-making capabilities, relying on external data obtained through their senses from their social environment. The social context gives them the main data filter, available to increase individual fitness [31]. Therefore, docility is a kind of attitude or disposition underlying those activities of cognitive niche construction, which are related to the delegation and exploitation of ecological resources. That is, docility is an adaptive response to (or a consequence of) the increasing cognitive demand (or selective pressure) on those information-gaining ontogenetic processes, resulting from an intensive activity of niche construction. In other words, docility permits the inheritance of a large amount of useful knowledge while lessening the costs related to (individual) learning.

In Simon's work, docility is related to the idea of *socializability*, and to altruism in the sense that one cannot be altruistic if he or she is not docile. In this perspective, however, the most important concept is docility and not altruism, because docility is the cognitive condition of the possibility of the emergence of altruism. I strongly believe, in the light of the coalition enforcement hypothesis, that moral altruism can be correctly seen as a sub-product of (or at least intertwined with) the violent behavior needed to "morally" defend and enforce coalitions. I have said that groups need to detect, control and punish social parasites, that for example do not share meat (food as

---

[4] On the moral/violent nature of gossip and fallacies cf. Bardone and Magnani [1].

nutrient as hard to acquire), by killing or injuring them (and any cooperators who refuse to carry out punishment) and to this aim they have to gain the cooperation of other potential punishers.

Research on chimps' behavior shows that punishment can be seen as altruistic for the benefit of the other members of the group (and to the aim of changing the future actions of the individual being punished), often together with "the function of keeping the top ranking male, or coalition of males, at top, or preserving the troop-level macrocoalition that disproportionately serves the interests of those on top" (cf. Rohwer [29, p. 805]). The last observation also explains how altruistic punishment can serve individual purposes (and so it can be captured by individual selection models): in chimps it reflects the desire to maintain status, that is a new high position in the hierarchy. Rohwer's conclusion is that "altruistic punishment need not have originated by group selection, as the Sober and Wilson model assumes. Seen through the lens of the linear dominance hierarchy, it is reasonable to suspect that altruistic punishment may have originated primarily through individual selection pressures" (cit. p. 810)[5].

I have said above that groups need to detect and punish social parasites by killing or injuring them (and any cooperators who refuse to carry out punishment) and to this aim they have to gain the cooperation of other potential punishers. This explains altruistic behavior (and the related cognitive endowments which make it possible, such as affectivity, empathy and other non violent aspects of *moral* inclinations) which can be used in order to reach cooperation. To control free-riders inside the group and guard against threat from other alien groups, human coalitions – as the most gregarious animal groups – have to take care of the individuals who cooperate. It is from this perspective that we can explain, as I have said above, quoting Bingham, why contemporary humans are not only violent but *also* very docile and "[...] unique among top predators in being relatively placid"[6].

The problem of docility is twofold. First, people delegate tasks of data acquisition to their experience, to external cultural resources and to other

---

[5] On the puzzling problem of the distinction between individual and group selection for altruism in the framework of multilevel selection cf. Rosas [30]: multilevel selection theory claims that selection operates simultaneously on genes, organisms, and groups of organisms. A history of the debate about altruism is given by Dugatkin [10].

[6] On the intrinsic moral character of human communities – with behavioral prescriptions, social monitoring, and punishment of deviance – for much of their evolutionary history cf. Wilson [47, p. 62] and Boehm [5]. [18] further emphasize the evolutionary adaptive role of morality (and so of cooperation) as "group stability insurance". The exigence of morality as group stability would explain the "viscosity" of basic aspects of the morality of a group and why morality is perceived as having an air of absolutism. I also add that this viscosity favors the "embubblement" I will illustrate below in subsection 4.3, which explains the obliteration – at the level of the agent awareness – of the actual violent outcomes of his moral actions.

individuals. Second, people generally put their trust in one another in order
to learn. A big cerebral cortex, speech, rudimentary social settings and primitive material culture furnished the conditions for the birth of the mind as a
"universal machine" – to use Turing's expression. I contend that a big cortex can provide an evolutionary advantage only in the presence of a massive
storage of meaningful information and knowledge on external supports that
only a developed (even if small) community of human beings can possess. If
we consider high-level consciousness as related to a high-level organization
of the human cortex, its origins can be related to the active role of environmental, social, linguistic, and cultural aspects. It is in this sense that docile
interaction lies at the root of our social (and neurological) development. It
is obvious that docility is related to the development of cultures, morality,
actual cultural availability and to the quality of cross-cultural relationships.
Of course, the type of cultural dissemination and possible cultural enhancements affect the chances that human collectives have to take advantage of
docility and thus to potentially increase or decrease their fitness.

As I have already noted, the direct consequence of coalition enforcement is
the development and the central role of cultural heritage (morality and sense
of guilt included): in other words, I am hinting to the importance of cultural
cognitive niches as new ways of arriving at diverse human adaptations. In
this perspective the long-lived and yet abstract human sense of guilt represents a psychological adaptation, *abductively* anticipating an appraisal of a
moral situation to avoid becoming a target of violent coalitional enforcement.
Again, we have to recall that Darwinian processes are involved not only in the
genetic domain but also (with a looser and lesser precision) in the additional
cultural domain, through the selective pressure activated by modifications in
the environment brought about by cognitive niche construction. According
to the theory of cognitive niches, coercive human coalition as a fundamental cognitive niche constructed by humans becomes itself a major element
of the selective environment and thus imposes new constraints (designed by
extragenetic information[7] on its members).

Some empirical evidence (presented for instance by Bingham in [4]) seems
to support the coalition enforcement hypothesis: fossils of *Homo* (but not
of Australopithecines) show, on observation of skeletal adaptations, how selection developed an astonishing competence in humans relating to the controlled and violent use of projectile and clubbing weapons (bipedal locomotion, the development of *gluteus maximus* muscle and its capacity to produce
rotational acceleration, etc.). The observed parallel increase in cranial volume
can be related to the increased social cooperation based on the reception,
use and transmission of "extragenetic information". Moreover, physiological, evolutionary, and obstetric constraints on brain size and structure indicate that humans can individually acquire a limited amount of extragenetical

---

[7] Cf. chapter 6 of my [22].

information, that consequently has to be massively stored and kept available in the external environment.

Usually it is said that Darwinian processes operating on genetic information produce human minds whose properties somehow include generation of the novel, complex adaptive design reflected in human material artifacts *sui generis*. However, following Bingham, these explanations

> [...] fail to explain human uniqueness. If building such minds by the action of Darwinian selection on genetic information were somehow possible, this adaptation would presumably be recurrent. Instead, it is unique to humans. Before turning to a possible resolution of this confusion, two additional properties of human technological innovation must be recalled. First, its scale has recently become massive with the emergence of behaviorally modern humans about 40,000 years ago. Second, the speed of modern human innovation is unprecedented and sometimes appears to exceed rates achievable by the action of Darwinian selection on genetic information. [3, p. 145]

Hence, a fundamental role in the evolution of "non" genetic information has to be hypothesized. Appropriately, coalition enforcement implies the emergence of novel extra-genetic information, such as large scale mutual information exchange – including both linguistic and model-based[8] communication between unrelated kin. As mentioned above, it is noteworthy that extra-genetic information plays a fundamental role in terms of transmitted ideas (cultural/moral aspects), behavior, and other resources embedded in artifacts (ecological inheritance). It is easy to acknowledge that this information can be stored in human memory – in various ways, both at the level of long-lived neural structures that influence behavior, and at the level of external devices (cognitive niches), which are transmitted indefinitely and are thus potentially immortal – but also independently of small kinship groups. Moreover, transmission and selection of extragenetic information is at least partially independent of an organism's biological reproduction.

3 Coalition Enforcement through Abduction: The Moral/Violent Nature of Language

In a study concerning language as an adaptation the cognitive scientist and evolutionary psychologist Pinker says: "[...] a species that has evolved to rely on information should thus also evolve a means to exchange that information. Language multiplies the benefit of knowledge, because a bit of know-how is useful not only for its practical benefits to oneself but as a trade good with

---

[8] Examples of model-based cognition are constructing and manipulating visual representations, thought experiment, analogical reasoning, etc. but it also refers to the cognition animals can get through emotions and other feelings. Charles Sanders Peirce already acknowledged the fact that all inference is a form of sign activity, where the word sign includes various model-based ways of cognition: "feeling, image, conception, and other representation" [29, 5.283].

others". The expression "trade good" seems related to a moral/economical function of language: let us explore this issue in the light of the coalition enforcement hypothesis I have introduced in the first section [27, p. 28].

Taking advantage of the conceptual framework brought up by Thom's catastrophe theory on how natural syntactical language is seen as the fruit of social necessity (cf. [22, chapter 8]), its fundamental function can only be clearly seen if linked to an intrinsic *moral* (and at the same time *violent*) aim which is basically rooted in a kind of *military intelligence*. Thom says language can simply and efficiently transmit *vital* pieces of information about the fundamental biological oppositions (life – death, good – bad): it is from this perspective that we can clearly see how human language – even at the level of more complicated syntactical expressions – always carries information (pregnances, in Thom's terms) about moral qualities of persons, things, and events. Such qualities are always directly or indirectly related to the survival needs of the individual and/or of the group/coalition.

Thom too is convinced of the important role played by language in maintaining the structure of societies, defending it thanks to its moral and violent role: "information has a useful role in the stability or 'regulation' of the social group, that is, in its defence" [37, p. 279]. When illustrating "military" and "fluid" societies he concludes:

> In a military type of society, the social stability is assured, in principle, by the imitation of the movement of the hierarchical superior. Here it is a question of a slow mechanism where the constraints of vital competition can impose rapid manoeuvres on the group. Also the chief cannot see everything and has need of special informers stationed at the front of the group who convey to him useful information on the environment. The invention of a sonorous language able to communicate information and to issue direction to the members of the group, has enabled a much more rapid execution of the indispensable manoeuvres. By this means (it is not the only motivation of language), one can see in the acquisition of this function a considerable amelioration of the stability of a social group.
>
> If language has been substituted for imitation, we should note that the latter continues to play an important role in our societies at pre-verbal levels (cf. fashion). In addition, imitation certainly plays a primary part in the language learning of a child of 1 to 3 years. [37, pp. 235–236]

I have illustrated in section 2 that in human or pre-human groups the appearance of coalitions dominated by a central leader quickly leads to the need for surveillance of surrounding territory to monitor prey and watch for enemies who might jeopardize the survival of the coalition.

This is an idea shared by Thom who believes that language becomes a fundamental tool for granting stability and favoring the indispensable manipulation of the world "thus the localization of external facts appeared as an essential part of social communication" [37, p. 26], a performance that is already realized by naming[9] (the containing relationship) in divalent structures: "X is in Y is a basic form of investment (the localizing pregnance of Y invests X). When X is invested with a ubiquitous biological quality (favorable or hostile), then so is Y" (*ibid.*). A divalent syntactical structure of language becomes fundamental if a *conflict* between two outside agents has to be reported. The trivalent syntactical structure subject/verb/object forges a salient "messenger" form that conveys the pregnance between subject and recipient. In sum, the usual abstract functions of syntactic languages, such as conceptualization, appear strictly intertwined with the basic *military* nature of communication.

I contend that this military nature of linguistic communication is intrinsically "moral" (protecting the group by obeying shared norms), and at the same time "violent" (for example, killing or mobbing to protect the group). This basic moral/violent effect can be traced back to past ages, but also when we witness a somehow *prehuman* use of everyday natural language in current mobbers, who express strategic linguistic communications "against" the mobbed target. These strategic linguistic communications are often performed thanks to *hypothetical reasoning*, abductive or not. In this case the use of natural language can take advantage of efficient hypothetical cognition through gossip, fallacies and so on, but also of the moral/violent exploitation of apparently more respectable and sound truth-preserving and "rational" inferences. The narratives used in a dialectic and rhetorical settings qualify the mobbed individual and its behavior in a way that is usually thought of by the mobbers themselves (and by the individuals of their coalition/group) as moral, neutral, objective, and justified while at the same time hurting the mobbed individual in various ways. Violence is very often subjectively dissimulated and paradoxically considered as the act of performing just, objective moral judgments and of persecuting moral targets. In sum, *de facto* the mobbers' coordinated narratives harm the target (as if she was just being *stoned* in a ritual killing), very often without an appreciable awareness of the violence performed.

This human linguistic behavior is clearly made intelligible when we analogously see it as echoing the anti-predatory behavior which "weaker" groups of animals (birds, for example) perform, for example through the use of suitable alarm calls and aggressive threats. Of course such behavior is mediated in humans through socially available ideologies (differently endowed with moral ideas) and cultural systems. Ideologies can be seen as fuzzy and ill-defined cultural mediators spreading what Thom calls pregnances that invest all those who put their faith in them and stabilize and reinforce the coalitions/groups: "[...] the follower who invokes them at every turn (and even out of turn) is demonstrating his allegiance to an ideology. After successful uses the ideological concepts are extended, stretched, even abused", so that

---

[9] It is important to stress that pregnant forms, as they receive names, loose their alienating character.

their meaning slowly changes in imprecise (and "ambiguous", Thom says)[10] ways, as we have seen it happens in the application of the archetypical principles of mobbing behavior. That part of the individual unconscious we share with other human beings – i.e. *collective unconscious*[11] – shaped by evolution – contains archetypes like the "scapegoat" (mobbing) mechanism I have just mentioned.

In this cognitive mechanism, a paroxysm of violence focuses on an arbitrary sacrificial victim and a unanimous antipathy would, mimetically, grow against him. The process leading to the ultimate bloody violence (which was, for example, widespread in ancient and barbarian societies) is mainly carried out in current social groups through linguistic communication. Following Girard [13, 14] we can say that in the case of ancient social groups the extreme brutal elimination of the victim would reduce the appetite for violence that had possessed everyone just a moment before, leaving the group suddenly appeased and calm, thus achieving equilibrium in the related social organization (a sacrifice-oriented social organization may be repugnant to us but is no less "social" just because it is rudimentary violent).

This kind of archaic brutal behavior is still present in civilized human conduct in rich countries, almost always implicit and unconscious, for example in the racist and mobbing behavior. Let me reiterate that, given the fact that this kind of behavior is widespread and partially unconsciously performed, it is easy to understand how it can be implicitly "learned" in infancy and still implicitly "pre-wired" in an individual's cultural unconscious (in the form of ideology as well) we share with others as human beings. I strongly believe that the analysis of this archaic mechanism (and of other similar *moral/ideological/violent* mechanisms) might shed new light on what I call the basic equivalence between engagement in morality and engagement in violence since these engagements, amazingly enough, are almost always hidden from the awareness of the human agents that are actually involved.

Recent evolutionary perspectives on human behavior, taking advantage of neuroscience and genetics (cf. Taylor [34])[12] have also illustrated the so-called *otherisation* – which decisively primes people for aggression – as a process grounded in basic human emotions, i.e. our bias towards pleasure and avoidance of pain. Perceiving others as the "others" causes fear, anger or disgust, universal "basic" responses to threats whose physiological mechanisms are relatively well understood. It is hypothesized that these emotions evolved to

---

[10] From this perspective the massive moral/violent exploitation of equivocal fallacies in ideological discussions, oratories, and speeches is obvious and clearly explainable, as I will illustrate below (cf. below, section 4).

[11] I have illustrated the speculative hypothesis of collective unconscious in chapters 4 and 5 of [22]).

[12] The book also provides neuroscientific explanations on how brains process emotions, evoke associations, and stimulate reactions, which offer interesting data – at least in terms of neurological correlates – on why it is reactively easy for people to harm other people.

enable our ancestors to escape predators and fight enemies. Of course the otherisation process continues when structured in "moral" terms, like for example in the construction of that special other that becomes a potential or actual scapegoat.

It is worth mentioning, in conclusion, the way Thom accounts for the social/moral phenomenon of scapegoating in terms of pregnances. "Mimetic desire", in which Girard roots the violent and aggressive behavior (and the scapegoat mechanism) of human beings [14] can be seen as the act of appropriating a desired object which imbues that object with a pregnance, "the same pregnance as that which is associated with the act by which 'satisfaction' is obtained" [37, p. 38]. Of course this pregnance can be propagated by imitation through the mere sight of "superior" individuals[13] in which it is manifest: "In a sense, the pleasure derived from looking forward to a satisfaction can surpass that obtained from the satisfaction itself. This would have been able to seduce societies century after century (their pragmatic failure in real terms having allowed them to escape the indifference that goes with satiety as well as the ordeal of actual existence)" (*ibid.*).

Recent cognitive research stresses the role of *intentional gaze* processing on object processing itself: objects falling under the gaze of others acquire properties that they would not display if not looked at. Observing another person gazing at an object enriches it of motor, affective, and status properties that go beyond its chemical or physical structure. We can conclude that gaze plays the role of transferring to the object the intentionality of the person who is looking at it. This result further explains why mimetic desire can spread so quickly among people belonging to specific groups (cf. Becchio [2])[14].

Grounded in appropriate wired bases, "mimetic desire" is indeed a sophisticated template of behavior that can be picked up from various appropriate cultural systems, available over there, as part of the external cognitive niches built by many human collectives and gradually externalized over the centuries (and always transmitted through activities, explicit or implicit, of teaching and learning), as fruitful ways of favoring social control over coalitions. Indeed mimetic desire triggers envy and violence but at the same time the perpetrated violence causes a reduction in appetite for violence, leaving the group suddenly appeased and calm, thus achieving equilibrium in the related social organization through a *moral effect*, that is at the same time a *carrier of violence*, as I have illustrated.

Mimetic desire is related to envy (even if of course not all mimetic desire is envy, certainly all envy is mimetic desire): when we are attracted to something the others have but that we cannot acquire because others already possess it (for example because they are rival goods), we experience an offense which generates envy. In the perspective introduced by Girard envy is

---

[13] Or through the exposure to descriptions and narratives about them and their achievements.

[14] On gaze cueing of attention cf. also Frischen *et al.* [11], who also establish that in humans prolonged eye contact can be perceived as aggressive.

a mismanagement of desire and it is of capital importance for the moral life of both communities and individuals. As a reaction to offense, envy easily causes violent behavior. In this perspective we can psychoanalytically add, according to Žižek, that "[...] the opposite of egotist self-love is not altruism, a concern for common good, but envy and resentment, which makes me act against my own interests. Freud knew it well: the death drive is opposed to the pleasure principle as well as to the reality principle. The true evil, which is the death drive, involves self-sabotage" [45, p. 76][15].

4   Fallacies as Distributed "Military" Intelligence

In section 2 above I have described the general framework of the so-called *coalition enforcement hypothesis*, a perspective which especially stresses the intrinsic *moral* (and at the same time *violent*) nature of language (and of abductive and other hypothetical forms of reasoning and arguments intertwined with the propositional/linguistic level). In this perspective language is basically rooted in a kind of *military intelligence*, as maintained by Thom [37]. Indeed we have to note that many kinds of hypothesis generation (from abduction to hasty generalization,[16] from *ad hominem* to *ad verecundiam*) are performed through inferences that embed formal or informal fallacies.

Of course not only language and model-based cognition carry morality and violence, motor actions and emotions: it is well-known that overt hostility in emotions is a possible trigger to initiate violent actions. It is well-known the "moral" role of emotions and I think the potential "violent" role of them is out of discussion. de Gelder *et al.* [12] indicate that observing fearful body expressions produces increased activity in brain areas narrowly associated with emotional processes and that this emotion-related activity occurs together with activation of areas linked with representation of action and movement. The mechanism of emotional (fear) contagion hereby suggested may automatically prepare the brain for action.

As I have already said above, in Thom's terms, language essentially efficiently transmits *vital* pieces of information about the fundamental biological opponents (life, death – good, bad) and so it is intrinsically involved in moral/violent activities. This conceptual framework can shed further light on some fundamental dialectical and rhetorical roles played by the so-called fallacies, which are of great importance to stress some basic aspects of human abductive cognition. In the following subsection I will consider some

---

[15] I have deepened the problem of envy, mimetic desire, and scapegoating mechanism, framing it in the perspective of religion, morality, and violence, in [23, chapter 6].

[16] This fallacy occurs when a person (but there is evidence of it also in animals, for example in mice, where the form of making hypotheses can be ideally modeled as a hasty induction) infers a conclusion about a group of cases based on a model that is not large enough, for example just one sample.

roles played by fallacies that have to be ideally related to the intellectual perspective of the coalition enforcement hypothesis illustrated above.

Of course the "military" nature is not evident in various aspects and uses of syntactilized human language[17]. It is hard to directly see coalition enforcement violent effect in the many *epistemic* functions of natural language, for example when it is simply employed to transmit scientific results in an academic laboratory situation, or when we pick up some information through the Internet – expressed in linguistic terms and numbers – about the weather. However, we cannot forget that even the more abstract character of the knowledge packages embedded in certain uses of language (and in hybrid languages, like in the case of mathematics, which involves considerable symbolic parts) still plays a significant role in changing the moral behavior of human collectives. For example, the production and the transmission of new scientific knowledge in human social groups not only transmits information but also implements and distributes roles, capacities, constraints, and possibilities of actions. This process is intrinsically moral because in turn generates precise distinctions, powers, duties, and chances, which can create new inter- and intra-group more or less violent conflicts or reshape the old ones.

New abstract biomedical knowledge about pregnancy and fetuses usually has two contrasting moral/social effects, 1) a better social and medical management of childbirth and related diseases; 2) the potential extension or modification of conflicts surrounding the legitimacy of abortion. In sum, even very abstract bodies of knowledge and more innocent pieces of information enter the semio/social process which governs the identity of groups and their aggressive potential as coalitions: deductive reasoning and declarative knowledge are far from being exempt from being accompanied by argumentative, deontological, rhetorical, and dialectic aspects. For example, it is hard to distinguish, in an eco-cognitive setting, between a kind of "pure" (for example deductive) inferential function of language and the argumentative or deontological one. For example, the first one can obviously play an associated argumentative role. However, it is in the arguments traditionally recognized as fallacious, that we can more clearly grasp the military nature of human language and especially of some hypotheses reached through the so-called fallacies.

Searle considers eccentric that aspect of our cultural tradition, according to which true statements, fruit od deductive sound inferences, that also describe how things are in the world, can never imply a statement about how they ought to be. Searle contends that to say something is true is already to say you ought to believe it, that is other things being equal, you ought not to deny it. This means that normativity is more widespread than expected. It is in a similar way that Thom clearly acknowledges the general and intrinsic "military" (and so moral, normative, argumentative, etc.) nature of language,

---

[17] The military/violent nature is obviously evident for example in hateful, racist, homophobic speech.

by providing a justification in terms of the catastrophe theory, as I have indicated in the previous section.

Woods contends that a fallacy is by definition considered "a mistake in reasoning, a mistake which occurs with some frequency in real arguments and which is characteristically deceptive" [44][18]. Traditionally recognized fallacies like hasty generalization and *ad verecundiam* are considered "inductively" weak inferences, while affirming the consequent is a deductively invalid inference. Nevertheless, when they are used by actual reasoners, "beings like us", that is in an eco-logical[19] and not merely logical – ideal and abstract – way, they are *no longer* necessarily fallacies. Traditionally, fallacies are considered mistakes that appear to be *errors, attractive* and *seductive*, but also *universal*, because humans are prone to committing them. Moreover, they are "usually" considered *incorrigible*, because the diagnosis of their incorrectness does not cancel their appearance of correctness: "For example, if, like everyone else, I am prone to hasty generalization prior to its detection in a particular case, I will remain prone to it afterwards" (*ibid.*).

Woods calls this perspective the traditional – even if not classical/Aristotelian – "EAUI-conception" of fallacies. Further, he more subtly observes

> [...] first, that what I take as the traditional concept of fallacy is not in fact the traditional concept; and, second, that regardless whether the traditional concept is or is not what I take it to be, the EAUI-notion is not the right target for any analytically robust theory of fallacyhood. [...] But for the present I want to attend to an objection of my own: whether the traditional conception or not, the EAUI-conception is not even so a sufficiently clear notion of fallacyhood. [...] If the EAUI-conception is right, it takes quite a lot for a piece of reasoning to be fallacious. It must be an error that is attractive, universal, incorrigible and bad. This gives a piece of reasoning five defences against the charge of fallaciousness. Indeed it gives a piece of erroneous reasoning four ways of "beating the rap". No doubt this will give some readers pause. How can a piece of bad reasoning not be fallacious? My answer to this is that not being a fallacy is not sufficient to vindicate an error of reasoning. Fallacies are errors of reasoning with a particular character. They are not, by any stretch, all there is to erroneous reasoning. [44, chapter 3]

If we adopt a sharp distinction between strategic and cognitive rationality, many of the traditional fallacies - hasty generalization for example - call for an equivocal treatment. They are sometimes cognitive mistakes and strategic successes, and in at least some of those cases, it is more rational to proceed strategically, even at the cost of cognitive error. On this view, hasty generalization or the fallacy of the affirming the consequence instantiate the

---

[18] Of course deception – in so far as it is related to *deliberate* fallaciousness – does not have be considered to be a part of the definition of fallacy (cf. Tindale [39]).

[19] That is when fallacies are seen in a social and real-time exchange of speech-acts between parties/agents.

traditional concept of fallacy (for one thing, it is a logical error), but there are contexts in which it is smarter to commit the error than avoid it.

According to Woods' last and more recent observations the traditional fallacies - hasty generalization included - do not really instantiate the traditional concept of fallacy (the EAUI-conception). In this perspective it is not that it is sometimes "strategically" justified to commit fallacies (a perfectly sound principle, by the way), but rather that in the case of the Gang of Eighteen traditional fallacies they simply are not fallacies. The distinction is subtle, and I can add that I agree with it in the following sense: the traditional conception of fallacies adopts – so to say – an *aristocratic* (ideal) perspective on human thinking that disregards its profound eco-cognitive character, where the "military intelligence" I have quoted above is fundamentally at play. Errors, in an eco-cognitive perspective, certainly are not the exclusive fruit of the so-called fallacies, and in this wide sense, a fallacy is an error – in Woods' words – "that virtually everyone is disposed to commit with a frequency that, while comparatively low, is nontrivially greater than the frequency of their errors in general".

My implicit agreement with the new Woods' "negative thesis" will be clearer in the light of the illustration of the general military nature of language I will give in the following subsections: 1) human language possesses a "pregnance-mirroring" function, 2) in this sense we can say that vocal and written language can be a tool exactly like a knife; 3) the so-called fallacies, are linked to that "military intelligence", which relates to the problem of the role of language in the so-called *coalition enforcement*; 4) this "military" nature is not evident in various aspects and uses of syntactilized human language. As already stressed above, it is hard to directly see the coalition enforcement effect in the many (too easily and quickly supposed to be universally "good") *epistemic* functions of natural language; also, we cannot forget that even the more abstract character of the knowledge packages embedded in certain uses of language – for example "logical" – still plays a significant role in changing the moral behavior and it is still error prone from many perspectives.

I have also to reiterate that, in this sense, it is hard to distinguish, in an eco-cognitive dimension, between a kind of "pure" (for example deductive) inferential function of language and the argumentative or deontological one, with respect to their error-making effects. So to say, there are many more errors of reasoning than there are fallacies. What I can bashfully suggest is that in the arguments traditionally recognized as fallacious, it is simply much more easier than in other cases, to grasp their error-proneness, for example when they constitute of assist hypothetical cognition. In sum, from an eco-cognitive perspective, when language efficiently transmits positive *vital* pieces of information through the so-called fallacies and fruitfully acts in the cognitive niche from the point of view of the agent(s), it is hard to still label the related reasoning performance as fallacious.

The so-called fallacies in practical *agent-based* and *task oriented* reasoning occurring in actual interactive cognitive situations, some important features immediately arise. In agent-based reasoning, the agent access to cognitive resources encounters limitations such as

- bounded information
- lack of time
- limited computational capacity[20].

4.1 Distributing Fallacies

It is now necessary to stress that the so-called fallacies and hypothetical reasoning which embeds them are usually exploited in a *distributed* cognitive framework, where moral conflicts and negotiations are normally at play. This is particularly clear in the case of *ad hominem*, *ad verecundiam*, and *ad populum*. We can see linguistic reasoning embedded in arguments which adopt fallacies, as distributed across

- human agents, in a basically visual and vocal dialectical interplay,
- human agents, in an interplay with other human agents mediated by various artifacts and external tools and devices (for example books, articles in newspapers, media, etc.)

From this perspective the mediation performed by artifacts causes additional effects: 1) other sensorial endowments of the listener – properly excited by the artifact features – are mobilized; these artifacts in turn 2) affect the efficacy of the argument: in this sense artifacts (like for example media) have their own highly variable cognitive, social, moral, economical, and political character. For example, the same *ad hominem* argumentation can affect the hearer in a different way depending on whether it is watched and heard on a trash television program or instead listened to in a real-time interplay with a friend. It is obvious that in the second case different cognitive talents and endowments of the listener are activated:

- positive emotional attitudes to the friend can be more active and other areas of the information and knowledge at disposal of the arguer – stored in conscious but also in unconscious memory – are at play. Both these aspects can affect the negotiatory interaction, which of course can also acquire
- the character of a dialectical process full of feedbacks, which are absent when an article in a newspaper is simply passively read (in this case the "rhetorical" effect prevails).

---

[20] On the contrary, logic, which Gabbay and Woods consider a kind of theoretical/institutional ("ideal", I would say) agent, is occurring in situations characterized by more resources and higher cognitive targets.

4.2 Aggressive Military Intelligence through Fallacies

Some aspects are typical of agent-based reasoning and are all features which characterize fallacies in various forms and can consequently be seen as good scant-resource adjustment strategies. They are:

1. Risk aversion (beings like us feel fear!),
2. guessing ability and generic inference,
3. propensity for default reasoning,
4. capacity to evade irrelevance, and
5. unconscious and implicit reasoning

Gabbay and Woods also contend that in this broader agent-based perspective one or other of the five conditions above remain unsatisfied, for example: i) fallacies are not necessarily errors of reasoning, ii) they are not universal (even if they are frequent), iii) they are not incorrigible, etc. Paradoxically, fallacies often *successfully* facilitate cognition (and hypothetical thinking) (*Abundance thesis*), even if we obviously acknowledge that actually beings like us make mistakes (and know it) (*Actually Happens Rule*). In sum, if we take into account the role of the so-called fallacies in actual human behavior, their cognitive acts show a basic, irreducible, and multifarious argumentative, rhetorical, and dialectic character. These cognitive acts in turn clearly testify that cognition can be successful and useful even in the presence of bounded information and knowledge and, moreover, in the absence of sound inferences. In this perspective deeper knowledge and sound inferences loose the huge privileged relevance usually attributed to them by the intellectual philosophical, epistemological, and logical tradition. "Belief" seems sufficient enough for human groups to survive and flourish, as they do, and indeed belief is more "economical" than knowledge as at the same time it simulates knowledge and conceals error.

The anti-intellectual approach to logic advanced by Woods' agent-based view is nicely captured by the *Proposition 8 (Epistemic Bubbles)*: "A cognitive agent X occupies an epistemic bubble precisely when he is unable to command the distinction between his thinking that he knows P and his knowing P" and *Corollary 8a*: "When in an epistemic bubble, cognitive agents always resolve the tension between their thinking that they know P and their knowing P in favour of knowing that P" [44]. Hence, we know a lot less than we think we do. Moreover, it is fundamental to stress that, when epistemic bubbles obviously change, the distinction between merely apparent correction and genuinely successful correction exceeds the agent's command and consequently the cognitive agent from his own first-person perspective favors the option of genuinely sound correction. In sum, detection of errors does not erase the appearance of goodness of fallacies.

This Humean skeptical conclusion is highly interesting because it shows the specific and often disregarded very "fragile" nature of the "cognitive"

Dasein[21] – at least of contemporary beings-like-us. I also consider it fundamental in the analysis of fallacies from the point of view of military intelligence.

Various studies address the problem of the intertwining between the feeling of knowing and knowing, which is of course related to our concept of epistemic bubbles and to the cognitive capacity to conceal error. Burton [7, p. 12] first of all notes that the feeling of knowing is commonly recognized by its absence and in any case strictly related to the widespread phenomenon of the so-called cognitive dissonance: "In 1957, Stanford professor of social psychology Leon Festinger introduced the term cognitive dissonance to describe the distressing mental state in which people 'find themselves doing things that don't fit with what they know, or having opinions that do not fit with other opinions they hold".

4.3 Moral Bubbles: Legitimating and Dissimulating Violence

A basic aspect of the human fallacious use of language regarding military effects is – so to say – the softness and gentleness which the constitutive capacity of fallacies to *conceal error* – especially when they involve abductive hypothesis guessing – can grant. Being constitutively and easily unaware of our errors is very often intertwined with the self-conviction that we *are not at all violently aggressive* in our performed argumentations (and in our possible related actions). Human beings use the so-called fallacies because they often work in a positive vital way – even if they of course involve violent aspects: if they are eco-cognitively fruitful, we cannot call them fallacies anymore. I contend that in this case we find ourselves inside a kind of *moral bubble*, that is very homomorphic with the epistemic bubble, illustrated in the previous subsection. A moral bubble relates to the fact that

- unawareness of our error is often accompanied by unawareness of the deceptive/aggressive character of our speeches (and behaviors).

Woods continues: "*Proposition 11 (Immunization)* Although a cognitive agent may well be aware of the Bubble Thesis and may accept it as true, the phenomenological structure of cognitive states precludes such awareness as a concomitant feature of our general cognitive awareness", and, consequently "Even when an agent X is in a cognitive state S in which he takes

---

[21] I am using here the word *Dasein* to refer to the actual and concrete existence of the cognitive endowments of a human agent. It is a German philosophical word famously used by Martin Heidegger in his magnum opus *Being and Time*. The word *Dasein* was used by several philosophers before Heidegger, with the meaning of "existence" or "presence". It is derived from da-sein, which literally means being-there/here, though Heidegger was adamant that this was an inappropriate translation of *Dasein*.

himself as knowing the Bubble Thesis to be true, S is immune to it as long as it is operative for X". In short, a skeptical conclusion derives that errors are unavoidable, their very nature lies embedded in their concealment; that is, in an epistemic bubble, "any act of error-detection and error-correction is subject in its own right to the concealedness of error" [44].

Furthermore, it can be argued the characteristic of morality defined as *viscosity* is a converging point between our *bottom up* perspective and that of sociobiology and evolutionary psychology, which is more of a *top down* one. This concept clarifies some other aspects of morality that further explain the insurgence of the moral bubble. Lahiti and Weinstein [18] introduce the concept of viscosity to provide an explanation of the gap between the absolutism of morality and the empirical evidence that moral regulations are often infringed with no major consequences neither for the whole moral system, nor for the very individual who performs the infractions. As a word borrowed from physics, viscosity refers to "the state of being thick, sticky, and semi-fluid in consistency, due to internal friction". To say that morality is viscous hints at its thickness and being glue-like, thus meaning its capability to be deformed, stressed, pulled apart and reassembled without showing decisive harm to its own stability and reproducibility. This is a trait displayed both at an inter-subjective and intra-subjective level, so that even major infringements of moral axiologies do not cause the moral to become self-defeating: such rapid regeneration of morality makes groups endorsing a morality "less likely to engage in rapid changes of commitment level that can compromise the efficacy of indirect reciprocity and ultimately threaten group stability. [Individuals belonging to those groups] will also be less likely to track drastic fluctuations in perceived group stability, decreasing susceptibility to sudden extrinsic threats" [18, p. 58].

From another perspective, though, this viscosity is a decisive trait of the moral disposition at the singular agent level: it allows a person to steal once and yet deeming robbery as morally wrong, to be unfaithful now and then and yet consider faithfulness as a fundamental moral virtue, to be violent and yet preach non-violence as the way to happiness and salvation. That is not a matter of hypocrisy, as Wilson [47] rightly points out in his analysis of Judaism in a group-serving perspective, but rather a necessary condition to the survival of morality itself by means of the moral bubble, whose scope is to avoid the cognitive breakdown that would be triggered by the constant appraisal of the major or minor inconsistency of our conduct with respect to our convictions. I have recently stressed in [21] how, since the Eighteenth Century, an ever-growing number of objects has been admitted to the category of those deserving moral treatment: women[22], animals, local and global environment, political entities, cultural artifacts and so on. That can be argued to further stress moral viscosity, insofar as our actions are susceptible to be evaluated as *good, right and righteous* or *bad, wrong and evil* in more

---

[22] By this I do not mean to endorse any misogyny but merely state a truth in the history of thought.

and more domains, which seems to encourage an hypertrophic growth of the *moral bubble*. The last consideration leads to the hypothesis, I have elaborated in [22, chapter 6], that a kind of overmorality (that is the presence of too many moral values attributed to too many human features, things, event, and entities), can be dangerous, because it furnishes more opportunities to trigger fresh violence by promoting a plenty of unresolvable conflicts.

Burton [7] contends that beliefs in general (and so also moral beliefs) are endowed with a kind or inertia, which further relates to shift between the mere belief to the idea of knowing, that is typical of the moral bubble: "The more committed we are to a belief, the harder it is to relinquish, even in the face of overwhelming contradictory evidence. Instead of acknowledging an error in judgment and abandoning the opinion, we tend to develop a new attitude or belief that will justify retaining it" [7, p. 12]. Furthermore, we consciously choose a false belief because it feels correct even when we know better, it seems that there is a gap between the specific sensation of feeling right or of being right. Paradoxically, we keep and trust a wrong belief even if we have a fresh knowledge which contradicts it. Similarly, embubbled, we adopt an axiological conviction (and we act according to it) even if we otherwise know the possible or actual violent outcomes of it.

I have said above, following Woods, that a skeptical conclusion derives that errors are unavoidable, their very nature lies embedded in their concealment; that is, in an epistemic bubble. Indeed, in a sense, there is nothing to correct, even if we are aware of the error in reasoning we are performing. Analogously, there is nothing to complain about ourselves, even if we are aware of the possible deceptive character of the reasoning we are performing. The awareness that something *has a priority* and that is *stable* is about the fact we are contending *our* opinions which are endowed with an intellectual value because they are ours and an intrinsic moral value because they fit some moral perspective *we* agree with as individuals. A perspective that we usually share with various groups we belong to but which can occasionally also be merely subjective and – obviously – can be seen as perverse from the perspective of others. Errors, but also deception and aggressiveness, are a constitutive "occluding edge" of agent-based linguistic acts and conducts. It is from this perspective that we can also grasp the effective importance reserved by humans for the so-called "intuition", where they simply reason in ways that are *typical* for them, and *typically* justified for them [44, chapter 4][23].

Finally, a quotation is noteworthy and self-evident

Proposition 12 (Inapparent falsity) The putative distinction between a merely apparent truth and a genuine truth collapses from the perspective of first-person awareness, i.e., it collapses within epistemic bubbles. *Corollary 12a.*

---

[23] It is interesting to note the recent attention to fraud, deception and their recognition in the area of multiagent systems (MAS), e-commerce, and agent societies, but also in logic (cf. [33]), for example by using the so-called experience-based reasoning (EBR), which also takes advantage of the concept of abduction.

As before, when in an epistemic bubble, cognitive agents always resolve the tension between only the apparently true and the genuinely true in favour of the genuinely true. *Corollary 12a* reminds us of the remarkable perceptiveness of Peirce [...] that the production of belief is the sole function of thought. What Peirce underlines is that enquiry stops when belief is attained and is wholly natural that this should be so. However, as Peirce was also aware, the propensity of belief to suspend thinking constitutes a significant economic advantage. It discourages the over-use of (often) scarce resources. (*ibid.*)

Hence, truth is "fugitive" because one can never attain it without thinking that one has done so; but thinking that one has attained it is not attaining it and so cognitive agents lack the possibility to distinguish between merely apparent and genuine truth-apprehension: "One cannot have a reason for believing P without thinking one does. But thinking that one has a reason to believe P is not having a reason to believe P". It can be said that fallibilism in some sense acknowledges the perspective above and, because of its attention to the propensity to guess (and thus also to abduce) and to error-correction, it does not share the error-elimination paroxysm of the syllogistic tradition.

4.4 Gossip, Morality, and Violence

If humans are so inclined to disregard errors it is natural to especially devote to the so-called fallacious reasoning a kind of natural, light military role which becomes evident when more or less vital interests of various kinds are at stake. In this case arguments that embed fallacies can nevertheless aim at 1) defending and protecting ourselves and/or our group(s) – normally, human beings belong to various groups, as citizens, workers, members of the family, friends, etc. – groups which are always potential aggressive coalitions; 2) attacking, offending and harming other individuals and groups. Here, by way of example, it is well-known that gossip takes advantage of many fallacies, especially *ad hominem, ad baculum, ad misericordiam, ad verecundiam, ad populum*, straw man, and begging the question[24].

From this perspective gossip (full of guessed hypotheses about everyone and everything, which exploit the so-called fallacies) contemplates

---

[24] Some considerations on the moral role of gossip are given by Magnani and Bardone in in [1]. The efficacious and presumptive "adaptive" role of language and of the theory of mind to justify punishment and bullying of weaker and/or endowed with low reputation individuals and groups by absent third parties is treated by Ingram, Piazza and Bering in [16]. The description of linguistic manipulation of reputation of others (and of the individual capacity to manipulate their own reputation in the mind of other individuals or groups) – through gossiping for example – and their role in recruiting the support of others are also analyzed. On the role of language to achieve some forms of symbolic status for a group, rather than to fundamentally achieve symmetrical cooperation, cf. also Dessalles [9].

1. the telling of narratives that exemplify moral characters and situations
   and so inform and disseminate the moral dominant knowledge of a group
   (or subgroup) (a teaching and learning role which enforces the group as
   a coalition) *possibly* favoring its adaptivity, while "at the same" time
   facilitating some disadvantage, persecution, and punishment

   a. of free riders inside the group (or inside the same subgroup as the
      arguer, or inside other subgroups of the same group as the arguer),
      and
   b. of alien individuals and groups presenting different moral and other
      conflicting perspectives.

Even if Wilson's evolutionary perspective in terms of *group selection* is questioned because of its strict Darwinist view of the development of human culture, a suggestion can be derived, especially if we reframe it in the theory of "cognitive niches": schemas of gossiping establish cognitive niches that can be adaptive but of course also *maladaptive* for the group [47]. Human coalitions produce various standard gossip templates which exploit fallacies and that can be interpreted in terms of conflict-making cognitive niches. However it is unlikely these "military" schemas, which embed both moral and violent aspects, can directly establish appreciable selective pressures.

The disseminating process of gossip shown above is *moral* and at the same time it secures the more or less *violent* persecution of free riders inside the group. Furthermore, the parallel process of protecting and defending ourselves and our groups, is *moral* and at the same time it secures aggression and *violence* against other, different, groups. In other words exploiting the so-called fallacies in gossip can be seen as *cooperative* because they carry moral knowledge, but it is at the same time *uncooperative* or even *conspiratory* because is triggers the violence embedded in harming, punishing and mobbing. It has to be said that the type of violence perpetrated through the so-called fallacies in these cases is situated at an intermediate level – in between sanguinary violence and indifference. If the "supposed to be" – from our intellectual viewpoint – fallacies are eco-cognitively fruitful for the individual or the coalition, can we still call them fallacies, if we take into account their own perspective?

Gossip proves to be an effective coalition-management tool also at a higher cognitive level, to keep within bonds hostilities among individuals and cliques (competitively aligned against each other) and reasserts once again the general values of the group. Gluckman states that: "[Gossip and scandal] control disputation by allowing each individual or clique to fight fellow-members of the larger group with an acceptable, socially instituted customary weapon, which blows back on excessively explosive users" [15, p. 13]. To make this point clearer, we might suggest a reasonable analogy: over the few past centuries, duels were a common method for settling arguments. Weaponry was not at the contenders' total discretion but was subject to a long-established regulation. Similarly, gossip can indeed be used as a social weapon; saying social, though, we do not only mean that it can be obviously used to damage other peers in the same coalition, but also that its very way of operating is determined by the coalition itself. Since the content of gossip is bounded by the evoked morals of the community, struggles are kept within the common set of shared norms, and any attempt to trespass these norms out of rage or indignation is discouraged by the very same mechanism. Thus, even internal struggles achieve the result of maintaining the group structured at a convenient degree.

The hypotheses generated by the so-called fallacies in a dialectic interplay (but also when addressed to a non-interactive audience) are certainly conflict-makers[25] but they do not have to be conceived absolutely as a priori "deal-breakers" and "dialogue-breakers", like Woods very usefully notes[26]. I would contend that the potential deceptive and uncooperative aim of fallacies can be intertwined with pieces of both information and disinformation, logical valid and invalid inferences, other typical mistakes of reasoning like perceptual errors, faulty memories and misinterpreted or wrongly transmitted evidence, but fallacious argumentations still can be – at first sight paradoxically – "civil" ways/conventions for negotiating. That is, sending a so-called deceptive fallacy to a listener is much less violent than sending a bullet, even if it "can" enter – as violent linguistic behavior – a further causal chain of possible "more" violent results. Also in the case of potentially deceptive/uncooperative fallacious argumentation addressed to a non-interactive audience, the listener is "in principle" in the condition to disagree with and reject what is proposed (like in the case of deceptive and fallacious advertising or political and religious propaganda). The case of a mobbed person is more problematic, often it is impossible to prevent the violent effects of mobbing performed through gossip: any reaction of the mobbed ineluctably tends to confirm the reason adduced by the mobbing agencies as right. This is one of the reasons that explains how mobbing is considered a very violent behavior.

In sum, when an argument (related or not to the so-called fallacies) "perpetrated" by the proposer(s), is accepted (for example when gossip full of *ad hominem* and *ad populum* arguments is fortunate and assumed inside a group), it has been proved successful and so must have been – simply – a good argument. When the argument is rejected, often, but not necessarily, it happens because it has been recognized as a fallacy and an error of reasoning: Proposition 12 (Error relativity), in [44], Woods clearly states: "Something may be an error in relation to the standards required for target attainment, in relation to the legitimacy of the target itself, or in relation to the agent's cognitive wherewithal for attaining it".

---

[25] The relationship between arguer and audience is analyzed by Tindale [38] in a contextual-based approach to fallacious reasoning.

[26] Several chapters of the new Woods' book [44] are devoted to an analysis of the argument *ad hominem* and contain a rich description of various cases, examples, and problems, which broaden my point.

4.5 Judging Violence: Abductive Fallacy Evaluation and Assessment

I recently watched a talk show on television devoted to the case of a Catholic priest, Don Gelmini, accused of sexual abuse by nine guys hosted in a home for people with addiction problems belonging to Comunità Incontro in Italy, a charitable organization now present worldwide, which he had founded many years before. Two pairs of journalists argued in favor and against Don Gelmini, by using many so-called fallacious arguments (mainly *ad hominem*) centered on the past of both the accused and the witnesses. I think the description of this television program is useful to illustrate the role of abduction in the filtration and evaluation of arguments, when seen as *distributed* in a real-life dialectical and rhetorical contexts. As an individual belonging to the audience, at the end of the program I concluded in favor of the *ad hominem* arguments (that I also "recognized" as fallacies) used by the journalists and so to the hypothesis which argued against Don Gelmini. Hence, I considered the data and gossip embedded in those fallacious reasoning describing the "immoral" and "judiciary" past of Don Gelmini more relevant than the ones which described the bad past of the witnesses. I was aware of being in the midst of a riddle of hypotheses generated by various arguments, and of course this was probably not the case of the average viewer, who may not have been trained in logic and critical thinking, but it is easy to see that this fact does not actually affect the rhetorical success or failure of arguments in real-time contexts, like it was occurring in my case. Indeed we are all in an "epistemic bubble" as listeners – compelled to think we know things even if we do not know them, a bubble that in this case forces you to quickly evaluate and pick up what you consider the best choice. I would like to put forward the idea that, at least in this case, the evaluation of the *ad hominem* arguments has to be seen as the fruit of an abductive appraisal, and that this abductive process is not rare in argumentative settings.

An analogy to the situation of trials in the common law tradition, as described by Woods [43], can be of help. Like in the case of the judge in the trial, in our case the audience (and myself, as part of the audience)[27] was basically faced with the *circumstantial* evidence carried by the two clusters of *ad hominen* arguments, that is, faced with evidence from which a *fact* can be reasonably inferred but not directly proven.

In a situation of lack of information and knowledge, and so of constitutive "ignorance", abduction is usually the best cognitive tool human beings can adopt to relatively quickly reach explanatory, non-explanatory, and instrumental hypotheses/conjectures[28]. Moreover, it is noteworthy that

---

[27] Furthermore, like the jury in trials, an audience is on average composed of individuals who are not experts able to "overt calibration of performance to criteria", but instead ordinary – reasonable and untutored – people reasoning in the way ordinary people reason, a kind of "intuitive and unreflective reasoning" [43].

[28] These various aspects of abductive cognition are fully explained in [22].

evidence – embedded in the *ad hominen* arguments – concerning the "past" (supposed) reprehensible behavior of the priest and of the witnesses were far from being reliable, probably chosen *ad hoc*, and deceptively supplied, that is, so to say, highly circumstantial for the judge/audience (and for me, in this case). I had to base my process of abductive evaluation regarding the fallacious dialectics between the two groups of journalists on a kind of sub-abductive process of *filtration* of the evidence provided, choosing what seemed to me the most reliable evidence in a more or less intuitive way, then I performed an abductive appraisal of all the data. The filtration strategy is of course abductively performed guided by various "reasons", the conceptual ones, for example being based on conceptual judgments of credibility. However, these reasons were intertwined with more or less other reasons as more or less conscious emotional reactions, based on various feelings triggered by the entire distributed visual and auditory interplay between the audience and the scene, in itself full of body gestures, voices, and images (also variably and smartly mediated by the director of the program and the cameramen). Along these lines, I might add that also the journalists fallaciously discussing the case were concerned with the accounts they could trust and certainly emotions played a role in their inferences.

In summary, I was able to abductively make a selection (selective abduction) between the two non-rival and incommensurable hypothetical narratives about the priest, – of course I could have avoided the choice, privileging indifference, thus stopping any abductive engagement – forming a kind of explanatory theory of them. The guessed – and quickly accepted without further testing – theory of what was happening in that dialectics further implied the hypothesis for guilt with respect to the priest. That is, the *ad hominen* of the journalists that were speaking about the priest's past (he was for example convicted for four years because of bankruptcy fraud and some acceptable evidence – data, trials documents – was immediately after provided by the staff of the television program) appeared to me convincing, that is, no more negatively biased, but a plausible acceptable argument. Was it still a fallacy from my own actual eco-cognitive perspective? I do not think so: it still was and is a fallacy only from a special subset of my own eco-cognitive perspective, the intellectual/academic one! The "military" nature of the above interplay between contrasting *ad hominem* arguments is patent. Indeed, they were armed linguistic forces involving "military machines for guessing hypotheses" clearly aiming at forming an opinion in my mind (and in that of the audience) which I reached through an abductive appraisal, quickly able to explain one of the two narratives as more plausible. In the meantime I became part of the wide coalition of the individuals who strongly suspect Don Gelmini is guilty and that can potentially be engaged in further "armed" gossiping[29].

---

[29] A vivid example of the aggressive "military" use of language is the so-called "poisoning the well" illiustrated by Walton, "[...] a tactic to silence an opponent, violating her right to put forward arguments on an issue both parties have agreed to discuss at the confrontation stage of a critical discussion" [40, p. 273]. "Poisoning the well" is often considered as a species of *ad hominem* fallacy.

1. In special contexts where the so-called fallacies and various kinds of hypothetical reasoning are at play, both at the rhetorical and the dialectic level, the assessment of them can be established in a more general way, beyond specific cases[30]. An example is the case of a fallacy embedding patently false empirical data, which can easily be recognized as false at least by the standard intended audience; another example is when a fallacy is structured, from the argumentative point of view, in a way that renders it impossible to address the intended audience and in these cases the fallacy can be referred to as "always committed".

2. Not only abduction, but also other kinds of (supposed to be) fallacious argumentation can be further employed to evaluate arguments in dialectic situations like the ones I have quoted above, such as for instance *ad hominem*, *ad populum*, *ad ignorantiam*, etc., but also hasty generalization and deductive schemes.

3. The success of a so-called fallacy and of the so-called fallacious hypothetical arguments can also be seen from the perspective of the arguer in so far as she is able to guess an accurate abductive assessment of her actual or possible audience's character. From this perspective an argument is put forward and "shaped" according to an abductive hypothesis about the audience, which the arguer guesses on the basis of available data (internal, already stored in memory, external – useful cues derived from audience features and suitably picked up, and other intentionally sought information). Misjudging the audience would jeopardize the efficacy of the argument, which would consequently be a simple error of reasoning/argumentation. Obviously also in assessment made by the audience inferences which are less complicated than abduction can be exploited, like hasty generalization, etc.

4. As is clearly shown by the example of the priest, arguments are not only distributed, as I have contended in the subsection 4.1 above, but they are also embedded, nested, and intertwined in self-sustaining clusters, which individuate peculiar global "military" effects.

Thom [37, pp. 281–282] usefully observes that there is, underlying the idea of information,

> [...] the existence of a seeker for whom it is profitable to learn this information, together with a giver, who gives, of his own free will in general, the information to the seeker. In all uses of the word information one is unable to identify the four elements: seeker, request, giver, advantage accruing to the seeker through knowledge of the information, one must suspect dishonesty in the use of the word.

Fallacies play a fundamental role in these cases, also when there has been no request and when information has little or no interest for the recipient, like in the case of a lot of publicity, which pretends it only offers information. Thom

---

[30] Examples are provided by Tindale [38].

concludes "in propagating the faith (*de propaganda fide*) at least the Church is aware of its aim and proclaims them". Another useful example provided by Thom concerns situations when the seeker and the giver are obscured, so that the violence involved if just of intellectual kind, such as in the case of the word information in the last decades of the previous century molecular biology, when overemphasizing the role of information leads to disguise (not necessarily intentionally) the almost-total ignorance in which we find ourselves in the study of the immensely complex unfolding of morphogenetic processes in embryology.

5 Conclusion

In this paper I have described the so-called coalition enforcement hypothesis, which sees humans as self-domesticated animals engaged in a continuous hypothetical activity of building cooperation through morality, incorporating punishing and violent policies at the same time. Indeed, I have tried to show how morality and violence can be seen as strictly intertwined with social and institutional aspects, implicit in the activity of cognitive niche construction. I have illustrated two key concepts that are in the background of the various argumentations contained in this paper: 1) the importance of abduction, as a way of guessing hypotheses, in moral reasoning and in describing the role of fallacies in moral and violent settings; 2) the philosophical usefulness of framing the problem of violence in a biological and evolutionary perspective, where – so to say – the "animality of human animals" is a tool for avoiding traditional and robust philosophical illusions, prejudices, and ingenuities about violence.

I have also shown that *hypothetical thinking* (and so *abduction*) is in turn very often embedded in various linguistic kinds of the so-called fallacious reasoning. Indeed, in evolution, coalition enforcement works through the building of social cognitive niches seen as new ways of diverse human adaptation, where guessing hypotheses plays a central role and, moreover, it happens just as it can, depending on the cognitive/moral options human beings adopt. I have further stressed the moral and violent effect played by human natural languages, focusing on the analysis of the relationships between language, logic, fallacies, and abduction. I have contended that this "military" nature of abductive hypothetical reasoning in linguistic communication (*military intelligence*) is intrinsically "moral" (protecting the group by obeying shared norms), and at the same time "violent" (for example, harming or mobbing others - members or not of the group – still to protect the group itself).

Finally, I have illustrated that not only language carries morality and violence, as the "military" power can be considered also active at the level of *model-based cognition*: taking advantage of the naturalistic perspective on abductive "hypothesis generation" at the level of both instinctual behavior and representation-oriented behavior, where nonlinguistic features drive a "plastic" model-based cognitive role, a wide range of cognitive performances gains a fundamental semiotic, eco-physical, and "military" significance, which furnished further insight into a kind of "social epistemology".

References

1. Bardone, E., Magnani, L.: The appeal of gossiping fallacies and its eco-logical roots. Pragmatics & Cognition 18(2), 365–396 (2010)
2. Becchio, C., Bertone, C., Castiello, U.: How the gaze of others influences object processing. Trends in Cognitive Sciences 12(7), 254–258 (2008)
3. Bingham, P.M.: Human uniqueness: A general theory. The Quarterly Review of Biology 74(2), 133–169 (1999)
4. Bingham, P.M.: Human evolution and human history: A complete theory. Evolutionary Anthropology 9(6), 248–257 (2000)
5. Boehm, C.: Hierarchy in the Forest. Harvard University Press, Cambridge (1999)
6. Boyd, R., Richerson, P.J.: Norms and bounded rationality. In: Gigerenzer, G., Selten, R. (eds.) Bounded Rationality. The Adaptive Toolbox, pp. 281–296. The MIT Press, Cambridge (2001)
7. Burton, R.A.: On Being Certain. St. Martin's Press, New York (2008)
8. Dellarosa Cummins, D.: How the social environment shaped the evolution of mind. Synthese 122, 3–28 (2000)
9. Dessalles, J.L.: Language and hominid politics. In: Knight, J.H.C., Suddert-Kennedy, M. (eds.) The Evolutionary Emergence of Language: Social Function and the Origin of Linguistic Form, pp. 62–79. Cambridge University Press, Cambridge (2000)
10. Dugatkin, L.A.: The Altruism Equation. Seven Scientists Search for the Origins of Goodness. Princeton University Press, Princeton (2008)
11. Frischen, A., Bayliss, A.P., Tipper, S.P.: Gaze cueing of attention. Psychological Bulletin 133(4), 694–724 (2007)
12. de Gelder, B., Snyder, J., Greve, D., Gerard, G., Hadjikhani, N.: Fear fosters flight: a mechanism for fear contagion when perceiving emotion expressed by a whole body. In: PNAS (Proceedings of the National Academy of Science of the United States of America) , vol. 101, pp. 16,701–16,706 (2004)
13. Girard, R.: Violence and the Sacred [1972]. Johns Hopkins University Press, Baltimore (1977)
14. Girard, R.: The Scapegoat [1982]. Johns Hopkins University Press, Baltimore (1986)
15. Gluckman, M.: Papers in honor of Melville J. Herskovits: Gossip and scandal. The American Economic Review 4(3), 307–316 (1963)
16. Ingram, G.P.D., Piazza, J.R., Bering, J.M.: The adaptive problem of absent third-party punishment. In: Høgh-Olesen, H., Bertelsen, P., Tønnesvang, J. (eds.) Human Characteristics: Evolutionary Perspectives on Human Mind and Kind, pp. 205–229. Cambridge Scholars, Newcastle-upon-Tyne (2009)
17. Koenigs, M., Tranel, D.: Irrational economic decision-making after ventromedial prefrontal damag: evidence from the ultimatum game. Journal of Neuroscience 27(4), 951–956 (2007)

18. Lahiti, D.C., Weinstein, B.S.: The better angels of our nature: Group stability and the evolution of moral tension. Evolution and Human Behavior 2, 47–63 (2005)
19. Magnani, L.: Abduction, Reason, and Science. Processes of Discovery and Explanation. Kluwer Academic Publishers, Dordrecht (2001)
20. Magnani, L.: Mimetic minds. Meaning formation through epistemic mediators and external representations. In: Loula, A., Gudwin, R., Queiroz, J. (eds.) Artificial Cognition Systems, pp. 327–357. Idea Group Publishers, Hershey (2006)
21. Magnani, L.: Morality in a Technological World. Knowledge as Duty. Cambridge University Press, Cambridge (2007)
22. Magnani, L.: Abductive Cognition. The Eco-Cognitive Dimension of Hypothetical Reasoning. Springer, Heidelberg (2009)
23. Magnani, L.: Philosophy of Violence. Morality, Religion, and Violence Intertwined (2010) (Forthcoming)
24. Moll, J., de Oliveira-Souza, R.: When morality is hard to like. Scientific American Mind 2/3, 30–35 (2008)
25. Moll, J., de Oliveira-Souza, R., Garrido, G., Bramati, I.E., Caparelli-Daquer, E.M., Paiva, M.L., Zahn, R., Grafman, J.: The self as a moral agent: Linking the neural bases of social agency and moral sensitivity. Social Neuroscience 2(3-4), 336–352 (2007)
26. Peirce, C.S.: Collected Papers of Charles Sanders Peirce. Harvard University Press, Cambridge (1931-1958), vols. 1-6, Hartshorne, C. and Weiss, P., eds.; vols. 7-8, Burks, A. W., ed
27. Pinker, S.: Language as an adaptation to the cognitive niche. In: Christiansen, M.H., Kirby, S. (eds.) Language Evolution, pp. 16–37. Oxford University Press, Oxford (2003)
28. Queiroz, J., Merrell, F. (eds.): Abduction: Between Subjectivity and Objectivity, vol. 153. De Gruyter, Berlin (2005) Special Issue of the Journal Semiotica
29. Rohwer, Y.: Hierarchy maintenance, coalition formation, and the origin of altruistic punishment. Philosophy of Science 74, 802–812 (2007)
30. Rosas, A.: Multilevel selection and human altruism. Biology and Philosophy 23, 205–215 (2008)
31. Secchi, D.: A theory of docile society: The role of altruism in human behavior. Journal of Academy of Business and Economics 7(2), 446–461 (2007)
32. Simon, H.A.: Altruism and economics. American Economic Review 83(2), 157–161 (1993)
33. Sun, Z., Finnie, G.: Experience-based reasoning for recognising fraud and deception. In: Ishikawa, M., Hashimoto, S., Paprzycki, M., Barakova, E., Yoshida, K., Köppen, M., Corne, D.W., Abraham, A. (eds.) Proceedings of the Fourth International Conference on Hybrid Intelligent Systems, HIS 2004, pp. 80–85 (2004)
34. Taylor, K.: Cruelty. Human Evil and the Human Brain. Oxford University Press, Oxford (2009)
35. Thagard, P.: How does the brain form hypotheses? Towards a neurologically realistic computational model of explanation. In: Thagard, P., Langley, P., Magnani, L., Shunn, C. (eds.) Symposium Generating explanatory hypotheses: Mind, computer, brain, and world. Proceedings of the 27th International Cognitive Science Conference, Cognitive Science Society, Stresa (2005) CD-Rom

36. Thagard, P.: Abductive inference: From philosophical analysis to neural mechanisms. In: Feeney, A., Heit, E. (eds.) Inductive Reasoning: Experimental, Developmental, and Computational Approaches, pp. 226–247. Cambridge University Press, Cambridge (2007)

37. Thom, R.: Esquisse d'une sémiophysique. InterEditions, Paris (1988). Translated by V. Meyer, Semio Physics: a Sketch. Addison Wesley, Redwood City (1990)

38. Tindale, C.W.: Hearing is believing: A perspective-dependent view of the fallacies. In: van Eemeren, F., Houtlosser, P. (eds.) Argumentative Practice, pp. 29–42. John Benjamins, Amsterdam (2005)

39. Tindale, C.W.: Fallacies and Argument Appraisal. Cambridge University Press, Cambridge (2007)

40. Walton, D.: Poisoning the well. Argumentation 20, 273–307 (2006)

41. Wilson, D.S.: Evolution, morality and human potential. In: Scher, S.J., Rauscher, F. (eds.) Evolutionary Psychology. Alternative Approaches, pp. 55–70. Kluwer Academic Publishers, Boston (2002)

42. Wilson, D.S., Timmel, J.J., Miller, R.R.: Cognitive cooperation. When the going gets tough, think as a group. Human Nature 15(3), 1–15 (2004)

43. Woods, J.: Ignorance, inference and proof: Abductive logic meets the criminal law. In: Proceedings of the International Conference Applying Peirce, Helsinki, Finlan (2009) (Forthcoming)

44. Woods, J.: Seductions and Shortcuts: Error in the Cognitive Economy (2010) (Forthcoming)

45. Žižek, S.: Violence [2008]. Profile Books, London (2009)

Different Cognitive Styles in the Academy-Industry Collaboration

Riccardo Viale

Abstract. Previous studies on obstacles in technology transfer between universities and companies emphasized the economic, legal, and organizational aspects, mainly focused in transfer of patents and licences. Since research collaboration implies a complex phenomenon of linguistic and cognitive coordination and attuning among members of the research group, a deeper cognitive investigation about this dimension might give some interesting answer to *academy-industry problem*. The main hypothesis is that there can be different cognitive styles in thinking, problem solving, reasoning and decision making that can hamper the collaboration between academic and industrial researchers. These different cognitive styles are linked and mostly determined by a different set of values and norms that are part of background knowledge. Different background knowledge is also responsible of bad linguistic coordination and understanding and of the difficulty of a successful psychology of group. The general hypotheses that will be inferred in this paper represent a research programme of empirical tests to control the effects on cognitive styles of different scientific and technological domains and geographical contexts.

1 Relationship between Background Knowledge and Cognitive Rules

The close relationship between background knowledge (BK) and implicit cognitive rules (ICR) [53, 81] is highlighted in the results of numerous studies on developmental psychology [63, 77] and cognitive anthropology [48, 77]. Our inferential and heuristic skills appear to be based on typical components of BK. Moreover, our reasoning, judgment, and decision-making processes seem to rely on principles that are genetically inherited from our parents.

Riccardo Viale
Fondazione Rosselli, Turin, Italy
e-mail: `riccardo.viale@fondazionerosselli.it`

L. Magnani et al. (Eds.): Model-Based Reasoning in Science & Technology, SCI 314, pp. 83–105.
springerlink.com   © Springer-Verlag Berlin Heidelberg 2010

As it was emphasized in Viale [76] dependence of ICR from BK is not justified by the cognitive theories that support an autonomous syntactic mental logic. According to these theories [3, 5, 56] the mind contains a natural deductive logic (which for Piaget is the propositional calculus) that allows to do some inference and not other. For example the human mind is able to apply the *modus ponens* and not *modus tollens*. In the same way, we could also presuppose the existence of a natural probability calculus, causal reasoning rule, risk assessment rule, and so on. Many empirical studies and some good theories give an alternative explanation that neglect the existence of mental logic and of other syntactic rules (for the pragmatic scheme theories: [10, 11, 12]; for the mental models theory: [24, 32]; for the conceptual semantic theory see [29]). The first point is that there are many rules that are not applied when the format is abstract, but which are applied when the format is pragmatic, that is when it is linked to every-day experience. For example the solution of the *selection task problem*, that is the successful application of *modus tollens*, is possible when the questions are not abstract but are linked to problems of everyday life [51, 52]. The second point is that, most of the time, the rules are implicitly learned through pragmatic experience [54, 14, 15]. The phenomenon of implicit learning seems so strong that it occurs even when the cognitive faculties are compromised. From recent studies [27] with Alzheimer patients it seems that they are able to learn rules implicitly but not explicitly. Moreover, the rules that are learnt explicitly in a class or are part of the inferential repertoire of experts are often not applied in everyday life or in test based on intuition (see the experiments with statisticians of Kaheneman and Tversky).

At the same time the pragmatic experience and the meaning that people give to the social and natural events are driven by background knowledge [57, 58, 61]. The values, principles, and categories of background knowledge, stored in memory, allow us to interpret reality, to make inferences, to act, that is to have a pragmatic experience. Therefore, background knowledge affects implicit learning and the application of the cognitive rules through the pragmatic and semantic dimension of reasoning and decision making[1]. The mental structure that connects background knowledge and cognitive rules can be represented by a *schema* (an evolution of the *semantic network* of Collins and Quillian, [16]), that is a structured representation that captures the information that typically applies to a situation or event [2]. They establish a set of relations that links properties. Thus the schema for a birthday party might include guests, gifts, cakes, and so on. The structure is that the guests give gifts to the birthday celebrant, and that everyone eats cake, and so on. What it is important is that the relationships within schemas and among different schemas allow us to make inferences, that is, they correspond to the

---

[1] It is not clear if the process is not linear but circular and recursive. In this case the cognitive rules might become part of the background knowledge and that could change its role in the pragmatic experience and in the reasoning and decision making processes.

implicit cognitive rules. For example consider our schema for glass. It specifies that if an object made of glass falls onto a hard surface, the object may break. This is an example of causal inference. Similar schemas can allow you to make inductive, deductive, analogical inferences, to solve problems and to take decisions [43, 55]. In conclusion the schema theory seems to be a good candidate to explain the dependence of ICR from BK.

2 Obstacles to Knowledge Transfer: Background Knowledge

What is the different background knowledge between university and industrial labs and how can this influence cognitive styles?

There were studies in the sociology of science that have focused on the values and principles that drive scientific and industrial research [76].

Academic research seems to be governed by a set of norms and values that are close to Mertonian *ethos* [44]. Communitarism, scepticism, originality, disinterestedness, universalism and so on were proposed by Robert Merton as the social norms of scientific community. He justified theoretically the proposal. Other authors like Mitroff [46] criticized the Mertonian ethos on an empirical base. He discovered that scientists follow often the Mertonian norms. Nevertheless there are cases in which they seem to follow the contrary of the norms. More recent studies [6] confirm most of the norms of Merton. The research should be Strategic, founded on Hybrid and interdisciplinary communities, able to stimulate the Innovative critique, Public and based on Scepticism (SHIPS). Recent studies [59, 78] confirm the presence of social norms that remind the Mertonian ethos. Scientist believe in the pursuit of knowledge per se, in the innovative role of critique, in the universal dimension of the scientific enterprise, in science as a public good. They believe in scientific method based on empirical testing, comparison of hypotheses, better problem solving and truth as representation of the world [78, pp. 216–219]. The fact that scientists have these beliefs don't prove that they act accordingly. The beliefs can be deviated by contingent interests and opportunistic reasons. They could also represent the pretended image of what they want to show to society. They also can vary from one discipline and specialization to another. Nevertheless the presence of these beliefs seems to characterize the cultural identity of academic scientists. Therefore they constitute part of their background knowledge and they can influence the implicit cognitive rules for reasoning and decision making. On the contrary industrial researchers are driven by norms that are contrary to academic ones. They can be summarized by the acronym PLACE [83]: Proprietary, Local, Authoritarian, Commissioned, Expert. The research is commissioned by the company that has the ownership of results, that can't be diffused, and are valid locally to improve the competitiveness of the company. The researchers are subjected to authoritarian decisions of the company and they develop a particular expertise valid locally. PLACE is a set of norms and values that characterizes the cultural identity of industrial researchers. They constitute part of their background knowledge and they may influence the inferential processes of reasoning and decision making.

To sum up, the state of art of studies on social norms in academic and industrial research seems insufficient and empirically obsolete. A new empirical study of norms contained in the background knowledge is essential. The study should control the main features characterizing cultural identity of academic and industrial researchers, established by previous studies. They can be summarized in following way:

- *Criticism vs. Dogmatism*: academic researchers follow the norm of systematic critique, scepticism, falsificatory control of knowledge produced by colleagues; industrial researchers aim at maintaining knowledge that works in solving technological problems.
- *Interest vs. Indifference*: academic researchers are not pushed in their activity mainly by economic interest but by epistemological goals; industrial researchers are pushed mainly by economic ends like technological competitiveness, commercial primacy, and capital gain.
- *Universalism vs. Localism*: academic researchers believe in a universal audience of peers, in universal criteria of judgement that can establish their reputation; industrial researchers think locally both for the audience and for the criteria of judgement and social promotion.
- *Comunitarism vs. Esclusivism*: academic researchers believe in the public and open dimension of pool of knowledge which they must contribute to increase; industrial researchers believe in the private and proprietary features of knowledge.

To the different backgrounds we should add also the different contingent features of the contexts of decision making (we refer to the decision-making context of research managers, that is heads of a research unit or of a research group) that become *operational norms*. The main features are related to time, results and funding [76].

In the pure academic context[2] the time for doing research is usually loose. There are some temporal requirements when one is working with funds coming from a public call, but, in a contract with a public agency or government department the deadline is usually not so strict, and the requested results are quite not well defined and not specified as to a particular product (e.g. a prototype or a new molecule or a new theorem). Therefore, time constraints don't press the reasoning and decision making processes of the researchers. On the contrary when an academic researcher works with an industrial contract, the time constraints are similar to those of the corporate researcher. Moreover, in a fixed given time a precise set of results must be produced and

---

[2] The analysis refers mainly to academic environment of Universities of Continental Europe.

presented to the company. According to private law the clauses of a contract with a company can be very punitive for the researcher and for the university that don't follow the signed expected requirements. In any case the effect of sub-optimal results for an academician working with a company are less punitive than for a corporate researcher. For him the time pressure is heavier because the results, in a direct or semi direct way, are linked to the commercial survival of the company. Sub-optimal behavior increases the risks for his career and also for his job. Therefore the great expectations on the fast production of positive concrete results press him in a heavier way. The different environmental pressure may generate a different adaptive psychology of time and a different adaptive ontology of what the result of the research might be. In the case of academic research, time might be less discounted. That is, future events tend not to be so underestimated as might happen in industrial research. The corporate researcher might fall into the bias of time discounting and myopia because of the overestimation of short term results. Even the ontology of an academic researcher in respect to the final products of the research might be different from the corporate one. While the former is interested in a declarative ontology that aims at the expression of the result in linguistic form (i.e. a report, a publication, a speech) the second aims at an object ontology. The results for him should be linked in a direct or indirect way to the creation of an object (i.e. a new molecule, a new machine, a new material, or a new process to produce them, or a patent that describe the process to produce them).

The third, different operational norm concerns financial possibilities. In this case it is not a problem of quantity of funding. The funding for academic research is usually less for each unity of research (or, better, for each researcher) than that in industrial research. But the crucial problem is the psychological weight of the funds. That is, how much the funds constraint and affect the reasoning and decision making processes of the researchers. In other words ceteris paribus the amount of money at disposal, how much the cognitive processes and in particular the attention processes refer to a sort of *value for money* judgment in deciding how to act. From this point of view it seems – but it is a topic to be investigated – that the psychological weight of money on academic researchers is less than on industrial researchers. Money is perceived with less value and therefore, influences decision making less. The reasons for this different mental representation and evaluation may come from: a) the way in which the funding is communicated and it can constitute a decision frame (with more frequency and relevance in the company because it is linked to the important decision of the annual budget); b) the symbolic representation of the money (with much greater emphasis in the company that has its *raison d'etre* in the commercial success of its products and in its increased earnings); c) from the social identity of the researchers linked more or less strongly to the monetary levels of the wage (with greater importance to the monetary level as an indicator of a successful career in a private company than in the university). The different psychological weight of the money has been analyzed by many authors, and in particular by Thaler [67].

To summarize the operational norms can be schematized in *loose time vs. pressing time*; *undefined results vs. well-defined results*; *financial lightness vs. financial heaviness*.

3   Obstacles to Knowledge Transfer: Cognitive Rules

How can the values in background knowledge and the operational norms influence the implicit cognitive rules of reasoning and decision making, and how are they an obstacle to the collaboration among industrial and academic researchers?

There are many aspects of cognition that are important in research activity. We can say that every aspect is involved, from motor activity to perception, memory, attention, reasoning, decision making and so on. Our aim however is to focus on the cognitive obstacles to the reciprocal communication, understanding, joint decision making and coordination between academic and corporate researchers and how that might hinder their collaboration.

I will analyze 6 dimensions of the interaction: language, group, thinking, problem solving, reasoning, and decision making [76].

1. It might be interesting to investigate the pragmatic aspects of **communication**. To collaborate on a common project means to communicate, mainly by natural language. To collaborate means to exchange information in order to coordinate one's own actions to pursue a common aim. This means "using language", as in the title of Clark's book [13], in order to reach the established common goal. Any linguistic act is at the same time an individual and a social act. It is individual because it is the subject that by motor and cognitive activity articulates the sounds that correspond to words and phrases and it is the subject that receives these sounds and makes the interpretation. Or in Goffman's [23] terminology about the linguistic roles, it is the subject that *vocalizes*, *formulates*, and *means* and it is another subject that *attends the vocalization*, *identifies the utterances* and *understands the meaning* [13, p. 21]. It is social because every linguistic act of a speaker has the aim to communicate something to one or two addressees (also in the case of *private settings* where we talk to ourselves because we ourselves play the role of an addressee). In order to achieve this goal there should be a coordination between the speaker's meaning and the addressee's understanding of the communication. But meaning and understanding is based on the knowledge, beliefs, and suppositions shared, that is, in shared background knowledge. Therefore the first important point is that it is impossible for two or more actors of a conversation to coordinate meaning and understanding without reference to their common background knowledge. "A common background is the foundation for all joint actions and that makes it essential to the

creation of the speaker's meaning and addressee's understanding as well" [13, p. 14]. A common background is shared by the members of the same cultural community.

A second important point is that the coordination between meaning and understanding is more effective when the same physical environment is shared (the same room in university or the same bench in a park) and the vehicle of communication is the richest possible. The environment represents a communicative frame that can influence meaning and understanding. Even more, gestures and facial expressions are rich in non-linguistic information and therefore are very important aids for coordination. From this point face-to-face conversation is considered the basic and most powerful setting of communication.

The third point is that the more simple and direct the coordination is the more effective the communication. There are different ways of making communication complex. The roles of speaking and listening (see above regarding linguistic roles) can be decoupled. Spokesmen, ghost writers, and translators, are examples of decoupling. A spokeswoman for a minister is only a vocalizer, while the formulation is the ghost writer's and the meaning is the minister's. Obviously, in this case, the coordination of meaning and understanding becomes more difficult (also because it is an institutional setting with many addressees). The non-verbal communication of the spokesman might be inconsistent with the meaning of the minister and the ghost writer might not be able to formulate correctly this meaning. Moreover in many types of discourse – plays, story telling, media news, reading – there is more than one domain of action. The first layer is the layer of the real conversation. The second layer is that of the hypothetical domain that is created by the speaker (when he is describing a story). By recursion there can be higher layers as well. For example the play requires three layers: the first is the real world interaction among the actors, the second is the fictional role of the actors, and the third is the communication with the audience. In face-to-face conversation there is only one layer and no decoupling. The role of vocalizing, formulating, and meaning is in the same person. And the domain of action identifies itself with the conversation. The coordination is direct without intermediaries. Therefore it is the most effective way of coordinating meaning and understanding with minor distortions of meaning and less misunderstandings. Academic and industrial researchers are members of different cultural communities, therefore they have different background knowledge. In the collaboration between academic and industrial researchers the coordination between meanings and understandings can be difficult if the background knowledge is different. When this is the case, as we have seen before, the result of the various linguistic settings will likely be the distortion of meaning and misunderstanding. When fundamental values are different (SHIPS vs. PLACE) and also when the operational norms of *loose time vs. pressing time*; *undefined product vs. well defined product*;

financial lightness vs. financial heaviness are different it is impossible to transfer the knowledge without losing or distorting shares of meaning.

Moreover, difficult coordination will increase in settings that utilize intermediaries between the academic inventor and the potential industrial user (*mediated settings* in [13, p. 5]). These are cases of an intermediate technology transfer agent (as in the case of the members of TTO of university or private of government TTA) that tries to transfer knowledge from the university to corporate labs. In this case, there is *decoupling* of speaking. The academic researcher is he who formulates and gives meaning to the linguistic message (also in a written setting), while the TT agent is only a vocalizer. Therefore, there may be frequent distortion of the original meaning, in particular when the knowledge contains a great share of tacit knowledge. This distortion is strengthened by the likely different background knowledge of the TT agent in respect to the other two actors in the transfer. The TT agents are members of a different cultural community (if they are professional from a TT private company) or from different sub-communities inside the university (if they are members of TTO). Usually they are neither active academic researchers nor corporate researchers. Finally, in the technology transfer there can be also the complexity of having more than one domain of action. For example, if the relation between an academic and industrial researcher is not face-to-face, but is instead mediated by an intermediary there is an emergent second layer of discourse. This is the layer of the story that is told by the intermediary about the original process and the techniques to generate the technology invented by the academic researchers. The story can also be communicated with the help of a written setting, for example patent or publication. All the three points show that a common background knowledge is essential for reciprocal understanding, and that face-to-face communication is a pre-requisite for minimizing distortion of meaning and misunderstanding that can undermine the effectiveness of knowledge transfer.

2. The second dimension of analysis is that of the **group**. When two or more persons collaborate to solve a common problem they elicit some interesting emergent phenomenon. In theory a group can be a powerful problem solver [28]. But to be so members of the group must share information, models, values and cognitive processes [28]. It is likely that the heterogeneity about skill and knowledge might be very useful for detecting more easily the solution. Some authors have analyzed the role of heterogeneity in cognitive tasks (e.g. the solution of a mathematical problem) and generation of ideas (e.g. the production of a new logo) and they have found a positive correlation between it and the success in these tasks [30]. In theory, this result seems very likely since finding a solution needs to look at the problem from different points of view. Different perspectives allow overcoming the phenomenon of *entrenched mental set*, that is, the fixation on a strategy that normally works well in solving

many problems, but that does not work well in solving this particular problem [66]. However the diversity that works is about cognitive skills or personality traits (Jackson, 1992). On the contrary when the diversity is about values, social categories, and professional identity it can hinder the problem solving ability of the group. In fact this heterogeneity generates the categorization of the differences and the similarities between the self and the others and the emergent phenomenon of the conflict/distance between *ingroup* and *outgroup* [72]. The relational conflict/distance of *ingroup vs. outgroup* is the most social expression of the negative impact of diversity of background knowledge on group problem solving. As it was showed by Manz and Neck [42] without a common background knowledge there is not sharing of goals, of social meaning of the work, of criteria to assess and to correct the ongoing activity, of foresight on the results and on their impact, and so on. As it is described by the theory of *teamthink* [42], the establishment of an effective group in problem solving relies on the common sharing of values, beliefs, expectations and a priori on physical and social world. For example academic and industrial researchers present a different approach concerning disciplinary identity. The academic has a strong faithfulness towards the *disciplinary matrix* [38], that is composed by the members of a discipline with their set of disciplinary knowledge and methods. On the contrary the industrial researcher tend to be opportunistic in using knowledge and in choosing peers. He doesn't feel to be member of *disciplinary invisible college* of peers and chooses *à la carte* which peer is helpful and what knowledge is useful to attain the goal of research. This asymmetry between academic and corporate researchers is an obstacle to the well functioning of the *teamthink*. The epistemological and social referents are different, therefore the communication becomes a dialogue between deafs. Lastly there is the linguistic dimension. As we have seen above, without a common background knowledge the coordination of meaning and understanding among the members of the group, that is the fundamental basis of collaboration, is impossible. Moreover without a common background knowledge, the pragmatic context of communication [25, 64] doesn't allow the generation of correct automatic and non automatic inferences between speaker and addressee. For example the addressee would not be able to generate proper *implicatures* [25] to fill the lack of information and the elliptical features of the discourse.

Lastly, different background knowledge influences problem solving, reasoning and decision making activity, in other words the implicit cognitive rules. Different implicit cognitive rules mean asymmetry, asynchrony, and dissonance in the cognitive coordination among the members of the research group. That means obstacle in the knowledge transfer, in the application of academic expertise and knowledge to the industrial goal, in the development of an initial prototype or technological idea towards a commercial end.

   Now I will tackle the hypothetical effect of values and operational
norms onto the implicit cognitive rules of academic and industrial
researchers.

3. The third dimension is about **thinking**. There are two systems of thinking that affects the way how we reason, decide and solve problem. The
   first is the *associative system* which involves mental operations based
   on observed similarities and temporal contiguities [60]. It can lead to
   speedy responses that are highly sensitive to patterns and to general tendency. This system corresponds to the system 1 of Kahneman [34]. The
   system represents the intuitive dimension of thinking. It is fast, parallel
   and mainly implicit. It is switched on by emotional and affective factors.
   Knowledge is mainly procedural. It is dominant on the second when the
   reasoning and decision making must be fast without the possibility to
   analyze all the details of a problem. The second is the *rule-based system*
   which involves manipulations based on the relations among symbols [60].
   It usually requires deliberate slow procedures to reach the conclusions.
   Through this system, we carefully analyze relevant features of the available data, based on rules stored in memory. It corresponds to system 2
   of Kahneman [34]. This system is slow, serial, mainly explicit. Knowledge is mainly declarative. It can be overridden by the first when there is
   time pressure, there are emotional and affective interferences and when
   the context of decision making doesn't pretend any analytical effort. The
   intuitive and analytical systems can give different results in reasoning
   and decision making. Generally all the heuristics are example of the first
   system. On the contrary the rational procedures of deductive and inductive reasoning are examples of the second. This system is switched on
   when there is epistemic engagement in reasoning and decision making
   and when the individual shows *need for cognition* [8]. Therefore the intuitive system is responsible of biases and errors of everyday life reasoning,
   whereas the analytical system allow us to reason according the canons
   of rationality. In reality often the first system is more adaptive than the
   second in many instances of everyday life [21]. The prevalence of one of
   the two systems in the cognitive activity of academic and industrial researchers will depend from contingent factors, as the need to end quickly
   the work, but also from the diverse styles of thinking. I can hypothesize
   that the operational norms of *pressing time, well defined results* and the
   social norm of *dogmatism* and *localism* will support a propensity to the
   activity of the intuitive system. On the contrary the operational norms
   of *loose time*, and *undefined results*, and the social norms of *criticism*,
   and *universalism* can support the activity of the analytical system. It
   is evident the role of time on activation of the two systems. Industrial
   researchers are used to follow time limits and to give value to time. Therefore this operational norm influences the speed of reasoning and decision
   making and the activation of the intuitive system. The contrary happens
   in academic labs. The other operational norm regarding the results seems

less evident. Who has not constraint of well defined results has the attitude to indulge in slow, and attentive way of analyzing the features of the variables and in applying rule based reasoning. Who should end with an accomplished work can't stop on analyzing the details and should go quickly to the final results. The social norm of criticism is more evident. The tendency to control and to criticize results produced by other scientists strengthens the analytical attitude in reasoning. Any form of control is a slow and precise analysis of the logical coherence, methodological fitness, and empirical support of a study. On the contrary in corporate labs the aim is to use good knowledge for practical results and not to increase the knowledge pool by overcoming previous hypotheses through control and critique. Finally the social norm of universalism vs. localism is less evident. Scientists believe in a universal dimension of their activity. The rules of scientific community should be clear and understandable by the peers. The scientific method, the reasoning style and the methodological techniques can't be understood and followed only by a small and local subset of scientists. Therefore they should be explicit in order to be diffused to the entire community. Thus the universality tends to strengthen the analytical system of mind. The contrary happens where there is no need of explicitness of rules and the evaluation is locally made by peers according to the working of the final product.

4. The fourth dimension is about **problem solving**. At the end of '50 Herbert Simon with some colleagues analyzed the effect of professional knowledge in problem representation. They discovered the phenomenon of "*selective perception*" [17], that is the relation between different professional roles and different problem representations. For example, in explaining the causes of a company crisis, the marketing manager will represent the problem mainly in terms of commercial communication, the staff manager mainly in terms of insufficient employment, and the book-keeper mainly in terms of an obsolete book-keeping and lack of liquidity. In the case of industrial and academic scientists I can suppose that the *selective perception* will be effective not only in relation with the professional and disciplinary roles but also with social values and operational norms. These norms and values might characterize the problem representation and therefore might influence reasoning and decision making. For example in representing the problem of a failure of a research programme, industrial researchers might point more to variables like cost and time whereas the academic scientists might be more oriented towards insufficient critical attitude and too local approach. Expert from novice are differentiated by different amount, organization, and use of knowledge in problem solving. What differentiates expert from novice is their schema for solving problems within their own domain of expertise [22]. The schemas of experts involve large, highly interconnected units of knowledge. They are organized according to underlying structural similarities among knowledge units. In contrast, the schemas of novices involve relatively small and

disconnected units of knowledge. They are organized according to super-ficial similarities [7]. Through practice in applying strategies experts may automatize various operations. The automatization involves consolidating sequences of steps into unified routines that require little or no conscious control. Through automatization experts may shift the burden of problem solving from limited-capacity working memory to infinite-capacity long-term memory. The freeing of their working memory capacity may better enable them to monitor their progress and their accuracy during problem solving. Novices in contrast, must use their working memory for trying to hold multiple features of a problem and various possible alternatives. This effort may leave novices with less working memory available for monitoring and evaluation. Another difference between expert and novice problem solvers is the time spent on various aspects of problems. Experts appear to spend more time determining how to represent a problem than do novices [40], but they spend much less time that do novices actually implementing the strategy for solution. Experts seem to spend relatively more time than do novices figuring out how to match the given information in the problem with their existing schemas. Once they find a correct match they quickly can retrieve and implement a problem strategy. Thus expert seems to be able to work forward from the given information to find the unknown information. In contrast novices seem to spend relatively little time trying to represent the problem. Instead, they choose to work backward from the unknown information to the given information. In the collaboration between academic and industrial scientists a cognitive dissonance might stem from asymmetric expertise in problem solving. Industrial researchers can be novice in aspects where academic scientists are expert and vice versa. If this is the case the opposite backward vs. forward approach and the different time in problem representation might produce cognitive dissonance and asymmetry. In any case it might be interesting to analyze the time spent by academic and industrial researchers in *problem representation*. The hypothesis is that time pressure together with intuitive system of thinking might bring the industrial researchers to dedicate less time in problem representation than academic researchers.

Time pressure can affect the entire *problem solving cycle* which includes [66]: *problem identification, definition of problem, constructing a strategy for problem solving, organizing information about a problem, allocation of resources, monitoring problem solving, evaluating problem solving*. In particular it might be interesting to analyze the effect of pressing vs. loose time in *monitoring and evaluation* phases. More time pressures could diminish the time devoted to these phases. Also dogmatism can accelerate the time spent for monitoring and evaluation whereas criticism might be responsible of better and deeper monitoring and evaluation of the problem solution.

Finally time pressure might have an effect also on *incubation*. In order to permit the old association resulting from negative transfer to weaken one needs to put the problem aside for a while without consciously thinking about it. You do allow for the possibility that the problem will be processed subconsciously in order to find a solution. There are several possible mechanisms for the beneficial effects of incubation [66]. The incubation needs time. Therefore the pressing time norm of industrial researcher might hinder the problem solving success.

5. The fifth dimension is about **reasoning**. Reasoning is the process of drawing conclusions from principles and from evidence. In reasoning we move from what is already known to infer a new conclusion or to evaluate a proposal conclusion. There are many features of reasoning that can differentiate academic and corporate scientists. I will concentrate on three aspects of reasoning that are crucial in scientific problem solving and that may affect the cognitive coordination between academic and industrial researchers.

The first is about **probabilistic reasoning** aimed to up-to-date an hypothesis according some new empirical evidence. In other words how the scientist deals with new data in order to strengthen or to weaken a given hypothesis. There is a canon of rationality, the Bayes theorem that prescribes how we should reason. The mathematical notation is the following:

$$P(H|D) = \frac{P(D|H)P(H)}{P(D|H)P(H) + P(D|\text{non}H)P(\text{non}H)}$$

This theorem tells us how to calculate the effect of new information on the probability of a thesis. Kahneman and Tversky [35] and Tversky and Kahneman [68, 70, 69] has experimentally proved that often we fall in *base rate neglect* that is we focus mainly in the new information and we neglect the prior probability. For example if we are controlling a theory T having prior probability $P(T)$ we tend to neglect it when we have new experimental data that change the prior probability in posterior probability $P(T|D)$. That is we give an excessive weight to new experiments and we forget the old ones compared to what it is prescribed by Bayes Theorem. Why do we forget prior probability and we give excessive weight to new data? According to Bar Hillel [1] we give more weight to new data because we consider them more relevant compared to the old ones. Relevance in this case might mean more affective or emotional weight given to the data and consequently stronger attentional processes on them. An opposite conservative phenomenon happens when the old data are more relevant. In this case we tend to ignore new data. In the case of industrial researchers an hypothesis may be that the time pressure, the financial weight, and well defined results tend to give more relevance to new data. New experiments are costly and they should be an important step towards the conclusion of the work. Therefore they are more relevant and privileged by the mechanisms of attention. On the contrary academic scientists without the influence of cost, time and the conclusion of the project can have a more balanced perception of relevance between old and new data.

The second is about *deductive reasoning* and in particular the hypothesis testing. It is well known in propositional logic the rule of *modus tollens* of conditional statements:

$$T \rightarrow d$$
$$\frac{\neg d}{\neg T}$$

If a theory T implies an experimental datum d and if d is falsified then the theory T is falsified. The only way to test the truth of a theory is modus tollens, that is trying to find its falsification. In fact it is wrong to test a theory in the following way, called the *Fallacy of Affirmation of the Consequent*:

$$T \rightarrow d$$
$$\frac{d}{T \quad \text{no}}$$

Modus tollens was popular in philosophy of science, mainly through the work of Karl Popper and in cognitive psychology mainly through the work of Peter Wason and Phil Johnson Laird. Wason [82] and Johnson Laird [24] have proved that in formal test people mistake the rule and tend to commit confirmation bias that corresponds to the Fallacy of Affirmation of the Consequent. More realistic tests [26] or tests linked to our pragmatic schemes [10, 11] improved the deductive performances. Also in science confirmation bias disappears when the falsificatory data are easy to produce, and non ambiguous [47, 24]. New studies that have analyzed the emotional and affective dimension of hypothesis testing have found that when individual is emotionally involved in a thesis he will tend to commit confirmation bias. The involvement can be various, economic (when one has invested money in developing an idea), social (because your social position is linked to the success of a project), organizational (because a leader that holds a thesis is always right) or biographical (because you have spent many years of your life in developing the theory). The emotional content of the theory causes a sort of regret phenomenon that pushes the individual to avoid falsification of the theory. From this point of view it is likely that financial heaviness and dogmatism together with other social and organizational factors would induce industrial researchers to commit more easily confirmation bias. Research is costly and it is fundamental for the commercial survival of company. Therefore their work should be successful an the results should be well defined in order also to keep or to improve their position. Moreover they don't follow the academic norm of criticism that prescribes the falsificationist approach towards scientific knowledge. Contrary to what happen to academic scientists that tend to be critic and should not be obliged to be successful in their research. It is likely that they are less prone to confirmation bias. The third aspect deals with **causal reasoning**. It is a fundamental aspect of reasoning in science and technology. Most of models, hypotheses and theories representing scientific and technological knowledge are causal. The main tenets of experimental method correspond to the technical evolution of Millian methods of *agreement* and *difference* [45]. It is not the place to deepen the epistemological discussion on causality and neither that on causal cognition (for a survey on the relationship between epistemological and cognitive dimension of causality see [73, 74, 75]). The aim of this paper is to single out potential different styles of reasoning between academic and corporate researchers. In this case a different approach to causal reasoning and causal models might have a strong effect in the cognitive coordination. According to Mackie [41] every causal reasoning is based on a *causal field*, that is the set of relevant variables able to cause the effect. It is well known that in front of the same event, for example a car accident or a disease, each expert will support a particular causal explanation (for a town planner the wrong design of the street, for a doctor the rate of alcohol of the driver, for an engineer the bad mechanics of the car, and so on). Usually once the expert have identified one of the suspected causes of a phenomenon he stops searching for additional alternative causes. This phenomenon is called *discounting error*. From this point of view the hypothesis may be that the different operational norms and social values of academic and corporate research may produce different discounting errors. Financial heaviness, pressing time, well defined results compared to financial lightness, slow time and ill defined results may limit different causal fields of the entire project. For example the corporate scientist can find the time as a crucial causal variable for the success of the project whereas the academic researcher doesn't care about it. In the same time the academic researcher can find crucial the value of universal scientific excellence of the results whereas the industrial researcher doesn't care about it. There is also the possibility of a deeper difference worth to be studied. One of the commonest bias in causal reasoning is to infer *illusory correlations* [9]. We confuse correlations with causal relations, that is we fall down in a sort of *magical thinking*. According to Johnson Laird and Wason [33] magical thinking happen for association based on contiguity, temporal asymmetry and resemblance between two events. The associative or intuitive system of thought is responsible of this phenomenon. As we know it is switched on when the time is little and there is no need of an articulated analysis of the problem. Consequently the values of dogmatism and localism and the operational norm of pressing time and well defined results of industrial researchers can be responsible of this causal bias. On the contrary the analytic or rule-based system of thought, more present in academic reasoning – because of social values of criticism and universalism and the operational norms of loose time – can neutralize the danger of illusory correlations and magical thinking.

6. The sixth dimension is about **decision making**. Decision making involve evaluating opportunities and selecting one choice over another. There are many effects and biases connected to decision making. I would focus on some aspects of decision making that can differentiate academic and industrial researchers.

The first deals with **risk**. In the psychological literature risk[3] is the multiplication of loss for the probability, whereas uncertainty is when an event is probable. Usually to the risk of loss is associated also the possibility of gain. In many cases to a bigger risk is associated a bigger gain (as in the case of gambling). People can have risk adversity when they don't want to take great risk in order to gain a big pay-off. They prefer to bet on red or black and not on a single number. On the contrary risk propensity exists when one takes bigger risk for bigger gain. For example betting on the least favored horse with a bigger listing. The risk behavior seems not linear. According to *prospect theory* [36, 71] risk propensity is stronger in situation of loss and weaker in situation of gain. A loss of 5$ cause a negative utility bigger than the positive utility caused by the gain of 5$. Therefore people react to the loss with risky choices aimed to recover the loss. Another condition that increases the risk propensity is *overconfidence* [19, 37] and *illusion of control* [39]. People often tend to overestimate the accuracy of their judgements and the probability of success of their performance. They believe to have better control of future events than the chance probability. This phenomenon is associated often to the *egocentric bias* of manager and to forms of *quasi-magical thinking* (like that of a dice player that throws the dices after having blown to them and thinks to have a better control on the results). Both the perception of loss and the overconfidence happen when there is competition, the decisions are charged of economic meaning, and have an economic effect. The operational norm of financial heaviness, and pressing time, and the social value of exclusivity, and interest of industrial researcher can increase the economic value of the choices, the perception of competitiveness, and consequently can increase the risk propensity. On the contrary the social values of communitarism, and indifference, and the operational norms of financial lightness and slow time of academic scientists may create an environment that doesn't induce any perception of loss and overconfidence. Thus the behavior tends to be more risk adverse.

A second feature of decision making is connected to **regret** and **loss aversion**. We see before that according to prospect theory individual doesn't like to loose and react with risk propensity. The loss aversion is

---

[3] In economics, risk is when we have probability assessment whereas uncertainty is when we have no probability assessment.

based on the regret that loss produce to the individual. The regret is responsible of many effects. One of the most important is the *irrational escalation* [65] in any kind of investment (economic, but also political and affective). When one is involved in an investment of money to reach a goal, as the building of a new prototype of missile or the creation of a new molecule to care the AIDS, has to consider the possibility of failure. One should monitor the various steps of the programme and especially when the funds are finished he has to analyze coldly if the project has some chance to succeed. In this case he should consider the moneys invested in the project as *sunk cost*, forget them and decide rationally. On the contrary people tend to be affectively attached to their project [49, 65]. They feel strong regret in admitting the failure and the loss of money and tend to continue the investment in an irrational escalation of wasted money to attain the established goal. The psychological mechanism is linked also to prospect theory and risk propensity under conditions of loss. The irrational escalation is stronger when there is stronger emphasis on the economic importance of the project. That is the typical situation of a private company that links the success of their technological projects to its commercial survival. The same industrial researchers have the perception that their job and the possibility of promotion are linked to the success of the technological projects. Therefore it is likely that they will tend more easily to fall in an irrational escalation compared to academic researchers that have the operational norm of financial lightness, and social norm of indifference and whose career is only loosely linked to the success of research projects.

The third aspect of decision making has to do with an irrational bias called *myopia* [18] or **temporal discounting**. People tend to strongly devaluate the gains in time. They prefer small gain at once than big gain in the future. Many behaviors of everyday life witness this bias. The small pleasure of a cigarette today is more than the big gain of being healthy after 20 years. The perceived utility of a choice of a certain job without perspective now is bigger than the choice of unstable work now with greater future professional perspectives. And so on. More recently these observations about discount functions have been used to study savings for retirement, borrowing on credit cards, and to explain addiction. Drug dependent individuals discount delayed consequences more than matched nondependent controls, suggesting that extreme delay discounting is a fundamental behavior process in drug dependence [4]. Some evidence suggests pathological gamblers also discount delayed outcomes at higher rates than matched controls [50]. All these phenomena show a complex risk behavior. People are risk adverse in the present, that is they want to have now a certain satisfaction (effect of drug, pleasure of gambling, certainty of a job), whereas they show high risk propensity for the future (high risk of death for drug, high risk of becoming poor for gambling, high risk of professional decline in the case of a job without perspectives). Usually this behavior is associated with overconfidence and illusion of control. Time discounter prefer the present because he thinks to be able to control the output, the results beyond any chance esteem. In the case of industrial researcher and of entrepreneurial culture, in general, the need to have results at once, to find fast solution to the problems, to assure the share holders and the market that the company is fine and is growing seems to match with the propensity with time discounting. The future results don't mind. What it is important is the "now", that is the ability to have new competitive products to commercially survive. Financial heaviness, pressing time, and well defined results might be responsible of the attitude to give more weight to the attainment of fast and complete results at once with the risk of products that in the future will be defective, little innovative and easily overcome by competing products. In the case of academic scientists the temporal discounting might be less strong. In fact the three operational norms – financial lightness, loose time, and undefined results – together with criticism and universalism might immunize them from myopic behaviors. Criticism is important because pushes the scientist not to be easily satisfied by quick and unripe results that can be easily falsified by the peers. Universalism is important because the academician wishes to pursue results that are not valid locally, but that can be recognized and accepted by the entire community and that can increase his scientific reputation. In academic community it is well known that reputation is built through a lengthy process, but it is destroyed in a fast way.

4   Conclusion

The present paper was a hypothetical deductive and analogical exercise to define potential interesting topics for empirical studies about the cognitive styles of academic and industrial researchers. My current proposals are general but the empirical studies should be made according different variable as disciplinary and technological domains; size of University/company; geographical context. Since the goal of the studies is to single out the obstacles to the academy-industry collaboration, the subjects of the test should be articulated in at least four categories: pure scientists, business oriented professor, academic entrepreneurs, and industrial researchers. In the case of technology transfer also the category of TTO officers should be included. The next passage will be to articulate the test for each of the cognitive variables and a questionnaire to control the presence of social values and operational norms [80].

In the meantime it has been made a pilot study composed of a questionnaire and a focus group [20, 79] aiming to deepen the risk behavior and the perception of time in academic and industrial researchers. It has been observed a distance between the two styles of decision making. From one side the pure academicians, having no contact with the business world, can maintain a decision making style that reflects the operational norm of financial lightness and the social value of disinterest for economic gain. From the other side the business oriented academicians have absorbed the social norm of economic interest and the operational norm of financial heaviness present among industrial participants. Therefore they have more regret for losses and are more risk inclined. Time perception and the operational norms loose time vs. pressing time differentiated business oriented academicians from entrepreneurial researchers. For the last ones time is pressing and it is important to find soon concrete results and not to waste money The answers show a clear temporal discounting. The charge of business participants to academicians was of looking too much ahead and not caring about the practical need of present. The different temporal perceptions were linked to the risk assessment. The need to obtain fast results to allow the survival of the company increased the risk perception of the money spent in the projects of R&D. On the contrary even if the academic participants were not pure but business oriented they didn't show the temporal discounting phenomenon and the risk was perceived in connection with the scientific reputation inside the academic community (the social norm of universalism). For them what was risky was the failure of scientific recognition and not that of a business (vestiges of academic values). They also were inclined more to comunitarism than exclusivity (vestiges of academic values). Knowledge should be open and public and not exclusive private property and monopole. For all participants misunderstandings about time and risk are the main obstacles to the collaboration. University members accuse company members to be too short minded and prudent in the development of new ideas; entrepreneurial participants charge university members with being too far minded and advanced in innovation proposal.

References

1. Bar-Hillel, M.A.: The base-rate fallacy in probabilistic judgements. Acta Psychologica 44, 211–233 (1980)
2. Barsalou, L.W.: Concepts: Structure. In: Kazdin, A.E. (ed.) Encyclopedia of Psychology, pp. 245–248. American Psychological Association, Washington (2000)
3. Beth, E., Piaget, J.: Etudes d'Epistemologie Genetique, XIV: Epistemologie Mathematique et Psichologie. PUF, Paris (1961)
4. Bickel, W.K., Johnson, M.W.: Delay discounting: A fundamental behavioural process of drug dependence. In: Loewenstein, G., Read, D., Baumeister, R.F. (eds.) Time and Decision, pp. 419–440. Russell sage Foundation, New York (2003)
5. Braine, M.D.S.: On the relation beween the natural logic of reasoning and standard logic. Psychological Review 85, 1–21 (1978)
6. Broesterhuizen, E., Rip, A.: No place for cudos. EASST Newsletter 3, 5–8 (1984)

7. Bryson, M., Bereiter, C., Scarmadalia, M., Joram, E.: Going beyond the problem as given: Problem solving in expert and novice writers. In: Sternberg, R.J., Frensch, P.A. (eds.) Complex problem solving, pp. 61–84. Erlbaum, Hillsdale (1991)

8. Cacioppo, J.T., Petty, R.E.: The need for cognition. Journal of Personality and Social Psychology 42, 116–131 (1982)

9. Chapman, L.J., Chapman, J.P.: The basis of illusory correlation. Journal of Abnormal Psychology 84(5), 574–575 (1975)

10. Cheng, P.W., Holyoak, K.J.: Pragmatic versus syntactic approaches to training deductive reasoning. Cognitive Psychology 17, 391–416 (1985)

11. Cheng, P.W., Holyoak, K.J.: On the natural selection of reasoning theories. Cognition 33, 285–313 (1989)

12. Cheng, P.W., Nisbett, R.: Pragmatic constraints on causal deduction. In: Nisbett, R. (ed.) Rules for Reasoning, pp. 207–227. Erlbaum, Hillsdale (1993)

13. Clark, H.H.: Using Language. Cambridge University Press, Cambridge (1996)

14. Cleeremans, A.: Implicit learning in the presence of multiple cues. In: Proceedings of the 17th Annual Conference of the Cognitive Science Society, pp. 298–303. Erlbaum, Hillsdale (1995)

15. Cleeremans, A., Destrebecqz, A., Boyer, M.: Implicit learning: News from the front. Trends in Cognitive Science 2, 406–416 (1998)

16. Collins, A.M., Quillian, M.R.: Retrieval time from semantic memory. Journal of Verbal Learning and Verbal Behaviour 8, 240–248 (1969)

17. Dearborn, D.C., Simon, H.A.: Selective perception: A note on the departmental identifications of executives. Sociometry 21, 140–144 (1958)

18. Elster, J.: Ulysses and the Syrenes. Studies in Rationality and Irrationality. Cambridge University Press, Cambridge (1979)

19. Fischhoff, B., Slovic, P., Lichtenstein, S.: Knowing with certainty: The appropriateness of extreme confidence. Journal of Experimental Psychology 3, 552–564 (1977)

20. Rosselli, F.: Modelli socio-cognitivi nel sistema della ricerca pubblica e nel mondo delle imprese (2008), http://www.fondazionerosselli.it

21. Gigerenzer, G.: Gut feeling. Viking, London (2007)

22. Glaser, R., Chi, M.T.H.: Overview. In: Chi, M.T.H., Glaser, R., Farr, M. (eds.) The nature of expertise, pp. xv–xxvii. Erlbaum, Hillsdale (1988)

23. Goffman, E.: Forms of Talk. University of Pennsylvania Press, Philadelphia (1981)

24. Gorman, M.: Simulating Science. Indiana University Press, Bloomington (1992)

25. Grice, H.P.: Studies in the Way of Words. Harvard University Press, Cambridge (1989)

26. Griggs, R.A., Cox, J.R.: The elusive thematic-materials effect in wason's selection task. British Journal of Psychology 73, 407–420 (1982)

27. Grossman, M., Smith, E.E., Koenig, P., Glosser, G., Rhee, J., Dennis, K.: Categorization of object descriptions in alzheimer's disease and frontal temporal demential: Limitation in rule-based processing. Cognitive Affective and Behavioural Neuroscience 3(2), 120–132 (2003)

28. Hinsz, V., Tindale, S., Vollrath, D.: The emerging conceptualization of groups as information processors. Psychological Bulletin 96, 43–64 (1997)

29. Jackendoff, R.: Language, Consciuosness, Culture. MIT Press, Cambridge (2007)

30. Jackson, S.: Team composition in organizational settings: Issues in managing an increasingly diverse work force. In: Worchel, S., Wood, W., Simpson, J. (eds.) Group Process and Productivity, pp. 138–173. Sage, Newbury Park (1992)
31. Johnson-Laird, P.: Mental Models. Cambridge University Press, Cambridge (1983)
32. Johnson-Laird, P.: How We Reason. Oxford University Press, Oxford (2008)
33. Johnson-Laird, P., Wason, P.C.: Thinking. Cambridge University Press, Cambridge (1977)
34. Kahneman, D.: Maps of bounded rationality. In: Frangsmyr, T. (ed.) Nobel Prizes 2002, pp. 449–489. Almquist and Wiksell, Stockolm (2003)
35. Kahneman, D., Tversky, A.: On the psychology of prediction. Psychological Review 80, 237–251 (1973)
36. Kahneman, D., Tversky, A.: Prospect theory: An analysis of decision under risk. Econometrica 47, 263–291 (1979)
37. Kahneman, D., Tversky, A.: On the reality of cognitive illusions. Psychological Review 103, 582–591 (1996)
38. Kuhn, T.: The Structure of Scientific Revolutions. Chicago University Press, Chicago (1962)
39. Langer, E.: Reduction of psychological stress in surgical patients. Journal of Experimental Social Psychology 11, 155–165 (1975)
40. Lesgold, A.M.: Problem solving. In: Sternberg, R.J., Smith, E.E. (eds.) The Psychology of Human Thought, pp. 188–213. Cambridge University Press, New York (1988)
41. Mackie, J.: The Cement of the Universe. A Study on Causation. Oxford University Press, Oxford (1974)
42. Manz, C.C., Neck, C.P.: Teamthink. beyond the groupthink syndrome in self-managing work teams. Journal of Managerial Psychology 10, 7–15 (1995)
43. Markman, A.B., Gentner, D.: Thinking annual review of psychology. Developmental Psychology 52, 223–247 (2001)
44. Merton, R.: The Sociology of Science. Theoretical and Empirical Investigations. University of Chicago Press, Chicago (1973)
45. Mill, J.S.: A System of Logic. Harper & Brothers, New York (1987)
46. Mitroff, I.I.: The Subject Side of Science. Elsevier, Amsterdam (1974)
47. Mynatt, R., Doherty, M.E., Tweney, R.D.: Confirmation bias in a simulated research environment: An experimental study of scientific inference. In: Johnson Laird, P.N., Wason, P.C. (eds.) Thinking. Cambridge Universty Press, Cambridge (1988)
48. Nisbett, R.: The Geography of Thought. The Free Press, New York (2003)
49. Nozick, R.: Newcomb's problem and two principles of choice. In: Moser, P.K. (ed.) Rationality in Action, Contemporary Approach. Cambridge University Press, New York (1990)
50. Petry, N.M., Casarella, T.: Excessive discounting of delayed rewards in substance abusers with gambling problems. Drug and Alcohol Dependence 56, 25–32 (1999)
51. Politzer, G.: Laws of language use and formal logic. Journal of Psycholinguistic Research 15, 47–92 (1986)
52. Politzer, G., Nguyen-Xuan, A.: Reasoning about promises and warnings: Darwinian algorithms, mental models, relevance judgements or pragmatic schemas? Quarterly Journal of Experimental Psychology 44, 401–421 (1992)

53. Pozzali, A., Viale, R.: Cognition, Types of 'Tacit Knowledge' and Technology Transfer. In: Topol, R., Walliser, B. (eds.) Cognitive Economics. New Trends, pp. 205–224. Elsevier, New York (2007)
54. Reber, A.S.: Implicit Learning and Tacit Knowledge. An Essay on the Cognitive Unconscious. Oxford University Press, Oxford (1993)
55. Ross, B.H.: Category learning as problem solving. In: Medin, D.L. (ed.) The Psychology of Learning and Motivation: Advances in Research and Theory. Academic Press, San Diego (1996)
56. Rumain, B., Connell, J., Braine, M.D.S.: Conversational comprehension processes are responsible for fallacies in children as well as in adults: If is not the biconditional. Developmental Psychology 19, 471–481 (1983)
57. Searle, J.: The Construction of Social Reality. Free Press, New York (1995)
58. Searle, J.: Philosophy in a new century. Cambridge University Press, Cambridge (2008)
59. Siegel, D., Waldman, D., Link, A.: Assessing the impact of organizational practices on the productivity of university technology transfer offices: An exploratory study, NBER Working Paper N. 7256 (1999)
60. Sloman, S.A.: The empirical case for two systems of reasoning. Psychological Bulletin 119, 3–22 (1996)
61. Smith, E., Kossylin, S.: Cognitive Psychology. Mind and Brain. Prentice Hall, Upper Saddle River (2007)
62. Sorensen, R.: Thought Experiments. Oxford University Press, Oxford (1992)
63. Spelke, E.S., Phillips, A., Woodward, A.L.: Infants' knowledge of object motion and human action. In: Sperber, D., Premack, D., Premack, A.J. (eds.) Causal Cognition, pp. 44–78. Oxford University Press, Oxford (1995)
64. Sperber, D., Wilson, D.: Relevance. Communication and Cognition. Oxford University Press, Oxford (1986)
65. Stanovich, K.: Who Is Rational: Studies of Individual Differences in Reasoning. Lawrence Erlbaum Associates, Inc., Mahwah (1999)
66. Sternberg, R.: Cognitive Psychology. Wadsworth, Belmont (2009)
67. Thaler, R.: Mental accounting matters. Journal of Behavioural Decision Making 12, 183–206 (1999)
68. Tversky, A., Kahneman, D.: Causal schemas in judgements under uncertainty. In: Fishbein, M. (ed.) Progress in Social Psychology, pp. 49–72. Erlbaum, Hillsdale (1980)
69. Tversky, A., Kahneman, D.: Evidential impact of base rate. In: Kahneman, D., Slovic, P., Tversky, A. (eds.) Judgement under Uncertainty: Heuristics and Biases, pp. 153–160. Cambridge University Press, Cambridge (1982)
70. Tversky, A., Kahneman, D.: Judgements of and by representativeness. In: Kahneman, D., Slovic, P., Tversky, A. (eds.) Judgement Under Uncertainty: Heuristics and Biases, pp. 84–98. Cambridge University Press, Cambridge (1982)
71. Tversky, A., Kahneman, D.: Advances in prospect theory: Cumulative representation of uncertainty. Journal of Risk and Uncertainty 5, 547–567 (1992)
72. Van Knippenberg, D., Schippers, M.C.: Work group diversity. Annual Review of Psychology 58, 515–541 (2007)
73. Viale, R.: Causality: Epistemological questions and cognitive answers. In: Costa, G., Calucci, G., Giorgi, M. (eds.) Conceptual Tools for Understanding Nature, pp. 89–109. World Scientific, Singapore (1997)
74. Viale, R.: Causal cognition and causal realism. International Studies in the Philosophy of Science 13(2), 151–167 (1999)

75. Viale, R.: Causality: Cognition and reality. In: Rossini Favretti, R., Sandri, G., Scazzieri, R. (eds.) Incommensurability and Translation, pp. 253–266. Edward Elgar, Aldershot (1999)

76. Viale, R.: Knowledge driven capitalization of knowledge. In: Viale, R., Etzkowitz, H. (eds.) The Capitalization of Knowledge: A Triple Helix of University-Industry-Government, Elgar, Cheltenham (2005)

77. Viale, R.: Local or universal principles of reasoning? In: Viale, R., Andler, D., Hirschfeld, L. (eds.) Biological and Cultural Bases of human Inferences, pp. 1–31. Erlbaum, Hillsdale (2005)

78. Viale, R.: Reasons and reasoning: What comes first. In: Boudon, R., Dmeleulenaere, P., Viale, R. (eds.) L'explication des normes sociales, pp. 215–236. Presses Universitaires de France, Paris (2005)

79. Viale, R.: Different cognitive styles between academic and industrial researchers. Columbia University, New York (2009)

80. Viale, R., Del Missier, F., Rumiati, R., Franzoni, C.: Different cognitive styles between academic and industrial researchers: an empirical study (2009), http://www.italianacademy.columbia.edu/ publications_working.html#0809

81. Viale, R., Pozzali, A.: Cognitive aspects of tacit knowledge and cultural diversity. In: Magnani, L., Li, P. (eds.) Model-Based Reasoning in Science, Technology, and Medicine, pp. 229–244. Springer, Berlin (2007)

82. Wason, P.C.: On the failure to eliminate hypotheses in a conceptual task. Quarterly Journal of Experimental Psychology 12, 129–140 (1960)

83. Ziman, J.M.: L'individuo in una professione collettivizzata. Sociologia e Ricerca Sociale 24, 9–30 (1987)

Abduction, Induction, and Analogy
On the Compound Character of Analogical Inferences

Gerhard Minnameier

Abstract. Analogical reasoning has been investigated by philosophers and psychologists who have produced different approaches like "schema induction" (Gick and Holyoak) or the "structure-mapping theory" (Gentner). What is commonplace, however, is that analogical reasoning involves processes of matching and mapping. Apart from the differences that exist between these approaches, one important problem appears to be the lack of inferential precision with respect to these processes of matching and mapping. And this is all the more problematic, because analogical reasoning is widely conceived of as "inductive" reasoning. However, inductive reasoning – in a narrow and technical sense – is not creative, whereas analogical reasoning counts as an important source of human creativity. It is C. S. Peirce's merit to have pointed to this fact and that induction can merely extrapolate and generalize something already at hand, but not the kind of reasoning that leads to new concepts. Indeed, inventive reasoning is usually identified with abduction, and consequently abduction should play at least some role in analogy. Peirce has claimed that analogy is a compound form of reasoning that integrates abduction and induction, but the intriguing question is still, how these two inferences are to be reconstructed precisely. In the proposed paper I hold that analogical reasoning can indeed be analyzed in this way and that this helps us to reach a much more precise and differentiated understanding of the forms and processes of analogical reasoning. In particular I hold that (at least) two forms of analogical reasoning have to be distinguished, because they

Gerhard Minnameier
RWTH Aachen University, Institute of Educational Research
e-mail: minnameier@lbw.rwth-aachen.de

L. Magnani et al. (Eds.): Model-Based Reasoning in Science & Technology, SCI 314, pp. 107–119.
springerlink.com

represent different inferential paths. The underlying inferential processed will be explicated in detail and illustrated by various examples.

1 Introduction: The Quest for Logic in Analogical Reasoning

Analogical reasoning is a very common form of thinking and problem-solving. Therefore, it is important to understand how it works and how it can be fostered in order to produce new knowledge – especially in the educational domain [1].

Existing psychological theories on analogical reasoning describe it as schema induction [2, 3, 4] or structure-mapping [5, 6, 7, 8]. Both approaches rest on the principle that certain more or less deep insights (schemata, structures) are transferred form a source to a target domain, and that this process comprises two characteristic sub-processes which might be called "matching" and "mapping". That is, first a target (the domain where a problem has to be solved) and a source (the domain from which the analogy is drawn) have to be matched, then the relevant features of the source have to be mapped onto the target.

A classic example is the radiation problem reported by Gick & Holyoak [2] that originally dates back to Duncker [9]: Subjects are first presented a story about a general who has to conquer a dictator's fortress. In order to succeed he has to use all his forces, but the access ways to the fortress are so small that his men cannot attack all at once. However, as there are different ways to access the fortress, the solution to the problem lies in splitting up his army and attach from different sides at the same time and with combined forces. After this story the subjects are presented another one, in which a surgeon intends to destroy a tumor in a human brain using radiation. The problem here is that a beam strong enough to destroy the tumor would also destroy all the healthy tissue along its trajectory. How to solve this problem? The idea here is that subjects match the two stories and remember that in the fortress story forces had to be split up and operate from different sides. Now this structure could be mapped to the target problem leading to the suggestion that several weaker beams might be applied from different sides that might converge in the spot where the tumor is located and thus destroy it while leaving the other parts of the brain unaffected.

Apart from the fact that only about 10 % of the subjects actually managed to use the analogy and solve the problem without further hints, the task and what is to be considered is very clear. However, what is as yet unclear, is how these processes of matching and mapping are to be understood precisely and in terms of an underlying logic that explains what kinds of inferences take place in individuals undergoing such analogical thought processes. Gentner

[10, p. 30] remarked there were processes of abduction and induction at work, but failed to make these inferential mechanisms explicit.

This is precisely the aim of the present paper. Based on the general view that all learning (at least in the cognitive realm) is inferential (see [1]), i.e. that all new knowledge is inferred from prior knowledge plus actual experience based on this prior knowledge, analogical reasoning has to be analyzable as a particular inferential form and therefore must have a logical structure in this sense. This structure is going to be revealed on the basis of the Peircean inferences of abduction, deduction and induction which are thought to cover reasoning as a whole – although various forms of these three basic inferences can be distinguished (see [11, 12, 13])[1].

In particular, I will build on C.S. Peirce's remark that analogical reasoning be a compound inference made up of abduction and induction. This theoretical reconstruction of analogy (at least one variant) will be presented in Section 2 after a short description of the three basic inferences and their dynamic interaction. The theoretical account is then going to be exemplified and deepened on the basis of a classic example: Johannes Kepler's attempt at explaining the movement of the planets (Section 3). In Section 4 I am going to proceed to a second form of analogical reasoning and its inferential structure, which contains one more inductive step than the previously described form. This also raises the question as to the status of the different instances of induction that are involved in the overall processes of analogical reasoning. These shall be analyzed in Section 5. In the conclusion (Section 6), I try to highlight also the prospects of applying this knowledge about the inferential functioning of analogies.

2 Inferential Reconstruction of Analogical Reasoning

Our analysis rests on an interpretation of Peirce's three inferences that the author believes expresses Peirce's own original view (as expressed in his later writings) and that fits well with up to date pragmatist approaches (see [14]).

According to this interpretation, abduction, deduction, and induction stand in a dynamic relationship, whereby abduction leads to new explanatory or technological hypotheses or concepts in order to solve a particular

---

[1] It should be noted however, that I do not approve of both Magnani's and Schurz's wide usage of abduction which includes "inference to the best explanation" (IBE) as what they call "selective abduction". In [14] I have argued that IBE can be reconstructed as "induction" in the Peircean sense (and without falling prey to Goodman's "new riddle of induction" (see [15]), to which Schurz [13, p. 202] alludes. Moreover, in my own approach the different forms are organized into the ordinary feed-forward inferences and in feed-back "theorematic" inferences (see [11]), and into explanatory versus technological inferences (see further below in the present paper).

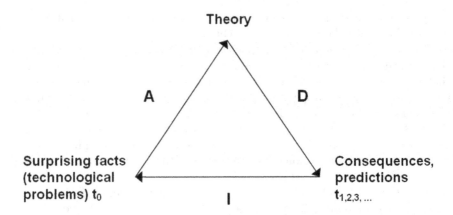

Fig. 1 The dynamic interaction of the three inferences.

explanatory or practical problem[2]. Deduction produces necessary consequences of the suggested approaches or concepts and induction attempts at answers to the question whether a theory (explanatory) or a strategy (technological) ought to be accepted or rejected. In Peirce's famous words:

> Abduction is the process of forming an explanatory hypothesis. It is the only logical operation which introduces any new idea; for induction does nothing but determine a value, and deduction merely evolves the necessary consequences of a pure hypothesis. Deduction proves that something *must* be; Induction shows that something *actually is* operative; Abduction merely suggests that something *may be*. Its only justification is that from its suggestion deduction can draw a prediction which can be tested by induction, and that, if we are ever to learn anything or to understand phenomena at all, it must be by abduction that this is to be brought about. [16, p. 106, CP 5.171]

Now let us take up Peirce's suggestion that analogy is a combination of abduction and induction. He said: "Analogy ... combines the characters of Induction and Retroduction" [17, p. 28, CP 1.65]. Unfortunately, he nowhere tells us, in what way (or ways) the two inferences should work together in

---

[2] It is well established that abduction is not restricted to explanatory problems. Some abductions do not aim at truths, but merely at strategies to achieve other goals like predictions, i.e. explaining observational data that are otherwise unexplainable. Newton's action at a distance is one such case, which serves this instrumental goal, even though it seemed highly implausible form an ontological point of view (see [18, pp. 118–119]). Gabbay and Woods [18] speak of "nonplausiblistic abduction", Magnani [12] speaks of "instrumental abduction". To my mind, this view of instrumental reasoning is still too narrow. Any practical problem – be it forecasting empirical data, e.g. concerning the weather, or be it an engineering problem – requires ideas on how to possibly solve this problem. And these reasoning processes could be labeled "technological abductions" as opposed to "explanatory abductions".

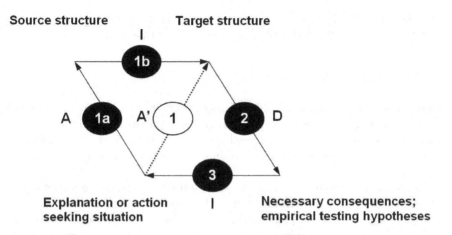

Fig. 2 Inferential model of analogical reasoning (variant 1).

analogical reasoning. Two routes seem possible, one rather direct one and another more indirect one. The direct variant is described in the following (see Figure 2), the second variant will be treated in Section 4.

As Figure 2 shows, analogy is conceived as a deviated form of abduction, in which one first abduces from the problem in the target domain to a possibly suitable structure in the source domain that might also do the job for the target. The crucial features are then projected (by an inductive inference) from source to target, i.e. the specific concepts of the source structure have to be replaced by suitable target concepts, and it has to be judged whether that transformation is possible and useful with respect to solving the original target problem. This second step of projection is said to be an induction, because induction is all about projection and extrapolation from known areas into unknown ones. The projection is valid and the analogy tenable (in the abductive sense of a plausible hypothesis worth of considering further), if all those structural features of the source can be mapped onto the target, that are needed to produce an abductively valid solution for the target problem.

What is important to understand the inferences, and hence analogy, better is to consider further that each inference is composed of three distinct sub-processes which Peirce calls "colligation", "observation", and "judgment" [19, pp. 267-269, CP 2.444]. Every logical reasoning, no matter what kind, starts from a set of premises that has to be *colligated* first. In the case of abduction, the premise consists of all the relevant information about the explanatory of technological problem. In the case of analogy the process of *observation* yields the connection the source structure. However, such ideas are always generated spontaneously as we reflect about the colligated premise. Therefore, the last process of *judgment* is needed to enable the individual to exert rational control over her reasoning. This is, one judges whether the observed idea can be rightly inferred from the premises, given the logical criterion of each inference (for an explication and formalization of these criteria see [11]; also [14] and [1]).

3 The Example of Kepler's "Vis Motrix"

Gentner [10] and Genter et al. [20] discuss the very interesting and illustrative example of Johannes Kepler's reasoning about a possible explanation of the motion of planets, in which he used an analogy with light. It provides a perfect foil for the inferential explication of analogical thinking according to variant 1.

Kepler started from the conviction that something drives the planets and at first considered two possibilities, first that every planet had a soul (*anima motrix*), second, that all planets were moved by one single soul located in the sun. Later on he changed the idea of soul into that of a force. Another crucial observation was that the outer planets moved slower than the inner ones, i.e. slower than could be expected on the basis of the relative length of the orbits. It is for this particular relationship that Kepler believed there must be only one single force located in the sun.

Thus, reasoning at this point begins with an abductive colligation. The abductive premise contains the explanatory problem that – in Kepler's understanding – something must move the planets so that the inner planets move faster and the outer ones slower in proportion to their distance from the sun. However, this raises not only the question, what kind of force this might be, but also of how a force that is emitted by the sun could possibly move distant planets without being in causal contact with them.

Starting from this *colligated* premise, Kepler got to the idea (*observation*) that light could be understood as a model to explain the movement of planets. *Judgment* would have to consider the explanatory qualities of light in relation to the problem at hand. These are that light is emitted by one source (the sun), travels – again, in Kepler's understanding – instantaneously, i.e. in no time, and is ineffective where there is nothing to be lit. Thus, according to Kepler, light acts at a distance, and – as is well known – light also gets weaker with distance as is spreads out into space (see Figure 3).

Hence the analogy proves fruitful at this first step 1a, since light bears characteristics that would, in principle, explain away the problems in the colligated abductive premise.

The next inferential step is to transfer the properties of light to the target problem of planetary motion. This inference starts with the colligated premise of the relevant properties of light and consists essentially in the replacement of all references to the source by references to the target (observation). Hence we might produce the following shift from:

1. "*Light*" is emitted by the sun and dispersed as shown in Figure 3.
2. "*Light*" is only operative, where there is an object to be lit.
3. "*Light*" is itself conceived as immaterial.

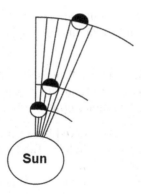

Fig. 3 Graphical illustration of the radiation of light (according to Kepler's own drawing in *Epitome Astronomiae Copernicae*, cited from [20, p. 24]).

to:

1. "*Vis motrix*" is emitted by the sun and dispersed as is shown for light in Figure 3.
2. "*Vis motrix*" is only operative, where there is an object to be moved.
3. "*Vis motrix*" is itself conceived as immaterial.

What we have to judge or evaluate in the context of induction is whether this shift is possible (i.e. makes sense) with all features relevant to the target problem. In this case it is possible, and therefore the induced properties yield an explanatorily possible approach for the original abductive problem. Therefore the induction is to be considered valid and therefore Kepler was also justified in assuming his analogy as a hypothesis for the explanation of the movement of planets which is worth to be considered and examined further (deductively and inductively in the sense of Figure 1).

4   A Second Variant of Analogical Reasoning

However, Kepler might have followed a slightly different route, which codes for a second form of analogical reasoning (which underlies many studies on analogical reasoning; see e.g. [21] for an overview). Instead of inferring straight from his explanatory problem in the target to the deep structure of the source, he could have observed and established similarity relations between the two domains, independently from the abductive problem-solving step. For instance, he could have observed the mere similarity that both light and the velocity of planets are the weaker the further away they are from the sun. This does not explain anything, but establishes a positive association (see Step 1a in Figure 4). This is another form of "matching", which does not yet include extraction of the deep structure of the source.

Revealing the deep structure would amount to an ordinary explanatory, i.e. abductive, move in the source domain (1b), explaining why light has these features:

1. *"Light"* is emitted by the sun and dispersed as shown in Figure 3.
2. *"Light"* is only operative, where there is an object to be lit.
3. *"Light"* is itself conceived as immaterial.

This, again, would then have to be followed by the final analogical move (1c), i.e. the induction of the properties of light to the target domain of planetary movement and the underlying force(s).

This second variant is based on more or less superficial relations between source and target that lead to the supposition that they might also be more deeply related. Such superficial similarities are especially exploited by young children up to 4 years of age, disregarding even deeper relational similarities, from which they are easily distracted (see [22]). However, this may seems to be due to the fact that these children lack the ability to extract abstract relational properties (see ibid, p. 251). It does not mean that older children, adolescents or adults would not use surface similarities to associate different objects or domains with each other.

In fact, surface similarities facilitate analogical problem solving of adults two. Craig, Nersessian, and Catrambone [23] report on an experiment with undergraduate students who had to solve Duncker's radiation problem (see above). After having reflected on the problem for 4 minutes (phase 1), they were presented with two different stories: an irrelevant one and which they could use for analogical transfer. This second story was varied for two different groups. The first group was given Gick's and Holyoak's [24] fortress

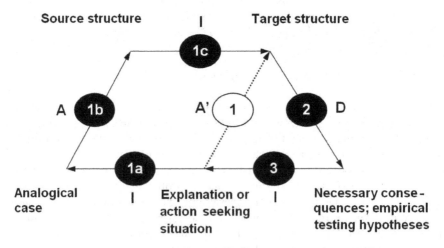

Fig. 4 Inferential model of analogical reasoning (variant 2).

story of a conqueror who had to use combined forces of small troops attacking simultaneously from different sides. The second group got the so-called "beehive story". It describes a man who wants to kill a queen in a beehive. Initially, he thinks of driving a big stick into the beehive, but then refrains from it, because this would set off the drones. Instead, he pushes in two smaller sticks from different sides that would meet in the middle. After these two stories, they were told the radiation problem again with the information that some time of incubation might help them to solve the problem (phase 2). Finally, they were given the radiation problem again, together with the hint that one of the two other stories might help them solving it.

With those subjects eliminated who produced the correct solution to the radiation problem right away, 20 out of 49 subjects in the beehive condition solved the radiation problem in phase 2, whereas only 9 out of 44 in the fortress condition were able to solve it. Of the remaining 29 subjects (beehive condition) 22 solved it in phase 3 as opposed to 15 out of 35 (fortress condition). This result may be understood as a rather clear indicator that surface similarities facilitate analogical problem solving, because the beehive story has bears many such surface similarities (sticks and beams, brain and beehive, tumor and queen), whereas the fortress story does not. The deep structure, however, is the same in both conditions: the convergent force schema[3].

5   Inferential Sub-processes and Different Forms of Induction in Analogical Reasoning

Following the above reconstruction of two forms of analogical reasoning, induction appears in different parts of the inferential processes and thus plays different roles in analogical reasoning. The present section attends to these differences in order to reveal them and to explain at the same time, why they are all still instances of one and the same type of inference, namely "induction".

To be able to explain both the inferential core and the differences, it is necessary to go one step further in the analysis of inferential reasoning, i.e. to

---

[3] Craig, Nersessian, and Catrambone [23, p. 179] believe that there also have to be closer structural similarities between beehive and radiation story compared with fortress and radiation story, because they think the hint in part 3 should have not differential influence if structural similarities are equal. However, if the surface similarities are not matched in the first place, the source structure has to be extracted independently from them, i.e. more in the sense of step 1a in variant 1 (see above, Figure 2). This may be already more difficult than inferring the deep structure from the matched surface features of the beehive story, because the surface similarities limit the space of features from which to abstract the deep structure. Moreover, the inductive transfer (1c in Figure 4) may also be facilitated by the concrete match of surface similarities, since the features to be matched are already present in concrete form.

explicate the sub-processes of inferential reasoning. These are independent of the type of inference and divide the whole process up into three distinctive parts, which Peirce calls "colligation", "observation", and "judgment" (as already mentioned above). Each inference starts from a collection of premises. Putting these together or representing them consciously is the process of "colligation". These colligated premises are then "observed" in order to produce a result, i.e. to infer something. However, since each such result is generated spontaneously and since an inference has to be justifiable in terms of its specific validity criterion (see [11]), a judgment is also needed to establish that the observed result is indeed valid or to reject the spontaneous idea. Peirce himself describes this process elsewhere with respect to abductive reasoning:

> When one contemplates a surprising or otherwise perplexing state of things (often so perplexing that he cannot definitely state what the perplexing character is) he may formulate it into a judgment or many apparently connected judgments; he will often finally strike out a hypothesis, or problematical judgment, as a mere possibility, from which he either fully perceives or more or less suspects that the perplexing phenomenon would be a necessary or quite probable consequence. [25, p. 177, CP 8.229]

The perplexing state of things here is the colligated premise, and contemplation means "observation" leading to the observed result (these are – a little misleadingly – called judgments). The eventual judgment leads to the acceptance of the abductive hypothesis as described in the cited passage.

Turning to induction, this inference is always about projecting certain properties from one set of cases to other sets of cases (see e.g. ibid, p. 178, CP 8.236), most typically from the observed ones to all relevant cases (which is generalization). Therefore, colligation consists in the conscious presentation of the cases from which to project. Observation refers to actually carrying out the projection, and a judgment is to be made as to whether the projection is valid. Let us analyze the inductive inferences included in Figure 4 according to this scheme.

With respect to (1a), colligation has to include both source and target, because they are meant to be associated just because they are both present (and not for any other specific reason)[4]. Then the surface similarity has to be observed, i.e. remarked, and finally established (judgment). In such an associative judgment the similarity-constituting features are mutually projected from one object/domain onto the other.

As for (1c) the situation is such that the deep structure of the source has been extracted, and it has to be colligated together with what is known about the target domain. This colligated premise has to be observed in order

---

[4] If only one is present, the process of activation would have to be reconstructed as a separate abduction. For instance when one looks at piece of wood or stone which reminds one of a face or anything else, this has to be understood as an abductive process of interpreting one's immediate perceptions, just as perception is thought to be abductive as well (see [12, pp. 268–276] and [26, pp. 307–308]).

find out, how the structure of the source would have to be mapped onto the target. Whether this mapping works and is valid is determined by the inductive judgment.

Finally, the conclusive inductive inference (3) leads to the acceptance – or rejection – of the hypothesis generated in (1a) through (1c). Colligation here refers to the consequences deduced from the hypothesis (2) and evidence from experiments or (previously acquired) experience. These empirical experiences have to be reflected upon in the light of the hypothesis under the various aspects that are relevant (observation). Eventually, the epistemic subject has to come to a conclusion (judgment) whether to accept or reject the hypothesis or whether the matter cannot be decided based on the evidence at hand. In the positive case the theoretical properties are projected as holding for the initial problematic case, for the tested cases and for all the untested cases in the past, present and future.

Distinguishing theses different kinds of induction is not only necessary in order to fully understand analogical inferences and to be able to initiate and guide inferential processes in learners (in an educational frame of reference), but also to understand the reach of the different inductive inferences. The analogical transfer (step 1c in Figure 4), e.g., merely yields abductive plausibility of the induced structure for the putative solution of the target problem, if the induction works. But it cannot be used as evidence in the sense of conclusive induction (step 3). However, this fallacy does not seem to be uncommon (see [27]).

6   Conclusion

Analogical reasoning is complex. This insight is not new, but what is new is that the inferential analysis revealed its specific compound character in relation to the three distinct inferences of abduction, deduction, and induction. Furthermore, the different steps that are involved in the overall process of analogical reasoning have been reconstructed and described in detail (down to the sub-processes of each inference), and it has been revealed that there are at least two different types of analogical inferences.

References

1. Minnameier, G.: Wissen und inferentielles Denken – Zur Analyse und Gestaltung von Lehr-Lern-Prozessen. Verlag Lang, Frankfurt/Main (2005)
2. Gick, M.L., Holyoak, K.J.: Schema induction and analogical transfer. Cognitive Psychology 15, 1–38 (1983)
3. Holyoak, K.J., Thagard, P.R.: A computational model of analogical problem solving. In: Vosniadou, S., Ortony, A. (eds.) Similarity and Analogical Reasoning, pp. 242–266. Cambridge University Press, Cambridge (1989)
4. Holyoak, K.J., Thagard, P.R.: Mental Leaps: Analogy in Creative Thought. Bradford, Cambridge (1995)

5. Gentner, D.: Structure-mapping: A theoretical framework for analogy. Cognitive Science 7, 155–170 (1983)
6. Gentner, D.: The mechanisms of analogical learning. In: Vosniadou, S., Ortony, A. (eds.) Similarity and Analogical Reasoning, pp. 199–241. Cambridge University Press, Cambridge (1989)
7. Gentner, D., Markman, A.B.: Structure mapping in analogy and similarity. American Psychologist 52, 45–56 (1997)
8. Gentner, D., Bowdle, B., Wolff, P., Boronat, C.: Metaphor is like analogy. In: Gentner, D., Holyoak, K.J., Kokinov, B.N. (eds.) The Analogical Mind – Perspectives from Cognitive Science, pp. 199–253. Bradford, Cambridge (2001)
9. Duncker, K.: A comparative (experimental and theoretical) study of productive thinking (solving of comprehensible problems). Journal of Genetic Psychology 33, 642–708 (1926)
10. Gentner, D.: Analogy in scientific discovery: The case of Johannes Kepler. In: Magnani, L., Nersessian, N.J. (eds.) Model-Based Reasoning: Science, Technology, Values, pp. 21–39. KluwerAcademic/Plenum Publishers, New York (2002)
11. Minnameier, G.: The logicality of abduction, deduction, and induction. In: Bergman, M., Paavola, S., Pietarinen, A.-V., Rydenfelt, H. (eds.): Applying Peirce (forthcoming)
12. Magnani, L.: Abductive Cognition: The Epistemological and Eco-Cognitive Dimensions of Hypothetical Reasoning. Springer, Berlin (2009)
13. Schurz, G.: Patterns of abduction. Synthese 164, 201–324 (2008)
14. Minnameier, G.: Peirce-suit of truth – why inference to the best explanation and abduction ought not to be confused. Erkenntnis 60, 75–105 (2004)
15. Goodman, N.: Fact, Fiction, and Forecast. Harvard University Press, Cambridge (1955)
16. Peirce, C.S.: Lectures on pragmatism. In: [28], vol. V, pp. 11–131 (1903)
17. Peirce, C.S.: Lessons from the history of science. In: [28], vol. I, pp. 19–49 (1896)
18. Gabbay, D.M., Woods, J.: A Practical Logic of Cognitive Systems. The Reach of Abduction – Insight and Trial, vol. 2. Elsevier, Amsterdam (2005)
19. Peirce, C.S.: The grammatical theory of judgment and inference. In: [28], vol. II, pp. 265–269 (1893)
20. Gentner, D., Brem, S., Ferguson, R.W., Markman, A.B., Levidow, B.B., Wolff, P., Forbus, K.D.: Analogical reasoning and conceptual change: A case study of Johannes Kepler. The Journal of the Learning Sciences 6, 3–40 (1997)
21. Holyoak, K.J.: Analogy. In: Holyoak, K.J., Morrison, R.G. (eds.) The Cambridge Handbook of Thinking and Reasoning, pp. 117–142. Cambridge University Press, New York (2005)
22. Richland, L.E., Morrison, R.G., Holyoak, K.J.: Children's development of analogical reasoning: Insights from scene analogy. Journal of Experimental Child Psychology 94, 249–273 (2006)
23. Craig, D.L., Nersessian, N.J., Catrambone, R.: Perceptual simulation in analogical problem solving. In: Magnani, L., Nersessian, N.J. (eds.) Model-Based Reasoning: Science, Technology, Values, pp. 167–189. Kluwer Academic/Plenum Publishers, New York (2002)
24. Gick, M.L., Holyoak, K.J.: Analogical problem solving. Cognitive Psychology 12, 306–335 (1980)
25. Peirce, C.S.: To Paul Carus, on illustrations of the logic of science. In: [28], vol. VIII, pp. 171–179 (1910)

26. Meyer, M.: Abduktion, Induktion – Konfusion. Zeitschrift für Erziehungswissenschaft 12, 302–320 (2009)
27. Ward, C., Gimbel, S.: Retroductive analogy: How to and how not to make claims of good reasons to believe in evolutionary and anti-evolutionary hypotheses. Argumentation 24, 71–84 (2010)
28. Peirce, C.S.: Collected Papers of Charles Sanders Peirce. Ed. by C. Hartshorne & P. Weiss (vol. 1-6) and A. Burks (vol. 7-8). Harvard University Press, Cambridge (1958)

Belief Revision vs. Conceptual Change in Mathematics

Woosuk Park

Abstract. In his influential book Conceptual Revolutions (1992), Thagard asked whether the question of conceptual change is identical with the question of belief revision. One might argue that they are identical, because "whenever a concept changes, it does so by virtue of changes in the beliefs that employ that concept". According to him, however, all those kinds of conceptual change that involve conceptual hierarchies (e.g., branch jumping or tree switching) cannot be interpreted as simple kinds of belief revision. What is curious is that Thagard's interesting question has failed to attract any serious response from belief revision theorists. The silence of belief revision theorists may be due to both wings of their fundamental principle of informational economy, i.e., the principle of minimal change and the principle of entrenchment. Indeed, Gärdenfors and Rott conceded that their formal theory of belief revision "is concerned solely with small changes like those occurring in normal science" [8]. In this paper, I propose to re-examine Thagard's question in the context of the problem of conceptual change in mathematics. First, I shall present a strengthened version of the argument for the redundancy of conceptual change by exploiting the notion of implicit definition in mathematics. If the primitive terms of a given mathematical structure are defined implicitly by its axioms, how could there be other conceptual changes than those via changing axioms? Secondly, I shall examine some famous episodes of domain extensions in the history of numbers in terms of belief revision and conceptual change. Finally, I shall show that there are extensive and intricate interaction between conceptual change and belief revision in these cases.

Woosuk Park
Humanities and Social Science Korea Advanced Institute of Science and Technology,
KAIST, Daejeon, South Korea
e-mail: woosukpark@kaist.ac.kr

L. Magnani et al. (Eds.): Model-Based Reasoning in Science & Technology, SCI 314, pp. 121–134.
springerlink.com

1 Introduction

In his influential book *Conceptual Revolutions* (1992), Thagard asked whether
the question of conceptual change is identical with the question of belief revision. He considered quite seriously the possibility that the answer is positive,
for one might argue that they are identical on the ground that "whenever a
concept changes, it does so by virtue of changes in the beliefs that employ
that concept" [28, p. 20]. According to him, however, all those kinds of conceptual change that involve conceptual hierarchies cannot be interpreted as
simple kinds of belief revision. In particular, *branch jumping* or *tree switching*
are much rarer events associated with conceptual revolutions [28, p. 37].

Though quite convincing, I count Thagard's discussion not so much a
final word as spadework in a new fertile ground. For, it seems rather obvious that there are many intriguing relationships between belief revision and
conceptual change left untouched. What is most curious is that Thagard's
interesting question has failed to attract any serious response from belief revision theorists. The lack of attention is especially curious in view of the fact
that theory of belief revision – with the AGM paradigm [1] at its core – has
over the past two decades expanded its scope far beyond epistemic logic and
philosophy of science to include computer science, artificial intelligence, and
economics[1]. As it has been so successful in such diverse fields, why shouldn't
belief revision theorists aim for an account of scientific revolution, thereby
countering Thagard's thesis of non-identity of conceptual change and belief
revision?

In some sense, it is not hard to understand why belief revision theorists
have not faced Thagard's question directly. Their silence may be due to
their fundamental principle of informational economy [23, p. 505]. Indeed,
Gärdenfors and Rott conceded that their formal theory of belief revision "is
concerned solely with small changes like those occurring in normal science"
[8, p. 39]. Is it, then, simply impossible to develop a theory of belief revision
that can cover more radical belief changes including scientific revolutions?

In this paper, I will re-examine Thagard's question in the context of the
problem of conceptual change in mathematics. Here, the choice of mathematics as the target area is deliberate. For, the problem of conceptual change in
mathematics can present an ideal opportunity to both Thagard and belief
revision theorists to consider the identity or non-identity of belief revision
and conceptual change. On the one hand, if some radical conceptual changes
in mathematics can be explained exclusively in terms of belief revision, then
an account of scientific revolution within the formal theory of belief revision may be on the horizon. On the other hand, it will be interesting to see
whether Thagard's theory of conceptual revolution in science is applicable
to mathematics at all. Since there is an on-going debate as to the existence

---

[1] One anonymous referee points out that after the publication of Thagard's book
[28] in 1992 there has been a revival of logics describing concepts, which are now
known as description logics. See [6, 3.8] for a brief history of description logics.

of revolution in mathematics [9], it could be a true test of Thagard's theory of conceptual revolution in science to see if it can do similar work in mathematics[2].

My strategy is as follows. In section 2 and 3, I shall briefly review Thagard's and Belief Revision theorists' positions on conceptual change respectively. I shall figure out in section 4 a strengthened version of the argument for the redundancy of conceptual change by exploiting the notion of implicit definition in mathematics. Section 5 will examine some famous episodes of revolutionary change in the history of numbers, and ask if it is possible to reconstruct those episodes in terms of belief revision and conceptual change respectively. The discussion will focus on the discovery of negative integers and complex numbers. Unlike the modern axiomatic number systems, the historical development of our conception of numbers seems to give a fair chance to both Thagard and belief revision theorists to make their cases. And, the results will be interesting because we seem to have here revolutionary changes that involve conceptual hierarchies such as *branch jumping* or *tree switching*. In section 6, I shall explore the implications that the extensive and intricate interaction between conceptual change and belief revision in numerical domain extensions have on the question of the identity or non-identity of conceptual change and belief revision.

2 Thagard's Challenge

In order for concepts and conceptual changes to be relevant for epistemology, Thagard thinks, it would be sufficient to demonstrate that the question of conceptual change is not identical to the question of belief change. But, at this stage, he concocts a possible argument for the redundancy of conceptual change:

> The issue of conceptual change is a red herring. Whenever a concept changes, it does so by virtue of changes in the beliefs that employ that concept (or predicate, if you are thinking in terms of sentences). For example, if you recategorize whales as mammals rather than fish, you have made an important change in the concept *whale*. But this amounts to no more than deleting the belief that whales are fish and adding the belief that whales are mammals. Your concept of mammal may also change by adding the belief that whales produce milk, but this merely follows from the other belief additions. So as far as epistemology is concerned, conceptual change is redundant with respect to the central question of belief revision. [28, p. 20]

---

[2] One might want to find such an application in [25]. However, their focus is on how "to use Darden's strategies for anomaly resolution to analyze developments in Greek mathematics following the discovery of the incommensurables" [p. 108] rather than the identity or difference between belief revision and conceptual change. I am indebted to one anonymous referee, who drew my attention to [25].

Thagard considers this argument quite seriously. According to him, however, all those kinds of conceptual change that involve conceptual hierarchies cannot be interpreted as simple kinds of belief revision. In particular, *branch jumping* or *tree switching* are much rarer events associated with conceptual revolutions[3].

What Thagard calls "branch jumping" refers to "shifting a concept from one branch of a hierarchical tree to another". Thagard invokes Copernican revolution and Darwinian revolution as typical examples of branch jumping:

> For example, the adoption of Copernican theory required the reclassification of the earth as a kind of planet, when previously it had been taken to be *sui generis*. Similarly, Darwin recategorized humans as a kind of animal, when previously they were taken to be a different kind of creature. [28, p. 36]

On the other hand, what Thagard calls "tree switching" is "the most dramatic kind of conceptual change" that "affect the organizing principle of a hierarchical tree". He finds examples of this kind of conceptual change in Darwin and Einstein:

> Darwin not only reclassified humans as animals, he changed the meaning of the classification. Whereas before Darwin *kind* was a notion primarily of similarity, his theory made it a historical notion: being of common descent becomes at least as important to being in the same kind as surface similarity. Einstein's theory of relativity changed the nature of part-relations, by substituting ideas of space-time for everyday notions of space and time. [28, p. 36]

What crucial differences are there between these cases of more radical changes and our belief change about whales? Exactly why can't we make sense of *branch jumping* and *tree switching* in terms of belief revision? Thagard claims that they require "adopting a new conceptual system", which is "more holistic than piecemeal belief revision". In the end, then, he claims that

> Conceptual change goes beyond belief revision when it involves the addition, deletion, or reorganization of concepts, or redefinition of the nature of the hierarchy. [28, p. 36]

---

[3] According to Thagard, we can distinguish between nine degrees of conceptual change: (1) Adding a new instance; (2) Adding a new weak rule; (3) Adding a new strong rule; (4) Adding a new part-relation; (5) Adding a new kind-relation; (6) Adding a new concept; (7) Collapsing part of a kind-hierarchy; (8) Reorganizing hierarchies by branch jumping; (9) Tree switching [28, p. 35]. Thagard claims that among these "(1)–(3) can be interpreted as simple kinds of belief revision". On the other hand, "(4)–(9) cannot, since they involve conceptual hierarchies" [28, p. 36].

3 The Conservatism of AGM Theory of Belief Revision

To the best of my knowledge, no belief revision theorist has tried to meet Thagard's challenge. In their discussion of the problem of belief revision in science, Gärdenfors and Rott merely express a disclaimer:

> The formal theory of belief revision we are going to expound in this chapter is concerned solely with small changes like those occurring in normal science. An essential feature of the present concept of belief revision is conservativity. [8, pp. 38-40]

What they call "conservativity" is represented in AGM theory by the so-called the principle of informational economy. They treated this principle as the principal philosophical (or methodological) rationale for their theory. Gärdenfors wrote:

> The criterion of informational economy demands that as few beliefs as possible be given up so that the change is in some sense a *minimal* change of K to accommodate for A. [7, p. 53]

Also, he wrote:

> Several of the postulates that have been formulated for contractions and revisions have been motivated by an appeal to the requirement that epistemic changes ought to be *minimal* changes necessary to accommodate the input. [7, p. 66]

At first blush, this principle is quite intuitive and irresistible. For, it is all too natural to avoid unnecessary loss of information when making changes in our earlier set of beliefs. However, this seemingly intuitive principle has been attacked severely by one of the most active researchers in belief change. In his article, "Two Dogmas of Belief Revision", Hans Rott calls into question the idea of informational economy, which he sets out in two distinct versions:

1. When accepting a new piece of information, an agent should aim at a minimal change of his old beliefs.
2. If there are different ways to effect a belief change, the agent should give up those beliefs which are least entrenched [23, p. 505].

Rott calls the first "the principle of minimal change", and the second "the principle of entrenchment". Contrary to their apparent individuality, however, Rott claims that these two maxims are "at root identical" and that "they can therefore be viewed as two incarnations of a unified idea of informational economy" [23, p. 506]. By his account, both (1) and (2) require the changes of belief to be minimal in a certain sense. What is notable is the fact that now Rott wants to understand (2), the principle of entrenchment, in terms of (1), the principle of minimal change. According to him, (2) requires one to respect minimality with respect to an ordering of priority or entrenchment between beliefs [24, p. 73].

Rott's criticism of the idea of informational economy as the basis of belief change consists of two observations, the gist of which is summarized by himself as saying that the idea of minimal change is difficult to formulate (observation 1) and that its application in the construction of revisions is ill-understood (observation 2) [23, p. 512]. From my point of view, however, much more important is to note that Rott's criticism is not well motivated in that it fails to hit the core of the idea of informational economy. According to Rott, he calls (1) and (2) *dogmas* "not because almost all researchers actually kept to these principles (quite the opposite is true), but because so many authoritative voices have *professed* that these principles are the principal philosophical or methodological rationale for their theories" [23, p. 507]. But, as he confesses, he himself has been one of them for a long period of time. If so, what is going on? According to his explanation, "many researchers (including the present author[Rott]) have believed in it and recited it time and again, *without actually keeping to it when building their theories*" [23, p. 521, my italics]. As long as Rott wants to save AGM theory, however, such an explanation is bound to be unconvincing and *ad hoc*.

In fact, much better argument is possible for motivating the radical attack on the both wings of the principle of informational economy, i.e., the principle of minimal change and the principle of entrenchment. For, even though it is hard to resist the intuitive appeal of the principle of informational economy, we seem to have some evidence in the form of Rott's testimony, and another (possibly stronger) intuition against the principle. Let me present that intuition in the following rhetorical form: *In order to achieve the growth of knowledge at the individual level or at the level of scientific progress, shouldn't our belief change be maximal?* The intuition or idea behind this invocation of maximal change must be that we indeed desire to achieve advancement as great as possible[4].

4   Toward a Strengthened Argument for the Redundancy of Conceptual Change

Is it impossible to develop a theory of belief revision that can cover more radical belief changes including scientific revolutions? At this stage, it seems meaningful to sketch a strengthened version of the argument for the redundancy of conceptual change. I will attempt to do so by exploiting the notion

---

[4] The ideas discussed in this section were more extensively presented in Park(2005). As one anonymous referee points out, there is room for further discussion as to whether we can take the principle of informational economy as one preventing large scale changes. For what the principle defends is that informational loss should be avoided *if possible* while accommodating the new information.

of implicit definition in mathematics[5]. Let us suppose that this notion truly captures the essence of mathematical method. The primitive terms of a given mathematical structure are defined implicitly by its axioms. If so, how could there be other conceptual changes than those *via* changing axioms? In order to flesh out the main idea of this argument, it would be all too natural to use number systems as examples. Above all, the cases of domain extension by the introduction of negative or imaginary number must be the best examples of the more radical changes in mathematics that involve changes in conceptual hierarchies. What kind of belief revision and conceptual change are at stake in domain extensions in mathematics? Thus, what we are left with are perfect testing grounds for any argument for (or against) the redundancy of belief revision and conceptual change.

Now, given the axiomatic method with implicit definition as its core, if we simply substitute "negative numbers" for "whales" and "number" for "mammals" in the previous argument, then we might present the strengthened version of the argument for the redundancy of conceptual change as follows:

> The issue of conceptual change in mathematics is a red herring. Whenever a mathematical concept changes, it does so by virtue of changes in the axioms that define implicitly that concept. For example, if you recategorize negative numbers as genuine numbers rather than numberlike pseudo or impossible entities, you have made an important change in the concept *negative number*. But this amounts to no more than deleting the axioms that make negative numbers as impossible entities and adding the axioms that make negative numbers as genuine numbers. Your concept of number may also change by adding the belief that subtraction of a larger number from a smaller one is allowed, but this merely follows from the other belief additions. So as far as epistemology is concerned, conceptual change is redundant with respect to the central question of belief revision.

The axiomatic system for rational positive integers and the axiomatic system for rational integers must be separate systems. Now, how to make sense of the transition from the former to the latter without resorting to belief revision? Yes, here we are supposed to have a more holistic rather than piecemeal belief revision that requires a redefinition of the hierarchy itself. For that very reason, we seem to need to adopt wholesale a new axiomatic system,

---

[5] The idea of implicit definition has never been clarified completely. We should still wait for a resolution of the famous Frege-Hilbert controversy. As a consequence, for too long we have been lacking a sound understanding of the role and function of definitions in mathematics. Further, as witnessed by Ulrich Majer's work, we still do not quite understand either the continuity or discontinuity of the traditional and Hilbertian axiomatic method. We even fail to decide whether the problem of implicit definition is merely a side issue, as Majer claims. See Majer(2002).

thereby defining all fundamental concepts implicitly. In other words, there would be no conceptual change without belief revision[6].

5 The Story of Negative and Imaginary Numbers

Now, let us examine some famous episodes of revolutionary change in the history of numbers. Specifically, let's look at the invention or discovery of negative integers and complex numbers. How are we to reconstruct those episodes in terms of belief revision and conceptual change respectively? Unlike the modern axiomatic number systems, the historical development of our conception of numbers seems to give a fair chance to both Thagard and belief revision theorists to make their cases. For, we seem to have here revolutionary changes that involve conceptual hierarchies such as *branch jumping* or *tree switching*.

Hilbert himself quite well summarized our problem situation when he alleged a difference in the *method* of investigation in the literature on the principles of *arithmetic* and on the axioms of *geometry*:

[2] Let us first recall the manner of introducing the concept of number. Starting from the concept of the number 1, one usually imagines the further rational positive integers 2, 3, 4 ... as arising through the process of counting, and one develops their laws of calculation; then, by requiring that subtraction be universally applicable, one attains the negative numbers; next one defines fractions, say as a pair of numbers—so that every linear function possesses a zero; and finally one defines the real number as a cut or a fundamental sequence, thereby achieving the result that every entire rational indefinite (and indeed every continuous indefinite) function possesses a zero. We can call this method of introducing the concept of number the *genetic method*, because the most general concept of real number is *engendered* [*erzeugt*] by the successive extension of the simple concept of number.

[3] One proceeds essentially differently in the construction of geometry. Here one customarily begins by assuming the existence of all the elements, i.e. one postulates at the outset three systems of things (namely, the points, lines, and planes) and then—essentially on the pattern of Euclid—brings these elements into relationship with one another by means of certain axioms—namely, the axioms of linking [Vernüpfung], of ordering, of congruence, and of continuity. The necessary task then arises of showing the *consistency* and the completeness of these axioms, i.e. it must be proved that the application of the given axioms can never lead to contradictions, and, further, that the system of axioms is adequate to prove all geometrical propositions. We shall call this procedure of investigation the *axiomatic method*. [11, pp. 1092–3]

---

[6] In passing, it is worthwhile to note that Thagard also leaves room for counting branch jumping and tree switching as a kind of belief revision. Anyway, they are "more holistic than piecemeal *belief revisions*" (emphasis mine) [28, p. 36, emphasis mine].

Hilbert goes on to ask whether the genetic method is the only suitable one for the study of the concept of number, and then gives a negative answer:

My opinion is this: Despite the high pedagogic and heuristic value of the genetic method, for the final presentation and the complete logical grounding [Sicherung] of our knowledge the axiomatic method deserves the first rank. [11, pp. 1093]

So, he tries to present an axiomatic system of arithmetic.

In other words, what we need to do in this section is nothing but reconstructing interpretations of the history of genetic method for domain extension of the theory of numbers in terms of belief revision and conceptual change. Among all the relevant episodes in the history of number concept, let us focus on those related to negative and imaginary numbers. No doubt we need to know (1) exactly when and why these were discovered, (2) exactly when and why they provoked most heated debates for their acceptance, and (3) exactly when and why they were finally granted the status as numbers. But it does not take much time to realize that our problems are not that easy to answer. In all these problems, there are intolerable ambiguity and vagueness. Also, both negative and imaginary numbers took a remarkably long time for their final acceptance.

Given the complex history of the acceptance of negative and imaginary numbers, it may be impossible to answer these three questions above with any degree of precision. We should be satisfied with a rough outline that identifies some of the most salient features in the history. Fortunately, Dunmore (1992) provides us with such an outline. According to Dunmore, answers to (2), seem to have a common pattern: new number-like entities "violated the rules of how numbers should behave" [4, p. 216]. Answers to (3) also seem to have one common characteristic: new number-like entities were accepted for their usefulness. But, Dunmore claims, "a further and more far-reaching revolution had to take hold for the adoption of negative and imaginary numbers to be completed". She understands this revolution as involving the modification of the "meta-level view of the nature of mathematics as a whole" [4, p. 216].

According to Dunmore, mathematical "community held a meta-level belief that mathematics was the science of magnitude and quantity, and that the purpose of the number concept was for measuring and counting"[7]. As long as the community sticks to this meta-level belief, negative and imaginary numbers cannot be accepted as legitimate numbers. "The revolutionary abandonment and replacement of this meta-level belief" was needed, and Dunmore identifies this revolution as the replacement of the view of

---

[7] Mancosu refers back to Giuseppe Biancani (1566-1624) for this belief: "According to Biancani, the objects of mathematics are quantities abstracted from sensible matter. Arithmetic and geometry, which together constitute pure mathematics respectively deal with discrete and continuous magnitude". Mancosu(1996), p. 16; See also, Park(2009).

mathematics as the science of magnitude by the modern view of "mathematics as a study of abstract structures" [4, p. 218].

Dunmore's interpretation of the episodes related to the acceptance of negative and imaginary numbers, as legitimate as positive and real numbers, not only makes clear many significant facts but also raises many intriguing further questions. First of all, the revolutionary character of the acceptance of negative and imaginary numbers involved more than a piecemeal revision of belief, but a tree switching that apparently cannot be explained by simple belief change. Secondly, it points to the fact that the revolutionary conceptual change took a long period of time to complete. Finally, it suggests that the conceptual revolution was not complete until the number systems were axiomatized. Each of these points deserves some further comments.

Let us consider what Pycior(1997) had to say about Girolamo Cardano:

> Cardano was one of the most consistent of the early supporters of the negative numbers. He not only divided numbers into two kinds, positive and negative, and regularly gave the negative solutions of equations, but even tentatively explored the consequences of working with imaginary roots, which he considered "sophistic" kinds of negative numbers. [21, p. 18]

"Dividing numbers into two kinds, positive and negative" must involve tree switching. For, unlike branching in a given tree, here we witness a switch from a tree consisting of only one branch to another tree consisting of two branches. The fact that we are dealing with the most radical conceptual change that involve conceptual hierarchies can be made manifest by the epithets used for the number-like entities newly introduced. For example, Dunmore quotes Euler's remark on imaginary numbers:

> Because all conceivable numbers are either greater than zero or less than zero or equal to zero, then it is clear that the square roots of negative numbers cannot be included among the possible numbers. Consequently we must say that these are impossible numbers. And this circumstance leads us to the concept of such numbers, which by their nature are impossible, and ordinarily are called imaginary or fancied numbers, because they exist only in the imagination. (quoted in Kline 1972, p. 594) [4, p. 216]

In order to appreciate Dunmore's second point, i.e., that the adoption of negative and imaginary numbers took a long period of time, Pycior's discussion of Wallis might be of help. For Pycior perceptively observes that for Wallis the negative and imaginary numbers were anomalies "explainable only by appeal to nonarithmetic considerations such as precedent, geometric interpretation, and usefulness" [21, p. 230]. In her discussion of Wallis' argument based on the usefulness of these numbers, Pycior wrote:

> He then went outside traditional formal mathematics and built a persuasive, rather than mathematically compelling, case for the negative and imaginary numbers. He would not define these numbers in any traditional way; rather he would persuade mathematicians to accept them on the basis of a series of cumulative arguments. In a bold and somewhat risky maneuver in chapter 66,

he tied the cause of the imaginaries to that of the negatives: he stressed that negative numbers as well as numbers involving $\sqrt{-1}$ were imaginary, according to the canons of arithmetic; then he argued that the usefulness of these numbers as well as geometric analogy sanctioned the mathematician's "supposing" that there were negative numbers and (the traditionally) imaginary numbers". [21, pp. 128-9]

Wallis' case fits perfectly with Dunmore's scheme of conceptual change in mathematics. With all his strenuous effort, Wallis failed to persuade the mathematical community to accept the negative and imaginary numbers. For the community was still under the spell of the meta-level belief about mathematics as the science of magnitude. Furthermore, it is evident that Wallis himself was sharing that meta-level belief. For that very reason, he appealed to persuasive arguments rather than purely mathematical ones.

As for the third point that the required replacement of the meta-level belief for the final acceptance of negative and imaginary numbers as legitimate numbers was nothing but the axiomatization of number systems, Dunmore does not provide much discussion. She just draws our attention to some landmarks such as "Gauss' first proof of the fundamental theorem of algebra", "the emergence of a formalist trend exemplified by Ohm and Peacock", "Hamilton's attempts to give meaning to the 'impossibilities of the so-called arithmetical algebra", and the fact that "in defining the real numbers, he[Hamilton] comes exceedingly close to giving all of what are now termed the field axioms" [4, p. 217].

Now, given the three significant facts highlighted by Dunmore's interpretation, we seem to face a puzzling situation. Apparently the introduction and acceptance of the negative or imaginary numbers needed radical conceptual changes like tree switching that cannot be explained by belief changes. But such a radical conceptual change that needed a long period of time was completed only when the newly introduced number system was axiomatized. In other words, though the introduction of new number requires a conceptual change more than mere belief change, such a conceptual change cannot be completed without another kind of belief change, i.e., changing the axioms.

6 The Relationship between Belief Revision and Conceptual Change in Mathematics

This puzzle suggests, I believe, there is an extensive and intricate interaction between conceptual change and belief revision in numerical domain extensions, no matter what conclusion we come to on the question of the identity or non-identity of conceptual change and belief revision. If we emphasize too much the novelty and superiority of the axiomatic method, as Hilbert does, then in some sense the axiom system for natural numbers and the axiom system for integers have nothing to do with each other. The concept of natural number must be implicitly defined by the axioms for the natural number system. When we accept the axioms for integers, the concept of integer is implicitly defined, and the concept of positive integer is thereby determined. The concept of natural number within the axiom system for integer must be entirely different from that within the axiom system for natural numbers.

It is interesting to note that Hilbert never counted Dedekind as exemplifying the axiomatic method. From Hilbert's point of view, Dedekind represents merely the most sophisticated genetic method. However, according to the recent characterization of Dedekind as a structuralist, Hilbert's view goes too far[8]. For example, Sieg and Schlimm find in Dedekind axiomatic as well as genetic approach. According to them, Dedekind's reflections on the differences between his own approach and that of Helmholtz and Kronecker as ultimately resulted in a "dramatic shift" [27, p. 122]. If so, Dedekind's contribution to modern axiomatic method cannot be too much emphasized. Furthermore, in view of our present puzzle, the so-called genetic aspect of Dedekind's method may not be a weakness but a strength. Unlike the purely axiomatic approach, we can see the connection between the system of natural numbers and that of integers.

Based on their careful study of Dedekind's *Nachlass*, Sieg and Schlimm provide us with a snap shot of how Dedekind struggled with the problem of how to generate the integers from the natural numbers:

> The first manuscript formulates at the outset basic facts regarding the series of natural numbers N: (1) N is closed under addition; addition is (2) commutative and (3) associative; (4) if $a > b$, then there exists one and only one natural number c, such that $b + c = a$, whereas in the opposite case, when $a \leq b$, no such number c exists. Dedekind notes that the fourth condition states a certain *irregularity* and raises the crucial question, whether it is possible to extend the sequence N to a system M (by the addition of elements or numbers to be newly generated) in such a way that M satisfies conditions (1)–(3) and also (4'), i.e., for any two elements a and b from M, there exists exactly one element c, such that $b + c = a$. And he asks, how rich must the *smallest* such system M be.

> In the following *Investigation*, which is also called *Analysis*, Dedekind assumes the existence of such a system M. He reasons that M must contain a unique element 0 (called zero), such that $a + 0 = a$; furthermore, for every element a in N there must be a new element a^* in M, such that $a + a^* = 0$. Thus, *any* system M satisfying (1)–(4') must contain in addition to the elements of N the new element zero and all the different new elements a^*. [27, p. 135]

---

[8] Here we seem to face a problem of how to find a way out from what I would call "the Hilbertian dilemma": Are we to emphasize the novelty of Hilbert's axiomatic method or the continuity of the history of axiomatic method? If we grasp the second horn, as Majer does, we might be giving up the hope for understanding the so-called second birth of mathematics in the 19th and 20th centuries. If we grasp the first horn, as logical positivists and Friedman do, we should make clear how the method of implicit definition truly works in scientific as well as mathematical practice.

In his letter to Keferstein, Dedekind explains in detail how he arrived at the so-called Dedekind-Peano axioms [3]. So, starting from the axiom system of natural numbers, we can understand how Dedekind generates the new numbers, i.e., the negative numbers to arrive at the system of integers.

7 Concluding Remarks

What is the implication of all this discussion to our starting point, i.e., Thagard's question as to whether the question of conceptual change is identical with the question of belief revision? Could we submit and examine a wild hypothesis that conceptual change and belief revision could be extensionally identical, even though there are intensional and/or procedural differences between them? I do not know. But both Thagard and belief revision theorists might learn some important lessons from examining such a hypothesis. Thagard may aim at extending his theory of conceptual revolution in science to mathematics. Belief revision theorists may extend their theory in such a way that it can cover not only normal science but also scientific revolutions. Especially, they may learn from Dedekind's exercises exactly when they can and should incapacitate their principle of informational economy in order to allow axiomatic belief change. Be that as it may, we cannot afford to ignore Thagard's timely and pertinent question as to the identity and non-identity of the question of conceptual change and the question of belief revision any more.

Acknowledgements. Many thanks are due to the encouragement and moral support of Lorenzo Magnani, Walter Carnielli, John Woods, Ahti-Veikko Pietarinen, Mary Keeler, Balakrishnan Chandrasekaran, and Paul Thagard. During the conference, Chandrasekaran raised an incisive question as to whether it is allowable to treat axioms simply as beliefs. I cannot discuss this problem in the present paper, though it may capture the whole point of Thagard's recent research on cognitive process even at the neural level. Finally, I am indebted to Jeffrey White's extensive comments on the penultimate draft.

References

1. Alchorron, C., Gärdenfors, P., Makinson, D.: On the logic of theory change: Partial meet contraction functions and their associated revision functions. The Journal of Symbolic Logic 50, 510–530 (1985)
2. Darden, L.: Theory Change in Science: Strategies from Mendelian Genetics. Oxford University Press, New York (1991)
3. Dedekind, R.: Letter to Keferstein. English translation by Stefan Bauer-Mengelberg and Hao Wang in van Heijenoort, pp. 98–103 (1967)
4. Dunmore, C.: Meta-level revolutions in mathematics. In: Gillies, D. (ed.) Revolutions in Mathematics, pp. 209–225. Oxford University Press, Oxford (1992)
5. Ewald, W.B. (ed.): From Kant to Hilbert. A Source Book in the Foundations of Mathematics, vol. 2. Oxford University Press, Oxford (1996)
6. Gabbay, D.M., Kurucz, A., Wolter, F., Zakharyaschev, M.: Many-Dimensional Modal Logics: Theory and Applications. Elsevier, Amsterdam (2003)

7. Gärdenfors, P.: Knowledge in Flux: Modeling the Dynamics of Epistemic States. MIT Press, Cambridge (1988)
8. Gärdenfors, P., Rott, H.: Belief Revision. In: Gabbay, D.M., et al. (eds.) Handbook of Logic in Artificial Intelligence and Logic Programming. Epistemic and Temporal Reasoning, vol. 4, pp. 35–132. Clarendon Press, Oxford (1995)
9. Gillies, D. (ed.): Revolutions in Mathematics. Oxford University Press, Oxford (1992)
10. Heidelberger, M., Stadler, F. (eds.): History of Philosophy and Science. Kluwer, Boston (2002)
11. Hilbert, D.: Über den Zahlbegriff. Jahresbericht der Deutschen Mathematiker-Vereinigung 8, 180–184, English translation in Ewald 1089–1096 (1996)
12. Manders, K.: Domain extension and the philosophy of mathematics. Journal of Philosophy 86, 553–562 (1989)
13. Muntersbjorn, M.M.: Naturalism, notation, and the metaphysics of mathematics. Philosophia Mathematica 7(2), 178–199 (1999)
14. Kvasz, L.: The history of algebra and the development of the form of its language. Philosophia Mathematica 14(3), 287–317 (2006)
15. Majer, U.: Hilbert's program to axiomatize physics (in analogy to geometry) and its impact on Schlick, Carnap and other members of the Vienna Circle. In: Heidelberger, M., Stadler, F. (eds.) History of Philosophy and Science, pp. 213–224. Kluwer, Boston (2002)
16. Mancosu, P.: Philosophy of Mathematics and Mathematical Practice in the Seventeenth Century. Oxford University Press, Oxford (1996)
17. Nagel, E.: Teleology Revisited and Other Essays in the Philosophy and History of Science. Columbia University Press, New York (1979)
18. Neal, K.: From Discrete to Continuous: The Broadening of Number Concepts in Early Modern England. Kluwer, Dordrecht (2002)
19. Park, W.: Belief revision in Baduk. Journal of Baduk Studies 2(2), 1–11 (2005)
20. Park, W.: The status of scientiae mediae in the history of mathematics. Korean Journal of Logic 12(2), 141–170 (2009)
21. Pycior, H.M.: Symbols, Impossible Numbers, and Geometric Entanglements: British Algebra Through the Commentaries on Newton's Universal Arithmetick. Cambridge University Press, Cambridge (1997)
22. Quine, W.V.O.: Two dogmas of empiricism. In: Quine, W.V.O. (ed.) From a Logical Point of View, pp. 20–46. Harvard University Press, Cambridge (1953)
23. Rott, H.: Two dogmas of belief revision. Journal of Philosophy 97, 503–522 (2000)
24. Rott, H.: Change, Choice and Inference: A Study of Belief Revision and Nonmonotonic Reasoning. Clarendon Press, Oxford (2001)
25. Rusnock, P., Thagard, P.: Strategies for conceptual change: Ratio and proportion in classical greek mathematics. Studies in History and Philosophy of Science 26, 107–131 (1995)
26. Sepkoski, D.: Nominalism and Constructivism in Seventeenth-Century Mathematical Philosophy. Routledge, New York (2007)
27. Sieg, W., Schlimm, D.: Dedekind's analysis of number: Systems and axioms. Synthese 147, 121–170 (2005)
28. Thagard, P.: Conceptual Revolutions. Princeton University Press, Princeton (1992)
29. Van Heijenoort, J. (ed.): From Frege to Gödel: A Sourcebook of Mathematical Logic. Harvard University Press, Cambridge (1967)

Affordances as Abductive Anchors

Emanuele Bardone

Abstract. In this paper we aim to explain how the notion of abduction may be relevant in describing some crucial aspects related to the notion of affordance, which was originally introduced by the ecological psychologist James J. Gibson. The thesis we develop in this paper is that an affordance can be considered an abductive anchor. Hopefully, the notion of abduction will clear up some ambiguities and misconceptions still present in current debate. Going beyond a merely sentential conception, we will argue that the role played by abduction is two fold. First of all, it is decisive in leading us to a better definition of affordance. Secondly, abduction turns out to be a valuable candidate in clarifying the various issues related to affordance detection.

1 The Environment: Constraint or Resource? Why a Theory of Affordance Matters

In the opening section of this paper we will try to set the stage for our discussion on affordance. We start off with a number of questions such as how and why may a theory of affordance contribute to shedding light on how humans distribute their cognition? What contribution can a theory of affordance provide in deepening some key issues concerning distributed cognition? What are these key issues?

In our view, one of the issues characterizing a theory of distributed cognition is related to how humans turn their environment into something potentially meaningful and beneficial for their survival and reproduction. For example, an animal, belonging to our environment, could be a threat to us and our property. A tiger, for instance, can attack humans causing them

Emanuele Bardone
Computational Philosophy Laboratory and Department of Philosophy,
University of Pavia, Italy
e-mail: `bardone@unipv.it`

L. Magnani et al. (Eds.): Model-Based Reasoning in Science & Technology, SCI 314, pp. 135–157.
springerlink.com © Springer-Verlag Berlin Heidelberg 2010

serious injuries and, sometimes, even death. However, human beings have domesticated some animals that now live in our niches and for which we have developed feelings of love and care. Scientists can even use animals as epistemic mediators and they have recently come up with some alarming implications of global warming by using bird migration as a source of clues.

Generally speaking, the environment and everything in it, presents two faces, on the one hand, it may threaten and constrain our activities while, on the other, the same threats and constraints can be overcome through resources provided by the environment itself.

Following the tradition of ecological psychology, the environment is turned into a source of chances when the human agent attains a stable, functional relationship with his surroundings (cf. Kirlik [20, p. 238]). Attaining a stable, functional relationship with one's surroundings – what Egon Brunswik [3] called *achievement* – basically refers to the way the human agent develops and sets up pre-determined associations with his environment so as to respond to its challenges by using the environment itself as a template. Recently, Clark [2] referred to *coupling* as the factor permitting an agent to simplify neural problem solving by making use of the environment as a kind of external representation conveying additional information and computational resources.

The theory of distributed cognition has placed great emphasis on how humans solve their problems by externalizing functions to the environment (Hutchins [17] Clark [6], Magnani [25]). That is, the environment is not neutral from the cognitive perspective, but it is ready for supporting and extending our cognition in a certain way [23]. Ecological psychology, for instance, insisted on the "invitation" or "demand" character of the environment (cf. Reed [38]). In this case, the emphasis is put on the way the agent and the environment may communicate or, even better, on how couplings may be established.

In order to look into the ambiguity of environment, let us consider this issue from an evolutionary perspective. Traditionally, evolution is considered as a response to modified selection pressures [51]. Accordingly, the environment is basically a source of constraints affecting the chances one organism has to live long enough to reproduce. So, environmental constraints caused by changing selection pressures make any organism experience some *adaptive lag*, namely, a mismatch between current selection pressures and behavior (cf. Laland and Brown [21]). Subsequently the organism tries to adapt to the environment so as to minimize the ecological mismatch.

This traditional view is partially erroneous or, at least, incomplete; it erroneously assumes that the causal arrow that goes from the environment to the organism points in one direction only. But this does not honor the facts; organisms adapt their environment and vice versa insofar as they alter the environment by constructing so-called ecological niches [8, 21]. That is, the adaptive process does not only regard the organism, but also the environment that "evolves" insofar as the selective pressures are modified by the organism's niche construction activity. It follows that the mismatch between current selection pressures and behavior can be minimized not only by the organism adapting to the new environment (and so still experiencing the environment as a source of constraints), but also the other way around. That is, the activity of niche construction appropriately selects some aspects of the environment so as to increase the organism's chances of survival and reproduction.

More generally, the adaptive complementarities of organism and environment point to the idea that the environment is not only a source of constraints, but also of resources and chances potentially benefiting the organism. The possibility to exploit environmental constraints as chances – and so enlarge one's behavioral repertoire – is granted to the organism that attains a stable relation with its environment, namely, achievement. According to the present argument, such an achievement is not a mere adaptation in which the human agent passively modifies his response to the environment. It is reached through the selective activity of niche construction in which the human agent aims at turning one part of the environment – the one that is meaningful to it – into a potential local representative (or, more simply, a clue) of the other – the one that is not meaningful yet. In this sense, the limits of the agent's adaptability correspond to his limits in creating less equivocal as possible indicators of distal events[1].

In this paper we will argue that a theory of affordance may clarify the way the environment is transformed from a source of constraints to a source of chances. More precisely, the theory of affordance is particularly helpful in understanding how the human agent looks at the environment as a source of chances. That is, affordances are the *locus* in which a stable and functional relationship between the agent and the environment is successfully achieved.

The notion of affordance does not only contribute to shedding light on how the environment turns into a source of chances for the agent, but it also introduces another important related question: the question of how to access environmental chances. If the environment provides the agent with opportunities for behavior, how does the agent come up with them? How could we describe such a process?

The question of access is central also for a theory of distributed cognition. The idea of distributed cognition arises to take into account the way external resources become part of one's cognitive system notwithstanding the fact that they are external to the brain. That said, the question of access is intimately related to distributed cognition as it explicitly addresses the problem of how the agent comes to be engaged in the exploitation of latent environmental resources in order to extend and shape pre-existing abilities. Here, the notion of affordance acquires a central meaning, as it offers a straightforward answer to the question of access. It states that the human agent has direct access to environmental chances. That is, the environment offers or furnishes chances

---

[1] Cf. Figueredo *et al.* [9]. We will come back to this issue in section 4.

that the agent directly detects and picks up as relevant and thus the external objects automatically become part of the agent's extended cognitive system.

The rest of the paper will proceed as follows, in the next section we will illustrate the notion of affordance dealing with main points that have recently emerged in the literature related to affordance. In section 3 we will present the most significant objection to a possible extension of affordance beyond the realm of visual perception. This is the objection that direct perception is a prerequisite for having affordances and therefore we have access to the affordances of the environment only by direct perception. By contrast, we will claim that the idea of "no direct perception, no affordance" is somehow misleading. In section 4 we will introduce the notion of abduction arguing that abduction may be of great help in understanding the problem of affordance detection. More precisely, we will present a particular interpretation of abduction that goes beyond the dualism between inference and perception. Such an interpretation will be central to our proposal. So, in section 5 we will contend that affordance detection can be illuminated by the notion of abduction, claiming that affordance informs us about *environmental symptomaticity*. In this sense, an affordance can be considered as an *abductive anchor*. The last section deals with an often neglected occurrence, that is, when the environment does not afford us.

2   What Are Affordances? The Received View

One of the most disturbing problems with the notion of affordance is that any examples provide different, and sometimes ambiguous insights on it. This fact makes very hard to give a conceptual account of it. That is to say, when making examples everybody grasps the meaning, but as soon as one tries to conceptualize it the clear idea one got from it immediately disappears. Therefore, we hope to go back to examples from abstraction without loosing the intuitive simplicity that such examples provide to the intuitive notion.

The entire debate during the last fifteen years about the notion of affordance is very rich and complicated, but also full of conflicts and ambiguities. This subsection aims at giving just an overview of some issues we consider central to introduce to our treatment.

Gibson defines "affordance" as what the environment offers, provides, or furnishes. For instance, a chair affords an opportunity for sitting, air breathing, water swimming, stairs climbing, and so on. By cutting across the subjective/objective frontier, affordances refer to the idea of agent-environment mutuality. Gibson did not only provide clear examples, but also a list of definitions [50] that may contribute to generating possible misunderstanding:

1. affordances are opportunities for action;
2. affordances are the values and meanings of things which can be directly perceived;
3. affordances are ecological facts;
4. affordances imply the mutuality of perceiver and environment.

We contend that the Gibsonian ecological perspective originally achieves two important results. First of all, human and animal agencies are somehow hybrid, in the sense that they strongly rely on the environment and on what it offers. Secondly, Gibson provides a general framework about how organisms directly perceive objects and their affordances. His hypothesis is highly stimulating: "[...] the perceiving of an affordance is not a process of perceiving a value-free physical object [...] it is a process of perceiving a value-rich ecological object", and then, "physics may be value free, but ecology is not" [12, p. 140]. These two issues are related, although some authors seem to have disregarded their complementary nature. It is important here to clearly show how these two issues can be considered two faces of the same medal. Let us start our discussion.

2.1 Affordances Are Opportunities for Action

Several authors have been extensively puzzled by the claim repeatedly made by Gibson that "an affordance of an object is directly perceived"[2]. During the last few years an increasing number of contributions has extensively debated the nature of affordance as opportunity for action. Consider for instance the example "stairs afford climbing". In this example, stairs provide us with the opportunity of climbing; we climb stairs because we perceive the property of "climbability", and that affordance emerges in the interaction between the perceiver and stairs [5, 43]. In order to prevent from any possible misunderstanding, it is worth distinguishing between "affordance property" and "what" and object affords [31]. In the former sense, the system "stairs-plus-perceiver" exhibits the property of climbability, which is an *affordance property*. Whereas in the latter the possibility of climbing is clearly *what* an object affords.

2.2 Affordances Are Ecological Facts

Gibson also argued that affordances are ecological facts. Consider, for instance, a block of ice. Indeed, from the perspective of physics a block of ice melting does not cease to exist. It simply changes its state from solid to liquid. Conversely, to humans a block of ice melting does go out of existence, since that drastically changes the way we can interact with it. A block of ice can chill a drink the way cold water cannot. Now, the point made by Gibson is that we may provide alternative descriptions of the world: the one specified by affordances represents the environment in terms of action possibilities. As Vicente [48] put it, affordances "[...] are a way of measuring or representing the environment with respect to the action capabilities of an individual [...] one can also describe it [a chair] with respect to the possibilities for action

---

[2] Cf. Greeno [15], Stoffregen [43], Scarantino [40], Chemero [5].

that it offers to an organism with certain capabilities". Taking a step further, we may claim that affordances are chances that are *ecologically rooted*. They are ecological rooted because they rely on the mutuality between an agent (or a perceiver) and the environment. As ecological chances, affordances are the result of a hybridizing process in which the perceiver meets the environment. The emphasized stress on the mutuality between the perceiver and the environment provides a clear evidence of this point.

2.3 Affordances Imply the Mutuality of Perceiver and Environment

Recently, Zhang *et al.* [54], also going beyond the ecological concept of affordance in animals and wild settings by involving its role in human cognition and artifacts, in an unorthodox perspective, connect the notion of affordance to that of distributed representation. They maintain that affordances can be also related to the role of distributed representations extended across the environment and the organism. These kinds of representation come about as the result of a blending process between two different domains: on one hand the internal representation space, that is the physical structure of an organism (biological, perceptual, and cognitive faculties); on the other the external representation of space, namely, the structure of the environment and the information it provides. Both these two domains are described by constraints so that the blend consists of the allowable actions. Consider the example of an artifact like a chair. On one hand the human body constrains the actions one can make; on the other the chair has its constraints as well, for instance, its shape, weight, and so on. The blend consists of the allowable actions given both *internal* and *external* constraints.

Patel and Zhang's idea tries to clarify that affordances result from a hybridizing process in which the environmental features and the agent's ones in terms of constraints are blended into a new domain which they call *affordance space*. Taking a step further, Patel and Zhang define affordances as allowable actions. If this approach certainly acknowledges the hybrid character of affordance we have described above and the mutuality between the perceiver and the environment, it seems however lacking with regard to its conceptual counterpart. As already argued, affordances are action-based opportunities.

2.4 Affordances as Eco-Cognitive Interactional Structures

Taking advantage of some recent results in the areas of distributed and animal cognition, we can find that a very important aspect that is not sufficiently stressed in literature is the dynamic one, related to designing affordances, with respect to their evolutionary framework (cf. Magnani and Bardone [27]): human and non-human animals can "modify" or "create" affordances by manipulating their cognitive niches. Moreover, it is obvious to note that human, biological bodies themselves evolve: and so we can guess that even the more basic and wired perceptive affordances available to our ancestors were very different from the present ones[3]. Of course different affordances can also be detected in children, and in the whole realm of animals. We will come back to this issue in sections 5 and 6.

3 No Direct Perception, No Affordance? The Problem of Affordance Detection

The theory of affordance potentially re-conceptualizes the traditional view of the relationship between action and perception according to which we extract from the environment those information which build up the mental representation that in turn guides action (cf. Marr [29]). From an ecological perspective, the distinction between action and perception is questioned. The notion of affordance contributes to shed light on that issue fairly expanding it.

We posit that the Gibsonian ecological perspective originally achieves two important results. First of all, human and animal agencies are somehow hybrid, in the sense that they strongly rely on the environment and on what it offers. Secondly, Gibson provides a general framework about how organisms directly perceive objects and their affordances. As already mentioned, his hypothesis is highly stimulating and worth considering: the perceiving of an affordance is not a process of perceiving a value-free physical object. The two issues are related, although some authors seem to have disregarded their complementary nature. It is important here to clearly show how these two issues can be considered two faces of the same medal.

In his research Gibson basically referred to "direct" perception, which does not require internal inferential mediation or processing by the agent. Donald Norman from Human Computer Interaction studies challenged the original Gibsonian notion of affordance claiming: "I believe that affordances result from the mental interpretation of things, based on our past knowledge and experience applied to our perception of the things about us" [32, p. 14]. Norman [33] then added that an interface designer should care more about "what actions the user perceives to be possible than what is true".

---

[3] The term "wired" can be easily misunderstood. Generally speaking, we accept the distinction between cognitive aspects that are "hardwired" and those which are simply "pre-wired". By the former term we refer to those aspects of cognition which are fixed in advance and not modifiable. Conversely, the latter term refers to those abilities that are built-in prior the experience, but that are modifiable in later individual development and through the process of attunement to relevant environmental cues: the importance of development, and its relation with plasticity, is clearly captured thanks to the above distinction. Not all aspects of cognition are pre-determined by genes and hard-wired components. For more information, see Marcus [28].

In this passage he meant that an affordance should be perceived or detected to be just that. This consideration later led Norman to dismiss the notion of affordance in interface design. As for Norman achievement, namely, the effective use of an affordance, is necessary to recognize that an affordance is present. Roughly speaking, if the user does not click on the button, then the button does not afford *clicking* or it is not *clickable*.

More generally, it seems clear that in some cases detecting an affordance depends on the organism's experience, learning, and full cognitive abilities, i.e. they are not independent of them, like Gibson maintained. For example infants at 12 to 22 weeks old already show complicated cognitive abilities of this type, as reported by Rader and Vaughn [35]. These abilities allow them to lean on prior experience of an object and therefore detect what Rader and Vaughn call "hidden affordances" (we will come back to this issue in the last section of the paper). Hidden affordances are those specified by information not available at the time of the interaction, but drawn from past experiences [35, p. 539]. The same event or place can provide different affordances to different organisms but also multiple affordances to the same organism. Following Norman's point of view, affordances suggest a range of *chances*: given the fact that artifacts are complex things and their affordances normally require a high-level of supporting information, it is more fruitful to study them following this view.

To give an example, perceiving the full range of the affordances of a door requires complex information about for example direction of opening or about its particular pull. Becoming attuned to invariants and disturbances often goes beyond the mere Gibsonian direct perception and higher representational and mental processes of thinking/learning have to be involved[4]. This means that for example in designing an artifact to the aim of properly and usefully exhibiting its full range of affordances we have to clearly distinguish among two levels: 1) the construction of the utility of the object and 2) the delineation of the possible (and correct) perceptual information/cues that define the available affordances of the artifact. They can be more or less easily be undertaken by the user/agent [10, 49, 30]: "In general, when the apparent affordances of an artifact match its intended use, the artifact is easy to operate. When apparent affordances suggest different actions than those for which the object is designed, errors are common and signs are necessary" (Cf. Gaver [10, p. 80]). In this last case affordances are apparent because they are simply "not seen". In this sense information arbitrate the perceivability of affordances, and we know that available information often goes beyond what it can be provided by direct perception but instead involves higher cognitive endowments.

---

[4] Turvey and Shaw [47] and Hammond *et al.* [16] pointed out that high-level organisms' cognitive processes like those referred to language, inference, learning, and the use of symbols would have to be accounted for by a mature ecological psychology.

Generally speaking, if certain affordances are hidden, and therefore require internal processing beyond "direct perception", should we still consider them to be affordances? The traditional view on affordance considers direct perception as prerequisite to an affordance. To put it simply, if we do not have direct perception, then we do not have an affordance. If this were correct, then the contribution of affordance for distributed cognition theory would be poor. In fact, it has to be said that of course it is impossible to think that direct perception can explain all cognitive phenomena, like many Gibsonian researchers maintain. In the next two sections we will try to put forward an alternative conception of affordance based on abduction. More precisely, we will try to show how we can still consider direct detection as an important aspect of affordance without, however, resorting to direct perception.

4 Abduction and the Hypothetical Dimension of Eco-Cognitive Interactions

Magnani [23] defines abduction as the process of *inferring* certain facts and/or laws and hypotheses that render some sentences plausible, that *explain* or *discover* some (new) phenomenon or observation; it is the process of reasoning in which explanatory hypotheses are formed and evaluated. An example of abduction is the method of inquiring employed by detectives: in this case we do not have direct experience of what we are taking about. Say, we did not see the murderer killing the victim. But we infer that given certain signs or clues, a given fact must have happened. More generally, we guess a hypothesis that imposes order on data.

According to Magnani [23], there are two main epistemological meanings of the word abduction: 1) abduction that only generates "plausible" hypotheses ("selective" or "creative") and 2) abduction considered as inference "to the best explanation", which also evaluates hypotheses. An illustration from the field of medical knowledge is represented by the discovery of a new disease and the manifestations it causes which can be considered as the result of a creative abductive inference. Therefore, "creative" abduction deals with the whole field of the growth of scientific knowledge. This is irrelevant in medical *diagnosis* where instead the task is to "select" from an encyclopedia of pre-stored diagnostic entities. We can call both inferences ampliative, selective and creative, because in both cases the reasoning involved amplifies, or goes beyond, the information incorporated in the premises [22].

Abduction can fairly account for some crucial theoretical aspects of hypothesis generation as well as manipulative ones. Accordingly, we may distinguish between two general abductions, *theoretical abduction* and *manipulative abduction* [23]. *Theoretical abduction* illustrates much of what is important in creative abductive reasoning, in humans and in computational programs. It regards verbal/symbolic inferences, but also all those inferential processes which are model-based and related to the exploitation of internalized models of diagrams, pictures, etc.

Theoretical abduction does not account for all those processes of hypothesis generation relying on a kind of "discovering through doing", in which new and still unexpressed information is codified by means of manipulations of some external objects (which Magnani [23] called *epistemic mediators*). These inferential processes are defined by manipulative abduction which, conversely, is the process of inferring new hypotheses or explanation occurring when the exploitation of environment is crucial. More generally, manipulative abduction occurs when many external things, usually inert from the semiotic point of view, can be transformed into "cognitive mediators" that give rise – for instance in the case of scientific reasoning – to new signs, new chances for interpretants, and new interpretations.

Therefore, manipulative abduction represents a kind of redistribution of the epistemic and cognitive effort to manage objects and information that cannot be immediately represented or found internally (for example exploiting the resources of visual imagery). If the structures of the environment play such an important role in shaping our semiotic representations and, hence, our cognitive processes, we can expect that physical manipulations of the environment receive a cognitive relevance.

This distinction contributes to going beyond a conception of abduction, which is merely logical, say, related only to its sentential and computational dimension, towards a broader semiotic dimension worth investigating (see Magnani [24]). Peirce himself fairly noted that the all thinking is in signs, and signs can be icons, indices, or symbols. In this sense, all *inference* is a form of sign activity, where the word sign includes "feeling, image, conception, and other representation" [29, 5.283]. This last consideration clearly depicts the semiotic dimension of abduction, which will be crucial in looking into the relationship between abduction and affordance.

We posit that abduction contributes to shed light on a wide range of phenomena – from perception to higher forms of cognition – which otherwise would not be appropriately considered and understood. In the previous section we have illustrated the problem with affordances, whether they are directly perceived or mediated by higher cognitive processes. Abduction clarifies this issue in the sense that it goes beyond the traditional conception, which contrasts automatic or spontaneous response and perception, on the one hand, with mediated inferences and more plastic and reasoned forms of cognition, on the other. Apparently, the difference does not seem to reside in the underlying processes (mediated or not), but the different cognitive endowments organisms (humans including) can take advantage and make use of. In this sense, it is better to distinguish two different aspects of the same problem: first of all, the word inference is not exhausted by its logical/higher cognitive aspects but is referred to the effect of various sensorial activities [24]. Second, the ability of making use of various cognitive endowments (instinctual or plastic), which is ultimately connected to abduction, is an evolving property. That is, it is open to improvements, modification, and evolution.

We maintain that *perceptual judgment* is the best example to illustrate that it is not only conscious abstract thought that we can consider inferential (cf. Magnani [24]). Indeed, perceptual judgment is meant to be a spontaneous and automatic response, which is connected to various cognitive endowments, which are "instinctual" and hard-wired by evolution. Seeing is meant to be a direct form of cognition, which does not need any kind of mediation, since it happens without thinking – almost instantaneously. However, what about the case in which stimuli appear to be ambiguous, say, they need to be disambiguated (for instance, a face which is half familiar half not)? Usually, as we get more and more clues, we get a clearer picture about who is the person approaching us. That is symptomatic of the fact that perceptual judgment is a sign-activity, namely, abduction. Indeed the perceptions we have cannot be deliberately controlled the way scientific inferences are. We do not have any conscious access to them. However, perceptions are always withdrawable, just like the case we are presenting: our perceptual judgment can be subjected to changes and modifications, as we acquire more clues/signs. In this sense, we may say that what we *see* is what our visual apparatus can, so to say, "explain" (cf. Rock [39], Thagard [45], Hoffman [18], Magnani [23])[5]. That is, people are very got used to impose order on various, even ambiguous, stimuli, which can be considered "premises" of the involved abduction [23, p. 107]. More generally, those forms of cognition just like perceptual judgment are still abduction, even though, as brilliantly Peirce noted, they tends "to obliterate all recognition of the uninteresting and complex premises" from which they was derived [29, 7.37]. In this sense, perceptual judgment is only apparently a not-mediated process: it is semi-encapsulated in the sense that it is not insulated from "knowledge" (cf. Raftopoulos [36, 37] and Magnani [23]).

Another example supporting the semi-encapsulated nature of perception is provided by those perceptual judgments made by the experts[6]. Consider a professional meteorologist: what a trained meteorologist sees when looking at the sky clearly goes beyond what ordinary people can really see. They can just see clouds, whereas a meteorologist would see various types and subtypes of clouds (cirrus clouds, altocumulus clouds, and many more) telling her a great deal about how the weather will evolve over the course of the next few hours. Thanks for the training they received, meteorologists as experts are able to spontaneously impose order over an additional range of signs coming from the

---

[5] Recently, Thagard [42] claimed that perception cannot be described as abductive. He argued that perception involves some neuropsychological processes that do not fit with the definition of abduction as the process of generation and evaluation of explanatory hypotheses.

[6] The so-called visual abductions (cf. Magnani [23]) are essential also in science in which new and interesting discoveries are generated through some kind of visual-model based reasoning. On this topic, see for instance Shelley [42] and Gooding [14]. On the education of perception, see Goldstone *et al.* [13].

sky, which are *learnt* and *knowledge-dependent*. Once got familiar with some particular set of signs, then perceptual judgment proceeds automatically and with no further testing, although they may involve sophisticated knowledge, which require a long period of learning and training.

5   Affordances as Abductive Anchors

As already mentioned, Gibson defines affordance as what the environment offers, provides, or furnishes. For instance, a chair affords an opportunity for sitting, air breathing, water swimming, stairs climbing, and so on. But what does that exactly mean from an abductive perspective we introduced so far?

Within an abductive framework, that a chair affords sitting means we can perceive some clues (for instance, robustness, rigidity, flatness) from which a person can easily say "I can sit down". Now, suppose the same person has another object O, and she/he can only perceive its flatness. He/she does not know if it is rigid and robust, for instance. Anyway, he/she decides to sit down on it and he/she does that successfully. Is there any difference between the two cases?

We claim the two cases should be distinguished: in the first one, the cues we come up with (flatness, robustness, rigidity) are *highly diagnostic* to know whether or not we can sit down on it, whereas in the second case we eventually decide to sit down, but we do not have any precise clue about. How many things are there that are flat, but one cannot sit down on? A nail head is flat, but it is not useful for sitting. That is to say, that in the case of the chair, the signs we come up with are "highly diagnostic". That is, affordances can be related to the variable (degree of) *abductivity* of a configuration of signs: *a chair affords sitting* in the sense that the action of sitting is a result of a sign activity in which we perceive some physical properties (flatness, rigidity, etc.), and therefore we can ordinarily "infer" (in Peircean sense) that a possible way to cope with a chair is sitting on it[7].

According to our perspective, the original Gibsonian notion of affordance deals with those situations in which the signs and clues we can detect prompt or suggest to interact with the environment in a certain way rather than others. In this sense, we maintain that detecting affordances deals with a (semiotic) inferential activity [52]. Indeed, we may be afforded by the environment, if we can detect those signs and cues from which we may abduce the presence of a given affordance.

There are a number of points we should now make clear. They will help us clarify the idea of affordances as abductive anchors. First of all, an affordance can be considered as a *hypothetical* sign configuration. An affordance is neither the result of an abduction nor the clues (in Gibsonian terms, the information specifying the affordance). We contend that an affordance informs

---

[7] For further information about the semiotic role played by abduction in affordance detection, see Magnani [24], Magnani and Bardone [27] and Bardone [1].

us about an *environmental symptomaticity*, meaning that through a hypothetical process we recognize that the environment suggests for us and, at the same time, enables us to behave a certain way. We can do something with the environment, we can have a certain interaction, we can exploit the resources ecologically available in a certain way. This is particularly interesting if we go back to the issue we introduced in the first section about how to transform the environment from a source of constraints to a source of resources. In that section we pointed out that the human agent (like any other living organism) tries to attain a stable and functional relationship with their surroundings. We now claim that affordances *invite* us to couple with the environment by informing us about possible symptomaticities. In doing so, affordances become *anchors* transforming the environment into an *abductive texture* that helps us establish and maintain a functional relationship with it.

The second point worth mentioning is related to how the human agent regulates his relationship with his environment. We have partly answered this question. Our idea is that the human agent *abductively* regulates his relationship with the environment. That is, the human agent is constantly engaged in controlling his own behavior through continuous manipulative activity. Such manipulative activity (which is eco-cognitive one) hangs on to abductive anchors, namely, affordances that permit the human agent to take some part of the environment as local representatives of some other. So, the human agent operates in the presence of abductive anchors, namely, affordances, that stabilize environmental uncertainties by directly signaling some pre-associations between the human agent and the environment (or part of it).

The third point is that an affordance should not be confused with a resource. We have just argued that an affordance is what informs us that the environment may support a certain action so that a resource can be exploited. Going back to the example of the chair, we contend that an affordance is what informs us that we can perform a certain action in the environment (*sitting*) in order to exploit part of it as a resource (*the chair*). This is coherent with the idea introduced by Gibson and later developed by some other authors who see an affordance as an "action possibility" (see section 2).

The idea that an affordance is not a resource but rather, something that offers information about one, allows it to be seen as anything involving some eco-cognitive dimension. That is, insofar as we gain information on environmental symptomaticity to exploit a latent resource, then we have an affordance.

The framework we have developed so far allows us to stress two additional important points (cf. Magnani and Bardone [27]). Being or not being afforded by external objects is something related to

1. the various instinctual or hard-wired endowments humans are already attuned to, and
2. the various plastic cognitive endowments they have attuned to by learning or, more generally, they can make use of by using past knowledge and/or more sophisticated internal operations.

In the first case, affordances are already available and belong to the normality of the adaptation of an organism to a given ecological niche. Thus, in most cases detecting an affordance is a spontaneous abduction because this chance is already present in the perceptual endowments of human and non-human animals. That is to say, there are affordances that are somehow pre-wired and thus neurally pre-specified like action codes which are activated automatically by visual stimuli. This is, indeed, coherent to what Gibsonian conception refers to.

Organisms have at their disposal a standard (instinctual) endowment of affordances (for instance through their wired sensory system), but at the same time they can extend and modify the range of what can afford them through the appropriate cognitive abductive skills (which are more or less sophisticated). That is to say, some affordance relies on prior knowledge, which cannot be available at the time of the interaction, as already mentioned.

The fact that complex affordances would require the appropriate cognitive skills and knowledge to be detected does not mean to say they cease to be affordances, strictly speaking. In this case, familiarity is a key component making easier the detection of an affordance. The term "familiarity" covers a wide range of situations in which affordance detection becomes almost automatic – coherently with the Gibsonian view – even if it requires an additional mediation of previous knowledge and even training.

That is the case of experts. Experts take advantage of their advanced knowledge within a specific domain to detect signs and clues that ordinary people cannot detect. Here again, a patient affected by pneumonia affords a physician in a completely different way compared with that of any other non medical person. Being abductive, the process of perceiving affordances mostly relies on a continuous activity of hypothesizing which is cognition-related. That A affords B to C can also be considered from a semiotic perspective as follows: A signifies B to C. A is a sign, B the object signified, and C the interpretant. Having cognitive skills (for example knowledge content and inferential capacities but also suitable wired sensory endowments) in a certain domain enables the interpretant to perform certain abductive inferences from signs (namely, perceiving affordances) that are not available to people not possessing those tools. To ordinary people a cough or chest pain are not diagnostic, because they do not know what the symptoms of pneumonia or other diseases related to cough and chest pain are. Thus, they cannot make any abductive inference of this kind.

Humans may take advantage of additional affordances, which are not *built-in*, so to speak, but have become stabilized. Now one important point should be added in. The possibility of displaying certain abductive skills depends also on those plastic endowments, which have been created and manufactured for a specific purpose. In this sense, the possibility of creating and then stabilizing affordances relies on the various transmission and inheritance systems that humans (but also other non-human creatures) display. That is, affordances are stabilized with recourse to a variety of means using, for instance, genetic but also behavioral and/or symbolic helpers. Environments change and so do perceptive capacities, when enriched through new or higher-level cognitive skills, which go beyond the ones granted by merely instinctual levels.

This dynamics explains the fact that if affordances are usually stabilized this does not mean they cannot be modified and changed and that new ones can be formed. Affordances are also subjected to changes and modifications. Some of them can be discarded, because new configurations of the cognitive environmental niche (for example new artifacts) are invented with more powerful offered affordances. Consider, for instance, the case of blackboards. Progressively, teachers and instructors have partly replaced them with new artifacts which exhibit affordances brought about by various tools, for example, slide presentations. In some cases, the affordances of blackboards have been totally re-directed or re-used to more specific purposes. For instance, one may say that a logical theorem is still easier to be explained and understood by using a blackboard, because of its affordances that give a temporal, sequential, and at the same time global perceptual depiction to the matter.

If framed within an evolutionary dimension, the difference between the two opposing views ceases to form a radical objection to the theory of affordance. As just mentioned, we manipulate the environment and thus go beyond the merely instinctual levels adding new or higher-level cognitive skills.

6   Why and When We Are Not Afforded

In the previous section we tried to illustrate how an abductive framework is particularly fruitful in understanding the process of being afforded. The aim of this closing section is, conversely, to survey a number of cases in which people are not afforded, and then to explain why not. The interest in spelling out the reasons why people might be not afforded resides on a very simple claim: a person not being afforded does not necessarily mean that there are no affordances to detect.

We already pointed out that an affordance is a symptomatic configuration of signs informing a person about a way of exploiting the environment, meaning that the environment enables her for cognitive coupling. If this is correct, then we should distinguish between two cases: 1) when the sign configuration is not symptomatic to him/her; 2) when there are no sign configurations at all.

6.1   Hidden, Broken, and Failed Affordances

In this subsection we will briefly illustrate several types of affordances: hidden affordances, broken affordances, and failed affordances. Let us start with the first type. A person might not be afforded because she cannot make use of certain signs. Basically, this happens because he/she lacks the proper knowledge in terms of abductive skills suitable for interpreting the signs or clues available in the environment. This is the case of *hidden* affordances previously introduced in section 3.

One of the simplest examples of hidden affordances regards babies. Babies are usually not afforded by their parents' environment, simply because they are still developing the repertoire of skills necessary for detecting even simple affordances. For example, infants of age 8-12 months are not yet able to detect the affordances of a spoon for eating [11]. They simply bang or wave the spoons they are given. Indeed, using a spoon for eating requires a number of perception-action skills that the infants are still developing.

From our perspective, the infants are not yet able to detect some affordances because their repertoire of abductive skills – pre-wired by evolution – requires a period of training. That means that the brain structures underlying certain abductive skills, even though already pre-determined at genetic level, have to be further specified by the relevant experiences the infant has in an enriched environment [28].

Another interesting case in which a person might be not afforded is provided by so called *broken* affordances[8]. Broken affordances regard all those people suffering from brain damage or some other injury. The impairment may be temporary or permanent, either way they simply miss some affordances just because they lost some crucial neural mechanism underlying those skills required to detect the affordances, which otherwise would be immediately available to them. For instance, patients affected by *optic ataxia* (dorsal stream impaired) are able to name an object appropriately and recognize its function, but remain unable to grasp and locate it to exploit its affordances. However, if allowed a delay between target presentation and movement execution, they demonstrate skilled pantomimed action, relying on their visual recollection of an object instead of its actual location [53]. What they lack is therefore the ability to unconsciously adjust ongoing movements something that seems to suggest the existence of an automatic pilot, which in this case is impaired [17, p. 2750]. In this case, damage to the dorsal stream prevents the patient from making use of a number of wired abductive skills – mainly manipulative and kinesthetic – which would have allowed him to detect the affordances available[9].

An affordance might not be detected because it is poorly designed so that a person can hardly make use of it, meaning that the sign configuration is poorly symptomatic. In this case we have what we call *failed* affordances. As already mentioned, Norman introduced the distinction between *potential* affordances and *perceived* affordances precisely because he wanted to draw a line of demarcation between merely potential chances and those that are actually exploited by the user [33]. He argued that only in the second case we do have affordances. However, it is worth noting that it is nearly

---

[8] We derive this expression from Buccino *et al.* [4].

[9] For more information about the relationship between dorsal stream and ventral stream and their role for affordance detection, see Magnani and Bardone [27].

impossible to predict whether or not a user will detect or perceive the affordances constructed by the designers.

Indeed, the cooperation between designers and users can be enhanced so as to allow designers, for instance, to characterize and thus predict the most likely user reaction. However, like any other activity involving a complex communicative act, there are always design trade-offs [44] between user needs and environment constraints identifying not the optimal solution, but the satisfying one. Design trade-offs simply indicate that some affordances intended by the designer cannot be optimally constructed, meaning that an affordance, like a sign configuration, might be ambiguous to the user. If so, then the distinction between potential affordances and perceived affordances starts blurring. Let us see how our conception of affordance may be helpful to solve this problem.

As already pointed out, an affordance informs us about environmental symptomaticity. That is, an affordance is a sign configuration that is totally or scarcely ambiguous so that we promptly infer that an environmental chance is available to us. Therefore, the ambiguity of a sign configuration can be a valuable indicator to demarcate the boundaries of what can be called an affordance and what cannot. However, it is worth noting that on some occasions the user fails to detect an affordance simply because of design trade-offs. For we propose to use the term failed affordances to indicate those situations in which a sign's configuration might favor some misunderstanding between designers and users as resulting from design trade-offs. Notwithstanding the fact that they are ambiguous – to a certain extent, failed affordances are still affordances, as their ambiguity is more a result of design trade-offs than of the absence of symptomaticity. Failed affordances are common, for instance, in HCI when designers simply fail to build a mediating structure – an interface – able to communicate the intended use of an object.

6.2 Not Evolved and Not Created Affordances

In this last subsection we will consider affordances with relation to creativity and evolution. Our main claim is that a person may not be afforded by the local environment in a certain way, just because there is nothing to be afforded by. This may happen for two reasons. First of all, because a configuration of signs supporting a certain activity has not been discovered (or invented) yet. Secondly, because that configuration of signs has not yet been selected by evolution; and maybe it never will be.

As regards the first category, the fact that there is nothing to be afforded by simply means that our abilities as eco-cognitive engineers are limited. As already mentioned in the first section, the limits of the agent's adaptability correspond to his limits in creating affordances. We have previously discussed failed affordances arguing that the detection of environmental symptomaticities may be partly impaired by ambiguity due to some design trade-offs. The ambiguity of a sign configuration may lead us to misunderstand how to exploit our local environment. It may take us more time to detect some hidden chances, but that does not mean that we are totally blind about them. By contrast, sometimes we simply lack any environmental symptomaticities to make use of, because they are not present.

More generally, we may view the creation of new affordances as part of the history of technology which humans are involved in as eco-cognitive engineers. Indeed, the history of technology is a very interesting field as it offers many examples concerning affordance innovation and creation. An interesting example is provided by the so-called *paperless world*. The digital revolution has drawn our attention to fascinating scenarios in which digital technologies – like PCs or tablets – would make paper documents useless or just something from the past. Actually, the history of what we might call "the quest for the paperless world" is extremely interesting when noting the amazing creativity of humans, but also some present limitations [41]. The advent of a paperless world has partly come about, as digital technologies have created new affordances. Basically, designers (and users) have unearthed chances for new interactions, in our terminology, new symptomaticities. For instance, the paperless office has provided new affordances related to archiving, browsing and searching through documents, etc. Designers have been able to make a screen touchable enabling users to underline and manage their notes in a completely new way.

However, some affordances are still missing or, at least, some affordances cannot be reproduced in the digital. An interesting case worth mentioning here is provided by post-it notes. Post-it notes are crucial in everyone's life for managing personal information. More precisely, post-its help us store "information scraps" [2]. Information scraps allow a person to hold a variety of personal information like for instance ideas, sketches, phone numbers and reminders, like to-do lists. Generally speaking, information scraps contain all the personal information that is too unexpected or miscellaneous to be encoded or stored in other forms than scribbled notes.

However simple they may seem, post-its are amazingly useful as they contain a number of affordances that allow us to have access to a kind of information that otherwise would be completely out of reach. For instance, a post-it is as ready-to-use as a corner of a sheet of paper. But unlike the corner it is 1) portable, as it can be put in a pocket, 2) it can be stuck on any other object like a computer monitor, a door, a window, etc., or 3) it can be handed to somebody else.

Although there are a number of computer programs that have been appropriately designed to help people take quick notes, the conclusion reached by Bernstein *et al.* [2] is that at the moment there are no computer applications able to surpass post-its. By definition information scraps are taken down on the fly, and usually a computer program – whatever it is – it takes too long.

For instance, even assuming that we have our laptop already working and available to take a note, it forces us, to make a type assignment or assign a category in order to make our note easily retrievable [2]. But this is precisely what we do not need to do when using a post-it. We do not need to think about how to categorize a certain piece of information or about how to set a deadline. We simply assign a particular meaning and a particular place to a particular pad of post-it so that it will be easily accessible and retrievable later on. Plain and simple, no computer applications afford us the way post-its do. In our terminology, designers have not yet figured out how to embed in the digital those symptomaticies signaling the affordances of a post-it. Here again the limits of being afforded are the limits of human beings as eco-cognitive engineers.

As already mentioned, there is an additional reason why we are not afforded, that is, when an affordance or a set of affordances have not been secured by evolution. In section 4 we argued that some affordances do not seem to require any mediation of knowledge, because they are already made available by evolution in terms of pre-wired and thus neurally pre-specified responses to the environment. Let us make a very simple example of affordances that are somehow secured by evolution, the case of the *opposable thumb*. In this case, some hominids developed a particular adaptation enabling them to oppose the thumb to the palmar side of the forefinger. That was a fundamental step in evolution: the opposable thumb made available a number of affordances that were simply unavailable before, like, for instance, grasping, handing, throwing, and so on. Those species that had not develop the opposable thumb simply lack certain affordances. For them, a tree branch is not graspable, they cannot pick something up and then throw it to hit an animal, and they cannot climb a tree.

Another example is that one of the human-chimpanzee common ancestors, the so-called *Ardipithecus*, had a *grasping foot*. However the hominid foot then evolved so that our later ancestors lost it. Indeed, that loss made a number of affordances unavailable, like those related, for example, with climbing trees or locomotion. In fact we still miss them and we can only use our feet for walking or standing, but not for grasping. This has a very simple consequence in that all the objects around us are designed bearing in mind that we do not push a button or type using our feet and we would simply miss such affordances, because they have been put out of reach by evolution.

More generally, what the two examples point to is that evolution makes an organism develop different affordances. That is, there are certain sign configurations that become meaningful for one organism, but not for others: evolution operates in such a way as to allow certain organisms not to be afforded by certain things. In this case, an achievement (that we can consider an adaptation) was not established, simply because it was not profitable for the organism.

Conclusion

In this paper we have been dealing with the notion of affordance. Our interest stems from the fact that it potentially contributes to understanding and further developing some central issues concerning the way human beings distribute their cognition. A theory of affordance posits that human beings have direct access to the chances that are ecologically made available by the activity of environmental selection, namely, niche construction.

The idea that we have direct access to environmental chances presents some problems that we have tried to discuss in this paper. According to the traditional conception of affordance, an affordance requires what Gibson called direct perception. The idea is that an affordance in order to termed as such should be perceived without mediation of any sort. We have argued in this paper that this idea is misleading, because it impedes the possibility to conceptually extend the notion of affordance beyond the domain of visual perception.

Our proposal relied on the notion of abduction. More precisely, we put forward an alternative concept of affordance based on the notion of abduction. We argued that an affordance can be considered as a hypothetical sign configuration from which one can infer that the environment supports a certain interaction rather than another. So, affordances can be considered as abductive anchors allowing us to establish a functional relationship with our environment. That is, our environment ceases to be simply a source of constraints to our activities, but it becomes a source of chances to exploit.

References

1. Bardone, E.: Seeking Chances. From Biased Rationality to Distributed Cognition (2010) (Forthcoming)
2. Bernstein, M., van Kleek, M., Karger, D., Schraefel, M.: Information scraps: How and why information eludes our personal information management tools. ACM Transactions on Information Systems 26(4), 1–46 (2008)
3. Brunswik, E.: Organismic achievement and environmental probability. Psychological Review 50, 255–272 (1943)
4. Buccino, G., Sato, M., Cattaneo, L., Roda, F., Riggio, L.: Broken affordances, broken objects: A TMS study. Neuropsychologia 47, 3074–3078 (2009)
5. Chemero, A.: An outline of a theory of affordances. Ecological Psychology 15(2), 181–195 (2003)
6. Clark, A.: Natural-Born Cyborgs. Minds, Technologies, and the Future of Human Intelligence. Oxford University Press, Oxford (2003)
7. Clark, A.: Supersizing the Mind. Embodiment, Action, and Cognitive Extension. Oxford University Press, Oxford/New York (2008)
8. Day, R.L., Laland, K., Odling-Smee, J.: Rethinking adaptation. The niche-construction perspective. Perspectives in Biology and Medicine 46(1), 80–95 (2003)

9. Figueredo, A., Hammond, K., McKiernan, E.: A Brunswikian evolutionary developmental theory of preparedness and plasticity. Intelligence 34, 211–227 (2006)
10. Gaver, W.W.: Technology affordances. In: CHI 1991 Conference Proceedings, pp. 79–84 (1991)
11. Gibson, E., Pick, A.: An Ecological Approach to Perceptual Learning and Development. Oxford University Press, Oxford (2000)
12. Gibson, J.J.: The Ecological Approach to Visual Perception. Houghton Mifflin, Boston (1979)
13. Goldstone, R., Landy, D., Son, J.: The education of perception. Topics in Cognitive Science 2, 265–284 (2010)
14. Gooding, D.: Seeing the forest for the trees: Visualization, cognition and scientific inference. In: Gorman, M., Gooding, D., Tweney, R., Kincannon, A. (eds.) Scientific and Technological Thinking, pp. 173–217. Lawrance Erlbaum Publishers, Mahwah (2004)
15. Greeno, J.G.: Gibson's affordances. Psychological Review 101(2), 336–342 (1994)
16. Hammond, K., Hamm, R., Grassia, J., Pearson, T.: Direct comparison of intuitive and analytical cognition in expert judgment. IEEE Transactions on Systems, Man, and Cybernetics, SMC 17, 753–770 (1987)
17. Himmelbach, M., Karnath, H.O., Perenin, M., Franz, V., Stockmeier, K.: A general deficit of the 'automatic pilot' with posterior parietal cortex lesions? Neuropsychologia 44(13), 2749–2756 (2006)
18. Hoffman, D.D.: Visual Intelligence: How We Create What We See. Norton, New York (1998)
19. Hutchins, E.: Cognition in the Wild. The MIT Press, Cambridge (1995)
20. Kirlik, A.: On Gibson's review of Brunswik. In: Hammond, K.R., Steward, T.R. (eds.) The Essential Brunswik. Beginnings, Explications, Applications, pp. 238–242. Oxford University Press, Oxford (2001)
21. Laland, K., Brown, G.: Niche construction, human behavior, and the adaptive-lag hypothesis. Evolutionary Anthropology 15, 95–104 (2006)
22. Magnani, L.: Abductive reasoning: philosophical and educational perspectives in medicine. In: Evans, D.A., Patel, V.L. (eds.) Advanced Models of Cognition for Medical Training and Practice, pp. 21–41. Springer, Berlin (1992)
23. Magnani, L.: Abduction, Reason, and Science. Processes of Discovery and Explanation. Kluwer Academic/Plenum Publishers, New York (2001)
24. Magnani, L.: Mimetic minds. Meaning formation through epistemic mediators and external representations. In: Loula, A., Gudwin, R., Queiroz, J. (eds.) Artificial Cognition Systems, pp. 327–357. Idea Group Publishers, Hershey (2006)
25. Magnani, L.: Morality in a Technological World. Knowledge as Duty. Cambridge University Press, Cambridge (2007)
26. Magnani, L.: Abductive Cognition. The Epistemological and Eco-Cognitive Dimensions of Hypothetical Reasoning. Springer, Heidelberg (2009)
27. Magnani, L., Bardone, E.: Sharing representations and creating chances through cognitive niche construction. The role of affordances and abduction. In: Iwata, S., Oshawa, Y., Tsumoto, S., Zhong, N., Shi, Y., Magnani, L. (eds.) Communications and Discoveries from Multidisciplinary Data, pp. 3–40. Springer, Berlin (2008)
28. Marcus, G.: The Birth of the Mind: How a Tiny Number of Genes Creates the Complexities of Human Thought. Basic Books, New York (2004)

29. Marr, D.: Vision. Freeman, San Francisco (1982)
30. McGrenere, J., Ho, W.: Affordances: clarifying and evolving a concept. In: Proceedings of Graphics Interface, Montreal, Quebec, Canada, May 15-17, pp. 179–186 (2000)
31. Natsoulas, T.: To see is to perceive what they afford: James J. Gibon's concept of affordance. Mind and Behavior 2(4), 323–348 (2004)
32. Norman, D.: The Design of Everyday Things. Addison-Wesley, New York (1988)
33. Norman, D.: Affordance, conventions and design. Interactions 6(3), 38–43 (1999)
34. Peirce, C.S.: Collected Papers of Charles Sanders Peirce. Harvard University Press, Cambridge (1931-1958). Vols. 1-6, Hartshorne, C. and Weiss, P., eds.; vols. 7-8, Burks, A. W., ed
35. Rader, N., Vaughn, L.: Infant reaching to a hidden affordance: evidence for intentionality. Infant Behavior and Development 23, 531–541 (2000)
36. Raftopoulos, A.: Is perception informationally encapsulated? The issue of theory-ladenness of perception. Cognitive Science 25, 423–451 (2001)
37. Raftopoulos, A.: Reentrant pathways and the theory-ladenness of perception. Philosophy of Science 68, S187–S189 (2001), Proceedings of PSA 2000 Biennal Meeting
38. Reed, E.: Encountering the World. Oxford University Press, Oxford (1996)
39. Rock, I.: Inference in perception. In: PSA. Proceedings of the Biennial Meeting of the Philosophy of Science Association, vol. 2, pp. 525–540 (1982)
40. Scarantino, A.: Affordances explained. Philosophy of Science 70, 949–961 (2003)
41. Sellen, A., Harper, R.: The Myth of the Paperless Office. The MIT Press, Cambridge (2002)
42. Shelley, C.: Visual abductive reasoning in archaeology. Philosophy of Science 63(2), 278–301 (1996)
43. Stoffregen, T.A.: Affordances as properties of the animal-environment system. Ecological Psychology 15(3), 115–134 (2003)
44. Sutcliffe, A.: Symbiosis and synergy? Scenarios, task analysis and reuse of HCI knowledge. Interacting with Computers 15, 245–263 (2003)
45. Thagard, P.: Computational Philosophy of Science. The MIT Press, Cambridge (1988)
46. Thagard, P.: How brains make mental models. SCI, vol. 314. Springer, Heidelberg (2010)
47. Turvey, M., Shaw, R.: Toward an ecological physics and a physical psychology. In: Solso, R., Massaro, D. (eds.) The Science of the Mind: 2001 and Beyond, pp. 144–169. Oxford University Press, Oxford (2001)
48. Vicente, K.J.: Beyond the lens model and direct perception: toward a broader ecological psychology. Ecological Psychology 15(3), 241–267 (2003)
49. Warren, W.: Constructing an econiche. In: Flach, J., Hancock, P., Caird, J., Vicente, K.J. (eds.) Global Perspective on the Ecology of Human-Machine Systems, pp. 210–237. Lawrence Erlbaum Associates, Hillsdale (1995)
50. Wells, A.J.: Gibson's affordances and Turing's theory of computation. Ecological Psychology 14(3), 141–180 (2002)
51. Williams, G.: Gaia, nature worship and biocentric fallacies. The Quarterly Review of Biology 67, 479–486 (1992)

52. Windsor, W.L.: An ecological approach to semiotics. Journal for the Theory of Social Behavior 34(2), 179–198 (2004)
53. Young, G.: Are different affordances subserved by different neural pathways? Brain and Cognition 62, 134–142 (2006)
54. Zhang, J., Patel, V.L.: Distributed cognition, representation, and affordance. Pragmatics & Cognition 14(2), 333–341 (2006)

A Model-Based Reasoning Approach to Prevent Crime

Tibor Bosse and Charlotte Gerritsen

Abstract. Within the field of criminology, one of the main research interests is the analysis of the *displacement of crime*. Typical questions that are important in understanding the displacement of crime are: When do hot spots of high crime rates emerge? Where do they emerge? And, perhaps most importantly, how can they be prevented? In this paper, an agent-based simulation model of crime displacement is presented, which can be used not only to *simulate* the spatio-temporal dynamics of crime, but also to *analyze* and *control* those dynamics. To this end, methods from Artificial Intelligence and Ambience Intelligence are used, which are aimed at developing intelligent systems that monitor human-related processes, and provide appropriate support. More specifically, an explicit domain model of crime displacement has been developed, and, on top of that, model-based reasoning techniques are applied to the domain model, in order to analyze which environmental circumstances result in which crime rates, and to determine which support measures are most appropriate. The model can be used as an analytical tool for researchers and policy makers to perform thought experiments, i.e., to shed more light on the process under investigation, and possibly improve existing policies (e.g., for surveillance). The basic concepts of the model are defined in such a way that it can be directly connected to empirical information.

1 Introduction

Within the field of criminology, one of the main research interests is the analysis of the *displacement of crime* [14, 20, 26]. Typically, certain locations

Tibor Bosse and Charlotte Gerritsen
Vrije Universiteit Amsterdam, Department of Artificial Intelligence,
De Boelelaan 1081a, 1081 HV Amsterdam, the Netherlands
e-mail: `tbosse@few.vu.nl, cg@few.vu.nl`

L. Magnani et al. (Eds.): Model-Based Reasoning in Science & Technology, SCI 314, pp. 159–177.
springerlink.com   © Springer-Verlag Berlin Heidelberg 2010

in a city seem to attract many criminal activities, but only for a short period. These locations where many crimes occur are called *hotspots*. Questions that are important in understanding the displacement of crime are: When do hot spots of high crime rates emerge? Where do they emerge? And, perhaps most importantly, how can they be prevented? In recent years, computational modeling and simulation have proved to be a useful instrument to answer such questions.

When investigating the literature on computational modeling of displacement of crime, different computational modeling approaches can be distinguished. Among the approaches that are applied, one can find agent-based modeling [6, 9, 13, 24], population-based modeling [9], cellular automata [19, 22], different spatial analysis techniques [18], and evolutionary computing techniques [24].

Although they all share the aim of investigating crime displacement, the perspectives taken in the above papers differ. For example, some authors try to develop simulation models of crime displacement in existing cities, which can be directly related to real world data (e.g., [22]), whereas others deliberately abstract from empirical information (e.g., [9]). The idea behind the latter perspective is that the simulation environment is used as an analytical tool, mainly used by researchers and policy makers, for thought experiments, to shed more light on the process under investigation, and perhaps improve existing policies (e.g., for surveillance) [16]. Also, some authors take an intermediate point of view (e.g., [6, 23]). They initially build their simulation model to study the phenomenon per se, but define its basic concepts in such a way that it can be directly connected to empirical information, if this becomes available.

This intermediate perspective is also taken in the current paper. Its main goal is to develop an agent-based simulation model of crime displacement, which can be used not only to *simulate* the spatio-/temporal dynamics of crime, but also to *analyze* and *control* those dynamics. This second aim distinguishes it from most existing approaches, which are mainly descriptive (instead of prescriptive).

To achieve this goal, we make use of techniques from Artificial Intelligence, and in particular from Ambient Intelligence (AmI). Ambient Intelligence [1, 2, 25] represents a vision of the future where humans will be surrounded by pervasive and unobtrusive electronic environments, which are sensitive, and responsive to their needs. In order to develop such intelligent environments, Bosse et al. [10] introduced a methodology to endow intelligent systems with the possibility to reason explicitly about the mental and physical states of humans. In the current paper, this methodology is reused in order to develop an intelligent system that reasons about crime displacement.

More specifically, this paper will first describe the development of an explicit *domain model* of crime displacement (which describes displacement in terms of states of the world over time, and transitions between these states). On top of that, model-based reasoning techniques (cf. [4]) will be applied to the domain model, in order to analyze which environmental circumstances result in which crime rates, and to determine which support measures are most appropriate (cf. [5]). Hence, both an *analysis model* and a *support model* will be developed.

The outline of this paper is as follows. First, in Section 2, some background information about the area of Ambient Intelligence will be provided. Next, Section 3 will introduce the basic methodology for the development of intelligent human-aware model-based systems that will be used in this paper. Based on this methodology, Section 4, 5 and 6 will introduce, respectively, the domain model, analysis model and support model for crime displacement. Section 7 will present some preliminary simulation results, and Section 8 will conclude the paper with a discussion.

2 Ambient Intelligence

Ambient Intelligence [1, 2, 25] represents a vision of the future where human beings will be surrounded by pervasive and unobtrusive electronic environments, which are sensitive, and responsive to their needs. Such an environment has a certain degree of awareness of the presence and states of living creatures in it, and supports their activities. It analyzes their behavior, and may anticipate on it. Ambient Intelligence (AmI) integrates concepts from ubiquitous computing and Artificial Intelligence (AI) with the vision that technology will become invisible, embedded in our natural surroundings, present whenever we need it, attuned to the humans' senses, and adaptive to them. In an ambient intelligent environment, people are surrounded by networks of embedded intelligent devices that can sense their state, anticipate, and when relevant adapt to their needs. Therefore, the environment should be able to determine which actions have to be undertaken in order to keep this state optimal.

For this purpose, acquisition of sensor information about humans and their functioning is an important factor. However, without adequate additional *knowledge* for analysis of this information, the scope of such applications is limited. As argued by Bosse et al. [10], AmI applications can show a more human-like understanding and base personal care on this understanding when they are equipped with knowledge about the relevant physiological, psychological, and/or social aspects of human functioning. For example, this may concern elderly people, patients depending on regular medicine usage, surveillance, penitentiary care, psychotherapeutical/selfhelp communities, but also, for example, humans in highly demanding tasks such as warfare officers, air traffic controllers, crisis and disaster managers, and humans in space missions; e.g., [17].

Within human-directed scientific areas, such as cognitive science, psychology, neuroscience and biomedical sciences, models have been and are being developed for a variety of aspects of human functioning. If such models of human processes are represented in a formal and computational format, and incorporated in the human environment in devices that monitor the physical and mental state of the human, then such devices are able to perform a more in-depth analysis of the human's functioning. This can result in an environment that may more effectively affect the state of humans by undertaking actions in a knowledgeable manner that improve their wellbeing and performance. For example, the workspaces of naval officers may include systems that, among others, track their eye movements and characteristics of incoming stimuli (e.g., airplanes on a radar screen), and use this information in a computational model that is able to estimate where their attention is focussed at. When it turns out that an officer neglects parts of a radar screen, such a system can either indicate this to the person, or arrange on the background that another person or computer system takes care of this neglected part. Note that for a radar screen it would also be possible to make static design changes, for example those that improve situation awareness (e.g. picture of the environment, [28]). However, as different circumstances might need a different design, the advantage of a dynamic system is that the environment can be adapted taking both the circumstances and the real-time behavior of the human into account.

In applications of this type, an ambience is created that has a better understanding of humans, based on computationally formalized knowledge from the human-directed disciplines. The use of knowledge from these disciplines in Ambient Intelligence applications is beneficial, because it allows taking care in a more sophisticated manner of humans in their daily living in medical, psychological and social respects. In more detail, content from the domain of human-directed sciences, among others, can be taken from areas such as medical physiology, health sciences, neuroscience, cognitive psychology, clinical psychology, psychopathology, sociology, criminology, and exercise and sport sciences.

Although it does not directly fit in the description of Ambient Intelligence, the system envisioned by the current paper has a number of similarities with the types of systems sketched above. That is, it will also take information about humans and their dynamics as input (namely the spatial distribution of individuals over the city, and information about crime rates), it will also be equipped with (formalized) knowledge from human-directed disciplines (in this case criminological knowledge about crime displacement), and it will also generate support measures as output (i.e. advice to reduce crime). Thus, in order to develop the intelligent system for reasoning about crime displacement, it makes sense to reuse approaches from the Ambient Intelligence area. In particular, the methodology from [10] is used, which is introduced below.

3 Methodology

In this section, the adopted approach to develop intelligent human-aware systems is presented in detail, cf. [10]. Here, human-aware is defined as being able to analyze and estimate what is going on in the human's mind (a form of mindreading) and in his or her body (a form of bodyreading). Input for these processes are observed information about the human's state over time, and dynamic models for the human's physical and mental processes. For the mental side, such a dynamic model is sometimes called a Theory of Mind (e.g., [3]) and may cover, for example, emotion, attention, intention, and belief. Similarly for the human's physical processes, such a model relates, for example, to skin conditions, heart rates, and levels of blood sugar, insulin, adrenalin, testosterone, serotonin, and specific medicines taken. Note that different types of models are needed: physiological, neurological, cognitive, emotional, social, as well as models of the physical and artificial environment[1].

A framework can be used as a template for the specific class of Ambient Intelligence applications as described. The structure of such an ambient software and hardware design can be described in an agent-based manner at a conceptual design level and can be given generic facilities built in to represent knowledge, models and analysis methods about humans, for example (see Figure 1):

- human state and history models
- environment state and history models
- profiles and characteristics models of humans
- ontologies and knowledge from biomedical, neurological, psychological and/or social disciplines
- dynamic process models about human functioning
- dynamic environment process models
- methods for analysis on the basis of such models

Examples of useful analysis methods are voice and skin analysis with respect to emotional states, gesture analysis, and heart rate analysis. The template can include slots where the application-specific content can be filled to get an executable design for a working system. The analysis method used in the current paper mainly addresses displacement of crime, i.e., it calculates how a certain distribution of persons over space would lead to movement of criminal activities.

A general approach for embedding knowledge about the interaction between the environment and the human(s) in Ambient Intelligence applications is to integrate dynamic models of this interaction (i.e. a model of the *domain*) into the application. This integration takes place by embedding domain models in certain ways within agent models of the intelligent application. By incorporating domain models within an agent model, the intelligent agent gets an understanding of the processes of its surrounding environment, which is a solid basis for knowledgeable intelligent behavior. Three different ways to integrate domain models within agent models can be distinguished.

---

[1] In this paper, the main focus is on social/environmental states and models, i.e., locations of persons, and decisions to move to other location. Nevertheless, the model is sufficiently generic to be extended with the other types of states as well.

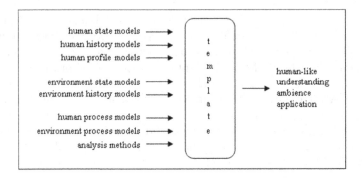

Fig. 1 Framework to develop intelligent human-aware systems.

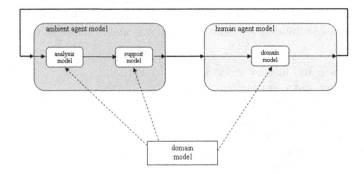

Fig. 2 Overview of the multi-agent system architecture.

A most simple way is to use a domain model that specifically models human behavior in the following manner:

- *domain model directly used as agent model*
  In this case a domain model that describes human processes and behavior is used directly as an agent model, in order to simulate human behavior. Note that here the domain model and agent model refer to the same agent.

Such an agent model can be used in interaction with other agent models, in particular with *ambient agent models* to obtain a test environment for simulations. For this last type of (artificial) agents, domain models can be integrated within their agent models in two different ways, in order to obtain one or more (sub)models; see Figure 2. Here the solid arrows indicate information exchange between processes (data flow) and the dotted arrows the integration process of the domain models within the agent models.

As shown in Figure 2, the following submodels can be obtained based on a domain model:

- *analysis model*
  To perform analysis of the human's states and processes by reasoning based on observations (possibly using specific sensors) and the domain model.

- *support model*
  To generate support for the human by reasoning based on the domain model.

Note that here the domain model that is integrated refers to one or more human agents, whereas the agent model in which it is integrated refers to an artificial agent (the intelligent system). In the following sections, this methodology will be applied to the domain of crime displacement. First a domain model is presented which represents the spatio-temporal dynamics of crime. Next, an analysis model is presented, which is able to reason about the domain model in order to predict crime rates for particular situations. And finally, a support model is presented, which is able to suggest to the user the most appropriate measures to reduce crime rates. For example, in case the analysis model predicts that the crime rates at the railway station will increase with 20% in the next year, and that these rates can be kept stable by increasing the amount of police by 5%, then it may propose to invest in 5% more police forces.

4 Domain Model

This section presents the domain model for crime displacement. The important concepts used are introduced in Section 4.1, and their formalization is described in Section 4.2.

4.1 Crime Displacement

As explained in the introduction, most large cities in the world contain a number of *hot spots*, i.e., locations where the majority of the crimes occur [15, 26]. Such locations may vary from railway stations to shopping malls. These hot spots usually have several things in common, among which the presence of many passers-by (which makes the location attractive for criminals) and the lack of adequate surveillance. However, after a while the situation often changes: the criminal activities shift to another location. This may be caused by improved surveillance systems (such as cameras) at that location, by an increased number of police officers, or because the police changed their policy.

Another important factor in explaining crime displacement is the *reputation* of specific locations in a city [20]. This reputation may be a cause of crime displacement, as well as an effect. For example, a location that is known for its high crime rates usually attracts police officers [15], whereas most citizens will be more likely to avoid it [27]. As a result, the amount of criminal activity at such a location will decrease, which affects its reputation again.

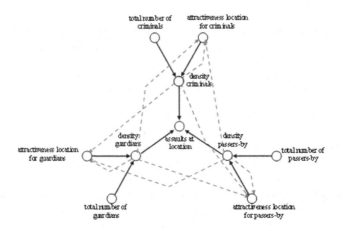

Fig. 3 Interaction between criminals, guardians, and passers-by.

To summarize, in order to model the process of crime displacement, several aspects are important. First, one should have information about the *total number* of agents in the different groups involved, i.e., the number of *criminals*, number of *guardians*, and number of *passers-by*. Next, it is assumed that the world (or city) that is addressed can be represented in terms of a number of different *locations*. It is important to know how many agents of each type are present at each location: the *density* of criminals, guardians, and passers-by. Furthermore, to describe the movement of the different agents from one location to another, information about the *reputation* (or *attractiveness*) of the locations is needed. This attractiveness is different for each type of agent. For example, passers-by like locations where it is safe, e.g. locations where some guardians are present and no criminals. On the other hand, guardians are attracted by places where a lot of criminals are present, and criminals like locations where there are many passers-by and no guardians. Finally, to be able to represent the idea of hot spots, the *number of assaults* per location is modeled. The idea is that more assaults take place at locations where there are many criminals and passers-by, and few guardians, cf. the Routine Activity Theory by [14].

The interaction between the concepts introduced above is visualized in Figure 3[2]. This figure depicts the influences between the different groups at

---

[2] Note that Figure 3 does not depict the influence of some basic attractiveness of a location for certain groups (i.e., an attractiveness that is independent of the distribution of agents at the location). For the sake of readability, this notion has been left out of the picture, but it often plays a role in reality. For instance, locations like a railway station will be visited more often by passers-by than other locations, simply because people need to go there to reach their desired destination. Therefore, the notion of basic attractiveness will also be considered in this paper.

Table 1 Variables used in the domain model.

Name	Explanation
c	Total number of criminals
g	Total number of guardians
p	Total number of passers by
$c(L, t)$	Density of criminals at location L at time t.
$g(L, t)$	Density of guardians at location L at time t.
$p(L, t)$	Density of passers-by at location L at time t.
$\beta(L, a, t)$	Attractiveness of location L at time t for type a agents: c (criminals), p (passers-by), or g (guardians)
$ba(L, a, t)$	Basic attractiveness of location L at time t for type a agents: c (criminals), p (passers-by), or g (guardians)
$assault\_rate(L, t)$	Number of assaults taking place at location L per time unit.

one location. Here, the circles denote the concepts that were mentioned above in italics, and the arrows indicate influences between concepts (influences on attractiveness have been drawn using dotted arrows to enhance readability).

4.2 Formalization

In order to build the domain model for crime displacement, the concepts that were introduced above (in italics) are formalized in terms of mathematical variables. The variable names that are used are summarized in see Table 1.

Next, a number of mathematical equations are introduced to represent the causal relations between these variables. Most of these ideas are taken over from [8] (and [7, 9]). First, the calculation of the number of agents at a location is done by determining the movement of agents that takes place based on the attractiveness of the location. For example, for criminals, the following formula is used:

$$c(L, t + \Delta t) = c(L, t) + \eta \cdot (\beta(L, c, t) \cdot c - c(L, t))\Delta t$$

This expresses that the density $c(L, t + \Delta t)$ of criminals at location L on time $t + \Delta t$ is equal to the density of criminals at the location at time t plus a constant η (expressing the rate at which criminals move per time unit) times the movement of criminals from t to $t + \Delta t$ from and to location L, multiplied by Δt. Here, the movement of criminals is calculated by multiplying the relative attractiveness $\beta(L, c, t)$ of the location (compared to the other locations) for criminals with the total number c of criminals (which is constant). From this, the density of criminals at the location at t is subtracted, resulting in the change of the number of criminals for this location. For passers-by, a similar formula is used:

$$p(L, t + \Delta t) = p(L, t) + \eta \cdot (\beta(L, p, t) \cdot p - p(L, t))\Delta t$$

However, as opposed to [9], the movement of the guardians is not (necessarily) modeled using this formula. Instead, to represent guardian movement, different strategies can be filled in (see Section 4).

Next, the attractiveness of a location can be expressed based on some form of reputation of the location for the respective type of agents. Several variants of a reputation concept can be used. The only constraint is that it is assumed to be normalized such that the total over the locations equals 1. An example of a simple reputation concept is based on the densities of agents, as expressed below.

$$\beta(L, c, t) = p(L, t)/p \qquad \text{for criminals}$$

$$\beta(L, p, t) = g(L, t)/g \qquad \text{for passers-by}$$

This expresses that criminals are more attracted to locations with higher densities of passers-by, whereas passers-by are attracted more to locations with higher densities of guardians. This definition of reputation is used in [9]. Although this definition is simple, which makes the model well suited for mathematical analysis, it is not very realistic. To solve this problem, in this paper, the following linear combinations of densities are used[3]:

$$\beta(L, c, t) = \beta_{c1} \cdot (1 - g(L, t)/g) + \beta_{c2} \cdot p(L, t)/p + \beta_{c3} \cdot \text{ba}(L, c, t)$$

$$\beta(L, p, t) = \beta_{p1} \cdot (1 - c(L, t)/c) + \beta_{p2} \cdot g(L, t)/g + \beta_{p3} \cdot \text{ba}(L, p, t)$$

This expresses that criminals are repelled by guardians, but attracted by passers-by. Similarly, passers-by are repelled by criminals, but may be attracted by guardians. In addition, for each type of agent some basic attractiveness can be defined. The weight factors (β_{xy}, which may also be 0) indicate the relative importance of each aspect. Again, for the guardians no formula is specified, since this depends on the guardian movement strategy that is selected.

Finally, to measure the assaults that take place per time unit, also different variants of formulae can be used (see [9]). In this paper, the following is used:

$$\text{assault\_rate}(L, t) = \max(c(L, t) \cdot p(L, t) - \gamma \cdot g(L, t), 0)$$

Here, the assault rate at a location at time t is calculated as the product of the densities of criminals and passers-by, minus the product of the guardian density and a constant γ, which represents the capacity of guardians to avoid an assault. The motivation behind this is that the maximum amount of assaults that can take place at a location is $c(L, t) \cdot p(L, t)$, but that this number

---

[3] Note that these attractiveness formulae are not normalized yet. To ensure that the values stay between 0 and 1, each attractiveness value is divided by the sum of the values over all locations. Moreover, the influence by agents from the same group is not considered.

can be reduced by the effectiveness of the guardians (which corresponds exactly to the Routine Activity Theory). In principle, this assault rate can become less than 0 (the guardians can have a higher capacity to stop assaults than the criminals have to commit them); therefore the maximum can be taken of 0 and the outcome described above. Based on this assault rate, the total (cumulative) amount of assaults that take place at a location is calculated as:

$$\text{total\_assaults}(L, t + \Delta t) = \text{total\_assaults}(L, t) + \text{assault\_rate}(L, t)\Delta t$$

Although the domain model is presented here in a purely mathematical notation, its actual implementation has been done in the agent-based modeling environment LEADSTO [11]. This environment is well suited for the current purposes, since it integrates both qualitative, logical aspects and quantitative, numerical aspects. The basic building blocks of LEADSTO are executable rules of the format $\alpha \twoheadrightarrow \beta$, which indicates that state property α leads to state property β. Here, α and β can be (conjunctions of) logical and numerical predicates.

5 Analysis Model

This section extends the domain model introduced in the previous section to an analysis model. The analysis model (and the support model, see next section) are created by taking the domain model as a basis, and applying model-based reasoning to it. In particular, two types of reasoning are applied (taken from [4]): forward and backward reasoning. In short, these types of reasoning make use of the following kinds of (simplified) rules (where X and Y are variables in a model, e.g. as in Figure 3):

- If we believe X and believe that Y depends on X, then we also believe Y.

  $\text{belief}(X) \wedge \text{belief}(\text{depends\_on}(Y, X)) \, \text{belief}(Y)$

- If we desire Y and believe that Y depends on X, then we also believe X.

  $\text{desire}(Y) \wedge \text{belief}(\text{depends\_on}(Y, X)) \, \text{desire}(X)$

To illustrate the idea, assume that we focus on an existing city, of which the average number of criminals, guardians, and passers-by at the different locations is known (to a certain extent). Thus, specific numbers can be assigned to the variables *density_criminals*, *density_guardians*, and *density_passers_by* in Figure 3 (which correspond to $c(L, t)$, $g(L, t)$, and $p(L, t)$ in Table 1). Then, via forward reasoning (the first rule shown above), the model can predict how the number of assaults will change over time.

One step further, instead of taking the actual densities of guardians at the different locations, the analysis model can also be used to investigate how the crime rates would change in case the densities of guardians were different. To this end, the analysis model is extended with the possibility to specify particular crime prevention *strategies*. The idea is that, in addition to the rules that govern the behavior of criminals and passers-by, the behavior of the guardians can be specified by selecting one out of multiple strategies.

In current practice, the crime prevention policies that are applied by law enforcement agencies are – mostly – reactive [12, 15]. That is, these agencies often only increase the level of guardianship at locations where crimes have been committed in the past. As a consequence, this often means that such a decision is made too late, because the damage has already been done. Instead, we hypothesize that a more anticipatory strategy (e.g., a strategy to invest in more guardians at locations where one predicts that a hot spot *will emerge*) may be more efficient.

To be able to investigate this, the analysis model in equipped with multiple strategies for movement of guardians (varying from reactive to anticipatory, and combinations of the two). The selected strategies are based on [7, 8], in which they were already tested against some initial scenarios. In total, the analysis model contains ten different strategies (see also Table 2):

1. The first strategy is a *baseline* strategy. In this case guardians do not move at all. Their density at the different locations remains stable over time.

2. The second strategy (called *reactive 1*) states that the amount of guardians that move to a new location is proportional to the density of criminals at that location.

3. The third strategy (*reactive 2*) states that the amount of guardians that move to a new location is proportional to the percentage of the assaults that have recently taken place at that location.

4. The fourth strategy (*reactive 3*) states that the amount of guardians that move to a new location is proportional to the percentage of all assaults that have taken place so far at that location.

5. The fifth strategy (*reactive 4*) states that the amount of guardians that move to a new location is proportional to the density of passers-by at that location.

6. In the sixth strategy (*anticipate 1*), the amount of guardians that move to a new location is proportional to the density of criminals they expect that location to have in the future.

7. In the seventh strategy (*anticipate 2*), the amount of guardians that move to a new location is proportional to the density of passers-by they expect that location to have in the future.

8. In the eighth strategy (*anticipate 3*), the amount of guardians that move to a new location is proportional to the amount of assaults they expect that will take place at that location in the future. This predicted amount of assaults is approximated by taking the average of the expected densities of criminals and passers-by.

9. The ninth strategy (*hybrid 1*) is a combination of *reactive 2* and *anticipate 2*. Here, the amount of guardians that move to a new location is the average of the amounts of guardians determined by those two strategies.

10. The tenth strategy (*hybrid 2*) is a combination of *reactive 3* and *anticipate 2*. Here, the amount of guardians that move to a new location is the average of the amounts of guardians determined by those two strategies.

To formalize these strategies, the following formula is used:

$$g(L, t + \Delta t) = g(L, t) + \eta \cdot \sigma(L, t) \Delta t$$

This formula is similar to the formulae used for criminals and passers-by, but the amount of guardians that move per time unit is indicated by the factor $\sigma(L, t)$, which depends on the chosen strategy. The different definitions of σ are shown in Table 2. For example, for the baseline strategy, $\sigma(L, t) = 0$, which means that the amount of guardians at time point $t + \Delta t$ is equal to the amount at t.

Table 2 Guardian movement strategies considered by the analysis model.

Strategy	Formalization of $\sigma(L, t)$
baseline	0
reactive 1	$(c(L, t)/c) \cdot g - g(L, t)$
reactive 2	$\text{aar}(L, t) \cdot g - g(L, t)$
reactive 3	$\text{taar}(L, t) \cdot g - g(L, t)$
reactive 4	$(p(L, t)/p) \cdot g - g(L, t)$
anticipate 1	$(c(L, t) + \eta 2 \cdot (\beta(L, c, t) \cdot c - c(L, t)) \cdot \Delta t)/c \cdot g - g(L, t)$
anticipate 2	$(p(L, t) + \eta 2 \cdot (\beta(L, p, t) \cdot p - p(L, t)) \cdot \Delta t)/p \cdot g - g(L, t)$
anticipate 3	$((c(L, t) + \eta 2 \cdot (\beta(L, c, t) \cdot c - c(L, t)) \cdot \Delta t)/c +$ $(p(L, t) + \eta 2 \cdot (\beta(L, p, t) \cdot p - p(L, t)) \cdot \Delta t)/p)/2 \cdot g - g(L, t)$
hybrid 1	$((\text{aar}(L, t) \cdot g - g(L, t)) +$ $(p(L, t) + \eta 2 \cdot (\beta(L, p, t) \cdot p - p(L, t)) \cdot \Delta t)/p \cdot g - g(L, t))/2$
hybrid 2	$((\text{taar}(L, t) \cdot g - g(L, t)) +$ $(p(L, t) + \eta 2 \cdot (\beta(L, p, t) \cdot p - p(L, t)) \cdot \Delta t)/p \cdot g - g(L, t))/2$

In the strategies *reactive 2* and *3*, the average assault rate $\text{aar}(L, t)$ and the total average assault rate $\text{taar}(L, t)$ are calculated by:

$$\text{aar}(L, t) = \text{assault\_rate}(L, t)/\Sigma_{X:loc}\,\text{assault\_rate}(X, t)$$

$$\text{taar}(L, t) = \text{total\_assaults}(L, t)/\Sigma_{X:loc}\,\text{total\_assaults}(X, t)$$

As can be seen from Table 2, the idea of the anticipation strategies it that the guardians use formulae that are similar to the formulae for movement of criminals and passers-by to predict how they will move in the near future. Obviously, these predictions will not be 100% correct, since they do not consider interaction between the different types of agents, but our assumption is that they may be useful means to develop an efficient strategy.

Furthermore, different values can be taken for the parameter $\eta 2$ in the anticipation strategies. This parameter represents the speed by which the criminals and/or passers-by move in the predicted scenario (or, in other words, the distance in the future for which the prediction is made). For example, by taking a very high value for $\eta 2$ in the *anticipate 1* strategy, guardians get the tendency to move to locations that are predicted to have a high density of criminals in the very far future.

As mentioned earlier, the idea of having different strategies is that the analysis model can test which one performs best. A question is however how to define the notion of a "good" strategy. One possibility is to look at effectiveness, e.g., by considering the strategy that results in the lowest crime rates (total_assaults) as the best. However, in reality also the *costs* of crime prevention play an important role. Various mechanisms to improve guardianship exist (e.g., adding and moving security guards, burglar alarms, fencing, lighting), but they all involve costs [12]. Thus, instead of only measuring the amount of assaults that result from each strategy, in the calculation of the "best" strategy one should compensate for the costs involved. For this reason, the following formula (which was not included in [4]) has been added:

$$\text{total\_costs}(t + \Delta t) = \text{total\_costs}(t) + \Sigma_{X:loc}\sigma(X, t) \cdot \varepsilon \Delta t$$

This formula counts the total costs that are spent on crime prevention (for all locations involved) during the simulation. Parameter ε represents the guardian movement costs per time step.

6 Support Model

On top of the analysis model presented above, also a support model for crime prevention has been developed. This model takes as input certain information about the future scenario for which the user desires support. Based on this information, it generates advices about which strategies are recommended to prevent crime in this scenario.

More specifically, the model first needs to have some information about the state of the world. In particular, the user needs to specify the geography of the city (i.e., which locations are relevant?), and the initial densities of the different types of agents for each location. In addition to this, the user needs to define a scenario, i.e., (s)he needs to indicate the total time span for which the system is to provide support, and to specify for each location how its basic attractiveness will change during this time span. For instance, in case a circus will temporarily come to town, the basic attractiveness of the location of the circus is likely to increase. Finally, the user has to specify the maximum amount of money (s)he desires to spend.

To summarize, the support model takes the following information as input (which needs to be entered by the user of the system):

- geography of the city (i.e., which locations are relevant?)
- initial densities of the different types of agents ($c(L, t)$, $g(L, t)$, $p(L, t)$) for each location
- total time span of the scenario
- basic attractiveness for the different types of agents (ba(L, c, t), ba(L, g, t), ba(L, p, t)) for each location over time
- maximum budget

On the basis of these settings, the support model requests the analysis model to perform simulations for all possible strategies, to determine for each of these strategies to which crime rates it would lead, and what its costs would be. After that, the support model selects the "best" strategies, and presents information about those strategies to the user. The strategies that are assessed as best are those strategies of which the costs are lower than the maximum budget. Moreover, concerning the remaining strategies, in case some strategy s1 turns out both more expensive and less effective than some strategy s2, then this strategy s1 is removed from the selection. Upon request, the model can also provide the user more detailed information about the dynamics of the effect of a particular strategy in the scenario.

7   Results

A prototype implementation of the model has been developed. To illustrate the behavior of the prototype, below (part of) the dynamics of an example execution are shown in detail.

This example addresses a scenario where there are three locations, and 3900 agents. The population considered consists of 600 (potential) criminals, 300 guardians, and 3000 passers-by. Initially, these agents are distributed equally over the three location (i.e., at each location, there are 200 criminals, 100 guardians and 1000 passers/by). Moreover, all locations start with the same basic attractiveness (= 0.33 on a [0, 1] scale). After 50 time steps the attractiveness of the locations changes: location 1 becomes very attractive (=0.6), location 2 becomes slightly less attractive (= 0.3), and location 3 becomes much less attractive (= 0.1). The scenario lasts 100 time steps and the maximum budget the user can spend is 100.

When executing the system based on these settings, for the analysis model would predict the dynamics of the scenario for each of the different strategies, as mentioned above. As an illustration, such a prediction is visualized for one particular strategy (in this case, the *reactive 2* strategy, see Table 2) in Figure 4. Figure 4 shows, from top left to bottom right, the assault rate, and the amount of criminals, guardians, and passers-by at the different locations. In all graphs, the red line indicates location L1, the green line indicates location L2, and the blue line indicates location L3. The black line in the upper left graph shows the total amount of assaults, i.e., the sum of the assaults at the three locations.

As can be seen in Figure 4, over the first 50 time points, the number of the different types of agents at the locations stays equal. After time point 50, the amounts change. The guardians move away from location 3 to location 1, which is the most attractive location. The criminals move away from location 1 because they want to move away from the guardians. The passers by move towards location 1 since they want to be at the safest location (i.e. the location with the highest amount of guardians and the lowest amount of criminals). In this case, the strategy used by the guardians seems to work well, because the total number of assaults (i.e., the black line in the upper left graph) grows

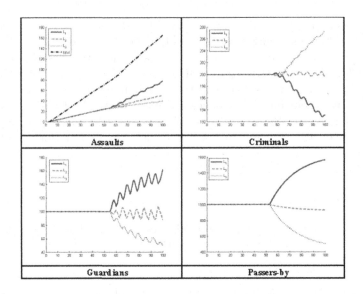

Fig. 4 Results of an example simulation run by the analysis model.

Table 3 Recommendation by the support model.

Recommended strategy	Predicted costs	Predicted % crime prevented
anticipate 1	30.38	7.7
anticipate 2	93.16	70.9
anticipate 3	45.10	40.6

not much faster than it did during the first 50 time points. When comparing this, for instance, with a baseline strategy in which the guardians are static (which is also tested by the analysis model but not shown in Figure 4), this turns out to be a significant improvement.

All in all, the analysis model tries out all possible strategies and provides the results to the support model. Based on this, the support model selects the most promising strategies (in the context of the user's preferences), and presents them as a recommendation to the user. Table 3 shows what this recommendation looks like for the current scenario.

As can be seen from Table 3, the system predicts that the three *anticipate* strategies are "best", i.e., they have costs that are below the budget of the user and are nevertheless effective. Moreover, the system predicts that strategy *anticipate 1* will be cheapest, but that strategy *anticipate 2* will be most effective.

Although this is only a single example scenario, it clearly illustrates that the model is able to generate an appropriate advice on police investment, which may actually be used by policy makes in order to reduce crime rates.

For a more detailed comparison between the different strategies in various scenarios, see [7, 8].

8 Discussion

This paper presented a model-based reasoning approach to analyze crime displacement. The approach was inspired by an existing methodology from Ambient Intelligence [1, 2, 25], which proposes that intelligent human-aware systems are composed of three separate components, namely a *domain model*, an *analysis model*, and a *support model* [10]. Although the context of crime displacement has some differences with standard AmI domains (e.g., we focus on group processes instead of individual processes, and use statistical data from police databases rather than sensor data), this methodology turned out very useful for our purposes. In the context of crime displacement, the role of the domain model was to simulate the dynamics of crime displacement, but on top of that, the analysis model proved useful to reason about such simulations for different settings, and the support model was able to generate advice on the basis of the results of this reasoning. The advice consists of a selection of guardian movement strategies that are recommended for a particular scenario, augmented with additional information about the costs and effectiveness of these strategies. A prototype version of the model has been implemented, and some initial tests have pointed out that the model provides realistic advices.

Despite these encouraging results, on should be careful not to over-generalize them. Currently, they were achieved in simulations that used several specific parameters and simplifying assumptions. Nevertheless, after further testing, the model may provide useful input for policy makers, in order to elaborate their thoughts about efficient strategies, and possibly improve existing surveillance policies.

References

1. Aarts, E., Collier, R., van Loenen, E., de Ruyter, B. (eds.): EUSAI 2003. LNCS, vol. 2875, p. 432. Springer, Heidelberg (2003)
2. Aarts, E., Harwig, R., Schuurmans, M.: Ambient intelligence. In: Denning, P. (ed.) The Invisible Future, pp. 235–250. McGraw Hill, New York (2001)
3. Baron-Cohen, S.: Mindblindness. MIT Press, Cambridge (1995)
4. Bosse, T., Both, F., Gerritsen, C., Hoogendoorn, M., Treur, J.: Model-based reasoning methods within an ambient intelligent agent model. In: Mühlhäuser, M., Ferscha, A., Aitenbichler, E. (eds.) Constructing Ambient Intelligence: AmI-07 Workshops Proceedings. Communications in Computer and Information Science (CCIS), vol. 11, pp. 352–370. Springer, Heidelberg (2008)

5. Bosse, T., Duell, R., Hoogendoorn, M., Klein, M.C.A., van Lambalgen, R., van der Mee, A., Oorburg, R., Sharpanskykh, A., Treur, J., de Vos, M.: A generic personal assistant agent model for support in demanding tasks. In: Proceedings of the Fourth International Conference on Augmented Cognition and 13th International Conference on Human-Computer Interaction, HCI 2009. LNCS, vol. 5638, pp. 3–12. Springer, Heidelberg (2009)
6. Bosse, T., Gerritsen, C.: Agent-based simulation of the spatial dynamics of crime: On the interplay between criminals hot spots and reputation. In: Proceedings of the Seventh International Joint Conference on Autonomous Agents and Multi-Agent Systems, AAMAS 2008, pp. 1129–1136. ACM Press, New York (2008)
7. Bosse, T., Gerritsen, C.: Comparing crime prevention strategies by agent-based simulation. In: Proc. of the 9th IEEE/WIC/ACM International Conference on Intelligent Agent Technology, IAT 2009, pp. 491–496. IEEE CS Press, Los Alamitos (2009)
8. Bosse, T., Gerritsen, C.: An agent-based framework to support crime prevention. In: Proceedings of the Ninth International Conference on Autonomous Agents and Multi-Agent Systems, AAMAS 2010. ACM Press, New York (in press, 2010)
9. Bosse, T., Gerritsen, C., Hoogendoorn, M., Jaffry, S.W., Treur, J.: Comparison of agent-based and population-based simulations of displacement of crime. In: Proceedings of the Eighth IEEE/WIC/ACM International Conference on Intelligent Agent Technology, IAT 2008, pp. 469–476. IEEE Computer Society Press, Los Alamitos (2008)
10. Bosse, T., Hoogendoorn, M., Klein, M.C.A., van Lambalgen, R., van Maanen, P.P., Treur, J.: Incorporating human aspects in ambient intelligence and smart environments. In: Mastrogiovanni, F., Chong, N.Y. (eds.) Handbook of Research on Ambient Intelligence and Smart Environments: Trends and Perspectives, IGI Global (in press, 2010)
11. Bosse, T., Jonker, C.M., van der Meij, L., Treur, J.: A language and environment for analysis of dynamics by simulation. International Journal of Artificial Intelligence Tools 16(3), 435–464 (2007) (in press)
12. Brand, S., Price, R.: The economic and social costs of crime, vol. 217. Home Office Research Study, London (2000)
13. Brantingham, P.L., Glässer, U., Singh, K., Vajihollahi, M.: Mastermind: Modeling and Simulation of Criminal Activity in Urban Environments. Technical Report SFU-CMPTTR-2005-01, Simon Fraser University (2005)
14. Cohen, L.E., Felson, M.: Social change and crime rate trends: A routine activity approach. American Sociological Review 44, 588–608 (1979)
15. Eck, J.E., Chainey, S., Cameron, J.G., Leitner, M., Wilson, R.E.: Mapping crime: Understanding hot spots. National Institute of Justice, U.S. Department of Justice (2005), http://www.ojp.usdoj.gov/nij/pubs-sum/209393.htm
16. Elffers, H., van Baal, P.: Spatial backcloth is not that important in simulation research: An illustration from simulating perceptual deterrence. In: Liu, L., Eck, J.E. (eds.) Artificial Crime Analysis Systems, pp. 19–34. IGI Global, Hershey (2008)
17. Green, D.J.: Realtime compliance management using a wireless realtime pill-bottle – a report on the pilot study of SIMPILL. In: Proc. of the International Conference for eHealth, Telemedicine and Health, Med-e-Tel 2005, Luxemburg (2005)

18. Groff, E.R.: The Geography of addressStreetJuvenile Crime Place Trajectories. Ph.D. Thesis, University of Maryland (2005)
19. Hayslett-McCall, K., Qui, F., Curtin, K.M., Chastain, B., Schubert, J., Carver, V.: The simulation of the journey to residential burglary. In: Liu, L., Eck, J. (eds.) Artificial Crime Analysis Systems: Using Computer Simulations and Geographic Information Systems, Information Science Reference, pp. 281–300 (2008)
20. Herbert, D.T.: The Geography of Urban Crime, Longman, Harlow, England (1982)
21. Liu, L., Eck, J. (eds.): Artificial Crime Analysis Systems: Using Computer Simulations and Geographic Information Systems. Information Science Reference (2008)
22. Liu, L., Wang, X., Eck, J., Liang, J.: Simulating crime events and crime patterns in RA/CA model. In: Wang, F. (ed.) Geographic Information Systems and Crime Analysis, pp. 197–213. Idea Group, Singapore (2005)
23. Malleson, N., Brantingham, P.: Prototype burglary simulations for crime reduction and forecasting. Crime Patterns and Analysis 2(1), 47–66 (2008)
24. Reis, D., Melo, A., Coelho, A.L.V., Furtado, V.: Towards optimal police patrol routes with genetic algorithms. In: Mehrotra, S., Zeng, D.D., Chen, H., Thuraisingham, B., Wang, F.-Y. (eds.) ISI 2006. LNCS, vol. 3975, pp. 485–491. Springer, Heidelberg (2006)
25. Riva, G., Vatalaro, F., Davide, F., Alcañiz, M. (eds.): Ambient Intelligence. IOS Press, Amsterdam (2005)
26. Sherman, L.W., Gartin, P.R., Buerger, M.E.: Hot spots of predatory crime: Routine activities and the criminology of Pplace. Criminology 27, 27–55 (1989)
27. Skogan, W.: Fear of crime and neighborhood change. In: Reiss Jr., A.J., Tonry, M. (eds.) Communites and Crime (Crime and Justice 8), pp. 203–229. Univ. of Chicago Press, Chicago (1986)
28. Wickens, C.D.: Situation awareness and workload in aviation. Current Directions in Psychological Science 11(4), 128–133 (2002)

Abducing the Crisis

Ricardo F. Crespo, Fernando Tohmé, and Daniel Heymann

Abstract. Macroeconomic crises are events marked by "broken promises" that shatter the expectations that many agents had entertained about their economic prospects and wealth positions. Crises lead to reappraisals of the views of the world upon which agents had based their expectations, plans and decisions, and to a reconsideration of theories and models on the part of analysts. A crisis triggers widespread efforts of abduction in search of new hypothesis and explanations. In this paper we will explore, in particular, the abductions that analysts may apply after a crisis and see how they reveal the prevalence of "wrong" abductions at the onset of the crisis. In order to carry out this exercise, we study the general role of abduction in economic analysis, both theoretical and practical. Economic theory generally proceeds by constructing models, that is, mental schemes based on mental experiments. They are often written in mathematical language but, apart from their formal expression, they use metaphors, analogies and pieces of intuition to motivate their assumptions and to give support to their conclusions. We try to capture all these elements in a formal scheme and apply the ensuing model of abduction to the analysis of macroeconomic crises.

Ricardo F. Crespo
IAE – Universidad Austral and CONICET
e-mail: RCrespo@iae.edu.ar

Fernando Tohmé
Departamento de Economía, UNS – CONICET
e-mail: ftohme@criba.edu.ar

Daniel Heymann
ECLAC – Universidad de Buenos Aires
e-mail: Daniel.HEYMANN@cepal.org

L. Magnani et al. (Eds.): Model-Based Reasoning in Science & Technology, SCI 314, pp. 179–198.
springerlink.com

1   Introduction

The extensive existing literature on the methodology of economics does not
mention abduction as a step of economic reasoning nor usually reflect on the
origin of economic hypotheses. However, given that abduction is a necessary
stage of every scientific development it is also used in Economics. Moreover,
in the case of economic phenomena (and more generally, in social contexts)
an additional role of abduction has to be taken into account. Abduction, as
a type of inference, not only plays a role in social science models building,
but also in the behavior of the agents whose actions are being modeled. That
is, economists (and other social scientists) have to abduce their models on
the basis of agents' abductions. Besides, these agents' abductions are subsequently based on their beliefs about other agents' abductions. For example,
in order to predict future events and thus do profitable investments, agents
should consider not only their beliefs but specially other people beliefs, the
average opinion. The analysts have to abduce hypotheses about behaviors
based on abductions of abductions. Keynes exemplifies this situation by a
very well-known (by economists) metaphor of newspapers beauty contests.
The competitors have to decide, from a large set of pictures, which are the
prettiest faces. The prize, however, is awarded to the competitor whose choice
is closest to the average decision (which would not necessarily be his actual
choice but the choice he suspects the others will do). In this case, Keynes
affirms, we reach "a third degree", and he adds that there are some who
practice the fourth, fifth and higher degrees [22]: we scientists, abduce what
agents abduce that other agents abduce and so on. This has only recently
been incorporated into the body of formal modeling tools used in the discipline [5].

While this is well-known among economists, the outbreak of large scale
crises usually poses hard questions about the predictive abilities of Economic
Theory as well as of the individual agents in real-world economies. It is in
such events that abduction plays a central role. Agents may have formed
wrong expectations about their future prospects, which leads to a number
of promises and compromises that are inevitably broken in a crisis. On the
other hand, analysts, seeing this, have to abduce a cogent explanation for the
outbreak of the crisis, in which the wrong abductions of the agents are a key
component.

The goal of this paper is to analyze the role of abduction in the explanation
of the causes of economic crises. In order to achieve this, we present a wider
discussion on the nature and role of abductive reasoning in Economics. The
plan of the paper is as follows: in Section 2 we will introduce the notion of abduction that we will use in the paper; Section 3 provides a general account on
how abduction operates in economics; Section 4 presents a formal framework
to show how abduction is performed and how important are certain guiding
criteria in the process. Finally, Section 5 presents the core of this paper, the
application to the abduction of crises.

2   Abduction and Inference to the Best Explanation

The aim of this Section is to present the notion of Abduction and "Inference to the Best Explanation" (IBE) that we will consider in the paper. We are aware that there are different possible interpretations and classifications of these notions. However, we do not intend to focus on these possible distinctions but to adopt one and to try to show how it applies in Economics. We would roughly affirm that our concept of Abduction is Peircean, as interpreted in Rescher [37], and that our notion of IBE is taken from Niiniluoto [33] and Lipton [29].

Although Aristotle discussed abduction under the name of **apagoge** (in *Posterior Analytics* **I**, 13), the modern view of abduction was firstly formulated by Charles S. Peirce. Two meanings can be discerned in his use of the term. First, he considers abduction as a type of logical inference. While deduction infers a result from a rule and a case, and induction infers a rule from the case and the result, abduction infers the case from the rule and the result. So understood, abduction can be identified with the fallacy of affirming the consequent. Thus, its result can only be conditionally accepted.

In a second sense, Peirce sees abduction as a way of arriving at scientific hypotheses. He formulates it as [36, 5.189]:

- The surprising fact C is observed.
- But if A were true, C would be a matter of course.
- Hence, there is reason to suspect that A is true.

The conception implied in this second formulation is more general than the first. A might be a case or a hypothetical rule (see [33]). However, Peirce states that, given that the fallacy of affirming the consequent remains, this procedure of discovery and postulation of hypotheses is only a first step of scientific research. That is, abduction in this second sense is a heuristic method assisted by some criteria formulated by Peirce. For him, the hypotheses should be explanatory [36, 5.171, 5.189, 5.197], economical [36, 6.395, 6.529, 8.43] and capable of being tested in experiments [36, 2.96, 2.97, 4.624, 5.597, 5.634, 8.740].

[36, 2.756–760] distinguishes three forms of induction: 1) crude induction, i.e., every day life empirical generalizations; 2) quantitative induction, i.e., statistical induction and, 3) qualitative induction, "the collaborative meshing of abduction and retroduction, of hypothesis conjecture and hypothesis testing" [37, 3]. This abduction corresponds to its second formulation and "retroduction [to] the process of eliminating hypotheses by experiential/experimental testing" [37, *ibid.*].

Aliseda [1] holds that

> Abduction is thinking from evidence to explanation, a type of reasoning characteristic of many different situations with incomplete information. Note that the word explanation – which we treat as largely synonymous with abduction – is a noun which denotes either an activity, indicated by its corresponding

verb, or the result of that activity. These two uses are closely related [...].
The process of explanation produces explanations as its products [...].

Boer ([4]) explores the meaning of explanation for Peirce concluding that we
can find in him elements of today prevailing models of explanation. It is clear
that explanation, in usual examples of abducting often points to causes: "we
want to know the cause" [36, 7.198; see also 2.204, 2.212, 2.213, 3.395, 3.690,
7.221]. Let us consider this put in [2]:

> You observe that a certain type of clouds (nimbostratus) usually precede
> rainfall. You see those clouds from your window at night. Next morning you
> see that the lawn is wet. Therefore, you infer a causal connection between the
> nimbostratus at night, and the lawn being wet.

There is a background knowledge that helps in identifying the explanation,
and consequently, the cause[1]. The final aim of scientific knowledge according to Peirce is, as Rescher remarks and argues, "the actual truth" [37][2].
Boersema, however, contends that Peirce does not take into account only
metaphysical (causal) aspects in his account of explanation but also epistemological and axiological elements, in the context of a broad theory of
inquiry. The need of retroduction as a second step of Peircean qualitative
induction means, however, that the second sense of abduction is a way to
suggest hypotheses or possible explanations pointing to true causes, but not
a sufficiently justified way to accept them. Nevertheless, Niiniluoto [33] indicates that Pierce considers "an extreme case of abductive inferences" that
are "irresistible or compelling" and come to us "like a flash" [36, 5.181]. In
these cases, Niiniluoto contends, "for Peirce [abduction] is not only a method
of discovery but also a fallible way of *justifying* an explanation" [33, italics
in the original]. That is, the strength of this flash would produce a change of
the epistemic state of the agent.

Niiniluoto thus distinguishes the procedure of suggesting hypotheses embodying a "weak conception" of abduction and the justification of the hypotheses as a "strong conception" of it. He equates this latter to an Inference
to the Best Explanation (IBE): "in the strong interpretation, abduction is
not only an inference to a potential explanation but to the *best explanation*"
[33, italics in the original]. In short, the weak conception is the best way of
arriving at hypotheses, but does not justify them. The strong conception,
in turn, is a fallible way of justifying explanations. This latter conception
implies a change of the epistemic state of the agent by which she accepts the
hypothesis, acquiring new knowledge [44]. Obviously, this acceptance does
not mean that the hypothesis is infallible: it is just an accepted hypothesis.

[29] considers IBE as a tool of exploration, generation and justification of
the hypotheses. Cresto [9, 10] proposes conceiving IBE as a complex process

---

[1] A subsequent problem would be to clarify what is the meaning of cause for Peirce.
For example, he criticizes the "grand principle of causation" [36, 6.68]. However,
it would go beyond the aim of this paper to deal with this topic.

[2] Peirce's notion of truth is another topic the paper will not deal with (see [15]).

which proceeds in two steps: the abductive stage and, after testing, the selective stage, in which the epistemic state of the agent changes. Developing Levi's [27, 28] expected epistemic utility theory, she applies this theory to the IBE, considering the epistemic virtues of simplicity (or parsimony), unification power, fertility, testability, economy, and accuracy as essential elements of her proposal. Similarly, Harman [16] proposes simplicity, plausibility and explanation power as criteria for judging the hypotheses while Thagard [44] considers consilience (how much a theory explains), simplicity and analogy. In turn, Lipton [29] mentions unification, elegance and simplicity as virtues leading to what he calls the "loveliest explanation". According to him, this "loveliest explanation" finally becomes the "likeliest explanation". In addition to empirical adequacy, which is required but not sufficient, other epistemic virtues enter into the play in the whole process of IBE. Each context indicates which virtue has more or less weight in the epistemic utility calculus. For example, as Keynes contends, vagueness may be more virtuous than precision when dealing with the complex social realm. For him, elegance and simplicity may be misleading and economy may be a vice instead of a virtue. This is compatible with Peirce's thought: for him "simplicity" does not imply a "simplified" hypothesis, but "the more facile and natural, the one that instinct suggests, that must be preferred" [36, 6.477].

The choice of these criteria is a key point to the process of postulating hypotheses (and, eventually, of justifying them). In Section 4 of the paper (formal framework) we will take into account **simplicity, unification power** (external coherence), **internal coherence** and **testability.**

3 Abduction in Economics

We contend that abduction is an essential component of economic analysis, theoretical and practical. Economic theory generally proceeds by constructing models [31], that is, mental schemes based on mental experiments [32]. They are often written in mathematical language but, apart from their formal expression, they use metaphors, analogies and pieces of intuition to motivate their assumptions and to give support to their conclusions (see [13]). In dealing with ongoing economic processes, agents and analysts must generally evaluate whether the situation resembles in a relevant way some instances observed or studied in the past, and whether this warrants applying somehow the "lessons" drawn from those experiences. The problem in judging "whether some pasts are good references for the future" becomes particularly severe when the economy is seen to undergo important changes as in the example that we will provide in Section 6. Simplicity in the Peircean sense, explanatory power, coherence and testability are rather unconsciously considered in this abduction of possible explanatory models.

The retroductive phase also involves problems implying abductive-like decisions. Although it sounds rather obvious, it must be recognized that there is a gap between the formulation of a question to be answered through measurement and the actual measurement providing the right answer. The difference arises from the fact that problems are qualitative while data are quantitative. In consequence, rough data (which certainly are the quantitative counterparts of qualitative concepts) must be organized according to the qualitative structure to be tested. That is, a correspondence between theory and data must be sought. So, for example, in economic theory there exists a crucial distinction between ordinal and cardinal magnitudes in the characterization of preferences. But once measurements are involved it is clear that the theoretical relational structure must be assumed to be homomorphic to a numerical structure [24]. This implies that if there exists a data base of numerical observations about the behavior of a phenomenon or a system, we might want to infer the properties of the qualitative relational structure to which the numerical structure is homomorphic. Of course, this is impaired by many factors:

- The syntactic representation of the qualitative structure can be somewhat ambiguous [3].
- Although the observations fall in a numerical scale, the real world is too noisy, allowing only a statistical approximation.
- The complexity of the phenomena may be exceedingly high. Then, only rough approximations may make sense.

These factors, which preclude a clear cut characterization of the observations, leave ample room for arbitrary differences. In this sense, the intuition and experience of the economist and the econometrician determine the limits of arbitrariness in an abductive-like fashion. As an example, consider the question "Did a specific economy grow in the last year?" To provide an answer, first, one has to define clearly what does it mean that an economy grows and which variables can be used to measure the phenomenon of growth. Economic theory states that economic growth means growth of the national income. But in order to answer the question an economist has to define what real world data will represent national income; i.e. she has to embed the available data into the framework given by the theory. In this case the national product is an available variable which is easy to measure and is considered (theoretically) equivalent to the national income. Therefore it is easy to check out whether the economy grew or not. But in the case where the question is something like "Did welfare increase in the last twenty years?" the procedure is far less simple. How do we define welfare and moreover, how do we make the concept operational? This is where the intuition of the economist is called in. Although theoretical concepts may be lacking, a set of alternative models of the notion of welfare and its evolution in time should be provided in order to check out which one fits better the real world data. When this question is settled it is possible to consider the development of a theory formalizing the properties satisfied in the chosen model. That is, when the abducting process is completed the theory-building phase can start.

The inferences that allow economists and econometricians to detect patterns in reams of data cannot be called statistical inductions. They are more a result of a detective-like approach to scarce and unorganized information, where the goal is to get clues out of unorganized data bases of observations and to disclose hidden explanations that make them meaningful. In other words: it is a matter of making guesses, which later can be put in a deductive framework and tested by statistical procedures. So far, it seems that it is just an "artistic" feat, which can only be performed by experts[3].

Let us give a sketchy description of the reasoning process in Economics. Economists have a background of general rules. When a surprising fact appears the first step is to try to come up with an explanation according to those rules. The best explanation obtains by delimiting the possible hypotheses until only one of them remains. In this process the economist uses information about similar situations as well as the features of the specific case to capture simple and coherent hypotheses and models.

We may distinguish the following steps in this process:

1. An abnormal event is detected, requiring an explanation.
2. The event is carefully described.
3. Some stylized facts are extracted from the description.
4. Situations sharing the same stylized facts are given particular attention.
5. Formal expressions, capturing the relations deemed essential in the explanation of the relevant stylized facts, are formulated.
6. Only those combinations of deductive chains and inductive plausibility that are both externally and internally coherent are chosen, discarding other possibilities.
7. This provides an original coherent explanation of the event.
8. The conclusions are tested.

Abduction is hidden in the whole process, but steps 5 and 6 are mostly deductive. On the other hand, step 8 is also inductive and retroductive. The whole process is a Peircean qualitative inductive process. Good economists have a guess instinct [36, 6.476–477] present in their scientific processes. This is not a mysterious miracle but an intellectual intuition, stemming from a theoretical framework or background knowledge, of experience, of hard work with theories, models and data. This leads good economists to foresee a set of probably successful models. Combining this gift with hard empirical work economists often overcome the problems of under-determination of theories by formulating local or context-dependent theories. Context-dependence is a characteristic feature of IBE [10, 11]. However, the economists always try to improve their models. This is because, given the fluctuating ontological condition of the economic material, a close relation with real situations is needed. The analogies sometimes work and sometimes not. Old or conventional theories may be misleading. Thus, economists need that special "gift for using

---

[3] This might be a reason for why formal logicians, until recently, did not intensively study abduction in contrast to the other forms of inference.

vigilant observation to choose good models" [21, p. 297]. This improvement, however, has a limit. On the one hand, the frequent urgency of decisions that cannot wait for further investigation, and the economy of research ([37, p. 65 ff.], extensively quoting Peirce), actually lead to accept the conclusions as fallible though reasonable inferences to the best explanation. On the other hand, the mentioned problems of quantification - conceptual, institutional, accuracy of data, calculation and even presentation -, also incline to accept a sufficiently probed fallible conclusion as a good one. Given this informal characterization we are interested in providing a formal framework for abduction in the next Section.

4 A Formal Framework for Abduction

What clearly separates abduction and IBE from statistical induction is that they require a previous meta-theoretical commitment. The study of the historical example of how Kepler derived the laws of planetary motion, allowed Peirce to clarify this point. Moreover, Peirce draw from this example some prescriptions on how to perform an abduction. First of all, data (or more generally information) had to be structured by means of Peirce's own classifications of signs [30]. Since in this view every set of data constitutes a sign, it can be classified according to Peirce's exhaustive taxonomy. The advantage of this approach is that there exists only a finite set of possibilities to match with the real world information. Once one of the possibilities becomes chosen, it is assumed to provide a clear statement of the kind of structure hidden in the data, although not necessarily as complex as a functional form.

Peirce's approach can be used as a heuristic guide for the formalization of abduction in economic analysis. This is because the pieces of information available to an economist cannot be all put in the same level. In fact, to classify a set of data in terms of the meaningfulness of the information conveyed is very useful in order to construct a testable hypothesis. While this is a hard task, the remaining chore is still harder: to work on the classified data base, trying to fit it to one of a bundle of possible functional forms.

We will try to make this discussion a bit more formal and develop an approach to qualitative model building in economics. In the first place, we should note that the meaning of **model** in this field is not the same as in mathematical logic. We will try to keep the meaning of the word as used in first-order logic, so in order to explain how abduction helps in economic model building, we need some previous definitions[4]:

Definition 1. *Given a first order language $\mathcal{L}$ a structure is $\Delta = \langle \mathbf{N}, \gamma, \mathcal{F}, \Pi \rangle$, where $\mathbf{N}$ is a set of individuals; γ is a function that assigns an individual to each constant of $\mathcal{L}$, $\mathcal{F}$ is a family of endomorphic functions on $\mathbf{N}$, while Π is a set of predicates on $\mathbf{N}$. An interpretation of any consistent set of well formed*

---

[4] For a precise characterization of these notions see [38].

formulas of $\mathcal{L}$, $\mathcal{T}(\mathcal{L})$ obtains through a correspondence of constants, function symbols and predicate symbols to Δ. A model of $\mathcal{T}(\mathcal{L})$ is an interpretation where every interpreted formula is true.

A structure can be thought of as a database plus the relations and functions that are, implicit or explicitly, true in it. An interpretation is a structure associated to a certain set of well-formed formulas (when deductively closed this set is called a *theory*). If, when replacing the constants by elements in the interpretation and the propositional functions by relations in the structure, all the formulas are made true in the interpretation, all the formulas become true in the interpretation, this structure is called a model. To say that abduction helps in model building means that it is a process that embeds the real-world information in a certain structure that is assumed to be the model of a theory or at least of a coherent part of one.

In economics it is usual to find that there is not a clear distinction between what is meant by "theory" and by "model". One reason is that for most applications, it is excessive to demand a theory to be deductively closed, which means that all its consequences should be immediately available. In the usual practice, statements are far from being deduced in a single stroke. On the other hand -and this explains clearly the confusion between theory and model- most scientific theories have an intended meaning more or less clear in its statements. This does not preclude the formulation of general and abstract theories, but their confrontation with data are always mediated by an intended model [41].

Another concern that may arise from our approach is whether any economically meaningful assertion can be embedded in a first-order language. The point is that most theories of sets, Zermelo-Frenkel and others, intended to provide a comprehensive foundation for mathematics, are first order [12]. Since most of the economic statement can be expressed as set-theoretic expressions, it seems that the previous definition of a structure is enough for our purposes. What we need is to translate the data base of observations into a formal structure such that [26]:

- Each element of interest in the data has a symbolic representation.
- For each (simple) relationship in the data, there must be a connection among the elements in the representation.
- There exist one-to-one correspondences between relationships and connections, and between elements in the data and in the representation.

This representation of the real world information, Λ, facilitates the abduction, by means of its comparison with alternative structures. The result of the abduction will determine an implicit representation of data, as we see if we consider the following definition:

Definition 2. *Given a set of structures $\{\Delta_i^{\mathcal{C}}\}_{i \in I}$ where I is a set of indexes, selected for satisfying a set of criteria $\mathcal{C}$, an abduction is the choice of one of them, say Δ^*, by comparison with Λ.*

In words, given a class of criteria, there might exist several (although we assume only a finite number) possible structures that may explain the data in Λ. To *abduce* Λ, is to choose one of them. We have to explain, on one hand, what those criteria might be and, on the other, how a single structure may be selected. With respect to the criteria, notice that in the case of Kepler's abduction he had at least one criterion in mind: trajectories of celestial bodies should be described by simple geometrical expressions. Under this criterion, Kepler had to choose one among a few structures comparing the movements implied by them with the behavior of a given set of real-world elements (the known planets of the solar system). Each of those structures was a simple geometric representation of the solar system. He finally chose the one that fitted the data best.

In general, the criteria represent all the elements that the scientist wants to find incorporated into the chosen structure. Given the criteria in $\mathcal{C}$ the set of structures that satisfies them is defined as follows:

Definition 3. *A criterion c_j defines a set of structures in which it is satisfied, $\{\Delta_i\}_{i \in I_j}$ (where I_j is a set of indexes corresponding to this criterion) . Then, $\mathcal{C} = \{c_j\}_{j \in J}$ defines a set of structures $\{\Delta_i^{\mathcal{C}}\}_{i \in I} = \cap_{j \in J}\{\Delta_i\}_{i \in I_j}$.*

In general, the number of criteria is reduced in order to ensure that the set of possible structures is not empty. The comparison of the structures with the data determines an order on $\{\Delta_i^{\mathcal{C}}\}_{i \in I}$:

Definition 4. *Given Λ, and two possible structures Δ_j, Δ_l we say that $\Delta_j \preceq \Delta_l$ if and only if $\mathbf{WFF}(\Delta_j) \bar{\cap} \Lambda \subseteq \mathbf{WFF}(\Delta_l) \bar{\cap} \Lambda$, where $\mathbf{WFF}(\cdot)$ is the set of well-formed formulas corresponding to a given structure and $\bar{\cap}$ is a satisfaction operator.*

To complete this definition, we have to provide a characterization of the satisfaction operator $\bar{\cap}$. Notice that if we had used only the set-theoretic intersection $\cap$ we would have missed the point of comparing Λ with the potential structures. Since Λ may just consist of a data base of numerical observations, a qualitative structure may not yield even a single one of those observations and still be meaningful. In order to address this question, we have to consider each relation $\mathcal{R}$ implicit in Λ. Then consider the collection of sets of observations in Λ, denoted 2^Λ. Then, an application of the Axiom of Choice for finite sets yields that:

Definition 5. *A proposition $\lambda_{\mathcal{R}}$ satisfies Λ if and only if for every finite subfamily sets in Λ there exists a choice S such that for every $a_1, \ldots, a_n \in S \subseteq 2^\Lambda$, $\mathcal{R}(a_1, \ldots, a_n)$.*[5]

Consider then, the family of the propositions $\lambda_{\mathcal{R}}$ for all relations $\mathcal{R}$ defined over Λ. These relations may represent the closeness of numerical values, or

---

[5] This follows in a logic defined over a hypergraph in which the observations constitute the nodes and sets of observations under the relation $\mathcal{R}$ the hyperedges [23].

the fact that they belong to a given interval or, closer to Peirce's aim, a hierarchy of observations, ones deemed more relevant than the others. In any case each of these formulas abstract away from the data base. But then:

Definition 6. *Given a structure* Δ, $\mathbf{WFF}(\Delta) \bar{\cap} \Lambda = \{\lambda_\mathcal{R} : \Delta \models \lambda_\mathcal{R}\}$, *where* $\models$ *is the classical relation of semantical consequence.*

That is, $\mathbf{WFF}(\Delta) \bar{\cap} \Lambda$ consists of those $\lambda_\mathcal{R}$ that are satisfied by Δ, and can be seen as well-formed formulas shared by the data base and the theory for which Δ is a model. Finally, the relation among structures $\preceq$ simply yields for every pair of structures, Δ_l, Δ_j a preference for the structure, say Δ_l, that satisfies not only the same formulas of the data base as Δ_j but also some more. Notice that the class of the wffs $\lambda_\mathcal{R}$ determine, as much as the candidate structures, the resulting order $\preceq$.

Even if this description is sound, in practice there exist serious difficulties associated with the detection of patterns and relations in a numerical database. This fact is well known by statisticians:[6] an approximate generalization is, according to any statistical test, indistinguishable from the form of a wrong generalization. Even if statistical inferences may preclude hasty generalizations, the fact is that qualitative data may not correspond directly to quantitative forms that can be statistically supported.

Other (non-statistical) methods lead to similar problems. Computational intelligence only provides rough approximations to the task of theory or model building. Systems like BACON (in any of its numerous incarnations) despite their claimed successes are only able to provide *phenomenological* laws [40]. That is, they are unable to do more than yield generalizations that involve only observable variables and constants. No deeper explanations can be expected to ensue from their use.

In the process of inquiry carried out by economists, the human side has a crucial task, not yet fully elucidated in the literature: the formation of concepts and the elicitation of qualitative relations. In fact, experts excel in detecting patterns and relations in disordered and noisy data. Of course, as it is well known in Combinatorics, more precisely in Ramsey Theory [14], with enough elements a regular pattern will exists, be it meaningful or not. In any case, an expert uses the patterns and relations he finds or imposes over the database and this is represented above by the procedure of selection $\mathcal{S}$.

Based on this possibility of finding expressions that "refine" the crude information in Λ we have the following result:

Proposition 1. *There exists a maximal structure* Δ^* *in the set* $\{\Delta_i^\mathcal{C}\}_{i \in I}$ *ordered under* $\preceq$.

Proof. First of all we will show that $\preceq$ is a partial order, i.e. that it verifies the following properties:

---

[6] See [39].

- **Reflexivity**: since $\mathbf{WFF}(\Delta_i)\bar{\cap}\Delta \subseteq \mathbf{WFF}(\Delta_i)\bar{\cap}\Delta$ then $\Delta_i \preceq \Delta_i$.
- **Transitivity**: if $\Delta_j \preceq \Delta_l$ and $\Delta_l \preceq \Delta_k$ then $\mathbf{WFF}(\Delta_j)\bar{\cap}\Delta \subseteq \mathbf{WFF}(\Delta_l)\bar{\cap}\Delta$ and $\mathbf{WFF}(\Delta_l)\bar{\cap}\Delta \subseteq \mathbf{WFF}(\Delta_k)\bar{\cap}\Delta$. By transitivity of $\subseteq$ it follows that $\mathbf{WFF}(\Delta_j)\bar{\cap}\Delta \subseteq \mathbf{WFF}(\Delta_k)\bar{\cap}\Delta$ i.e. that $\Delta_j \preceq \Delta_k$.

Since we assume that Δ is finite, $\langle \{\Delta_i^{\mathcal{C}}\}_{i\in I}, \preceq \rangle$ is bounded: for any Δ_i such that $\Delta \subseteq \mathbf{WFF}(\Delta_i)$ there is no other Δ_l such that $\Delta_i \prec \Delta_l$. Therefore, using Zorn's Lemma it follows that $\langle \{\Delta_i^{\mathcal{C}}\}_{i\in I}, \preceq \rangle$ has a maximal element. $\qquad\square$

A trivial case of a maximal structure Δ^* arises when $\Delta^* \models \Lambda$. That is, when all the observations in the data base are satisfied in the structure. But, as said, this is not only difficult to be found, but also undesirable, if the data base includes noisy and otherwise imprecise observations.

So far, many structures may be chosen. Sufficient conditions for uniqueness can be achieved if certain criteria are included in $\mathcal{C}$:

Definition 7. • $\mathbf{c}^{min}$ **(Minimality)**: *given two structures Δ_i, Δ_j, such that $\mathbf{WFF}(\Delta_i) \subseteq \mathbf{WFF}(\Delta_j)$ and $\mathbf{WFF}(\Delta_j) \not\subset \mathbf{WFF}(\Delta_i)$, select Δ_i.*
- $\mathbf{c}^{comp}$ **(Completeness w.r.t. Λ)**: *given two structures Δ_i, Δ_j, where $\Lambda \subseteq \mathbf{WFF}(\Delta_i)$ but $\Lambda \not\subset \mathbf{WFF}(\Delta_j)$, select Δ_i.*
- $\mathbf{c}^{conc}$ **(concordance w.r.t. Λ)**: *a given structure Δ is selected if for every $\lambda_{\mathcal{R}}$ derived from Λ, either $\lambda_{\mathcal{R}}$ or $\neg\lambda_{\mathcal{R}}$ belongs to $\mathbf{WFF}(\Delta)$.*

Then we have the following result:

Proposition 2. *If $\{\mathbf{c}^{min}, \mathbf{c}^{com}\} \subseteq \mathcal{C}$ and the set of possible structures is otherwise unrestricted, Δ^* is unique.*

Proof. There are two cases to consider. If $\{\Delta_i^{\mathcal{C}}\}_{i\in I} = \emptyset$ then we can define $\Delta^* = \emptyset$, which is trivially unique. If $\{\Delta_i^{\mathcal{C}}\}_{i\in I} \neq \emptyset$, according to $\mathbf{c}^{comp}$, $\Lambda \subseteq \mathbf{WFF}(\Delta_i^{\mathcal{C}})$ for all $i \in I$. On the other hand, according to $\mathbf{c}^{min}$ there is a minimal $\Delta_i^{\mathcal{C}}$ such that $\Lambda \subseteq WFF(\Delta_i^{\mathcal{C}})$. Since one of the possible structures is a Δ' such that $\Lambda = \mathbf{WFF}(\Delta')$ we can define $\Delta^* \equiv \Delta'$, which is unique. $\qquad\square$

Similarly:

Proposition 3. *If $\{\mathbf{c}^{min}, \mathbf{c}^{conc}\} \subseteq \mathcal{C}$ and the set of possible structures is unrestricted, Δ^* is unique.*

These results shows that a unique structure can be selected if the restrictions on possible structures obey to methodological criteria like minimality, completeness or concordance. This is not without a cost: if the only wffs in the chosen structure are the ones drawn from the database it is not possible to provide more than a description (data fitting) of the available information. This means in turn that if only methodological criteria are to be used, the result of the inference is the generation of a prototype, i.e. only a statistical inference is performed. In Economics these criteria are usually violated since sometimes inferences are drawn from partial samples from a bigger database (violation of c^{comp}), some observations are discarded as outliers (violation of c^{min}), or some information in the database is not used (violation of c^{conc}). Nevertheless they represent extreme case of very desirable properties: minimality involves **simplicity** while completeness and concordance approximate **unification power** (i.e. external coherence). On the other hand, the fact that the abduction yields a structure implies **internal coherence**. A final requirement, **testability**, is satisfied when the structure yields observable outcomes not found in Λ, that have to be checked out in the real world.

The chosen relational structure Δ^* is, as said, conceived as a model of a theory $\mathcal{T}$, which in economic parlance is the actual "model" sought for. To derive this $\mathcal{T}$, one might choose one from a collection of closed sets of wffs of $\mathcal{L}$, each one having Δ^* as a model. One candidate is just $\mathbf{WFF}(\Delta^*)$ itself as a bundle of first-order formulas. Other possible theories may involve information that is certainly not present in the data. In the case that $\mathcal{T} \equiv \mathbf{WFF}(\Delta^*)$, the theory is, as said, called *phenomenological*. Otherwise, the theory is said *representational* and involves to postulate non-observable properties and entities.

As seen above, the burden of the task of performing abductions is on the set of criteria used. Although this is true for every science, in human affairs it seems that the hidden assumptions account for a good deal of surprising results that lead, in turn, to policies affecting the lives of the members of entire societies. This is, in fact, the case of economic crises.

5 Abduction and Macroeconomic Crises

The meaning of the term crisis in the economic literature is not without ambiguity. However, there is a commonsense group of characteristics that we will retain here as defining a set of critical macroeconomic events: (*i*) they have a large scale, in the sense that they are reflected in wide swings in macroeconomic aggregates and affect the behavior and economic performance of a population as a whole, (*ii*) they are perceived by most agents as a sizeable disturbance in their economic life and prospects, (*iii*) they typically involve moments of abrupt economic change marked, for example by shifts in asset prices much sharper than normal measures of variability, or by "big news" like the failure of some large firm and (*iv*) they are memorable events, which lead many people (agents and perhaps analysts) to reconsider plans and beliefs or, at least, to engage in active after-the-fact learning in order to revise opinions and expectations. Economic depressions, financial crashes, hyperinflations and collapses of monetary and exchange regimes would belong to this category, although clearly they form a heterogeneous set in terms of the economic processes at work. We shall concentrate here on phenomena of the family that includes the ongoing worldwide crisis and the Great Depression (lumping together those events implies already a presumption of comparability despite their clear dissimilarities, but it may be admissible at this point because of the widespread use of the analogy and the related search for similarities and contrasts).

Concern for the study and the understanding of crises is actually older than macroeconomics as an established discipline and it has operated historically as a strong motivation to investigate in the field. Modern macroeconomic theory, on its side, has increasingly become committed to a set of analytical and procedural presumptions, which lead to look for representations of macroeconomic behavior as the result of well coordinated (except for some noise which acts as an additional constraint) optimal decisions of agents, equipped with rational expectations, that is with knowledge of the probability distributions relevant for their plans. These research criteria, sometimes elevated to the rank of methodological prescriptions, can be seen as the outcome of past debates on the theory of macroeconomic fluctuations and inflation, which generated dissatisfaction with earlier theories. At the same time, their application to the study of crises, as if they could claim a universal range of validity, has been subject to paradoxes and problems in the interpretation of salient facts, which seem to call for new searches of the abductive type.

Consider the following facts, which describe the onset of a crisis:

1. A large, relatively closed economy (or, to make the point more starkly, the world economy as a whole) undergoes a period of rapid expansion, marked by substantial technical changes and the emergence of new patterns of the division of labor, together with strong increases in aggregate demand.
2. The volume of credit rises strongly, through the issue of wide variety of instruments. Both consumers and firms show willingness to enter into debt, and higher asset prices indicate the strength of the demand by prospective lenders.
3. There are public discussions about the possibility that the economy has grown an unsustainable bubble, and that policy corrections may have to be applied. The authorities decline to do so, with the view that it is not clear that asset prices and credit flows are out of line and, if that was the case, policies can handle a potential correction without much disruption.
4. Eventually, doubts about the sustainability of asset prices induce some falls. Banks start showing worsening results due to increasing defaults on their loans. Evidence of those problems reduces asset demands and leads to cautious attitudes by households and firms in their demand for consumption and investment.
5. At some point, information emerges about a large mass of bad debts, which threatens banks with failure. The central bank provides massive assistance to troubled institutions. However, credit flows are disrupted, and asset prices fall precipitously.
6. The economy enters into a sharp recession, with substantial falls both in investment and consumption, and higher unemployment. Bad news about the real economy are followed by further declines in the demand for private bonds and equities. At the same time, interests on the public

debt drop to very low levels. The government decides substantial increases in its spending, and runs large deficits.

This scenario, highly simplified as it is, has still too much detail to lend itself to a precise analysis. Consider then, with the facts just described as background, the question of what would explain the wide swings in macroeconomic activity and in asset valuations that is, of what set of features of the economic process are crucial in determining those fluctuations, and thus deserve a closer exploration in view of advancing towards a general account of crises. This question requires some initial definition of a framework of analysis, and a narrowing down of the set of alternatives to be contemplated. Among the various possibilities we will contemplate the following arguments that could be used to rationalize the choice of the different hypotheses as basic elements of the approach to the subject:

- Δ_1: The statistical properties of macroeconomic aggregates (output, consumption, investment, employment) resulting from historical data can be represented to a reasonable degree of approximation by DSGE's (Dynamic Stochastic General Equilibrium Models) where the behavior of agents is described through the solution of dynamic programming problems in stochastic environments where the main impulses driving the system are random, exogenous, shocks to the aggregate productivity of the economy, or shifts in monetary policy which distort labor supply and demand decisions, and where the expectations of the agents are rational. The analysis should be based on such a construction, with the presumption that the severity of the macroeconomic swings would be determined by the extraordinary magnitude of the shocks hitting the system[7].

- Δ_2: The distinguishing feature of the episodes under consideration is the amplitude of the fluctuations in credit, from a phase of easy financing which promotes exaggerated valuations of assets and unsustainable buildups of credits, to a collapse of the bubble where the supply of lending falls dramatically. Monetary policy is the main regulator of credit conditions. Therefore, the crisis can be traced to an unduly lax attitude of monetary authorities, and a later tightening, associated with a strong credit contraction [43, 34, 35].

- Δ_3: All crises of this type are ultimately great swindles, where some groups of economic agents gain to the detriment of others. Thus, a crisis reveals deep problems of incentives, like rewarding bank officers for making loans that they know will not be repaid, or giving bonuses to executives for short – run profits when their actions will eventually lead to bankruptcy. In extreme cases, like the pyramid schemes posing as investment funds, even seemingly sophisticated traders were cheated. The behavior of trying to take advantage of other persons was also a decisive factor in asset valuation, as rational agents kept demanding stocks and bonds, knowingly

---

[7] This line of explanation has generated a significant literature on past episodes of economic depression [8, 20]. For the crisis started in 2008, see [6] and [7].

"riding the bubble" because they understood that they would be able to sell at prices before the collapse [17, 19].

- Δ_4: The marking characteristic of crises like the one being analyzed is the widespread frustration of economic plans and expectations, reflected particularly in large- scale defaults on financial commitments. Such crises involve as their crucial element a drastic fall in the wealth perceptions of substantial groups of agents, implying that previous decisions on production, spending and credit supply or demand had been based on wrong anticipations. Crises are memorable events, which lead to revisions of beliefs on the part of economic agents, policymakers and analysts: the search for lessons to learn from such episodes indicates that models of analysis and decision are in fact being reconsidered. Moreover, crises tend to occur after periods where changes in the present or prospective configuration of the economies (due to technological developments, policy reforms or a modified international environment) are apt to sustain expectations of future increases in real revenues and thus to promote a perception that a credit boom has fundamental underpinnings. Those swings in beliefs and expectations in economies undergoing changes in structure or performance should be one of the basic objects of analysis research and a necessary element in macroeconomic models seeking ranges of validity that include those episodes [25, 18, 42].

The previous arguments are not necessarily incompatible (one can for example insist in the central role of changes in expectations without denying the relevance of monetary policies or incentive effects in the actual course of events), but they can base different approaches to the subject, and also generate quite distinct policy implications. We shall maintain here that Δ_1 is actually the most general, in that it points to an essential component of the phenomena. The alternatives, as those briefly outlined above, would not qualify as cores of a theory of crises. Nevertheless, they qualify as explanations for a crisis of the types that have been usual in the world since 1994[8].

To abduce an explanation we need, as said in the previous section, a family of criteria. The following are adequate in this case:

- c^A: The hypotheses should satisfy the onset of the crisis (1 to 6, above).
- c^B: The explanations should be internally consistent.
- c^C: The border conditions of each hypothesis should be observable.

The cogency of these criteria is evident. c^A just captures the idea that, if we seek an explanation for a crisis, it should describe its process. On the other hand c^B indicates that inconsistent structures should be disregarded. Finally, c^C, indicates that any reference to external variables or conditions invoked in an explanation should be testable (otherwise, the explanation would hinge on unverifiable assumptions).

---

[8] That is, the Mexican, South-Asian, Russian, Brazilian, Turkish, Argentinian, Dot.com, and the Subprime crises.

To proceed, let us note that in principle, all four hypotheses satisfy $\mathbf{c}^A$. That is, a crisis that follows each of those prescriptions will undergo steps 1 to 6. But the DSGE models with rational expectations (Δ_1), even when augmented with financial propagation mechanisms beg the question about the impulse that would shock the system and about the ex-ante probability that agents may have assigned to a large disruption of their plans. That is, Δ_1 may fail to satisfy $\mathbf{c}^C$.[9]

As for the oscillations in monetary policies (Δ_2), it is true that very low interest rates would raise asset prices transitorily and also cause transitory increases in aggregate spending. However, if economic actors anticipate correctly an interest rate fluctuation and its consequences, it will not induce defaults or disturb the wealth perceptions of agents. A similar argument holds for malincentives and fraud (Δ_3). In particular, a too risky loan policy by bank managers, if perceived as such by the public, would lead to falls in the market valuation of those institution and, in the limit, to a refusal to buy the debt or the equity of those institutions before the bubble develops; the expected possibility of moral hazard through government bailouts would either draw a response of policy-makers or, if these are not willing to react, and that is well understood by the private sector, it would provoke a precautionary response of taxpayers and holders of government bonds who discount the coming fiscal burden. In sum, there seems to be no adequate alternative to establishing a close association between expectational errors and swings in asset prices and credit flows marked at some point by uncommonly large defaults and by sharp and widespread cuts in consumption. That is, Δ_2 and Δ_3 do not satisfy $\mathbf{c}^B$.

In summary, only Δ_1 and Δ_4 might satisfy the three criteria for abduction. Criteria like $\mathbf{c}^{min}$ or $\mathbf{c}^{com}$ are not strictly satisfied by either candidate, so the only remaining possibility is to establish an order $\preceq$ between Δ_1 and Δ_4. In this sense, we have that $\Delta_1 \preceq \Delta_4$, since Δ_4 may yield an explanation for a crisis even in the absence of external shocks.

The previous argument restricts the classes of theories among which the search for a representation of crisis-type events would proceed to those for which Δ_4 might be part of their models. It may be noted in this regard that some knowledge about the probability of behaviors leading to crises would be of great importance. In any case, an essential part of an applicable model would be a specification of how agents form expectations about their future incomes and about the returns of various assets, in economies with evolving configurations. In one way or another, this would require addressing concretely how people determine their representations of economic conditions to come and, consequently, how they understand their environment and project its trends.

Δ_4 raises some important questions in this sense. The most important arises from the realization that the behavior of the agents starts from

---

[9] [6] claims that the large increases in the price of commodities during the 2000's act as impulses for the financial crisis started in 2008.

ex-ante abductions on the overall prospects of the economy that end up being obviously mistaken. Why does this happen? It might be because they do not use the right criteria or because they disregard important data. Thus, in fact, the analyst would be involved in a "second order" abduction when trying to understand expectation formation. This certainly seems a difficult matter for inquiry, but one that cannot be left aside when trying to develop workable theories of phenomena of the social importance of macroeconomic crises.

6 Conclusions

The example presented in the last section shows that meta-theoretic assumptions may lead in an otherwise innocent looking abduction to a result that may affect not only our understanding of the behavior of economies but also the actual behavior through the application of economic policies. In this matters it seems that abductive reasoning must abandon the realm of implicit activity to become an open activity, that may be discussed with the same seriousness as the values of statistical estimates.

An abductive inference should be reported providing:

- The set of criteria to be considered, precisely stated.
- The alternative hypotheses that are postulated (obeying to the criteria). Each should be represented by a system of relations, which constitutes a necessary condition for the respective hypothesis.
- The tests showing which of the hypothesis is accepted. The acceptation criteria should be already stated in the set of general criteria.

Therefore, any discussion on the inference can be based either on the criteria used or on the set of postulated hypotheses. In the first case, the criteria may be wrong, biased, insufficient, etc. In the second case, any new hypotheses added to the list may conform to the originally stated criteria. Both types of discussion may enliven the scientific evaluation of the available information.

This analysis, as it has been revealed in the case of macroeconomic crises, must also cover the abductions performed by the economic agents. The results of this line of research may have important practical applications, starting with the design of more effective policies and regulations on processes in which uncertainty is pervasive.

Acknowledgements. We acknowledge useful comments by Eleonora Cresto and Alfredo Navarro as well as the permanent encouragement to pursue this topic by the late Ana Marostica. We also received thought-provoking questions and suggestions after the presentation of a draft version of this paper at the meeting on *Model-Based Reasoning in Science and Technology. Abduction, Logic, and Computational Discovery*, held in Campinas, Brazil (December 2009). The usual caveat applies.

References

1. Aliseda, A.: Logics in scientific siscovery. Foundations of Science 9, 339–363 (2004)
2. Aliseda, A.: Abductive Reasoning. Logical Investigations into Discovery and Explanation. Springer, Dordrecht (2006)
3. Barwise, J., Hammer, E.: Diagrams and the concept of logical system. In: Allwein, G., Barwise, J. (eds.) Logical Reasoning with Diagrams, pp. 49–78. Oxford University Press, Oxford (1994)
4. Boersema, D.: Peirce on explanation. The Journal of Speculative Philosophy 17, 224–236 (2003)
5. Brandenburger, A.: The power of paradox: Some recent developments in interactive epistemology. International Journal of Game Theory 35, 465–492 (2007)
6. Caballero, R.J., Farhi, E., Gourinchas, P.O.: Financial Crash, Commodity Prices and Global Imbalances. NBER Working Paper 14521 (2008)
7. Cochrane, J.H.: Lessons from the financial crisis. Regulation 32, 34–37 (2010)
8. Cole, H., Ohanian, L.: The Great Depression in the US from a neoclassical perspective. Federal Reserve Bank of Minneapolis Quarterly Review 23, 2–24 (1999)
9. Cresto, E.: Creer, inferir y aceptar: una defensa de la inferencia a la mejor explicación apta para incrédulos. Revista Latinoamericana de Filosofía 28, 201–229 (2002)
10. Cresto, E.: Inferring to the Best Explanation: A Decision-Theoretic Approach. Ph. D. Thesis, Columbia University (2006)
11. Day, T., Kincaid, H.: Putting inference to the best explanation in its place. Synthese 98, 271–295 (1994)
12. Devlin, K.: The Joy of Sets. Springer, New York (1993)
13. Frigg, R.: Models in Science, Stanford Encyclopedia of Philosophy (2006), http://plato.stanford.edu/
14. Graham, R., Spencer, J., Rothschild, B.: Ramsey Theory. Wiley and Sons, New York (1990)
15. Haack, S.: Two fallibilists in search of truth. Proceedings of the Aristotelian Society 51, 73–84 (1977)
16. Harman, G.: The inference to the best explanation. The Philosophical Review 74, 88–95 (1965)
17. Hart, O., Zingales, L.: How to avoid a financial crisis, Working Paper, Chicago University (2009)
18. Heymann, D.: Macroeconomics of broken promises. In: Farmer, R. (ed.) Macroeconomics in the Small and the Large, pp. 74–98. Elgar Publishing, Cheltenham (2008)
19. Kane, E.: Incentive Roots of the Securitization Crisis (2009), http://www.voxeu.org
20. Kehoe, T., Prescott, E. (eds.): Great Depressions of the Twentieth Century. Federal Reserve Bank of Minneapolis, Minneapolis (2007)
21. Keynes, J.M.: A treatise on probability (1921). In: The Collected Writings of John Maynard Keynes VIII, MacMillan, London (1973)
22. Keynes, J.M.: The general theory and after: Part II. Defence and Development. In: The Collected Writings of John Maynard Keynes, vol. XIV. MacMillan, London (1973)

23. Kolany, A.: On the logic of hypergraphs. In: Mundici, D., Gottlob, G., Leitsch, A. (eds.) KGC 1993. LNCS, vol. 713, pp. 231–242. Springer, Heidelberg (1993)
24. Krantz, D., Luce, R.D., Suppes, P., Tversky, A.: Foundations of Measurement, vol. I. Academic Press, New York (1971)
25. Leijonhufvud, A.: Stabilities and instabilities in the macroeconomy (2009), http://www.voxeu.org
26. Levesque, H.: Foundations of a functional approach to knowledge representation. Artificial Intelligence 23, 155–212 (1984)
27. Levi, I.: Decisions and Revisions. Cambridge University Press, Cambridge (1984)
28. Levi, I.: Inductive expansions and non-monotonic reasoning. In: Rott, H., Williams, M. (eds.) Belief Revision, pp. 7–56. Kluwer, Dordrecht (2001)
29. Lipton, P.: Inference to the Best Explanation. Routledge, London (2004)
30. Marostica, A.: Abduction: The creative process. In: Gorna, R., Hausden, B., Posner, R. (eds.) Signs, Search and Communication: Semiotic Aspects of Artificial Intelligence, pp. 134–150. Walter de Gruyter, Berlin (1993)
31. Morgan, M., Morrison, M.: Models as mediating instruments. In: Morgan, M., Morrison, M. (eds.) Models as Mediators, pp. 10–37. Cambridge University Press, Cambridge (1999)
32. Nersessian, N.: How do scientists think? Capturing the dynamics of conceptual change in science. In: Giere, R. (ed.) Cognitive Models of Science, pp. 5–22. University of Minnesota Press, Minneapolis (1992)
33. Niiniluoto, I.: Defending abduction. Philosophy of Science 66 (suppl.), S436–S451 (1999)
34. Obstfeld, M., Rogoff, K.: Global imbalances and the financial crisis: Products of common causes. Working Paper, University of California, Berkeley (2009)
35. O'Driscoll, G.: Money and the present crisis. Cato Journal 29, 167–186 (2009)
36. Peirce, C.S.: Collected Papers Vols. 1–8. In: Hartshorne, C., Weiss, P., Burks, A. (eds.). The Belknap Press of Harvard University Press, Cambridge (1931–1958)
37. Rescher, N.: Peirce's Philosophy of Science. University of Notre Dame Press, Notre Dame (1977)
38. Shoenfield, J.: Mathematical Logic. Addison-Wesley, New York (1967)
39. Simon, H.: On judging the plausibility of theories. In: Van Rootselaar, B., Staal, J. (eds.) Logic, Methodology and Philosophy of Science, vol. III, pp. 439–459. North-Holland, Amsterdam (1968)
40. Simon, H.: Computer modeling of scientific and mathematical discovery processes. Bulletin of the American Mathematical Association (New Series) 11, 247–262 (1984)
41. Stigum, B.: Toward a Formal Science of Economics. The MIT Press, Cambridge (1990)
42. Schiller, R.: The Subprime Solution. Princeton University Press, Princeton (2008)
43. Taylor, J.: The financial crisis and the policy response: An empirical analysis of what went wrong. Working Paper, Stanford University (2008)
44. Thagard, P.: The best explanation: Criteria for theory choice. The Journal of Philosophy 75, 76–92 (1978)

Pathophysiology of Cancer and the Entropy Concept*

Konradin Metze, Randall L. Adam, Gian Kayser, and Klaus Kayser

Abstract. Entropy may be seen both from the point of view of thermodynamics and from the information theory, as an expression of system heterogeneity. Entropy, a system-specific entity, measures the distance between the present and the predictable end-stage of a biological system. It is based upon statistics of internal characteristics of the system. A living organism maintains its low entropy and reduces the entropy level of its environment due to communication between the system and its environment. Carcinogenesis is characterized by accumulating genomic mutations and is related to a loss of internal cellular information. The dynamics of this process can be investigated with the help of information theory. It has been suggested that tumor cells might regress to a state of minimum information during carcinogenesis and that information dynamics are integrally related to tumor development and growth. The great variety of chromosomal aberrations in solid tumors has limited its use as a variable to measure tumor aggressiveness or to predict prognosis. The introduction of Shannon's entropy to express karyotypic

Konradin Metze
Department of Pathology, Faculty of Medicine, interdisciplinary working group "Analytical Cellular Pathology" and National Institute of Photonics applied to Cellular Biology, University of Campinas, Campinas, Brazil
e-mail: kmetze@fcm.unicamp.br

Randall L. Adam
Institute of Computing, interdisciplinary working group "Analytical Cellular Pathology" and National Institute of Photonics applied to Cellular Biology, University of Campinas, Campinas, Brazil

Gian Kayser
Institute of Pathology, University Freiburg, Freiburg, Germany

Klaus Kayser
UICC-TPCC, Institute of Pathology, Charite, Berlin, Germany

* Supported by FAPESP 2007/52015-0 and CNPq 479074/2008-9.

L. Magnani et al. (Eds.): Model-Based Reasoning in Science & Technology, SCI 314, pp. 199–206.
springerlink.com   © Springer-Verlag Berlin Heidelberg 2010

diversity and uncertainty associated to sample distribution has overcome this problem. During carcinogenesis, mutations of the genome and epigenetic alterations (e.g. changes in methylation or protein composition) occur, which reduce the information content by increasing the randomness and raising the spatial entropy inside the nucleus. Therefore, we would expect a raise of entropy of nuclear chromatin in cytological or histological preparations with increasing malignancy of a tumor. In this case, entropy is calculated based on the co-occurrence matrix or the histogram of gray values of digitalized images. Studies from different laboratories based on various types of tumors demonstrated that entropy derived variables describing chromatin texture are independent prognostic features. Increasing entropy values are associated with a shorter survival. In summary, the entropy concept helped us to create in a parsimonious way a theoretical model of carcinogenesis, as well as prognostic models regarding survival.

1 Introduction

Schrödinger [16] applied the thermodynamic laws to living organisms and postulated that for biological processes a continual in-flow of negative entropy is essential, which is equivalent to an exportation of entropy to its environment.

Later, Prigogine pointed out that living systems may be seen as thermodynamic stationary, non-equilibrium states, where the activity is characterized by entropy production or energy dissipation [13, 12].

In the following text we will show, how the entropy concept can help in a parsimonious way to create a theoretical model of carcinogenesis and prognostic models regarding survival. We are looking at entropy both from the view point of thermodynamics and from information theory, as an expression of system heterogeneity.

2 Entropy in Normal Cells and Tissues

Entropy may be seen both from the view point of thermodynamics, as measure of nonreversible energy, and of information theory, i.e. as an expression of system heterogeneity or the information content of a message [8]. A living system can be described by features that define its present stage and its expected development. Entropy, a system-specific entity measures the distance between the system's present stage and its predictable end-stage and is based on statistics of internal characteristics, such as thermodynamical, geometrical, or biochemical features.

For the analysis of entropy in living systems, the following assumptions are stated [8, 5]:

(a) A living organism has structures which can recognize certain objects in its environment and tries to import these objects applying a recognition process. Some biological structures may act as entropy-diminishing machines.
(b) It has defined spaces (macro-stages) of lower entropy compared with their environment.
(c) These spaces are equipped with mechanisms ("entropy diminishing machines"), which transform information into physicochemical changes. Exchange mechanisms allow intracellular storage and translation of information as negative entropy and export of thermodynamic entropy as heat and reaction creating a transmembrane entropy gradient in an open system far from equilibrium [5].

   This process requires free energy. The entropy-diminishing machines obey general biological laws, i.e., they will become less efficient with increasing age.
(d) All products (molecules, energy, entropy) must pass through a surface, e.g. a membrane.
(e) The transmembrane entropy-gradient is essential for life. Its catastrophic loss will provoke cell death.

Thus, living organisms maintain a lower entropy level and, in addition, lower the entropy level of its environment due to communication between the system and its environment. The system will decrease the number of degrees of freedom (or macro-stages) in the environment, and thus lower its entropy. Any part of a living organism which can be described by a non-random probability function may be a source of information. This information is generated by synthesizing "ordered" structures, e.g. macrocolucules or transmembrane gradients of ions. The information content of biological system may be obtained from the probability density functions describing the distribution of observable variables [5].

Classical information theory [17] permits to calculate informational entropy.

The concepts of thermodynamic and information entropy are closely related in living organisms. Information enables biological structures to counterbalance the effects of the second law of thermodynamics, which predicts increasing total system disorder (entropy) with time [5]. The information content increases when entropy (disorder) decreases. Shannon's entropy and the classical thermodynamic entropy are closely related, since at least the equivalent amount of entropy must be lost to acquire a given number of bits of Shannon information. (by multiplying the number of bits by Boltzmann's constant) [5]. Living organisms, which are open systems, retain information within the system, but simultaneously export the increase in thermodynamic entropy into the environment through heat flow or excretion of metabolic products.

3 Entropy in Neoplasias

At the moment there are two main paradigms regarding the pathophysiology of carcinogenesis: according to the most popular paradigm, carcinogenesis is usually accompanied by accumulating genomic mutations (as well as epigenetic modifications) so that most cells in malignant neoplasias contain a large number of mutations. During cancer progression, tumor cells undergo several additional genomic changes. Mutations that enhance tumor progression are most likely to be selected and the cells carrying these mutations tend to be dominant in the tumor. Intracellular entropy increases, which makes a permanent export necessary in order to maintain the highly organized structures of life. This may provoke an energetic overload, which can enhance genetic instability and thus induce further mutations.

Hauptmann [7] postulated that in this context cancer may be interpreted as an adaptative phenomenon, i.e. as a response to cellular stress. Adaptative mechanisms could be aneuploid polyploidization or the change in chirality of proteins and carbonhydrates, because the higher intrinsic energy of enantiomers would be able to reduce entropy of the cell.

Tumor progression may take various routes in the genome. A great number of different genetic and epigenetic alterations may provoke the same type of cancer, so that it is nearly impossible to establish for many solid malignancies a "prototypical cancer genotype". This phenomenon is interpreted as indicative of an underlying stochastic non-linear dynamics [5]. In this way, the evolution of cancer is related to a loss of internal cellular information

Shannon's Entropy-gray level histogram

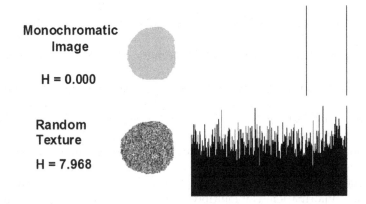

Fig. 1 In a gray value histogram with 256 gray values (8-bit image) Shannon's entropy of a monochromatic image (represented by a very narrow peak in the histogram on the right) will be 0. When all possible gray values are present (with random numbers representing the frequency pf each gray value) H will approximate the maximum of 8.

through mutations and epigenetic alterations. Thus, carcinogenesis appears to be related to genomic information degradation and the dynamics of this process can be investigated with the help of information theory. Gatenby and Frieden [5] corroborated these concepts applying two different simulation approaches. They came to the conclusion that tumor cells might regress to a state of minimum information during carcinogenesis and that information dynamics are integrally related to tumor development and growth.

In the alternative concept of carcinogenesis postulated by Duisberg et al [3], emphasizing chromosomal alterations instead of genetic mutations, entropy plays also a central role. According to Duesberg's hypothesis "cancer is caused by chromosomal disorganization, which increases karyotypic entropy" [3]. Because of the lack of chromosomal symmetry in aneuploidy, every time an aneuploid cell divides, the genome will be more disorganized. Thus the entropy of the genome increases with each cell division [14].

4 Applications for Prognosis

The great variety of chromosomal aberrations in solid tumors, which make often each patient an "unique case", has limited their use as a variable to measure tumor aggressiveness or to predict prognosis or for differential diagnosis. Thus, this degree of data heterogeneity has obscured the identification of a possible set of recurrent chromosome abnormalities specific for a certain tumor type. In order to overcome these difficulties Castro et al [2] described the karyotypic diversity of solid tumors applying Shannon's entropy concept. Tumor aggressiveness was correlated with karyotypic diversity estimated by Shannon's information entropy. After confronting different models, the authors suggested that the process which causes chromosome aberrations is neither deterministic nor totally random.

The absence of any unique pattern on the genotype suggests that a better description of the copy number data can be achieved if individual characteristics of each sample are considered. Freire et al [4] presented the entropy method as an initial scan of the copy number data, quantifyimg aberrations as a deviation from centrality without the bias of untested assumptions. This procedure is rapid, robust and does not need parameter calibration. Alfano [1] created a mathematical model that describes cellular phenotypic entropy as a function of cellular proliferation, survival, cellular transformation and differentiation. Genes may express splice variants, called alternative splicing, which can be quantified with the help of Shannon's entropy [15]. Severe alterations of this phenomenon are found in neoplasias. Alternative splicing showed highly significant gains of Shannon's entropy in many types of cancers and was correlated with estimated cellular proliferation rates.

Normal interphase nuclei reveal an organization of DNA, histones and nonhistone proteins which define the chromatin structure. During carcinogenesis, besides mutations of the genome, many epigenetic alterations (e.g. changes

Blasts – Acute Lymphoid Leukemia (B-ALL)

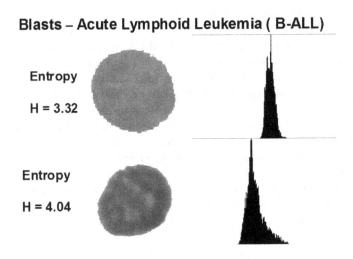

Entropy

H = 3.32

Entropy

H = 4.04

Fig. 2 Two examples of the gray-value histogram entropy of B-ALL blasts (May-Grünwald-Giemsa staining).

in the methylation or protein composition) may occur, which reduce the information content by increasing the randomness and raising the spatial entropy inside the nucleus. Therefore, we would expect an increase in the entropy of nuclear chromatin in cytological or histological preparations with increasing malignant potency of a tumor [8]. In this case, entropy is calculated based on the co-occurrence matrix or the histogram of the gray values of digitalize images [10].

Indeed, studies from different laboratories based on different types of cancer demonstrated that entropy-derived variables describing chromatin texture are independent prognostic features regarding survival, with increasing entropy values associated with shorter survival [8].

We would like to illustrate this concept with the following example [11]: Bone marrow smears of patients with acute B-lymphoid leukemia (B-ALL) stained by May-Grünwald-Giemsa were captured by a digital camera, the images gray-value transformed (8 bits) and the nuclei segmented for further analysis. Shannon's entropy was calculated for the gray value histogram of each nucleus according to the formula: $H = -\Sigma p(i) \log_2 p(i)$, with $p(i)$ the probability space containing the events. In this study, the minimum theoretical value for H was 0 for a monochromatic image and 8, when all gray values were equally present in the image of the nucleus (figure 1 and figure 2). When analyzing the survival of the patients in a prognostic model, Shannon's entropy of the nuclear texture was an independent prognostic factor for overall survival in a multivariate Cox regressionm with higher values suggestive of a worse outcome. It had been show earlier that in more aggressive cases of B-ALL, an increased number of DNA methylation changes can be found [6].

More methylation changes provoke accentuated alterations of the nuclear architecture in the stained slides, since the Giemsa staining pattern corresponds to that of the methylated CPG islands. Thus the association between higher entropy of the Giemsa staining pattern and worse prognosis of the patients can be easily explained.

The entropy concept is also of prognostic value when applied to cellular structures or functions visualized by immunohistochemistry. Furthermore, structural entropy, which is a measure of heterogeneity of neighboring events within a tumor, as well as its flow, have also shown to be of prognostic relevance [9].

5 Conclusions and Perspectives

The entropy concept helped us to create in a parsimonious way a model of the very heterogeneous and complex process of carcinogenesis. Furthermore, it enabled us to build up simple prognostic or predictive models based on genetic or morphologic analysis of neoplasias. Future studies should investigate the advantages of new kinds of entropy, such as Pincus or Tsallis' entropy in this context [14].

References

1. Alfano, F.D.: A stochastic model of oncogene expression and the relevance of this model to cancer therapy. Theoretical Biology and Medical Modelling 3, 5 (2006)
2. Castro, M.A.A., Onsten, T.T.G., De. Almeida, R.M.C., Moreira, J.C.F.: Profiling cytogenetic diversity with entropy-based karyotypic analysis. Journal of Theoretical Biology 234, 487–495 (2005)
3. Duesberg, P., Li, R., Fabarius, A., Hehlmann, R.: The chromosomal basis of cancer. Cellular Oncology 27, 293–318 (2005)
4. Freire, P., Vilela, M., Deus, H., Kim, Y.W., Koul, D., Colman, H., Aldape, K., Bogler, O., Yung, W., Coombes, K., Mills, G.B., Vasconcelos, A.T., Almeida, J.S.: Exploratory analysis of the copy number alterations in glioblastoma multiforme. PLoS ONE 3, e4076 (2008)
5. Gatenby, R.A., Frieden, B.R.: Information dynamics in carcinogenesis and tumor growth. Mutation Research 568, 259–273 (2004)
6. Gutierrez, M.I., Siray, A.K., Bhargava, M., Ozbek, U., Banavali, S., Chaudhary, M.A., Soth, H.E.I., Bathia, K.: Concurrent methylation of multiple genes in childhood all: Correlation with phenotype and molecular subgroup. Leukemia 17, 1845–1850 (2003)
7. Hauptmann, S.: A thermodynamic interpretation of malignancy: Do the genes come later? Medical Hypotheses 58, 144–147 (2002)
8. Kayser, K., Kayser, G., Metze, K.: The concept of structural entropy in tissue-based diagnosis. Quantitative Cytology and Histology 29, 296–308 (2007)

9. Metze, K., Adam, R.L., Silva, P.V., Carvalho, R.B.D., Leite, N.J.: Analysis of chromatin texture by pinkus' approximate entropy. Cytometry Part A 59A, 63 (2004)

10. Metze, K., Ferreira, R.C., Adam, R.L., Leite, N.J., Ward, L.S., de Matos, P.S.: Chromatin texture is size dependent in follicular adenomas but not in hyperplastic nodules of the thyroid. World Journal of Surgery 32, 2744–2746 (2008)

11. Metze, K., Mello, M.R.B.D., Adam, R.L., Leite, N.J., Lorand-Metze, I.G.H.: Entropy of the chromatin texture in routinely stained cytology is a prognostic factor in acute lymphoblastic leukemia. Virchows Archiv. 451, 114 (2007)

12. Nicolis, G.: Self-Organization in Nonequilibrium Systems. Wiley, New York (1977)

13. Prigogine, I.: Thermodynamics of Irreversible Processes. Wiley, New York (1967)

14. Rasnick, D.: Aneuploidy theory explains tumor formation, the absence of immune surveillance, and the failure of chemotherapy. Cancer Genetics and Cytogenetics 135, 66–72 (2002)

15. Ritchie, W., Granjeaud, S., Puthier, D., Gautheret, D.: Entropy measures quantify global splicing disorders in cancer. PLoS Computational Biology 4, e1000,011 (2008)

16. Schrödinger, E.: What is Life? Cambridge University Press, Cambridge (1944)

17. Shannon, C., Weaver, W.: The Mathematical Theory of Communication. University of Illinois Press, Chicago (1949)

A Pattern Language for Roberto Burle Marx Landscape Design

Carlos Eduardo Verzola Vaz and Maria Gabriela Caffarena Celani

Abstract. Patterns were developed by Christopher Alexander [2] to synthesize rules of good design practice. Although he does not tell us where he took his patterns from, it is possible to infer that they are the result of his sensible observation of existing situations in European cities. However, these solutions are not necessarily true for situations in other countries, with different climates, economies and societies.The Brazilian landscape designer Roberto Burle Marx is considered to have achieved the highest level of excellence and success in his designs for private gardens and public open spaces. In other words, there is no doubt about his being considered a "specialist", in the AI (artificial Intelligence) sense, in his field. The present paper proposes a systematization of the knowledge present in the work of Brazilian landscape designer Marx as "patterns" that can be used by students to overcome their difficulties related to the lack of professional experience.

Introduction

The fields of artificial intelligence and cognitive sciences acknowledge that inference power alone is not enough for solving certain types of problems, such

Carlos Eduardo Verzola Vaz
School of *Civil Engineering*, Architecture and Urban Design (FEC),
State University of Campinas, Campinas, Brazil
e-mail: cevv00@gmail.com

Maria Gabriela Caffarena Celani
School of *Civil Engineering*, Architecture and Urban Design (FEC),
State University of Campinas Campinas
e-mail: celani@fec.unicamp.br

L. Magnani et al. (Eds.): Model-Based Reasoning in Science & Technology, SCI 314, pp. 207–219.
springerlink.com　　　　　　　　　　　　　　　© Springer-Verlag Berlin Heidelberg 2010

as wicked and multi-criteria problems. Expert systems have been developed to help solving them. Expert systems consist of a data base of previously developed solutions in a specific field, and a set of inference rules, which can help finding the best solution for a specific situation.

In architecture schools, it is possible to notice how even the smartest novice students often struggle with their design exercises due to their lack of professional experience. No matter how hard they study, the solution to their problems cannot be found on books.

Christopher Alexander's *A Pattern Language* [2] was conceived as a data base intended to help architects find good solutions to design problems, based on the experience compiled in the book. In order to make easy for readers to consult the book, Alexander's patterns were organized in 253 numbered entries.

Each pattern might be seen as an in-the-small handbook on a common, concrete architectural domain. Entries intertwine these "problem space", "solution space", and "construction space" issues in a simple, down-to-earth fashion, so that each evolve concurrently when patterns are used in development. The patterns have five different parts [9]:

1. Name. A short familiar, descriptive name or phrase, usually more indicative of the solution than of the problem or context;
2. Context. Delineation of situations under which the pattern applies. Often includes background, discussions of why this pattern exists, and evidence for generality;
3. Problem. A description of the relevant forces and constraints, and how they interact. In many cases, entries focus almost entirely on problem constraints that a reader has probably never thought about. Design and construction issues sometimes themselves form parts of the constraints;
4. Solution. Static relationships and dynamic rules (microprocess) describing how to construct artifacts in accord with the pattern, often listing several variants and/or ways to adjust to circumstances.

Each pair of problem and solution, which forms a pattern is presented in a format that consists of: (see Table 1) [6]

The objective of the present research is to propose a design solution database system based on Alexander's pattern language, but more effective and easier to be consulted by architecture students. The work will result in a new format to each pattern and a different and more visual categorization of the knowledge system that all patterns are capable to form. It proposes a systematization of the knowledge present in the work of Brazilian landscape designer Marx as "patterns" that can be used by students to overcome their difficulties related to the lack of professional experience.

Table 1 Alexander pattern format.

A PHOTOGRAPH	showing an example of the pattern in use
AN INTRODUCTORY PARAGRAPH	which sets the pattern in the context of other, larger scale patterns
THE HEADLINE	an encapsulation of the problem (one or two sentences)
THE BODY	of the problem (this can be many paragraphs long)
THE SOLUTION	Always stated in the form of an instruction
A DIAGRAM	Shows the solution in the form of a diagram
A CLOSING PARAGRAPH	Shows how this pattern fits with other, smaller patterns

1 Patterns as a Method

Although Alexander [2] does not tell us where he took his patterns from, it is possible to infer – for example, looking at the photographs in the book – that they are the result of his sensible observation of existing situations in European cities. However, these solutions are not necessarily true for situations in other countries, with different climates, economies and societies, so if we use pattern language as a method, rather than as rigid standards, we can probably get better results.

In Brazil, for example, landscape designer Roberto Burle Marx is considered to have achieved the highest level of excellence and success in his designs for private gardens and public open spaces. In other words, there is no doubt about his being considered a "specialist", in the AI (artificial Intelligence) sense, in his field. However, he has passed on and we cannot ask him any question; we can only look at his work.

Extracting patterns from Burle Marx's constructed work, which proved successful, may help novice designers using the master's knowledge to overcome their lack of experience. The advantage of this approach is the fact that, by using Burle Marx's patterns, and not directly his designs, students will not run into the common error of copying existing solutions *"ipsis literis"*; they will have to invent their own designs (in terms of geometry, materials, and so on), based on general rules of good practice.

1.1 Materials and Method

The study started with an analysis of Alexander's [2] book *A Pattern Language – towns, buildings, construction*, and a selection of all the patterns that were directly related to landscape design. Fifty two patterns were selected.

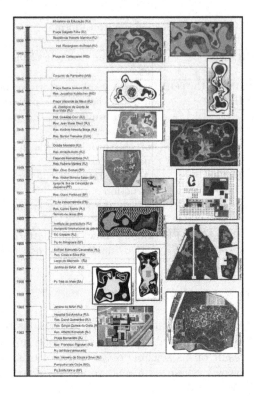

Fig. 1 Burle Marx's designs organized chronologically.

Next, images of virtually all of the projects developed by Burle Marx were collected, and organized chronologically (Figure 1). By looking at his designs, it was possible to infer a hierarchy of fields in which his decisions were taken (Figure 2).

His first level of decision was probably the overall zoning of the space, which is typical of the modernist movement. The next level refers to the definition of typical elements of a garden or urban open space: circulation, passive leisure, active leisure, green area, water, focus element and relationships with the context. Each of these elements were further subdivided into more specific sub-elements. For example, circulation was subdivided in vehicle circulation and pedestrian circulation, and so on.

The next step in the study consisted of taking the previously selected landscape design-related patterns from Alexander and categorizing them in terms of Burle Marx's hierarchical levels of decision (Figure 2).

Next, the relationships between Alexander's selected patterns, as pointed by the author himself, were mapped by means of a graphic representation (Figure 3). It was possible to infer that the patterns that have more relationships, i.e., the ones that receive the greatest number of connecting lines, are the most important ones. These patterns correspond to Burle Marx's circu-

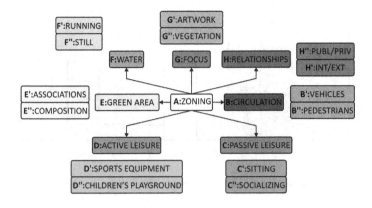

Fig. 2 A hierarchy of elements inferred from Burle Marx's designs.

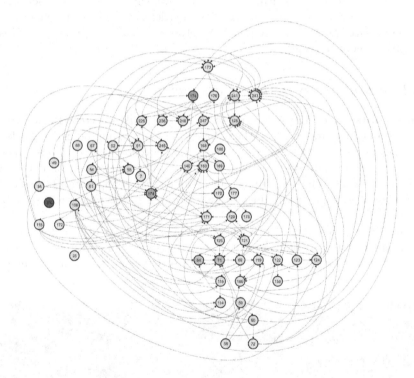

Fig. 3 Alexander's landscape design-related patterns categorized according to Burle Marx's levels.

lation and passive leisure categories, which, coincidently, are the minimum necessary elements of any garden or public open space (Figure 4).

The final step in the process, consisted of looking at Burle Marx's work and trying to infer how he would typically solve problems present in Alexander's

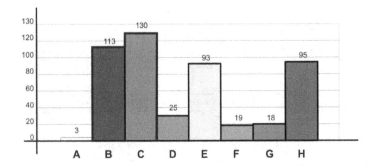

Fig. 4 Graphic representation of the relationships between Alexander's selected patterns.

selected patterns. In his writings Burle Marx was more worried in describing the correct way to use the vegetation as a garden compositional element, so the only source that we had to study the other categories (Figure 2) were his designs plans or photographs.

1.2 Examples of Other Categories

Garden seats, one of Alexander's patterns, was typically solved by Burle Marx with two main characteristics: there was always something behind a garden seat, such as a raised flowerbed, to serve as a "psychological" sit-back. Also, seats were almost always L-shaped, in such a way that people could sit and talk and look at each other. Images of different seat configurations are show in Figure 5. From it is possible to infer a design pattern to create specific places to seat.

Another example is paths. Alexander proposed a few patterns for pedestrian paths, but none of them talked about graphic compositions on them. Burle Marx typically used graphic compositions on the floor to direct

 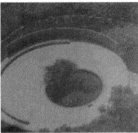 

Fig. 5 Images of garden seats from Burle Marx's work.

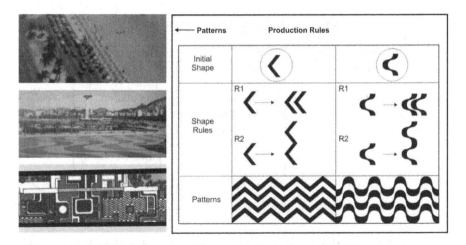

Fig. 6 A pedestrian path pattern inferred from Burle Marx's work.

pedestrian walking and to give people a sense of scale though the repetition of graphic elements. A graphic rule for this pattern is shown in Figure 6.

However, *A pattern language* doesn't take into account the fact that the same element may be seen from different points of view when considered at different moments of the process or at different scales. The introduction of different types of *categorizations* helped to organize the design elements in smaller groups, for faster access. Besides, the network of patterns is too complex to follow. A *graphic visualization system* would make the database easier to understand. Finally, as pointed by Knight [8], Alexander's patterns verbal descriptions were open to personal interpretations. A *graphic representation of application rules,* such as those used in Stiny and Gips's [11] shape grammars, could make the instructions more rigorous but at the same time easier to understand.

2   Visual Database System

2.1   Categorizations

In *A pattern language*, design elements are treated as concrete things. However, we know that during the design process architects think about design elements according to different categorizations, with different abstraction levels.

For example, at the very beginning of a project a designer may think about architectural elements as geometric primitives, such as points and lines. As the design evolves and changes scale, lines may be turned into walkways and points into sitting areas. Next, as the process evolves, a sitting area may be changed into a children's playground, due to functional requirements. At the

Table 2 The three categories proposed.

Category type	Phase	Categories
Geometry	Conceptual phase, small scale	Point, line, plane, volume
Function	Intermediate phase, medium scale	Circulation, resting, passive leisure, active leisure, ornamentation, contemplation
Material	Final phase, larger scale	Hard, fluid, vegetal

end of the process, the designer may consider different materials for paving the walkway and the playground, such as sand, a fluid material, brick, a hard material, or grass, a plant material.

In summary, there are different ways of categorizing design elements, each of them being more appropriate for a specific phase of the design process. Table 1 shows the three types of categorizations proposed in the present work.

In order to illustrate the three types of categorizations above, an example based on kitchen objects is presented in Figure 7. There are 5 items: a knife, a plate, a glass, a teacup and a cooking pan. The items are grouped together in different manners in each category. In certain cases, the same item may appear in two categories. For example, according to Function, the knife can be classified both for Eating and for Cooking. At certain moments a knife may be seen as a linear element. At other moments its material or function may be of more interest.

2.2   Data Representation

When *A pattern language* was published as a website[1] there were hyperlinks between each pattern and the ones related to it, making the look up process more dynamic. Hipertext can represent sequences of links linearly, but network structures cannot be properly visualized or understood.

The issue of visualizing complex inter-related information has been addressed by many researchers, such as Stouffs and Krishnamurti [14]. More recently, a new type of commercial dynamic representation software has been developed, the so-called "mind map" applications. This type of software is based on Buzan's [4] diagrams, which combine words and images connected by branches in a hierarchical or a network structure. These diagrams are also called "concept maps" and "content maps". Although mind maps can

---

[1] http://www.patternlanguage.com

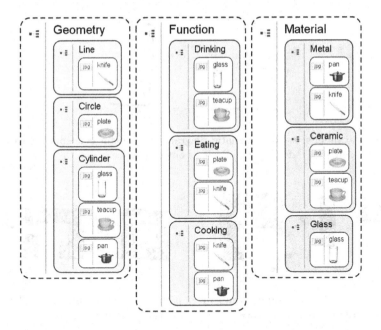

Fig. 7 Kitchen items organized according to the three categories proposed.

be drawn by hand, the advantage of computer implementations is that they allow mixing text, digital images, Internet addresses, and references to files stored in the computer. Besides, they can perform automatic searches on online databases.

There are many examples of implementations, such as Tufts University's Visual Understanding Environment, VUE, which is a free product. Besides, VUE allows representing ontologies, with the specification of relationships between parts. For this reason, VUE was chosen to build a prototype of the proposed design consulting system.

2.3   Rules

Another important issue that will be addressed in the proposed system is the use of graphic rules, instead of verbal descriptions, to represent the design solutions. In A pattern language most of the patterns were illustrated by photographs of examples of good and bad situations. Some of the patterns had also schematic diagrams that were used to explain concepts (Figure 8). Only one pattern had graphic rules to explain what could be done in specific cases (Figure 9).

Shape grammars rules are based on Post's production rules [10]. They are represented by a shape on the left hand side, an arrow in the middle, and a transformed shape in the right hand side. Whenever the shape on

Fig. 8 Alexander's use of schematic diagrams to explain concepts (convex and non-convex shapes; Pattern 106 – Positive outdoor space.

Fig. 9 The only two graphic rules in *A pattern language*, used to explain the transformation of negative spaces into positive spaces (Pattern 106).

the left hand side of the rule is identified, it may be replaced by the shape on the right hand side of the rule. In other words, graphic rules explain in which situations a transformation may be used. Therefore, they can guide the correct application of the good solutions stored in the database.

The advantage of using rules to guide the use of design solutions is that they can be diagrammatic, i.e., they can treat elements in a more or less abstract way. This means that they can be applied to different situations, whereas specific examples may induce direct copy. Rules can be combined to guide the designer in a refinement process: they can represent design elements at an abstract level, for further specification on a more advanced step in the design process.

11 shows some examples of landscape design rules. Rule 1 inserts a vertical plane along a pathway. Rules 2a and 2b further specify the vertical plane: Rule 2a substitutes it by a wall covered with vines, and Rule 2b by a sequence of palm trees.

In the design process, usually a designer starts by thinking about compositional elements, but always keeping function and material in mind. The dialogue below illustrates a possible conversation of a landscape architect with himself during a design process:

I want a visual bareer along this pathway.
What geometric element could be used as a bareer?
A vertical plane!
What is this vertical plane made of?
A wall covered with vines.

But this is too closed! I want something more open that can still give the feeling of enclosure.
I could use a sequence of palm trees!

Figure 11 shows examples of the application of rules 2a and 2b in actual landscape design projects developed by Burle Marx.

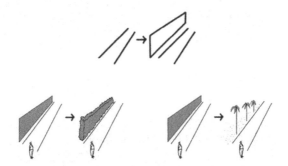

Fig. 10 Rule 1 (top): Inserting a vertical plane along a pathway; Rule 2a (bottom left): Substituting the vertical plane by a wall covered with vines; Rule 2b (bottom right): Substituting the vertical plane by palm trees.

Fig. 11 Examples of the application of Rules 2a (left) and 2b (right) in landscape projects developed by Burle Marx.

3   Making the Prototype

In order to test the proposed system, a prototype is being developed. The database will be filled with Roberto Burle Marx's landscape design solutions. Burle Marx is one of the best well-know Brazilian landscape architects worldwide. He gained international reputation for designing hundreds of gardens in Brazil and abroad, and is considered an expert in the field [1] and [5]. The prototype is being developed in VUE software, and will include verbal descriptions, graphic rules and photographs of actual gardens designed by Burle Marx.

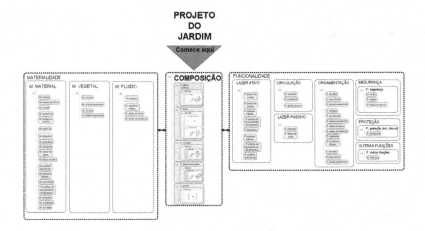

Fig. 12 The prototype's general schema.

A collection of images showing examples of Burle Marx's good design so-lutions has already been prepared. Next, graphic rules will be developed to explain the situations in which each design solution can be applied. Finally, the images and rules will be stored in the categories already created in VUE's visual map, shown in Figure 12. A network of connections will then be estab-lished in the environment, which will allow dynamic search of design solutions, jumping from one category to the other.

4   Conclusion

After mapping Burle Marx's design patterns onto Alexander's landscape de-sign related patterns, the next step in this research will consist of testing the efficacy of patterns in a more visual categorization for teaching landscape design. An experiment with novice undergraduate architecture students will be carried out. Improvement may be needed to clarify some of the patterns according to students understanding. These experiments will confirm if our initial hypothesis – that design patterns can be used to help novice students make better designs – was correct.

References

1. Adams, W.H.: Roberto Burle Marx: landscapes reflected. Princeton Architec-tual Press, New York (2000)
2. Alexander, C., et al.: A Pattern Language – Towns, Buildings, Construction. Oxford University Press, New York (1977)
3. Bittencourt, G.: Inteligência Artificial: Ferramentas e teorias. Editora UFSC, Florianópolis (2006)

4. Buzan, T.: The Mind Map Book. Penguin Books, London (1996)
5. Fleming, L.: Roberto Burle Marx: Um Retrato. Editora Index, Rio de Janeiro (1996)
6. Fincher, S.: Analysis of design: An exploration of patterns and pattern languages for pedagogy. Journal of Computers in Mathematics and Science Teaching: Special Issue CS-ED Research 18(3), 331–348 (1999)
7. Kalay, Y.: Architecture's New Media: Principles, Theories, and Methods of Computer Aided Design. MIT Press, Cambridge (2004)
8. Knight, T.W.: Transformations in Design: A Formal Approach to Stylistic Change and Innovation in the Visual Arts. Cambridge University Press, Cambridge (1994)
9. Lea, D., Alexander, C.: An introduction for object-oriented designers. Software Engineering Notes 19(1), 39–46 (1994)
10. Stiny, G., Gips, J.: Algorithmic Aesthetics. University of California Press, Berkeley (1978)
11. Stiny, G.: Introduction to shape and shape grammars. Environment and Planning B: Planning and Design 7, 343–351 (1980)
12. Mitchell, W.J.: The Logic of Architecture. MIT Press, Cambridge (1990)
13. Theodoridis, S.: Pattern Recognition. Academic Press, USA (1999)
14. Stouffs, R., Krishnamurti, R.: Data views, data recognition, design queries and design rules. In: Proceedings of the First International Conference on Design Computing and Cognition, Cambridge, Massachusetts, USA, pp. 219–238 (2004)
15. Terzides, K.: Algorithmic Architecture. MIT Press, Cambridge (2006)

A Visual Model of Peirce's 66 Classes of Signs Unravels His Late Proposal of Enlarging Semiotic Theory

Priscila Borges

Abstract. In this paper I will present the visual model of Peirce's 66 classes of signs, which I call the Signtree Model, and show how the model helps on developing the enlarged semiotic system that Peirce left unfinished. Peirce's best-known classification is that of 10 classes of signs. However, in his later years, when developing the sign process in much greater detail, Peirce proposed a classification of no less than 66 classes of signs. In contrast to the first classification, Peirce never worked out the details, making it a difficult topic that has received little attention from semioticians. For a better understanding of the 66 classes, I built the Signtree Model, which makes clear that the 66 classes work together composing a single dynamic system. As the Signtree describes all the 66 classes and visually shows how they are related in a dynamic system, the model can be a powerful tool for semiotic analysis, revealing details of a complex process composed of many elements and multiple relations emphasizing semiosis and the growing of signs. More than that, the Signtree gives clues about philosophical issues such as the relation between semiotic and pragmatism, between semiotic and metaphysics, and the relation among the three branches of semiotic: speculative grammar, critical logical and methodeutic.

1 Introduction

Peirce conceived of his semiotic as a logical discipline, an abstract and general theory for the mapping, classification, and analysis of sign processes. His best known and most thoroughly elaborated general classification of signs consists of ten main classes, but Peirce went further in his reflections on

Priscila Borges
Pontificia Universidade Católica de São Paulo, PUC-SP and FAPESP,
São Paulo, Brazil
e-mail: `primborges@gmail.com`

L. Magnani et al. (Eds.): Model-Based Reasoning in Science & Technology, SCI 314, pp. 221–237.

the classification of signs until he arrived at a system of sixty-six classes. However, he did not elaborate this system of sixty-six classes in a detailed manner, and it remained an unfinished project, which became a controversial topic in Peircean scholarship.

I have been dedicating my research to this controversial topic for a good number of years. Two years ago I proposed a visual model to represent the 66 classes of signs, which I called Signtree. The idea was to build a model that could visually describe the 66 classes and to show the complexity of the sign system. With that purpose, I created a 2D model and 3D model. The 2D model was used as a guide to build the side views and then create the 3D model. Both diagrams describe the logical structure of all the 66 classes of signs and present them together as a complex system. But only the 3D model made explicit the dynamic of this system and the relations between semiotics and Peirce's philosophy.

The purpose of the visual model was to provide a detailed graphical representation and to contribute to a better understanding of this system. Graphical diagrams are very useful in making complex and abstract conceptual systems more clear. They streamline the work with the large number of classes that are related to each other in many different ways, because in the visual model one can see the whole set of classes at the same time. On a written text the classes of signs have to be described one after the other, and, no matter how much one tries to explain their relations, they can never appear together.

The proposed graphical model has been repeatedly subjected to revision and testing. Since its first appearance, the Signtree has gone through some improvements specially an important change on its design. The change was not on its logical structure, but on its visual form, making the Signtree model more accurate and representative of Peirce's semiotics. The details of the change will be presented later on.

On this paper I will present some diagrams to show the creation process of the model and I will demonstrate that the 66 classes of signs as displayed in the Signtree can help on developing a method to apply all these classes on semiotic analysis and also that observing and experiencing this complex semiotic system shows some clues about philosophical issues.

Peirce defines semiotics as the 'science of the necessary laws of thought' [17, 1.444]. The study of signs begins with the observation of the signs characteristics that are well known and continues, in processes of abstraction and inferences, with the elaboration of a more comprehensive general system of all possible types of signs. Any classification is fallible and must be subject to a critical reexamination in processes of abstraction and learning from observation and experience. Abstraction involves mental diagrams as useful tools in the discovery of conceptual structures. By means of observing a mental diagram, new insights about the domain under scrutiny may be obtained [17, 2.227]. The present study follows these Peircean guidelines in its proposal of a graphical representation of the sixty-six classes of signs and in its revision and testing.

The model was created according to the logic of the phenomenological categories applied to the ten trichotomies, which produces the 66 classes of signs. I will briefly present the phenomenological categories and the sign trichotomies so that you can understand how the Signtree represents these concepts. I am not going deeper on its explanation for the sake of not drifting from the course of this presentation.

2 The Phenomenological Categories

Peirce's three phenomenological categories of firstness, secondness, and thirdness, are the foundation of his semiotics [17, 8.328]. According to their definition, "the First is that which has its being or peculiarity within itself. The Second is that which is what it is by force of something else. The Third is that which is as it is owing to other things between which it mediates" (W5: 229). The three categories are interrelated as follows: firstness is independent of any other category; secondness depends on firstness; and thirdness depends on secondness and firstness. They are represented in the model as circles, squares and triangles respectively (Fig. 1).

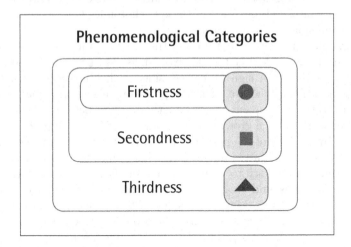

Fig. 1 The phenomenological categories.

3 The Sign Trichotomies

Among Peirce's many definitions of the sign is the following: "A REPRESEN-TAMEN is a subject of a triadic relation TO a second, called its OBJECT, FOR a third, called its INTERPRETANT, this triadic relation being such

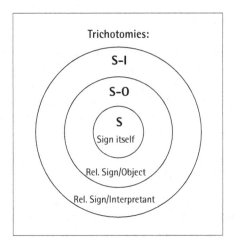

Fig. 2 The 3 sign trichotomies.

that the REPRESENTAMEN determines its interpretant to stand in the same triadic relation to the same object for some interpretant" [17, 1.541].

In his earlier classification of signs, Peirce considers only three trichotomies: the sign in itself, the sign in relation to its object, and the sign in relation to its interpretant. Each of these trichotomies belongs to one of the three phenomenological categories.

The first sketch in the elaboration of the 2D diagram was a tree diagram with upward branches for the ten classes of signs. The growth of a tree indeed evinces an affinity with sign processes, since each bifurcation of a branch results in a triadic structure. The temporal order in the sequence of the antecedent to the subsequent evinces another affinity between the growth of signs in semiosis and the growth of the branches of a tree (Fig. 2).

Inspired by the idea of the parallelism between the growth of trees and the growth of signs, the Signtree adopted the diagrammatic image of tree rings used in dendrochronology to count the age of the trees by counting their annual growth rings and to derive insights into climate changes over the centuries from their size. More than signs of time, tree rings are indices of influences between ecological systems. The growth of tree rings has affinities with the process of semiosis. The first trichotomy lies in the center of all rings; the next trichotomy begins with the second ring, and so on.

Peirce derived his ten main classes of signs from the logic of his phenomenological categories. Thus, if the first constituent of the trichotomy is of the nature of firstness, it can only determine relations of this very category. If the first constituent of the trichotomy is an existent, which is of the nature of secondness, then it can determine as its second constituent a relation of mere possibility (firstness) or existence (secondness). Finally, if the ground of the sign

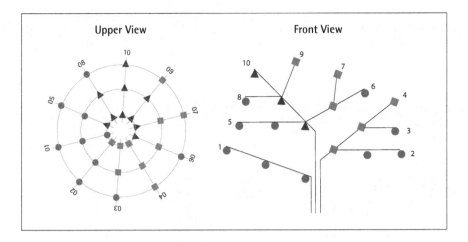

Fig. 3 The reduced Signtree Model (10 classes).

is a law (thirdness), the relation between sign and its object can be one of a possibility (firstness), existence (secondness), or law (thirdness) (Fig. 3).

It is well known that Peirce expanded the system of sign relations first by introducing the additional subdivision of the object into the immediate and dynamical object and then by introducing the subdivision of the interpretant into the immediate, the dynamical and the final one. The immediate object is the way in which the dynamical object is represented within the sign. The dynamical object is the object that is outside the sign and which the sign intends to represent. To represent it, the sign must determine an interpretant, which also represents the object of the sign. That is possible because within the sign there is the immediate interpretant, which has the power of determining an interpretant outside the sign, that is, the dynamical interpretant. This dynamical interpretant is an interpretant produced in an interpreting mind. A sign can determine more than one dynamical interpretant since all dynamical interpretants are potentially contained in the immediate interpretant. The final interpretant is the interpretative result to which every interpreter might come when the semiotic process is sufficiently developed [24, pp. 493–94] (Fig. 4).

The system of the 66 classes obeys the same logical rules, which determine the system of the ten classes of signs. When three trichotomies are considered, the structure of each sign must be described in three stages; with ten trichotomies, each class must be described in ten stages (Fig. 5).

4 The 10 Trichotomies and Their Determining Order

The logical premises valid for the elaboration of the ring shaped diagram proposed in this paper initially suggested that the central ring should represent

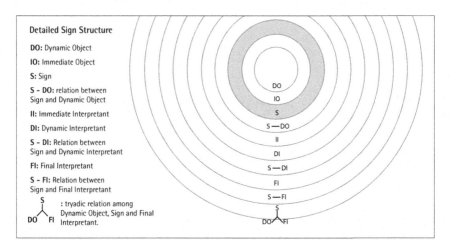

Fig. 4 The 10 sign trichotomies.

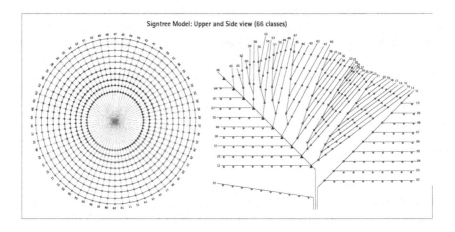

Fig. 5 The Signtree Model (66 classes).

the ground of the sign, since this is where a process of semiosis has its origin. A diagram following this line of thought would have two advantages: first, it would be in accordance with Peirce's own description of the processes of semiosis and his order of the ten trichotomies of 1908; second, it would bring into relief the relation of the sign with its object, showing that the cognition and semiosis begins with the sign.

However, a diagram constructed in this way would not sufficiently comply with Peirce's premises concerning the possible relations between the ground of the sign and its immediate object [17, 8.353–365]. According to the logic of the ring diagram, the rings evince relations of determination in the order from their inclusion. Hence, since the object determines the sign and not vice-versa, it is necessary to put the dynamical object in the central ring, followed by the immediate object and the ground of the sign.

The diagram complies with Peirce's premise that everything we can tell about the object is what the sign exhibits of it; this is why the representation of human knowledge appears in the middle and not in the beginning of the semiotic process in its diagrammatic representation. To represent the object centrally as the starting point of the process of semiosis is then a good diagrammatic method of showing that the origin of knowledge is not the human being. Representing the dynamical object in the center of the diagram also allows showing that the process of semiosis is more encompassing than the human mind and that humans on their own will never have full but only approximate cognizance of either the origin or the endpoint of this process. To place the object at the center, which represents the starting point of the process of semiosis, is also in agreement with Peirce's premise that there is indeed a reality, which does not depend on what we think of it. Furthermore, it is in accordance with the theory of semiotic growth, which takes the concept of intelligence far beyond the limits of the human mind. Last, but not least, the central circle can also represent the backward movement exerted by the dynamical object, whereas the ring of its periphery can represent the infinite possibilities of semiosis, both representing a temporal order similar to the one represented by the tree diagram.

The Signtree takes uncertainty and the theory of continuum into account. Its center represents the incomplete knowledge of the dynamical object in the uncertainty of its beginning. Its line of circumference represents the growth of rings as a growth of signs in time. The triadic relations represented by the last circumference of the circle stand for the growth of semiosis in its form of overlapping rings, in which the subsequent ring does not annihilate its antecedent ring, showing that both represent the growth of ideas. The interpretation of a sign is a process in which further signs are created with the same potential, which in turn accounts for the logical possibility of infinite semiosis; its further implications will be discussed below.

5 Further Implications of the Diagram

What are the relations between Peirce's philosophy and his semiotics that can be elucidated by the diagram of his thought? The 2D diagram served as a guide for the design of the ground plan and the side view further to be elaborated as the 3D diagram. These diagrams offer a detailed representation of the logical structure of the sixty-six classes of signs. They show a complex and coherent system without isolating any of its elements. However, only the 3D model is able to shed light on the relation between semiotics and Peirce's philosophy. Represented in the form of a tree with root and braches, the diagram of the system of signs has ecological implications of growth. (To see the 3D model please access: http://www.youtube.com/watch?v=O4iRL1kFSLk)

Let us consider how the roots are formed. As shown above, the dynamical object of semiosis always goes back in time in relation to the sign; it is never fully apparent in the sign in all of its implications. The sign can represent it in many different ways, but always only partially, never completely. Since it is impossible to have full access to the dynamical object, one might say, it is withdrawing itself. Its movement of withdrawal is represented in the axle z by the direction indicated by the negative sign. Since the dynamical object is located in the central ring, we can imagine that its movement of withdrawal forms the trunk and roots of the tree.

To see how the branches grow, it is necessary to consider the exterior rings. The last three rings show the final interpretant, the relation between the sign and its final interpretant, and the relation between the dynamical object, the sign, and its final interpretant. Since the final interpretant is not an existent, but a possible representation created by the sign, the end of the semiotic process is unattainable; the goal of semiosis is always in the future *ad infinitum*. The ring that represents the relation between the sign and its final interpretant points to the description of the process of semiosis in its complete way: the triadic relation between the object, the sign, and its interpretant.

Two processes are going on simultaneously in semiosis. On the one hand, the dynamical object withdraws in the direction of the ground, forming the trunk and the roots. This movement makes the object more complex and impedes the possibility of the full representation of the sign. On the other hand, the triadic sign relation involves a process of mediation, which can also be understood as a way of thought. This process indicates the growth of signs, represented by the branches in the diagram. The nature of these two processes justify the assumption that the movement of withdrawal of the dynamical object represents a link between semiotics and metaphysics, whereas the representation of the growth of semiosis, evident from the insight that the final interpretant is in the future, constitutes a link between semiotics and pragmatism. These hypotheses concerning the connections between semiosis and the sciences are supported by ideas which Peirce elaborates in his "critical analysis of logical theories" under the title of *Minute Logic* [17, 2.1–118]. In a passage from this treatise, Peirce emphasized that metaphysics is possible only if "founded on the science of logic" [17, 2.36], that is, on semiotics.

Taking into account the trichotomy order and the imagined movement of the dynamical object forming the roots, I decided to draw roots on the diagram (Fig. 6). To do that, instead of representing the dynamic and imme- diate object trichotomies, that are the first two in the center of the diagram, on the branches, I represented them on the roots of the tree. This change emphasized that human experience begins with the sign, which is the first trichotomy to appear in the branch. Although the sign represents and give

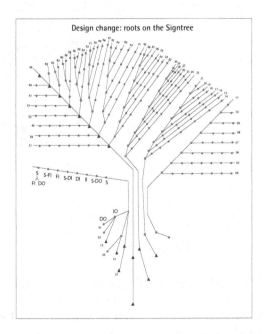

Fig. 6 Design change: roots on the Signtree.

information about the object, the sign is determined by the object, which is underneath the ground.

How metaphysics and pragmatism are founded on semiotics will become evident if one considers the starting and end point of the diagram. Why are there on the first trichotomy only one class of firstness, 10 of secondness and 55 of thirdnessof the diagram and that on the last trichotomy are 55 classes of firstness, 10 of secondness and 1 of thirdness? And what that means?

6 Semiotics and Metaphysics

In the ring representing the dynamical object, there are fifty-five classes of collective signs, ten classes of signs of occurrence and one class of abstractive sign (sign of possibility) (Fig. 7). If the dynamical object is what determines the sign and appears only by mediation of the sign, it might then be the real. Since metaphysics searches for the reality below appearance, the withdrawal of the object appears to be congruent with the goals of metaphysics. To obtain a better understanding of this idea, one has to clarify what reality is in the context of Peirce's philosophy; it is that "which is as it is independently of how we may think it to be"[17, 7.659].

Reality is not the same as existence. Reality is not restricted to a single instance of experience defined as a brute fact but it embraces a temporal dimension allowing the facts to be perceived in their regularity. This is in

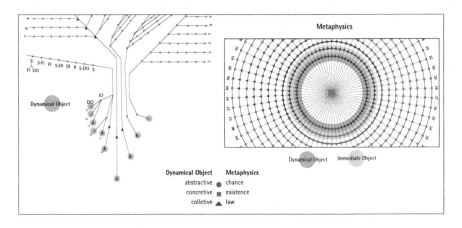

Fig. 7 Metaphysics.

accordance with the realist view that generals are real, and it explains the dominance of the classes of collective signs in reality. The impossibility of having any complete access to reality can be explained by the fact that no particular can express any general.

From these premises follows the conclusion that generals are real, but not that they are existent, for existence is the domain of individuals. It follows that if reality is grounded in the trichotomy of the dynamical object, existence has to be grounded in the trichotomy of the immediate object, for "the immediate object is the object *as it appears* at any point *in* the inquiry or semiosis process" [20]. Thus, the relation between Peircean semiotics and metaphysics is represented in the center of the diagram.

Considering that the phenomenological categories are expressed in Peirce's metaphysics as chance, existence, and law, and that chance is pure possibility, existence requires occurrence, and law indicates a necessity, it is possible to see the correspondence of these categories with the ones, which Peirce ascribes to the dynamical object. Chance relates to the class of abstractive signs or signs of possibility (red circle); existence relates to signs of occurrence or concretive signs (green squares), and law relates to signs of collection or collective signs (blue triangles).

7 Semiotic and Pragmatism

What happens at the other end seems to be exactly the opposite, since there is predominance of the category of firstness in the classes of signs and only one class expressed by thirdness (Fig. 8). But the detailed analysis of this other side of the diagram will show that there is a fusion between them. Considering that the triadic relation expressed in the last trichotomy is seen as the description of the process of thought, one can inquiry into what the

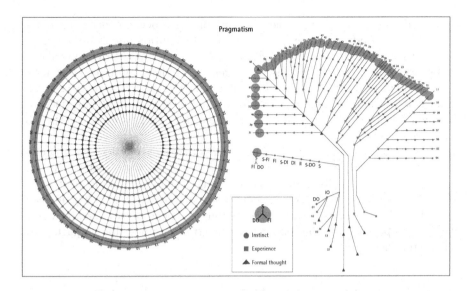

Fig. 8 Pragmatism.

function of thought is as the main goal of pragmatism. Therefore, the passage from the Peircean theory of signs to his pragmatism, or pragmaticism, is represented by the thinnest and most remote branches of this tree. According to Peirce, the ultimate purpose of thought is the development of an idea, rather than action itself.

What is found in the last trichotomy, the one that describes the triadic relation between the sign, its dynamical object and the final interpretant, are thoughts of three kinds: instinct, experience, and form. Of the sixty-six classes of signs, fifty-five are expressed by thinking in the form of instinct, ten by experience and one by formal thought. It seems that the possibility of attaining the pragmatic ideal (concrete reasonableness) is contained in this sole class of signs resulting in formal thought.

The concrete reasonableness considers self-control for the acquisition of new habits. Therefore, it is not strange that this ideal be situated in the only class of signs entirely composed of relations of thirdness, which will be essential in this system. Reason does not lead to complete determination, to a final thought, nor does it lead to any truth as conceived by common-sense. Reason is thought at the level of thirdness and, since semiosis means the creation of ever new signs indefinitely since the final interpretant will always be in the future. A thought of reason must be capable of giving rise to other such thoughts equally capable of the same *ad infinitum*. Reason thus does not point towards any certainty or determined thought, but to the possibility of the creation of thoughts.

As sentiments, pleasure, will, and desire are not self-controlled, reason is the only self-controlled quality, the only that can be freely developed by human doings. But as an incipient and becoming process, it needs to materialize and embody something. In a process of evolution, ideals do not grow by themselves; existents embody classes of ideals, so that their coming about transforms the very ideals themselves. This means that reason has to congregate existent elements, which make it concrete so that it can be developed. And it is through instinctive thought, understood as habitual thought that thought becomes concrete. Thus, the whole classes of signs work together to the growing of ideas.

I could go further on the relations of semiotics, metaphysics and pragmatism, but since the purpose of this paper is to show how a visual diagram can illuminate many disciplines and not going deep on each topic, next I will show how I started using the diagram for semiotic analysis. For that, instead examining how the classes are manifested in one trichotomy, I examined how the classes pass trough the ten trichotomies trying to describe its particular characteristics.

8   Comparison between the Two Classifications

One way to describe the 66 classes of signs is by comparing the ten classes of signs with the 66 classes of signs. The way signs are arranged in the small diagram and the detailed description Peirce gave of the ten classes exemplifies the fundamental logical relations among the sign constituents, and it may serve as a guideline to make the 66 classes diagram more comprehensive.

Since the 66 classes of signs system is an extension of the 10 classes system, which has been deeply discussed and described, the study on the 66 classes begins with a comparison between these two systems (Fig. 9). The first step for the comparison is to identify on the Signtree the 3 trichotomies that compose the 10 classes of sign.

Comparison:

By comparing the two systems of signs we can define 10 well known groups of signs (Fig. 10). All the classes of one group might share qualities that characterize them as being part of the group. The group type is the first characteristic required to start describing the 66 classes. It is a general description, but very useful because it shows the dependence relations among classes and the function of the classes in the semiotic system. To proceed with a semiotic analysis using the Signtree one might know that a sign can have characteristics of the first, the second and the third categories. So, instead of searching for a branch that describes the sign process, one should regard how semiosis follow the branches growing. Semiotic analysis should describe a sign in its most details, for that it is necessary to consider the classes working together, and the meaning of the sign getting more sophisticated as the branches are going higher (that means describing relations of the third category).

The classes of sign describe the conditions of signification. Then, a semiotic analysis should begin describing the most fundamental qualities of the sign

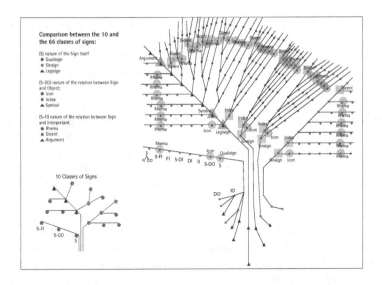

Fig. 9 Comparison between the 10 and the 66 models.

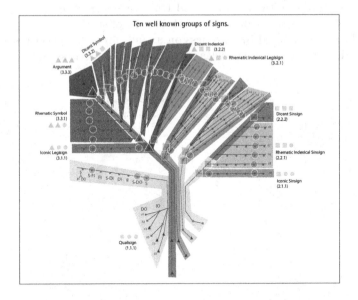

Fig. 10 Groups of signs on the Signtree model.

(qualisign), which can be only possible qualities and are given in the lowest branch. The next three branches (sinsigns) containing 24 classes describe the existent qualities, facts and instances of law that are present in the sign. Finally, the highest branch (legisign) describes the conventionality of the sign, what rules it.

This highest branch is composed of symbolic classes that are signs of law. They depend on its replicas to express information. The classes that follow from the symbol on the Signtree triadic branch are not concerned with the conditions of the sign to signify, but with the knowledge-producing value, or with its communicative significance.

Then, we can draw some other analogies on this last Signtree branch. Seven classes of signs follow from the symbol bifurcation. They describe the conditions of symbols to signify. Therefore, they concern to *Speculative Grammar*. Relations of firstness that follow on this branch represent the possibility of the symbol to be interpreted, which produces a hypothesis. And relations of secondness following it represents the actual test of predictions based on hypothesis.

The next bifurcations establish the conditions of reaching truth conclusions from a reasoning form (Fig. 11). To determine a valid argument it is necessary to guarantee the validity of its leading principles. "If one can guarantee the validity of a leading principle, then, given true premises, the conclusion will have a guarantee of being either necessarily or probably true, depending on the type of argument" [17, 2.464]. These last classes of signs seem to be related to the second and third branch of semiotics: *Critical Logic and Methodeutic*. The former concerns the accuracy and truth of information, and the latter establishes the formal conditions for attainment of truth.

The branch composed by thirdness on its interpretants concern to the validity of thought. The various twigs with relations of firstness and secondness

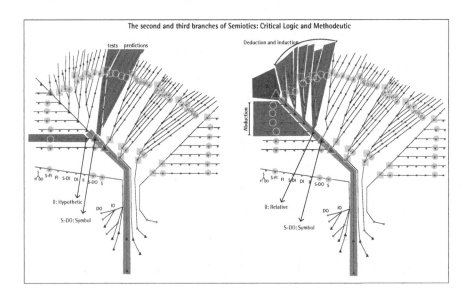

Fig. 11 The second and third branches of Semiotics: Critical Logic and Methodeutic.

that follow from the thirdness branch shows that valid reasoning is based on experience. According to Peirce, "deduction proves that something must be, Induction shows that something actually is operative, Abduction merely suggests that something may be" [13, 2.216]. Then, it seems that the classes that have only relations of firstness on the interpretants might be related to abduction. The classes composed by relations of secondness to induction, and the classes composed by relations of firstness and secondness would represent deduction.

9 Final Considerations

Semiotics is not an isolated discipline; it is the ground of Peirce's philosophical framework. The representation of this system in the form of a tree diagram reveals its complexity. Since the growth of the branches of this tree represents the evolution of thought, and the growth of the roots the complexities of reality, the diagram of Peirce's system of thought finds its ground in the tree of knowledge connecting the world of thought with the one of reality. Pragmatism and metaphysics are the sciences whose relevance to the dimensions of the diagram becomes apparent. Besides that, the Signtree model evinces a semiotic philosophy able to specify its constituents without isolating them in the course of analysis. The concept of continuum is able to integrate all constituents of Peirce's philosophy into a unified framework and the sixty-six classes of signs are a result of Peirce's effort of defining the points of relevance on this continuum.

References

1. Bergman, M.: Peirce's Philosophy of Communication, New York, London. Continuum Studies in American philosophy (2009)
2. Clark, G.: New light on Peirce's iconic notation for the sixteen binary connectives. In: Houser, N., Roberts, D., Evra van, J. (eds.) Studies in the Logic of Charles Sanders Peirce, pp. 304–333. Indiana University Press, Bloomington and Indianapolis (1997)
3. Freadman, A.: Peirce's second classification of signs. In: Colapietro, V.M., Olshewsky, T. (eds.) Peirce's Doctrine of Signs: Theory, Applications, and Connections, pp. 143–160. Mouton de Gruyter, Berlin (1995)
4. Houser, N.: A peircean classification of models. In: Anderson, M., Merrell, F. (eds.) On Semiotic Modeling, pp. 431–439. Mounton de Gruyter, Berlin (1991)
5. Houser, N., Roberts, D., van Evra, J. (eds.): Studies in the Logic of Charles Sanders Peirce. Indiana University Press, Bloomington (1997)
6. Kent, B.: The interconnectedness of Peirce s diagrammatic thought. In: Houser, N., Roberts, D., van Evra, J. (eds.) Studies in the Logic of Charles Sanders Peirce, pp. 445–459. Indiana University Press, Bloomington and Indianapolis (1997)
7. Ketner, K.: Elements of Logic: An Introduction to Peirce's Existential Graphs. Texas tech University Press, Lubbock (1990)

8. Liszka, J.: A General Introduction to the Semeiotic of Charles Sanders Peirce. Indiana University Press, Bloomington and Indianapolis (1996)
9. Merrel, F.: Peirce, Signs, and Meaning. University of Toronto Press, Toronto (1997)
10. Müller, R.: On the principles of construction and the order of Peirce's trichotomies of signs. Transactions of the Charles S. Peirce Society 30(1), 135–153 (1994)
11. Nöth, W.: Handbook of Semiotics. University of Toronto Press, Toronto (1995)
12. Ormiston, G.: Peirce's categories: Structure of semiotic. Semiótica 19(3/4), 209–231 (1977)
13. Peirce, C.: The Essential Peirce: Selected Philosophical Writings, vol. Indiana University Press, Bloomington (1839-1914), Edited by P.E. Project
14. Peirce, C.: The Essential Peirce: Selected Philosophical Writings, vol. 1. Indiana University Press, Bloomington (1867-1893), Edited by P.E. Project
15. Peirce, C.: The New Elements of Mathematics, vol. 1. Mouton Publishers/Humanities Press, Hague (1976)
16. Peirce, C.: Semiotic and Significs. The Correspondence between Charles S. Peirce and Lady Victoria Welby. Indiana University Press, Bloomington (1977)
17. Peirce, C.S.: Collected Papers, 8 vols. Harvard University Press, Cambridge (1931)
18. Pietarinen, A.: Peirce's magic lantern of logic: Moving pictures of thought (2003), http://www.helsinki.fi/science/commens/papers.html
19. Pietarinen, A.: Signs of Logic: Peircean Themes on the Philosophy of Language, Games, and Communication. Springer, Berlin (2006)
20. Ransdell, J.: Some leading ideas of Peirce's semiotic. Semiotica 19(3/4), 157–178 (1977)
21. Sanders, G.: Peirce's sixty-six signs? Transactions of the Charles S. Peirce Society 6(1), 3–16 (1970)
22. Santaella, L.: Teoria Geral dos Signos. Como as Linguagens Signicam as Coisas. Pioneira, São Paulo (2000)
23. Santaella, L.: O Método Anticartesiano de C.S. Peirce. UNESP, São Paulo (2004)
24. Short, T.: Peirce's Theory of Signs. Cambridge University Press, New York (2007)
25. Skagestad, P.: Peirce's semeiotic model of the mind. In: Misak, C. (ed.) The Cambridge Companion to Peirce, pp. 241–256. Cambridge University Press, New York (2004)
26. Spinks, C.: Diagrammatic thinking and the portraiture of thought. In: Anderson, M., Merrell, F. (eds.) On Semiotic Modeling, pp. 441–448. Mounton de Gruyter, Berlin (1991)
27. Stjernfelt, F.: Diagrammatology. An Investigation on the Borderlines of Phenomenology, Ontology, and Semiotics. Springer, Berlin (2007)
28. Waal de, C.: On Pragmatism. Wadsworth Publishing, New York (2005)
29. Weiss, P., Burks, A.: Peirce's sixty-six signs. Journal of Philosophy 42(14), 383–388 (1945)
30. Zellweger, G.S.: Untapped potential in Peirce's iconic notation for the sixteen binary connectives. In: Houser, N., Roberts, D., Evra van, J. (eds.) Studies in the Logic of Charles Sanders Peirce, pp. 334–386. Indiana University Press, Bloomington and Indianapolis (1997)

31. Zeman, J.: Peirce on the indeterminate and on the object: Initial reflexions. Grazer Philosophische Studien 32, 37–49 (1988)
32. Zeman, J.J.: Peirce logical graphs. Semiótica 12, 239–256 (1974)
33. Zeman, J.J.: Peirce philosophy of logic. Transactions of the Charles S. Peirce Society 22(1), 1–22 (1986)

The Role of Agency Detection in the Invention of Supernatural Beings

An Abductive Approach

Tommaso Bertolotti and Lorenzo Magnani

Abstract. Over the last decade, a multidisciplinary approach (merging cognitive science, anthropology and evolutionary psychology) has stressed the fundamental importance of cognitive constraints concerning the complex phenomenon defined as "religion". The main feature is the predominant presence of belief in agent-concepts that display a strong counterfactual nature, in spite of a minor degree of *counterintuitiveness*. Consistently with the major trend in cognitive science, we contend that agents populating religious beliefs were generated by the same processes involved to infer the presence of ordinary agents. Coherently with the Peircean framework, in which all cognitive performance is considered as a sign-mediated activity, our main point is that those processes of agency detection are characterized at all levels – from the less conscious to the higher ones – by the inferential procedure called *abduction*. Hence, the very invention of supernatural agents seems to be the result of a particular series of abductive processes that served some other purposes before (i.e. the detection of predators and preys) and whose output was coherent with that of other abductive patterns. Eventually, they would be externalized and recapitulated in the well-known figures of deities, spirits, and so on: thoughts concerning supernatural beings, at first rather vague, were *embodied* in material culture with the result of fixing them in more stable and sharable representations that could be manipulated and acquired back in the mind in the definitive form.

Tommaso Bertolotti
Department of Philosophy, University of Pavia, Pavia, Italy
e-mail: mahatma.tom@gmail.com

Lorenzo Magnani
Department of Philosophy, University of Pavia, Pavia,
Italy and Sun Yat-sen University, Guangzhou, P.R. China
e-mail: lmagnani@unipv.it

L. Magnani et al. (Eds.): Model-Based Reasoning in Science & Technology, SCI 314, pp. 239–262.
springerlink.com   © Springer-Verlag Berlin Heidelberg 2010

1 Introduction

Each and every day, we deal with a number of entities that do not belong to our everyday ecologies: our minds are full of items that we could not see nor perceive in the outside world. This description does not only match entities such as daemons, fairies, spirits, aliens, gods and deities, but also square roots, quarks and gluons, gravity fields, sinusoids: in brief, not only typically religious entities, but also many scientific objects cannot be detected by our standard human sensory apparatus and hence be grasped "naturally" by our minds, and therefore belong to the broad category of *counterintuitive* (see for instance Atran and Medin's book for a wide approach to the relationship between nature, folk-sciences and culture [2]).

A meaningful difference, though, keeps scientific concepts separate from religious ones: science may deal with what is counterintuitive *insofar as its understanding requires a consistent intellectual effort*, while religion's objects are counterintuitive *because ontologically counterfactual*.

Both of those categories regard things which are not immediate for us, as we said, objects that do not belong to our normal ecologies. Still, counterintuitive objects like an imaginary number (i.e. the square root of -2) or the process of replication of DNA or a chemical reduction-oxidation reaction, in spite of their being extremely unappealing to ordinary human minds, they can be integrated in a more complex system, analyzed within the proper artifactual framework, they can be the object of controlled and reproducible experiments and their *existence* can be eventually confirmed. Conversely, human minds seem particularly eager to absorb as many counterfactual objects as they can: such objects can be defined as those entities that violate (more or less systematically and at different levels of magnitude) common expectations about the world around us.

As mentioned above, counterfactual entities are amazingly appealing to human minds notwithstanding the factual impossibility to investigate them on a scientific basis and the lack of plainly empirical evidence on their account: systematic skepticism about counterfactual belief is in fact a distinctive trait of a scientific state of rationality, which is rather uncommon among average human beings. The unattainability of an empirical proof as for counterfactual elements is intrinsic to their definition, and it is not due to the rarity of episodes in which they are involved: over the last two decades, the popular distribution of portable camcorders first, and then the advent of digital media devices would have been expected to cause a massive rise in sightings and consequent documentation of UFOs, ghost and paranormal activities, Big Foot's, Loch Ness monsters and so on, but this was clearly not the case.

The seemingly unstoppable diffusion of religion among human beings can be regarded as the best example of the success that counterfactual beliefs met within our minds. Scott Atran stresses how this commitment to counterfactual entities should have proved particularly maladaptive for mankind: "[...] to take what is materially false to be true (e.g. people think and laugh and cry and hurt and have sex after they die and their body disintegrate) and to take what is materially true to be false (e.g., people just die and disintegrate and that's that) does not appear to be a reasonable evolutionary strategy" [1, chapter 1]. The key to solve this issue seems to be that humans usually know when to believe (and behave) "as if" counterfactual entities really belonged to their ecologies and when to suspend this belief. Let us consider an example from Christian faith: in the Gospels, Jesus is reported to have said "I tell you the truth, if you have faith as small as a mustard seed, you can say to this mountain, 'Move from here to there' and it will move. Nothing will be impossible for you." (Matthew 17:20, NIV) and "And these signs will accompany those who believe: in my name they will drive out demons; they will speak in new tongues; they will pick up snakes with their hands; and when they drink deadly poison, it will not hurt them at all" (Mark 16:17-18), but the great majority of Christians, even if it might sometimes believe that to be true, do not behave in their everyday life as if they could move a mountain by the power of faith or try to handle poisonous snakes fearless of the likely injures (or death) they might incur into. There are some exceptions to this *moderate* and widespread approach to Christian belief, but they all fall down within the laws of Nature: we can consider for instance the practice of snake-handling, typical of a Pentecostal minority in the eastern United States [40]. Those believers take literally the verses of the Gospels we just quoted, and manage bare-handedly venomous snakes, recurring to prayer in case of bites: laws of physiology cannot be changed by mere conviction and, as a matter of fact, such practice has resulted in 70-80 reported deaths over the last century.

Over the next sections we are going to deal with a possible origin of the supernatural entities that populate our beliefs: our explanation will hint at a solution of the paradox opposing religion's persistence and its likely negative effects with respect to individuals that sincerely believe in things that cannot be proven to exist, and that cannot be proven to affect our lives.

2 Hypotheses about the Naturalness of Religion

The cognitive origins of religion have become a hotly debated topic over the last decade: the attempt to *naturalize* religion and investigate it as any other *natural* phenomenon overflowed the boundaries of strictly academic debate. Scholars (coming from various fields, i.e. anthropology, psychology, biology, etc.) were prompted to write widely successful books that would convey to a broader public the latest findings in the domain: the most famous example of this trend is probably Daniel Dennett's *Breaking the Spell* [10].

This literature can provide us with useful introductory hints for our research, as we move to a level of higher details. Cognitive anthropologist Pascal Boyer, whose book *Religion Explained* [7] can be regarded as one of the forerunners of Dennett's best seller, pointed out how religion was *natural* to human beings, and conversely science itself appears to be one of the most *unnatural* activities they ever undertook.

Seemingly in conformity with this claim about the *unnatural* character of science, Dixon – in his much appreciable *Introduction* to the confrontation between science and religion – stated that many of the issues starting with Galileo's clash with the Church can be brought back to the problem of scientific *realism*, that is believing (or not) that we are able, through science, to provide an account of phenomena that cannot be observed, and therefore counterintuitive insomuch as *not* intuitable [12].

Boyer highlighted how religion, as a complex phenomenon, originates from a wide range of stages of the mind, starting from the most automatic and unconscious levels. According to Boyer's imagery, religion is natural because most of the cognitive activity crucial for religion takes place in the mind's *basement* [7, chapter 9], that is to say, hidden from the believer's awareness. Basements are gloomy, damp, poorly lit: the basement is a topical setting for horror and mystery fiction, because what happens in the basement is not usually clear to the upper floors of the house (otherwise there would be no mystery to solve at all!).

The aim of the basement metaphor is to discharge the naïf conception about how our perceptions and our cognitive performances inform our beliefs: oddly enough, it can be said that the perceptions we take more for granted (such as the visual ones) come *from the basement* and not simply from the outside world, as if from a window. There is no *homunculus*, no self-in-the-head who perceives the outer world as if senses were openings in the body shell. If we truly want to investigate the formation of our beliefs (any belief, and hence of religious ones, too), it is important to accept the neuroscientific findings suggesting that there is no such thing as a central controller that makes sense of the different perceptual stimuli: it is rather just the different perceptual stimuli that make sense into the global picture that we are aware of, thanks to a sort of *self-assessment* operated by each different system [13]. This is the kind of work going on in the less conscious layers of the mind, so crucial for the very possibility of religion. On the other hand, science is called unnatural because most of the work seems to take place in the fully conscious levels of the mind: science is about which method should be used to obtain the best knowledge, even perception and attention are strictly controlled and directed to a target: this positive conception of science can be the result of a favorable bias, but the higher standards of science (and in a lesser measure, philosophy until it was outclassed by modern science as the prime provider of knowledge) are actually aimed towards a critical analysis of the immediate self-assessment of perceptual channels.

It might be argued that few things are less intentional and consciously ordered than religious hierarchies, for instance, but it is important to bear in mind that we do not mean to deal with one particular religion, nor with religion *as a whole*: to make use of Boyer's useful distinction, we are not interested in theological religion but in practical religion, that is to say that our investigations focus on those beliefs that are "[...] not concerned with the general question of God's existence or powers, but more understandably with practical questions: what to expect now and what do do next" [7, chapter 9]. Nearly any contemporary religion rests on a theology: a theology is a corpus of ordered beliefs and (recommended) inferential rules and the very etymology of theology is the same as for biology, geology, neurology. A *logos* can constitute only in the upper, well illuminated floors on the mind. Conversely, we are interested in what actual human agents perceive in their conscience when their behavior can be filed as "religious". Of course, theology is accepted also because of these *naturally originated* beliefs, and vice versa those beliefs are partly informed and shaped by theology, but it is important to stress how theology and practical religion are not the same thing and, hence, the former cannot be used to invalidate explanations concerning the latter.

So far, so good: still, it is clear that, by now, one might wonder what kind of "work" is exactly ingoing in the recesses of the basement of the mind, that Boyer describes chiefly in an operative fashion. Such description is perfectly suitable to his purpose, but we shall attempt to get out of the analogy and shed further light on those underground mental activities in order to acquire a better understanding of the cognitive traits that make religions one of the constitutive characteristics of *homo sapiens sapiens*.

3   Abduction as a Multilevel Model for Perception

We want to briefly introduce our main take that human cognition chiefly displays an *abductive* nature: we shall describe how, within an abductive framework, the human brain operates in a similar way over a wide range of stimuli, thus drawing a continuous line from the most basic levels of perception to the origins of belief in supernatural agents and, *a fortiori*, of the complex series of phenomena we label as religion.

Abduction, as understood within the Peircean framework, can be accounted for as the process of inferring certain facts and/or laws and hypotheses that render some sentences plausible, that explain (and also sometimes discover) some (eventually new) phenomenon or observation: it is the process of reasoning in which hypotheses are formed and evaluated (for a full account of abduction and, in a broader sense, of abductive cognition see Magnani's extensive monograph [24]). It must not be regarded as a merely sentential inferential process, concerned with the discovery of physical laws or resolving mysteries (Doctor Watson makes a constant mistake in Conan Doyle's novels by complimenting Holmes for his brilliant deduction: his hypotheses are actually abductions from clues to the situation that originated them). Abduction opens a much wider field of investigation: survival, for any animate organism, is a matter of coping with the environment and the relationship with the environment is mediated by a series of cues the organism must make sense of in order to generate, even if tacitly, some knowledge it did not possess before.

This *making sense of signs* is an abductive activity that human beings share with any organism endowed with a nervous system or, on an even bigger perspective, any organism capable of reacting actively to modifications of its environment (from bacteria to *homo sapiens sapiens*), as shown in Magnani's paper about animal abduction [24].

As hinted in the previous section, our brain cannot operate directly on the outside world, it needs operating on inner representations of the outside world. As for all cognitive processing, both data and hypotheses can have a full range of verbal and sensory representations, involving words, sights, images, smells... but also kinesthetic experiences and other feelings such as pain, and so for all sensory modalities: all representations are brain structures and that abduction certainly is a neural process in terms of transformations of neural representations. We can also re-conceptualize abduction neurologically as a process in which one neural structure representing the explanatory target generates another neural structure that constitutes a hypothesis.

It is important to focus on how this *reduction* to a neural representation necessarily applies both to lower and higher cognitive processes: as a digital memory storage unit contemplates only one's and zero's in a bistable activation state, so a human brain can only process input *encoded* as neural structures. A neural structure can be seen as a set of neurons, connections, and spiking behaviors, and their interplay, and the behavior of neurons as patterns of activation (as maintained by the connectionist tradition) also endowed with an important exchange of chemical information. In this perspective it is clear we can say all representations are brain structures and abduction is a neural process in terms of transformations of neural representations. Such a description means to transcend a mere computational analogy, it is useful insofar as to negate that everything in our brain exists as and *only as* a structure of neural activation would be like breaking a computer open and insisting that we can actually *see* images, physically *write* on documents and *hear* the sounds that are just stored in its memory. Now, let us consider a consequence of this, crucial for our next steps: different kinds of cognition, be it sentential, visual, emotional, and so on are encoded within the same framework.

Neural processing of externally originated input is clearly not a random activity: we are not contending that the brain could be seen as a unique and undifferentiated abductive inferential machine. Conversely, we are about to provide a brief recapitulation of scientific hypotheses that suggest the exact opposite: our brain seems to display an extreme specialization in the accomplishment of basic and more complex tasks. The unicity of the neural and abductive ground, though, might encourage both the hypotheses that those neural systems are able to influence one another and that each specialized pattern within the brain displays an abductive endowment.

The patterns of neural activation we have just mentioned could be comprehended by the definition of *mental modules*. Mental modules were introduced in literature by Fodor [16, 12] who argued that mental phenomena are resulting from the operation of multiple distinct processes. Basically, according to Fodor modules are reflex-like, hardwired devices that process information in a specific, domain dependent, and stereotyped ways. Such multiple specialized processes have two basic features: the first is *domain specificity*, whereas the second is *encapsulation*. Domain specificity refers to the fact that a domain is specialized to accept as input pre-determined and pre-defined classes of information. Encapsulation refers to the fact that any activity of information processing is influenced by nothing external to the process, but only by what the system accepts as "legal inputs" [4]. To have a better idea of how a module works, it is been advocated the idea of a pipe: as a piece of information is accepted, then it is processed according to procedures that cannot be externally re-engineered.

Later on, Cosmides and Tooby [46] have developed the idea of modules claiming that they evolved during evolution in a Darwinian way, namely, descent with modification. Accordingly, modules would be described in terms of specialized functions designed to adapt to the relevant aspects of the ancestral environment humans lived in.

The theory of mental modules (or massive modularity) has been hotly questioned during the last two decades, and it still results highly controversial (Carruthers provides a quick yet complete review in [8]). The main objection raised against modularity is concerning the presence of a "central" system and its ability to be flexible in handling information from multiple source and for which there is no specific mechanism designed.

Recently, Barrett [3] has put forward an alternative perspective on modularity that is potentially able to avoid the objections usually raised by those who do not support any modular view of the mind: he suggested that specificity and encapsulation should be re-defined respectively in terms of access and processing. Both access (the information accepted as input) and process (the procedures devoted to handling the information) can be specific or general, so that some mechanisms, for instance, might "have access to large amounts of information in the mind but only process information that meets its input criteria" [4, p. 631], and vice versa. This alternative perspective on the role of modularity is less narrow than the original one allowing to escape from the pipe-like trap. For instance, it has not to assume complete isolation from other systems regarding the kind of information a single module can accept and then process. In fact, a module can guarantee flexibility and specificity at the same time. From an epistemic point of view, our abductive perspective on human cognition can explain many characteristics of the modular mind theory, but with the advantage of providing a coherent account of low-level brain functions as well.

Let us consider a very simple cognitive performance: vision. Vision is not an immediate perceptual performance: as we stated before, vision is simple only if we accept the nave conception of a inner controller peeking through our eyes, mere openings in the head. Vision is a complex series of physical, chemical and neural reactions that turn a raw mixture of signs into something that is readable by other systems in the brain[1]. As Jacob and Jannerod illustrate in their monograph about visual cognition [21]:

> [T]here are many stages on the way from the retina through the optic nerve to the higher levels of information-processing in the visual cortex. Each such stage carries some information about the distal stimulus and about everything the distal stimulus stands in some non-accidental correlation with. However, neither the retina nor the optic nerve represent everything they carry information about. [21, p. 5]

Being mediated, sensorial perceptions are the result of an inferential activity: according to Peirce, perceptions are abductions, and thus they are hypothetical and it is always possible to withdraw them. Moreover, given the fact that judgments in perception are fallible but indubitable abductions, we are not in any psychological condition to conceive that they are false, as they are unconscious habits of inference: this is precisely what informs our naïve conception of an immediate sensorial perception. Peirce considers perception a fast and uncontrolled knowledge-production process. Perception is a kind of vehicle for the instantaneous retrieval of knowledge that was previously assembled in our mind through inferential processes: "[. . .] a fully accepted, simple, and interesting inference tends to obliterate all recognition of the uninteresting and complex premises from which it was derived" [31, 7.37].

Perception is abductive in itself: as Peirce would say, "[a]bductive inference shades into perceptual judgment without any sharp line of demarcation between them" [30, p. 224]. As we maintain that perception is an inferential activity, it is easy to understand why the distinction between retrieving and producing knowledge becomes fuzzy: working on a set of signs, the result of perception is always something new and different from the initial set of signs. Therefore, many types of perception display the characteristics of *semi-encapsulation*, insofar as they work as both a *bottom-up* mechanism but are endowed with some *top-down* cognitive affections as well (Raftopoulos' contribution to this research is fundamental, for instance in [37, 36]). This is not only true for human vision, we can for instance consider animal sensorial perception such as a dog's ability to make sense of olfactive cues, or a cat's perception of vibrations through its whiskers, or a bat noticing the presence of a moth through its ultrasonic echolocation.

In all of these examples, perceptions are the result of an "inferential" (in the Peircean sense) activity, more or less complex, always mediated: what they all have in common is the striving to make sense of *per se* unreadable, raw signs, and make the output available to other cognitive performances. In fact, as Peirce himself had stated, "[a] mass of facts is before us. We go through them. We examine them. We find them a confused snarl, an

---

[1] "[D]ifferent visual attributes of objects are processed in separate cortical areas in the visual brain of primates: neurons in area V3 respond to moving shapes; neurons in area V4 respond to colors; neurons in areas MT and V5 are specialized for the processing of motion" [21, p. 8].

impenetrable jungle. We are unable to hold them in our minds. [...] But suddenly, while we are poring over our digest of the facts and are endeavoring to set them into order, it occurs to us that if we were to assume something to be true that we do not know to be true, these facts would arrange themselves luminously. That is abduction [...]"[2]. When contending that all perceptual activity is inferential, and hence a low-level yet mediate form of new-knowledge generation, we clearly suggest that the meaning of the word inference is not exhausted by its *logical* aspects but is referred to the effect of various sensorial activities.

If awareness, whether propositional or perceptual, is semiotic, then all awareness involves the interpretation of signs, and all such interpretation is inferential: semiosis not only involves the interpretation of linguistic signs, but also the interpretation of non-linguistic signs, and this will be of crucial importance for our account of the origins of belief in supernatural agents.

Going beyond a conception of abduction which is merely logical, say, related only to its sentential and computational dimension, we move towards a broader semiotic dimension worth investigating. Peirce himself fairly noted that the all thinking is in signs, and signs can be icons, indices, or symbols. In this sense, all inference is a form of sign activity, where the word sign includes "feeling, image, conception, and other representation" [31, 5.283].

Thesen and his colleagues provide an interesting review of neurological data showing how neural inputs coming from different sensorial system are merged at very early stages of cognition: "[t]raditionally, multisensory processing in the cortex has been assumed to occur in specialized cortical modules relatively late in the processing hierarchy and only after unimodal sensory processing in the so-called 'sensory-specific' areas", while recent imaging studies hint that "senses influence each other even at the earliest levels of cortical processing, that is, at the level of the primary sensory cortices" [43, p. 85]. Pettypiece's team explored the interaction of visual and haptic sensorial input and highlighted how the interplay is not a fixed one but varies with each task [33]. Emotions seem to play a crucial role in those processes as well: Phelps and her colleagues reported that their behavioral experiments were the first able to demonstrate how emotions influence not only superior faculties but also processes as low as early vision are influence by them [34].

Our claim that abductive inferences display a multi-modal tendency does not contradict the presence of pre-wired patterns for neural activation – each dealing with its proper input and output – within the shared internal semiotic environment. Even beliefs and desires, that traditional philosophy interpreted as merely propositional attitudes, can be usefully seen as brain structures (stages of neural activation), and moreover, in this extended framework the concept of inference can be reinterpreted to include non-verbal representations from all sensory modalities and their hybrid combinations, going beyond its merely logical meaning in terms of arguments on sentences. Such a

---

[2] Cf. Peirce's "Pragmatism as the logic of abduction", in [30, pp. 227–241], the quotation is from footnote 12, pp. 531–532.

perspective can really help us understand how the nature of abduction is indeed multimodal on the basis of its very neurological ground: on that account it is possible to understand how abductive inferences are not exclusively sentential, or iconic, or model-based, and so on, but all of these aspects can meddle in the neural circuits that make up the basement of our mind, so that emotions (such as fear of predators), desires (i.e. sexual desire) are processed together with a neural structure denoted by a different origin (visual, for instance), in order to achieve an output that can be emotional, visual or neither of the two, and represented in still another formulation.

Human beings, just like much any organism, do not *contemplate* the environment uninterestedly, but as *survival machines* they are actively involved in continuos problem-solving activities and the surroundings are hardly ever uniformly relevant to us while, on the contrary, our scope (on the short or on the long term, if it is the case of a complex plan) leads us to select which traits of the environment are relevant[3]. This is extremely important as far as a process like the detection of other agents is concerned: as it will emerge in the next section, actual inferences concerning the presence of other agents in the surroundings are extremely dependent upon emotions, contextual features and previously acquired knowledge.

4 From the Detection of Natural Agency to the "Invention" of Supernatural Agents

Nearly any cognitive account of the origins of religion takes as a fundamental assumption the predominance of agent concepts: as a matter of fact, such predominance is displayed not only by religion but by nearly all fictional and folkloric templates as well. We ourselves are agents, we consider our fellow humans (and hominids) as agents and we would file under "agency" our preys and predators as well. Atran highlights three fundamental characteristics connected with the identification of others as agents [1, chapter 3]: "observable, short-term productions of complex design", evidence of "internal motivations" directing behavior, and finally the display of teleological acting. A fundamental characteristic of an agent's behavior are, in fact, "telic" actions: not only they imply the possibility of reaching that discrete goal but also the eventuality of stopping the particular action[4].

All these aspects are picked up by human brains as *signs*. Our semiotic brains [23] perform a series of abductive operations upon the vast amount

---

[3] A quite famous experiment was conducted on *inattentional blindness*: subjects were asked to observe a certain performance (a ball being passed) in a movie and, as they were focused on it, they would miss the appearance of a gorilla beating his chest among the actors [41].

[4] *The tiger is attacking the goat* describes a telic action, because it is oriented toward the discrete goal of killing the goat but the action (attacking) could be interrupted by the tiger, if needed.

of signs surrounding us: what is sometimes referred to as agency-detection module (as said by Atran [1]) or device (in Barrett's formulation [5]) could in fact be only a pre-disposed abductive pattern, as maintained in the previous section. Picking up certain signs and cues, our brain would infer the most likely subject that originated them. What we are referring to is the capacity to detect the presence of another agent from a series of clues in the surrounding environment, i.e. another human (or hominid), a friend or a foe, or animals that could be predators or preys: such an ability is clearly of extreme importance, especially when humans had not reached their ecological dominance yet [15]. Yet, even once the ecological dominance was achieved, that is to say, when other human beings became the main major threat for the survival of an individual, the detection of agents remained of crucial importance: we just became more sensitive to recognize cues signaling complex conscious volition, moral behavior, for instance something happening to someone who just committed a mischief, *as if* she was being punished. [5]

Of course, the cues hinting to the presence of another agent are not picked up randomly: any sign is not any sign. They must respond to certain patterns in order to be received and acknowledged by the cognitive systems dedicated to the detection of agents: we admit the existence of a proper domain on which such inferences are operated, so that they can provide us with a sensible output [42]. Such proper domains comprehend the signs that are effectively produced by agents. When inferring from signs that are fairly symptomatic of the presence of an agent, the quality of our abduction is likely to be elevated. The call of a blackbird or the howling of a coyote are fairly symptomatic of the presence of such creatures. Similarly, spearheads, artistic artifacts, evident manipulations of the environment and similar signs displaying complex design are symptomatic of the presence of other human beings. This inferential activity displays an instinctual and subconscious, hence self-assessing, nature: it is deeply integrated in our neural wiring and we usually assume its output to be correct and reliable.

It might be interesting to notice how this ability in agency detection became highly developed in human beings, and its integration with high level, conscious abductive inferences surely contributed to our becoming dominant in many different ecologies, yet it is a feature shared by at least the vast majority of animals and insects. We can take a look at it in a perspective useful to our scope, in the fashion of *reverse engineering* and thus quickly survey what happens when an organism manages to avoid being identified as an agent or interferes with the exact recognition, for instance in animal camouflage and imitation. The base situation could be described as this:

> Property G matters to the survival of the animal (e.g. a sexually active male competitor or an insect to capture). The animal's sensory mechanism, however, responds to instantiations of property F, not property G. Often enough in the animal's ecology, instantiations of F coincide with instantiations of G.

---

[5] For further reference about anthropomorphism see Gebhard's [18] and other essays in Clayton and Opotow's edited book about psychological relevance of nature [9].

So detecting an F is a good cue if what enhances the animal's fitness is to produce a behavioral response in the presence of a G. [21, p. 8]

The presence of an agent who detains the property G is *abduced* on the basis on one or more properties F that usually signify the relevant property. If an organism is hunted as a prey or avoided as a predator because of a property G, it must try to reduce the occurrences of the properties signaling their characteristic, and this varies widely from organism to organism[6]. Cats have been known since the Fifties to see monochromatically and perceive objects as long as they are in motion [19, 27]: when a lizard is hunted by a cat, it may recognize the predator and freeze before its eyes, thus becoming invisible to the eyes of the cat who suspends its *belief* according to the location and presence of the lizard: the lizard makes itself irrelevant to the cat's neural patterns who process the cue signaling surrounding agency. Further examples are possibles: many insects are shaped like leaves or small branches and move *as if* displaying nonbiological but rather atmospheric movement, thus they are not perceived by the agent detection systems and do not come to the practical awareness of the organism: a typical example of this is the *Phasmatodea*, know as "stick insect" [6, 35].

Some examples are even more thought-provoking because organisms sometimes do not just try to fall out of the domain of other organisms' agent-detectors, but *mess* with them. Let us think about *mimicry*: insects, but animals too, developed visual, auditory or chemical characteristics [11] of other species, usually poisonous or dangerous ones (flies looking like bumblebees or innocuous snakes looking like venomous ones). By mimicry, those creatures exhibit a trait, say a property F, that is picked by surrounding organisms as signals of a property G which the signaling animal does not actually possess: the organism thus tricks the agent-detection system of its potential predator by inducing a *false belief* in it.

Animals can deceive other animals' agent-detecting systems in even more cunning ways: many flying insects (i.e. butterflies) or fishes display big eyes depicted on their body or wings [30]. This is perhaps the most interesting case, as long as those trait provoke in the potential predators the formulation of a (temporary, at least) *belief* about the presence of another predator, or of a big prey that they would not be able to outdo, thus giving to the animal a chance to escape. This peculiar camouflage technique could be said to make active use of the way the predator's agency-detector receives cues from the environment and makes use of them to provide sensible output: big eyes looking at me *automatically* mean another predator is tracking me or a big prey is aware of my presence.

---

[6] The situation we just described could also be accounted for recurring to Thom's semiophysical perspective [44, 45] (also explained in [24, chapter 8]): the *salient* trait is the displayed property, which hits an organism's sensorial apparatus as separated from something else – and hence detectable – while those very saliences are invested with the *pregnance* of the relevant property (i.e. being food, predator, sexual mate etc.).

As far as a non-human animal or a human being are concerned, if the quality of the output inference is is low and defective, e. g. when we fail to detect a predator, or we think that a tiger is actually a goat, or we fail in distinguishing a friend for a foe, death can be a most likely result.

Consequently with what we just described about animal camouflages techniques, it is important to add that our systems had not only to detect ordinary signs of other agents but also to infer their presence when they were *actively* trying to hide it by concealing its signs or producing incoherent ones. This explains why our mental systems for the detection and inference of other agents in the surrounded could develop to be so "touchy and hypersensitive", as Barrett claims:

> In our evolutionary past our best opportunities for survival and reproduction and our biggest threats were other agents, so we had to be able to detect them. Better to guess that the sound in the bushes is an agent (such as a person or tiger) than assume it isn't and become lunch. If you reacted unnecessarily (e.g., because of the wind blowing in the brush), little is lost. [5, p. 85]

Still, the distinction between what our mind-brain *should* process and what is *actually* processed is rather fuzzy and we often commit errors and we take one kind of sign for another, different one (i.e. we see some dust moving and we think we saw an insect). Similarly to the animal cases we described before, magnitude can be a common cause for such – so to say – "abductive error": that is, a similar *kind* of sign but with different intensity such as a very loud noise or the movement of large bodies, such as stones and clouds. Such misperception are likely to happen, especially if we consider our ancestors' impossibility to rely on the scientific knowledge we are accustomed to: if we know nothing about what a geyser is, its shrieking and hissing is likely to be cognitively *filed* as the hissing of a snake or the shrieks of a bird.

The only problem is magnitude: dimensions divert from what we are used to process. It is clearly the case with natural conformations of stone that resemble human artifacts, if not for the size. At any rate, once these signs are picked up by agent-dedicated abductive systems, then they are *necessarily* the signs of an agent which originated them.

The common scientific approach to religion, inherited from the 19th century, claimed that "myths [and subsequently supernatural beings] prevailed in an early, usually the earliest, stage in an evolutionary scheme, or that myths were the result of primitive mythopoeic man's attempt to explain such natural phenomena as the rising and setting of the sun." [14, p. 3]. We argue, though, that such an explanation quite does not grasp the essence of the very beginning of belief in supernatural and can only apply to a later, more structured evolution of "religion": most of all, it makes sense only as far as we benefit of an alternative, that is scientific knowledge, to match with religious belief.

Our point is that supernatural beings were *invented* indeed, but according both to the actual meaning of the word, i.e. to *create or design (something that had not existed before)* and the original latin etymology, that is *in venire*: to *come upon*, to stumble on. The generation of belief in something *super*natural, inferred from certain signs, is just as creative and non *theoretical* as the generation of the belief in an antelope hiding in the bushes: it is the same kind of inferential pattern, just operating on different kinds of signs. The first time our ancestors felt the cognitive need to invent-and-discover the existence of supernatural agents, they were not behaving as theologians. They were not engaging in highly speculative reasoning about the essence of what goes beyond our reason. The first glimpses of belief in supernatural agents might not have involved words, let alone a specific *logos* like the ones we are accustomed to, when dealing about religious and mystical matters. Historically, supernatural agents proved indeed to be ideal components of theories explaining different puzzling aspects of our world, but we maintain that their very origin did not display such theoretical and consciously fictional character. That is to say, at the very beginning of belief in supernatural, those super-agents were held to be as real as everyday agents, animal and human.

> Many events lead to agency detection without a known agent as a possible candidate. Suppose a woman walking alone through a deep gorge rounds a bend in the trail and rocks tumble down the steep wall and nearly hit her. HADD [Hypersensitive Agency Detection Device] might reflexively search for the responsible agent. A man hiking through an unfamiliar forest hears something behind a nearby shrub. HADD screams, 'Agent!' If, after detecting agency in these sorts of cases, a candidate superhuman agent concept is offered and seems consistent with the event, belief could be encouraged. Similarly, when a god concept is already available as a good candidate, events that HADD might have overlooked become significant. [5, p. 86]

The reason why this kind of inference trying to make sense of the surroundings is not critically questioned once it gets to higher conscience is that it rests on the same pre-assumptions that inform higher conscience. As Barrett puts it, we can stop and consider the evidence,

> But the evidence (if available) is always filtered and distorted by the operation of mental tools. We never have direct access to evidence but only processed evidence – memories. When asked 'how many colors are in a rainbow?' I might recall the last time I saw a rainbow and what it looked like. But this 'evidence' has already been tainted by non-reflective beliefs. [5, p. 81]

Non-reflective beliefs are the base components of folk-physics, folk-psychology, folk-biology and so on. They embody many kinds of regularities that human beings witness during the development of their mind-brain system, and they become the arguments of low-level abductive inferences, similar to:

1. Situation X causes Effect Y
2. *Hence, Effect Y is likely to be symptomatic of Situation X*
3. I notice Effect Y
4. *Therefore, I must be in presence of Situation X*

If we transpose this simple model into our exemplar narrative about the woman witnessing some falling rocks, we can obtain something like this:

1. An animal climbing on a cliff causes some gravel and rocks to move and fall when he treads over them
2. *Hence, falling rocks are likely to be symptomatic of an animal stepping up hill*
3. I notice rocks falling down
4. *Therefore, I must be in presence of an animal stepping uphill*

As said, shifts in magnitude allow the same interpretation: if an animal-agent can cause gravel to fall, a big and heavy stone must have been displaced by a mighty powerful agent, but if those stones were seen falling, then the existence of such powerful agent is not minimally questioned. It requires a powerful institutional agent[7] such as a scientific enterprise to transcend the dimension of the singular observer and break our intuitive pre-assumptions, that is our non-reflective beliefs about the world. Consider the movement of the Sun: we know and we believe, because we were taught so, that the Earth revolves around the Sun and still in our everyday life we believe and we behave according to our intuitive assumption that the Sun revolves around the Earth.

Non-reflective beliefs cannot be erased or overwritten no matter how strong and convincing – from a conscious point of view – the evidence against them can be. This can explain why, in spite of the amount of positive scientific knowledge we can rely on, beliefs concerning supernatural agents are extremely resilient: as a matter of fact, many of us still fear presences in the dark, or pray and put their trust in almighty supernatural beings whose might cannot be – at least *scientifically* – tested.

We are not concerned with cultural refinements of god-concepts and other religious objects at the moment: what needs to be stressed at this moment is how beliefs in the existence *supernatural* agents are automatically produced because of the very way our mind-brain system deals with simply *natural* objects. Hence, the first invention of supernatural agents is not a matter of being credulous, and far less an epistemic struggle to provide an explanation to the mysteries of the universe: this is not to say that religion never deals with such aspect, but they are not the heart of the matter. Conversely, we may say that if our ancestors had not been able to respond to external stimuli with the generation of supernatural concepts, that would have been a sign of poor capacity to cope with their natural and physical environment, and this could have had a dramatic evolutionary impact.

---

[7] By introducing the term "institutional agent" we mainly mean to stress the difference between the approach on reality displayed by a first person agent and an agency (such as *Logic*) that by method, distribution of knowledge and commitment manages to transcend the first person dimension. For a further discussion of John Wood's concept of *institutional agent* see for instance Magnani's monograph on abduction [24, chapter 7] and Wood's fore-coming book [48].

To make a clear example, North and South American native populations would draw a supernatural agent from the phenomena usually connected with thunderstorms: our unreflective beliefs about the world (that is, those informing folk-physics) suggest us that an effect must always have a cause. As stated before, to our mind-brain system, cause means agent doing something. If we experience an effect in our ecology, then an agent must have caused it. The bigger the effect, the more powerful the agent *must* be. It is important to stress once again how these universalizing inferences from the cause-effect situation do have an abductive nature: if any cause has a discrete effect, then anything perceived as a discrete effect hints to a cause.

Thus, agent-detecting abductive processes would pick up some of the signs originated by the storms as if originated by an agent, and the resulting *super*agent would be elaborated as a Thunderbird, whose enormous wings stirred the wind and whose powerful cry was thunder itself, as the South American tradition held:

> According to the Ashluslay Indians of the Paraguayan Chaco, thunder and lightning are produced by birds who have long, sharp beaks and who carry fire under their wings. The thunder is their cry and lightning the fire which they drop over the earth. They were also the owners of fire and their enmity against mankind began after they had been deprived of that element. [26, p. 132]

North American Indian myth provide a very similar account of the Thunderbird. The cultural relationship between the two supernatural agents is more of an anthropologist's matter, and hence we will not deal with it; conversely, it is important to notice how in both cultures the same kind of signs – related to the weather and the phenomena of sky and air – are processed *as if* symptomatic of an avian super-agent.

> When it is stormy weather the Thunderbird flies through the skies. He is of monstrous size. When he opens and shuts his eyes, he makes the lightning. The flapping of his wings makes the thunder and the great winds. Thunderbird keeps his meat in a dark hole under the glacier at the foot of the Olympic glacial field. That is his home. When he moves about in there, he makes the noise of thunder there under the ice. [38, p. 320]

Adding to our example, once the aforementioned Thunderbird-concept is ready, it can be inserted in another explanation, which makes sense to those who are already comfortable with the original Thunderbird concept. In this case, also the avalanche is considered as an effect of the Thunderbird's actions:

> Some men were hunting on Hoh mountains. They found a hole in the side of the mountain. They said, "This is Thunderbird's home. This is a supernatural place". Whenever they walked close to the hole they were very afraid. Thunderbird smelled the hunters whenever they approached his place. He did not want any person to come near his house. He caused ice to come out of the door of his house. Whenever people came near there, he rolled ice down

the mountain side while he made the thunder noise. The ice would roll until it came to the level place where the rocks are. There it broke into a million pieces, and rattled as it rolled farther down the valley. Everyone was afraid of Thunderbird and of the thunder noise. No one would sleep near that place over night. [38, p. 320]

The inferential process leading to the *invention* of the Thunderbird can be reassumed as follows, within a clearly abductive structure:

1. A bird taking flight or landing raises dust whirlwinds and emits a distinctive sound flapping its wings
2. *Hence, whirlwinds and flapping sounds are symptomatic of a the presence of a bird*
3. I notice impressive whirlwinds accompanied by what sounds like a deafening flapping sound
4. *Therefore, I must be in presence of some huge and mighty bird, that is the* Thunderbird

In brief, we share with the traditional view on religion the idea that the origins of supernatural rest in its ability to *explain*, but we contend that its very genesis has a much less intentional, conscious and theoretic nature than commonly thought. It is an *explanation* indeed, but one that belongs to the cognitive urge of human beings to constantly make sense of the surrounding environment. Of course it quickly merged and was structured within human beings' *constitutive* curiosity, and within the strife for knowing causes and origins of natural phenomena: still, in a diachronic perspective, the very first pulse of belief in supernatural agents might have rather sprouted from an essentially wired[8] (and mostly unconscious), neuronal-cognitive processing of environmental signs.

5 Embodying Supernatural Agents by Disembodying Cognition

If so far we have attempted to provide a mind-based explanation about the origin of everything loosely "supernatural" within our brains, the question "How did supernatural entities get in our mind *as we know them*, if they do

---

[8] The term *wired* can be easily misunderstood. Generally speaking, we accept the distinction between cognitive aspects that are *hardwired* and those which are simply *pre-wired*. By the former term we refer to those aspects of cognition which are fixed in advance and not modifiable. Conversely, the latter term refers to those abilities that are built-in prior the experience, but that are modifiable in later individual development and through the process of attunement to relevant environmental cues: the importance of development, and its relation with plasticity, is clearly captured thanks to the above distinction. Not all aspects of cognition are pre-determined by genes and hardwired components. See also Barrett's considerations on the issue in [4].

not exist in our ecologies?" is still unanswered. How did humans come up with the precise graphic representation of angels, and the same with Egyptian gods, dragons, fairies and so on? This section will deal with the processes of distribution of "raw" beliefs about supernatural agents into material culture, which allowed a verbal and/or iconic blending of characteristics from different cognitive domains, and eventually lead to the recapitulation of the supernatural concept *in the final form* shared by human minds.

In order to proceed in this investigation, we must make clearer a concept we have already mentioned: *semiotic brains*. That is, brains that make up a series of signs and that are engaged in making or manifesting or reacting to a series of signs: through this semiotic activity they are at the same time engaged in being minds and so in thinking intelligently.

Several studies [28, 29, 20, 22] in cognitive paleoanthropology – even if rather speculative – contend that reflective and high-level consciousness in terms of thoughts about our own thoughts and about our feelings (that is, consciousness not merely considered as raw sensation) is intertwined with the development of modern language (speech) and material culture. 250.000 years BP several hominid species had brains as large as those of modern day humans, but their archaeological findings did not provide any major evidence of art or symbolic behavior, so far. If we consider high-level consciousness as related to a high-level organization of human cortex, its origins can be related to the active role of environmental, social, linguistic, and cultural aspects.

As a matter of fact, the production of new artifacts (such as *hand axes* as in Mithen's account [29]) had to rely on two crucial factors:

1. a good degree of *fleeting consciousness* (thoughts about thoughts).
2. the exploitation of *private speech* (i.e. speaking to oneself) to allow an overall supervision and appraisal of the various activities involved in the development of an artifact (as for hand axes, private speech served to trail between planning, fracture dynamic, motor control and symmetry). In children as well we may witness a kind of private muttering which makes explicit what is implicit in the various abilities.

It is extremely important to stress that material culture is not just the product of this massive cognitive chance but also its cause. "The clever trick that humans learnt was to disembody their minds into the material world around them: a linguistic utterance might be considered as a disembodied thought. But such utterances last just for a few seconds. Material culture endures", as stressed by Mithen [29, p. 291]. Fleeting consciousness and rudimental private speech provide a kind of blackboard where previously distinct cognitive resources can be exploited all together and in their dynamic interaction. The result of this synthesis can be similar, from the phenomenological point of view, to the isolated application of the single components, but from the psychological and semiotic perspective it sparks a revolution, because it allows the blending of several cognitive domains, thanks to a distribution of cognitive tasks into the external environment.

From this perspective the semiotic expansion of the minds is in the meantime a continuous process of disembodiment of the minds themselves into the material world around them. In this regard the evolution of the mind is inextricably linked with the evolution of large, integrated, material cognitive semiotic systems.

It may take a little effort to find this argument compelling (and especially the following one strictly concerning supernatural beings), because of our modern-humans' linguistic bias. That is to say we believe the highly-organized mind of modern-day human beings has already been pre-wired *according to* the effects of such cognitive distributions. Most of all, our semiotic brains make use of a powerful symbolic language and an advanced private speech: we are thus allowed to rehearse in our symbolic imagination cognitive distributions and blending we would otherwise externalize on material supports.

Nevertheless, when we have to infer a meaning from a set of data we cannot understand, we often rely on *model-based reasoning* in form of recurring schemas, diagrams or other kinds of visual manipulations: similarly, the need for conceptualizing agents that went beyond mere biological ones sparked the need to manipulate and hybridize already known features by distribution over external supports.

A wonderful example of meaning creation through disembodiment of mind is the carving of what is probably a mythical being from the last ice age, 32.000 years ago: a half human/half lion figure carved from mammoth ivory found at Hohlenstein Stadel, Germany, often displayed as an example of the Aurignacian culture[9].

> An evolved mind is unlikely to have a natural home for this being, as such entities do not exist in the natural world, the mind needs new chances: so whereas evolved minds could think about humans by exploiting modules shaped by natural selection, and about lions by deploying content rich mental modules moulded by natural selection and about other lions by using other content rich modules from the natural history cognitive domain, how could one think about entities that were part human and part animal? Such entities had no home in the mind. [29, p. 291]

A mind consisting of different separated intelligences cannot come up with such entity. The only way is to extend the mind into the material word, giving the environment a primitive organization and exploiting in a semiotic way external materials (such as stone, ivory, *etc.*) and various techniques to impress a modification on them: "[...] artifacts such as this figure play the role of anchors for ideas and have no natural home within the mind; for ideas that take us beyond those that natural selection could enable us to possess" [29, p. 291].

---

[9] To appreciate the importance of this finding, consider that the first subsequent evidences of human-animal hybrid go back to the third millennium BC: that is, more than 25.000 years after the Aurignacian "Lion-Man".

In the case of our figure we deal with an anthropomorphic thinking created by the material representation serving to semiotically anchor the cognitive representation of supernatural being. In this case the material culture disembodies thoughts, that otherwise would soon disappear – without being transmitted to other human beings – and realizes a systematic semiotic delegation to the external environment. The early human mind possessed two separated intelligences for thinking about animals and people. Through the mediation of the material culture the modern human mind can manage to think *internally* about the new concept of animal and human at the same time. But the new meaning occurred over there, in the external material world from where the mind picked it up.

In this perspective we acknowledge that material artifacts are tools for thoughts as language is: tools (and their related new signs) for exploring, expanding, and manipulating our own minds. In this regard the evolution of culture is inextricably linked with the evolution of consciousness and thought.

Through the mediation of the material culture the modern human mind can arrive to internally think the new meaning of animals and people at the same time. This process involves two fundamental representational activities:

- *external representations* are formed by external materials that express (through reification) concepts and problems already stored in the brain or that do not have a natural home in it;
- *internalized representations* are internal re-projections, a kind of recapitulations (learning), of external representations in terms of neural patterns of activation in the brain. They can sometimes be internally manipulated like external objects and can originate new internal reconstructed representations through the neural activity of transformation and integration.

As for the Hohlenstein Stadel "Lion-Man", the *external representation* stage concerns the single parts of the figurine, the human body and the lion head, while the *internalized representations* comprehends the Lion-Man as a meaningful whole.

It is plausible to imagine that most anthropomorphic and zoomorphic hybrid deities – and supernatural creatures in a wider sense – were generated in a similar way. This is not to say that a Hindu should not be scandalized if we labeled his pantheon a series of primitive iconic hybrids: as human beings progressed in the use of their mind as a "semiotic sketchpad", they could use it as a *virtual* support reproducing part of the tasks they would face in their environment. Imagination supported by advanced language and full consciousness is affected by a much lesser number of constraints than material culture [24, chapter 3, in particular subsection 4.2].

As we saw in the previous section, the new hybrid super agent – once internalized in the mind of human being – becomes the explanation of the original signs accepted not as the best (from an epistemological point of view) but as the most satisfactory: further occurrences of the same signs will mechanically lead to the identification of the super-agent as causing the phenomenon:

the processes we just analyzed undergo a similar confirmation bias. Once we have the result, the externalization, blending and recapitulation process is obliterated and the new hybrid concept is accepted within the human brain as if it had been there *originally*. Cryptozoology and akin *borderline sciences* seem to rely on the naïve assumption that if we have the concept of some creature in our minds and our cultures, and furthermore we have depictions of it, then it probably exists (or at least it existed sometimes in the past). Human beings exhibit a similar behavior very often when, for instance, after watching an horror or sci-fi movie they expect the monster, alien, undead or vampire to sneak up in a dark hallway at night in their own house, even though they consciously know they just witnessed a production of human fiction.

With this respect we can draw the last, but not least, consideration of our investigation. Once we externalize and embody our thoughts about supernatural beings into the material culture, a dramatic increase in *manipulability* resulted. By associating a supernatural agent with its representation the agent became immediately present *there and then*. It could be seen, contemplated, touched, transported from one place to another, shown to other people who in turn possessed different ones: we can easily imagine a never ending continuity stretching from the *Lion Man* to present day Crucifixes.

The abductive theoretical framework on which we have been relying from the beginning is useful once again, with the concept of *manipulative* abduction [24, p. 1.6]: the externalization and embodiment of agent-concepts in artifacts were the result of an hypothetical eco-cognitive distribution in the environment, so that thinking and discovery were achieved *through doing*. The possibilities brought about by the externalization and materialization of thoughts exceed those of the same thought when it was *just* in the mind.

The manipulation of (what would become) the religious artifact generates *new* knowledge concerning it, which is promptly re-absorbed in the manipulator's mind: the process that produced the new knowledge is obliterated and this information gathers with what was already known about the supernatural agent before its materialization: thus, in a self-reinforcing dimension, what was as a matter of fact the consequence of the externalization process is considered to be the very knowledge that permitted it.

6 Conclusion

Our philosophical analysis might lead us to a partial revision of our theories about the development of religion in human societies. If belief in supernatural agents has a constitutive origin within our mind-brain system, we can imagine religion to have a much less generative role as far as the extramundane is concerned.

That is, from our cognitive perspective, it might be argued that, at least at the very beginning, religion did not create beliefs in supernatural beings, but conversely religion might be seen as a cultural tool aimed at answering the emergence of such entities in our minds: thus, the issue at stake here is not about religion being maladaptive[10] *per se* but rather religion being a way to regulate and control potentially dangerous beliefs about *super*natural entities conceived by our minds.

The topic should be further developed to shed light on the processes that led to the development of *religion* in its cognitive and logic diachronic dimension. The multidisciplinary approach to the matter, that emerged to full academic decency over the last two decades, could greatly contribute to other fields (not only academical ones) in which the relationship with religion is a difficult one: studies exploring the *naturalness* of religion to the human being could have a noteworthy impact on moral conflicts and other ones, such as the enduring science-religion dichotomy.

References

1. Atran, S.: In Gods We Trust: The Evolutionary Landscape of Religion. Oxford University Press, Cambridge (2005)
2. Atran, S., Medin, D.: The Native Mind and the Cultural Construction of Nature. MIT Press, Cambridge (2008)
3. Barrett, H.C.: Enzymatic computation and cognitive modularity. Mind & Language 20(3), 259–287 (2005)
4. Barrett, H.C., Kurzban, R.: Modularity in cognition: framing the debate. Psychological Review 113(3), 628–647 (2006)
5. Barrett, J.: Cognitive science, religion and theology. In: Schloss, J., Murray, M.J. (eds.) The Believing Primate, pp. 76–99. Oxford University Press, Oxford (2009)
6. Bedford, G.O.: Biology and ecology of the Phasmatodea. Annual Review of Entomology 23(1), 125–149 (1978)
7. Boyer, P.: Religion Explained. Vintage U.K. Random House, London (2001)
8. Carruthers, P.: Simple heuristics meet massive modularity. In: Carruthers, P., Laurence, S., Stich, S. (eds.) The Innate Mind. Culture and Cognition, vol. 2, Oxford University Press, Oxford (2007)
9. Clayton, S., Opotow, S. (eds.): Identity and the Natural Environment: The Psychological Significance of Nature. The MIT Press, Cambridge (2003)
10. Dennet, D.: Breaking the Spell. VIKING, New York (2006)
11. Dettner, K., Liepert, C.: Chemical mimicry and camouflage. Annual Review of Entomology 39(1), 129–154 (1994)

---

[10] Our use of adaptive and maladaptive can be said to be *loosely darwinian*. We did not wish to enter the debate on wether religion is an evolutionary adaption: its mal-*adaptiveness* would just signify a negative impact on the fitness and the welfare of the concerned individuals. Arguments about the contended evolutionary role of religion tend to focus rather on the social side than on the cognitive one: for further reference on this subject see Wilson's *Darwin's Cathedral* [47]; reflections on the evolutionary weight of religion can also be found in the rich edited book *The Believing Primate* [39] and of course in Atran's monograph [1].

12. Dixon, T.: Science and Religion: A Very Short Introduction. Oxford University Press, Oxford (2008)
13. Driver, J., Spence, C.: Crossmodal attention. Current Opinion in Neurobiology 8(2), 245–253 (1998)
14. Dundes, A.: Introduction. In: Dundes, A. (ed.) Sacred Narrative: Readings in the Theory of Myth, pp. 1–3. University of California Press, Berkley (1984)
15. Flinn, M.V., Geary, D.C., Ward, C.V.: Ecological dominance, social competition, and coalitionary arms races: Why humans evolved extraordinary intelligence. Evolution and Human Behavior 26, 10–46 (2005)
16. Fodor, J.: The Modularity of the Mind. The MIT Press, Cambridge (1983)
17. Fodor, J.: The Mind Doesn't Work That Way. The MIT Press, Cambridge (2000)
18. Gebhard, U., Nevers, P., Bilmann-Mahecha, E.: Moralizing trees: anthropomorphism and identity in children's relationships to nature. In: Clayton, S., Opotow, S. (eds.) Identity and the Natural Environment: The Psychological Significance of Nature, pp. 91–111. The MIT Press, Cambridge (2003)
19. Gunter, R.: The absolute threshold for vision in the cat. Journal of Physiology 114, 8–15 (1951)
20. Humphrey, N.: The Mind Made Flesh. Oxford University Press, Oxford (2002)
21. Jacob, P., Jeannerod, M.: Ways of Seeing: The Scope and Limits of Visual Cognition. Oxford University Press, Oxford (2003)
22. Lewis-Williams, D.: The Mind in the Cave. Thames and Hudson, London (2002)
23. Magnani, L.: Mimetic minds. In: Loula, A., Gudwin, R., Queiroz, J. (eds.) Artificial Cognition Systems, pp. 327–257. Idea Group Publishing, London (2006)
24. Magnani, L.: Animal abduction. Studies in Computational Intelligence 64, 3–28 (2007)
25. Magnani, L.: Abductive Cognition: The Epistemological and Eco-Cognitive Dimensions of Hypothetical Reasoning. Springer, Heidelberg (2009)
26. Métraux, A.: South American Thunderbirds. The Journal of American Folklore 57(224), 132–135 (1944)
27. Meyer, D.R., Miles, R.C., Ratoosh, P.: Absence of color vision in cat. Journal of Neurophysiology 17(3), 289–294 (1954)
28. Mithen, S.: The Prehistory of the Mind: A Search for the Origins of Art, Religion, and Science. Thames and Hudson, London (1996)
29. Mithen, S.: Handaxes and ice age carvings: hard evidence for the evolution of consciousness. In: Hameroff, A., Kaszniak, A., Chalmers, D. (eds.) Toward a Science of Consciousness III: The Third Tucson Discussions and Debates, pp. 281–296. MIT Press, Cambridge (1999)
30. Neudecker, S.: Eye camouflage and false eyespots: chaetodontid responses to predators. Environmental Biology of Fishes 25(1-3), 143–157 (1989)
31. Peirce, C.S.: Collected Papers of Charles Sanders Peirce. Harvard University Press, Cambridge (1931-1958), Vols. 1-6, Hartshorne, C., Weiss, P. (eds.); vols. 7-8, Burks, A.W (ed)
32. Peirce, C.S.: The Essential Peirce: Selected Philosophical Writings. Indiana University Press, Bloomington (1992-1998); vol. 1 (1867-1893) ed. by N. Houser and C. Kloesel; vol. 2 (1893-1913) ed. by the Peirce Edition Project
33. Pettypiece, C.E., Goodale, M.A., Culham, J.C.: Integration of haptic and visual size cues in perception and action revealed through cross-modal conflict. Experimental Brain Research 201(4), 863–873 (2010)

34. Phelps, E.A., Ling, S., Carrasco, M.: Emotion facilitates perception and poten-tiates the perceptual benefits of attention. Psychological Science 17(4), 292–299 (2006)
35. Purser, B.: Jungle Bugs: Masters of Camouflage and Mimicry. Firefly Books, New York (2003)
36. Raftopoulos, A.: Is perception informationally encapsulated? The issue of theory-ladenness of perception. Cognitive Science 25, 423–451 (2001)
37. Raftopoulos, A.: Reentrant pathways and the theory-ladenness of perception. Philosophy of Science 68, S187–S189 (2001), Proceedings of PSA 2000 Biennal Meeting
38. Reagan, A.B., Walters, L.V.W.: Tales from the Hoh and Quileute. The Journal of American Folklore 46(182), 297–346 (1933)
39. Schloss, J., Murray, M.J. (eds.): The Believing Primate. Oxford University Press, Oxford (2009)
40. Scott, S.L.: "They don't have to live by the old traditions": saintly men, sinner women, and an Appalachian Pentecostal revival. American Ethnologist 21(2), 227–224 (1994)
41. Simons, D.J., Chabris, C.F.: Gorillas in our midst: sustained inattentional blindness for dynamic events. Perception 28, 1059–1074 (1999)
42. Sperber, D., Hirschfeld, L.A.: The cognitive foundations of cultural stability and diversity. Trends in Cognitive Sciences 8(1), 40–46 (2004)
43. Thesen, T., Vibell, J.F., Calvert, G.A., Österbauer, R.A.: Neuroimaging of mul-tisensory processing in vision, audition, touch, and olfaction. Cognitive Process-ing 5(2), 84–93 (2004)
44. Thom, R.: Stabilité Structurelle et Morphogénèse. Essai d'une théorie générale des modèles. InterEditions, Paris (1972). Translated by D. H. Fowler, Structural Stability and Morphogenesis: An Outline of a General Theory of Models, W. A. Benjamin, Reading, MA (1975)
45. Thom, R.: Modèles mathématiques de la morphogenèse. Christian Bourgois, Paris (1980); Translated by W. M. Brookes and D. Rand, Mathematical Models of Morphogenesis, Ellis Horwood, Chichester (1983)
46. Tooby, J., Cosmides, L.: The psychological foundations of culture. In: Barkow, J., Cosmides, L., Tooby, J. (eds.) The Adapted Mind. Oxford University Press, Oxford (1992)
47. Wilson, D.S.: Darwin's Cathedral. Chicago University Press, Chicago and Lon-don (2002)
48. Woods, J.: Seductions and Shortcuts: Error in the Cognitive Economy (2010) (Forthcoming)

Part II
Formal and Computational Aspects of Model Based Reasoning

Does Logic Count?
Deductive Logic as a General Theory of Computation

Jaakko Hintikka

Abstract. What is the relation of the ordinary first-order logic and the general theory of computation? A hoped-for connection would be to interpret a computation of the value b of a function $f(x)$ for the argument a as a deduction of the equation $(b = f(a))$ in a suitable elementary number theory. This equation may be thought of as being obtained by a computation in an equation calculus from a set of defining equations plus propositional logic plus substitution of terms for variables and substitution of identicals. Received first-order logic can be made commensurable with this equation calculus by eliminating predicates in terms of their characteristic functions and eliminating existential quantifiers in terms of Skolem functions. It turns out that not all sets of defining equations can be obtained in this way if the received first-order logic is used. However, they can all be obtained if independence-friendly logic is used. This turns all basic problems of computation theory into problems of logical theory.

1 Deductive Logic vs. Theory of Computation

Deductive logic and the theory of computability are twins separated at birth. The theory of computability was born in the context of studies of logical provability like the famous *Entscheidungsproblem* (decision-problem for first-order logic). Its first major achievements concerned the limitations of computational methods in logic. Gödel showed that the truths of elementary arithmetic are not recursively enumerable and Church showed that the logical truths of the received first-order (RFO) logic do not form a recursive set. Yet each of the two twins has its own identity which is especially important to keep in mind when we discuss the scope and the limits of what one of

Jaakko Hintikka
Department of Philosophy, Boston University, Boston, USA
e-mail: hintikka@bu.edu

L. Magnani et al. (Eds.): Model-Based Reasoning in Science & Technology, SCI 314, pp. 265–274.
springerlink.com   © Springer-Verlag Berlin Heidelberg 2010

them can do, absolutely or to help the other one. Most of the discussion in the earlier literature has been about what computational (recursive) methods can do in logic, not vice versa. For instance, the results of Gödel and Church are not about what logic can do in arithmetic or in logic, respectively. These results do not reveal any limitations to what logic, mathematics, axiomatic method, or the human mind can accomplish. They only show what computers cannot do. This massive fact is almost universally neglected not only in popular expositions but in philosophical discussions.

This illustrates the fact that the relations of the two twins are as complicated as the relations of two family members often are. This paper is an attempt to sort out some of those relations and hopefully enhance the cooperation of the two. My main (but not only) question is not what computation theory can do for logic, but what logic can do for the theory of computability. In other words, does logic count?

The family relations here are apparently complicated by the fact that logic and computability theory are on the face of things conceptually incommensurable. Computability means computability of functions whereas the nonlogical primitives of applied first-order logic are predicates (properties and relations). This difference cuts deeper than what is usually realized. Usually, logicians treat functions as predicates of a certain kind. A function $f(x)$ is on this view merely an alter ego of a predicate $F(x, y)$ that satisfies the two conditions

(1.1) $(\forall x)(\exists y)F(x, y)$

(1.2) $(\forall x)(\forall y)(\forall u)(((F(x, y) \& F(x, u)) \supset y = u)$

Then we apparently can set

(1.3) $(\forall x)(\forall y)((f(x) = y) \leftrightarrow F(x, y))$

However, there is an important theoretical catch here. When we try to move from predicates to functions, the assumptions (1.1)–(1.2) have to be imported as ad hoc additional truths. They are not logical truths that could be seen from the notation itself. In the simplest possible sense, the logic of functions is not strictly speaking a part of the logic of predicates, although it can be, is still treated as a part of first-order logic of quantifiers.

2 Quantifiers as Dependence Indicators

Fortunately, there is a far deeper connection between first-order predicate logic and functions. An important part of the conceptual job description of quantifiers is to serve to express relations of dependence (and by implication independence) of their variables. Such de facto dependence between quantified variables is expressed by the formal dependence of a quantifier, say $(Q_2 y)$, on another, say $(Q_1 x)$. Now in RFO logic this formal dependence is expressed by the fact that (Q_2y) occurs within the scope of (Q_1x). This scope is marked by parentheses, as in

(2.1) $(Q_1x)(\ldots\ldots(Q_2y)(\ldots\ldots)\ldots\ldots)$

These dependencies are of course governed by certain functions. These functions have a name: they are the Skolem functions of the sentence in question.

Skolem functions are the lifeblood of first-order logic. Quantifiers can be viewed as being little more than proxies for Skolem functions. Trafficking in functions rather than in predicates thus does not distinguish computation theory from logic, since both do so.

This role of Skolem functions is brought out by their role in truth-conditions for quantificational propositions. In a manner of speaking, they *are* the truth conditions. In game-theoretical semantics (GTS), a sentence S is true iff there exists a winning strategy for the verifier in the correlated game $G(S)$. The Skolem functions of S codify such winning strategies. Here the obvious truth-condition for S is the existence of (a full set of) its Skolem functions.

This can be made even more intuitive in a different way. On our natural understanding of truth is a quantificational discourse, a sentence S is true if and only if suitable "witness individuals" that vouchsafe the truth. For instance,

(2.2) $(\forall x)(\exists y)F[x, y]$

is true if and only if for each value a of x there exists a value of y satisfying $F[a, y]$. As this example shows, witness individuals can depend on other individuals, as a function of which they are given. For instance, in (2.2) y is a function f of x. The truth of S is thus tantamount to the existence of a full set of such "witness functions". For instance, (2.2) is equivalent with

(2.3) $(\exists f)(\forall x)F[x, f(x)]$

But a moment's thought shows that the "witness functions" of S are precisely its Skolem functions.

3   The Equational Calculus

After these prolegomena, how can we re-establish a cooperation of the lost twins? How can we bring together computations and deductions? At first, this might seem to be unproblematic. The basic idea is to construe any computation of the value (say b) of a function f for the arguments a as a deduction of the equation $(f(a) = b)$ in a suitable system of number theory, and likewise for functions with more than one argument.

This at first seems easy enough. It is in fact trivially easy in the theory of primitive recursive equations. There we are dealing with what can be computed from a finite set of recursion equations of two rules:

(SI)   Substitutivity of identicals
(TS)   Term substitution: A variable may be replaced by a term.

In the case of primitive recursive functions, it is possible to allow replacement only by constant terms. All variables are thought of as being bound to initial universal quantifiers. In recursion theory the two computation rules (Sl) and (TS) are treated as rules of computation. However, they are ipso facto also valid rules of logical inference. Hence we have a duck-rabbit ambiguity here. A computation of a value of a primitive recursive function is automatically also a deduction of the desired equation from the defining recursion equations of the function in question.

But primitive recursive functions do not exhaust the class of computable functions. Hence the question arises how to generalize this connection between computations and deductions. One of the first ideas here is to allow also the use of propositional (truth-functional) logic. After all, propositional inferences are purely tautological. Quantificational logic enters the picture only in the guise of (Sl) and (TS). Actually, it also enters tacitly or strictly speaking structurally authorizing the formation of new functions by function composition and by the so-called projection functions. Such a function can e.g. map each given sequence of arguments $< a_1, a_2, \ldots, a_i, \ldots, a_j >$ to a_i.

The system we can thus reach is worth some attention in its own right. It will turn out to be capable of serving as the missing link between computation and deduction. It will be called the equational calculus. On the one hand, this equational calculus can be considered as a system of logic that can for instance be compared with the usual first-order logic. It has no predicates, only functions, and the atomic expressions are all equations. It comprises propositional logic. There are no existential quantifiers, only individual variables that are thought of as being bound to initial universal quantifiers. Since negation signs can be pushed deeper and deeper into formulas, we do not need to consider negation if we allow atomic formulas to be inequalities in addition to identities.

4 How to Compute a Function

But the very same equational calculus can serve as a system of computation. This is possible in that its nonpropositional rules of inference are (Sl) and (TS) which are also rules of computation. A derivation of an equation $(b = f(a))$ can thus be construed as a computation of the value of f for the argument a. The only rules of inference besides the tautological inference rule of propositional logic are (Sl) and (TS).

Used as a system of deductive logic, the equational calculus deals with questions as to what follows from a given propositional combination (truth-functions) of equations. This reduces to the question of what is implied by a conjunction of equations, some of them negated. Such combinations of equations (negated or unnegated) will be called defining equations (or sets of such equations.) It is easily seen that the equational calculus considered as a deductive logic is semantically complete.

Now from the condition $F[x, y]$ a number of sentences (for instance, from a set of equations involving x and y) of the form $F(a, b)$ can be deduced corresponding to the computation of the value b of f for the argument a. Hence the remaining question here is how to ascertain that (4.1)–(4.2) are satisfied.

In the equational calculus, any condition $F[x, y]$ will define a (total) function of one variable such that

(4.1) $(f(x) = y) \leftrightarrow F[x, y]$

assuming that the following conditions are satisfied:

(4.2) $(\forall x)(\exists y)F[x, y]$

(4.3) $(\forall x)(\forall y)(\forall u)((F[x, y] \,\&\, F[x, u]) \supset (y = u))$

An explanation is that from the condition $F[x, y]$ a number of (for instance, from a set of equations involving x and y) a number of sentences of the form $F(a, b)$ can be deduced corresponding to the computation of the values b of f for the argument a. Hence the first question here is how to ascertain that (4.2)–(4.3) are satisfied.

In the case of primitive recursion equations (4.2) and (4.3) are automatically satisfied. In general, it may of course happen that (4.2) fails in the sense that no equation of the form $(f(a) = b)$ can be deduced from a given set of equations. But this only means that f is a partial function.

Now (4.3) is satisfied for a condition $F[x, y]$ expressible in the equational calculus (i.e. in the sole terms of functions) as soon as the condition is consistent. For otherwise we would have a contradiction like $f(a) = b_1, f(a) = b_2$, $b_1 \neq b_2$. Such generalized computations come to an end in a finite number of steps if the initial equations and other conditions are like the defining equations for primitive recursive functions in that they determine the value of a function for a given argument in terms of its values for smaller arguments.

Hence the equational calculus can serve as a theory of computation. Admittedly, from the condition $F[x, y]$ we cannot tell whether or not it is in fact consistent because this question is recursively unsolvable.

This problem does not turn out to be difficult to by-pass. What we have to do is to find a class of functions corresponding to defining conditions $F(x, y)$ which are known to be consistent and which at the same time allow for definitions of all computable functions. This is what is achieved in one of the main forms of the general theory of computability, recursive function theory. The class of functions defined there is the class of all (general) recursive functions. It equals the class of computable functions definable by other means, such as the theory of Turing machines, the lambda calculus etc.

In recursion theory the satisfaction of these requirements is guaranteed by requiring that the function forming operations that go beyond primitive recursive functions are restricted to minimization operations. They take a condition $g(x, y) = 0$ and turn it to a requirement of a minimality on x for a given y by requiring that

(4.4) $(g(x, y) \neq 0) \vee (x \geq f_g(y))$

Here f_g is the new function defined by means of g. It is worth noting that (4.4) involves the propositional connectives for negation and disjunction. For its use we therefore need propositional logic over and above such values as (Sl) and (TS). The condition (4.4) is like the equations for primitive recursive functions in that the value of $f_g(y)$ is determined by it in terms of the values of $g(x, y)$ with $x = f_g(y)$. This also shows the consistency of the defining conditions of general recursive functions.

The only additional qualification needed is that it is computationally impossible always to ascertain that (1.1) is true. But this only means that recursion theory deals with partial functions and not only with total functions.

5 Logic as Equational Calculus

This turns the general theory of computability into the study of the equational calculus which is at the same time a part of deductive logic. Hence what remains to be done is to study the relation of the equational calculus to the usual forms of first-order logic. Is the equational calculus a part of first-order logic, as might first seem to be the case? If so, how can we deal with partial functions that are also studied in the theory of computability but not in RFO logic?

Now the RFO logic is usually is formulated in terms of predicates as non-logical primitives instead of functions. Accordingly, the theory of first-order logic both proof theory and modes theory are usually couched in the same terms. Hence in order to bring these rich theories to bear on the study of computability, we must see what happens when we try to reformulate the usual (RFO) logic in terms of functions only and to dispense with existential quantifiers in favor of equations.

Once again nothing seems easier. Given a usual first-order sentence or theory, what can obviously be done is the following:

(i) Predicates are replaced by their characteristic functions

This is merely a matter of interpretation. It does not affect the logic.

(ii) Each existential quantifier formula $(\exists x)G[z]$ occurring in a context

$S[\ldots\ldots (\exists x)G[x] \ldots\ldots]$

is replaced by

(5.1) $F[f(y_1, y_2, \ldots)]$

where $(Q_1y_1), (Q_2y_2), \ldots$ are the quantifiers on which $(\exists x)$ depends in S. (It is assumed here for notational convenience that all formulas are in the negation normal form.) Such functions f are what we have been calling Skolem functions of S. In view of the role of Skolem functions in first-order logic we can appreciate the naturalness of this translation of RFO logic in the equational calculus

The result is a reduction to the same equational calculus as we met before. In this way, the entire RFO logic is turned into a kind of universal algebra where there are no existential instantiations, only symbolic (and numerical) calculations with equations. The only nonpropositional rules are (SI) and (TS). It would therefore be a prime candidate for the role of a universal algebra, or at least the role of the true "algebra of logic". (A couple of important qualifications to these claims will be registered later in this paper.)

This morphing of first-order logic into an equation calculus is often useful for actual deductions. For instance one can formulate the axioms of group theory in quantificational terms without introducing any new function symbols. Such a treatment turns into a morass of complications in no time at all, however. The advantages of an equational approach have been pointed out earlier by Alfred Tarski and Stig Kanger.

The translation of RFO logic into the equational calculus means turning any logical argument in the usual first-order logic into a symbolic computation. Without any loss of generality, it can be assumed that all the formulas involved are in a negation normal form and that the logical argument in question follows tableau-type rules. In this correspondence, each existential instantiation in the original corresponds to an introduction of a new Skolem function term (term consisting in combinations of Skolem functions). In particular, there is a one-one correspondence between individual constants introduced in a deduction by existential instantiation and constant Skolem function terms introduced in the correspondent computation.

Hence the answer to my title question is in one interpretation an emphatic "yes". Not only can we think of deductions as symbolic computations. This is all that there is to our basic first-order logic.

6   Limitations of the Received Logic in Computation Theory

What has been established so far still leaves a big question unanswered. All computations can be thought of as taking place in the equational calculus. All deductions can be carried out as equational computations. But are all equational computations such translations? Can all computations be interpreted as hidden deductions?

If the logic we are thinking of is the received first-order logic, the answer is *no*. Not all sets of equations that can be used to define a recursive function can be obtained from formulas of the RFO logic. This is because the sets of functions figuring in such equations must satisfy certain conditions that are imposed on them by the structure of RFO formulas. Such formulas have a labeled tree structure. Each Skolem function comes from an existential quantifier somewhere in the tree. Its location in the tree is indicated by the selection of its arguments. These arguments are the variables bound to the quantifiers on which the Skolem-function generating existential quantifier depends. In RFO logic notation, they are all the numerical quantifiers lower in the same branch. Hence the argument sets of the Skolem functions of a given formula must have the same tree structure in terms of class-inclusion. For one thing, the argument sets must be partially ordered, with all upwards chains linearly ordered.

There is no reason why the functions in arbitrary recursion equations should do so, and it is easy to find actual examples to the contrary. The simplest ones are perhaps recursion equations for parallel processing. There the parallelism is intuitively equivalent to the branching of the quantifier structure. Thus logic can "count", but not all "counting" (that is all computation) can be construed as deduction in RFO logic.

This result shows important things about the relation of RFO logic and the equational calculus considered as a system of deductive logic. The equational calculus is stronger in representative power than the received first-order logic. There are functions that can be defined in the former but not in the latter.

7   IF Logic to the Rescue

This diagnosis of the reasons why all computations cannot be construed as deductions reveals also a cure. The underlying reason is that in RFO logic the formal dependences between quantifiers is expressed by the nesting of their scopes. This is a very special kind of relation, among other things transitive and antisymmetric. This nesting relation creates the tree structure that limited RFO logic as a theory of computation. It does not make it possible to express in RFO logic all possible patterns of dependence and independence between variables. Since Skolem functions are the means of codifying those dependencies, not all patterns of Skolem functions are expressible.

Now this is precisely the restriction that is removed where we move from RFO logic to independence-friendly (IF) first-order logic. There we can express the independence of a quantifier $(Q_2 y)$ of another quantifier $(Q_1 x)$ in whose formal scope it occurs by writing it as $(Q_2 y / Q_1 x)$. In its simplest variant, only independencies of existential quantifiers of universal ones are considered. (As always, formulas are assumed to be in a negation normal form.) This is the IF logic considered in this paper. It can immediately be extended by admitting sentence-initial contradictory negations. It will be called extended independence-friendly (EIF) logic.

Unextended IF logic is as strong as the $\sum_1^1$ fragment of second-order logic. In fact, the existence of the Skolem functions is expressed by a $\sum_1^1$ sentence.

The converse translation takes us from the $\sum_1^1$ fragment of second-order logic to IF first-order logic.

For instance, a "branching quantifier" formula like

(7.1) $(\exists f)(\exists g)(\forall x)(\forall y)F[x, f(x), y, g(y), h(x, y)]$

has no equivalent in RFO logic However, in IF logic it translates as

(7.2) $(\forall x)(\forall y)(\exists z/\forall y)(\exists u/\forall x)F[x, z, y, u, h(x, y)]$

More can be said here, however. When a formula of the equational calculus is considered as a formula of a suitable first-order logic, it is seen that it contains (in its negation normal form) only universal quantifiers, not existential ones. It is hence the negation of a purely existential formula. Hence it is equivalent with a $\prod_1^1$ formula, a mirror formula of a $\sum_1^1$ one. Hence the equational calculus is not strictly speaking equivalent with IF first-order logic, but its mirror image, the other half of EIF logic which is equivalent with the $\prod_1^1$ fragment of second-order logic.

This difference matters, for this $\prod_1^1$ half of EIF logic has a complete proof procedure even though IF logic in the narrow sense does not. The mirroring relation implies that IF logic does have a complete disproof procedure. This can perhaps be seen most easily by proving that IF logic is compact, but this is merely a reflection of the semantical meaning of suitable attempted proofs of logical truth is frustrated attempts to construct a counter-example. The other side of the coin is that the $\prod_1^1$ half does not have a complete disproof procedure.

EIF logic could claim to be the true universal algebra, if it did not allow for the further generalization that removes all restrictions from the use of contradictory negation.

In EIF logic, we can among other things formulate a descriptively complete axiomatization of elementary number theory. The only unusual axiom would express that there are no infinite descending chains of numbers. This could be expressed by

(7.3) $\neg (\forall x)(\forall z)(\exists y/\forall z)(\exists u/\forall x)((((x=z) \leftrightarrow (y=u))\ \&\ (x > y)\ \&\ (z > u))$

It is IF first-order logic that is equivalent with the unrestricted equation calculus and hence can serve as the medium of recursion theory (general theory of computation). In IF logic, the law of excluded middle does not hold for the primary negation. Hence truth-value gaps and partiality are automatically admitted. For instance, the minimization operation in recursion theory that takes us from a function $g(x, y)$ to the function $f(y)$ that yields the smallest value x such that $g(x, y) = 0$ can now be implemented by defining f as the characteristic function of a predicate of the form

(7.4) $G(x, y, o) \& (\forall u)(G(u, y, o) \supset (u \geq x))$

(cf. (4.4) above.)

8   EIF Logic as a Framework for a General Theory of Computability

But the failure of tertium non datur seems to create a serious obstacle to the use of IF logic as a framework of computation theory. It implies (as was pointed out) that there is no complete axiomatization for IF first-order logic. Hence there does not seem to be any hope of construing all computations as formal deductions.

This does not matter, however. It suffices for the purposes of computability theory to have a complete disproof procedure. For then a computation of the value $f(a)$ of a function $f(x)$ for the argument a from a set of equations E can be captured by a disproof of the sentence

$$(8.1)\quad E \,\&\, (\forall y) \,\neg\, (f(a) = y)$$

A successful disproof will produce a value of y, say b, such that the equation $(f(a) = b)$ is either true or neither true-nor-false. The latter possibility is in keeping with the fact that we are in the general theory of computation dealing with partial functions.

Thus the formula

$$(8.2)\quad E \,\&\, (\forall y) \,\neg\, (f(x) = y)$$

can serve as the defining condition for the function f in the sense of sec 4.

References

1. Hintikka, J.: The Principles of Mathematics Revisited. Cambridge University Press, Cambridge (1996)
2. Hintikka, J., Sandu, G.: What is the logic of parallel processing? International Journal of Foundations of Computer Science 6, 27–49 (1995)
3. Phillips, I.C.C.: Recursion theory. In: Abramsky, S., Gabbay, D.M., Maibaum, T.S.E. (eds.) Handbook of Logic in Computer Science, vol. 1, pp. 79–187. Clarendon Press, Oxford (1992)
4. Rogers Jr., H.: Theory of Recursive Functions and Effective Computability. McGraw-Hill, New York (1967)
5. Sandu, G.: Independence-Friendly Logic. Cambridge Unversity Press, Cambridge (forthcoming)

Causal Abduction and Alternative Assessment: A Logical Problem in Penal Law

Claudio Pizzi

Abstract. Epidemiological investigations very often allow saying with certainty that there is a relation between a macrophenomenon F and a certain value of increase or decrease of a certain pathology P. The abductive inference which leads to such a conclusion, however, does not allow establishing which cases of the pathology P are actually caused by cases of F and which are not. Given that in order to establish penal responsibility in most Western countries the law requires that there is a causal relation among token – events (which here we will identify with so-called Kim-events) it is frequently argued that in such cases no causal relation, and a fortiori no penal responsibility, can be properly established. The problem will be examined with the tools of quantified conditional logic. The aim of the paper is to argue that identifying a causal relation in which causes and effects are at a different level of determination does not prevent establishing penal responsibilities.

1 The Semmelweis' Case

Epidemiology is a branch of medicine whose object is the distribution of diseases in populations and the study of their causal determinants. It was born about the middle of the XIX century, and since the beginning it was able to reach noteworthy results. It had an impressive development after the second World War, extending its realm to the area of clinico-pharmacological inquiries. In the Sixties a new paradigm of clinical behavior was offered by so called Evidence Based Medicine, defined as "the conscientious, explicit and judicious use of current best evidence in making decisions about the care of individual patients"[1]. In other words it was theorized that

---

Claudio Pizzi

Dipartimento di Filosofia e Scienze Sociali, Università di Siena, Siena, Italy

e-mail: `pizzic@msn.com`

[1] See [12, p. 1].

L. Magnani et al. (Eds.): Model-Based Reasoning in Science & Technology, SCI 314, pp. 275–289.

individual diagnoses should take into account epidemiological evidence. In the last decades the importance of investigations about pollution widened the borders of medical epidemiology giving rise to new disciplines which are called *eco-epidemiology* and *socio-epidemiology*.

The Semmelweis case offers an interesting example of a successful epidemiological inquiry, even if its protagonist was an obstetrician. The episode is well known and is frequently quoted by philosophers of science as a paradigmatic application of experimental method[2].

Semmelweis was impressed by the meaningful difference in the numbers of cases of death from childbed fever (puerperal fever) which were recorded, beginning from 1840, among parturient women hospitalized in two Maternity Divisions of his hospital: in 1845 they were 11.4% in the first clinic and 2.7% in the second. The problem of understanding what "made a difference" was intriguing, since apparently there was no manifest difference among the properties of the women of the two classes. Semmelweis entertained various conjectures, not excluding psychological ones. The first clinic had the reputation of being a "clinic of horrors". Among other frightening aspects it should be mentioned that a priest preceded by an attendant ringing a bell used to bear the last sacrament to dying women: this was supposed to provoke a debilitating choc which might have been a possibile cause of puerperal fever. But the conjecture had to be rejected: after eliminating this disagreeable liturgy mortality did not decrease. Also the idea that the treatment received by male students working in the First Division could have been different from the one received by the midwives working in the Second had to be rejected, since the treatment turned out to be exactly the same. However Semmelweis observed, but was unable to explain, that hospitalized women who had childbirth in the street, before reaching the hospital, turned out to have a better chance of avoiding puerperal fever.

The solution of the puzzle was reached thanks to a serendipian intuition. During the autopsy of a colleague named Kollethcka, who died of septicemia from a wound to his finger produced during an autopsy, Semmelweis observed an analogy between this infection and the course of the disease registered in puerperian fever. After this, he observed that the male students followed courses which put them in contact with "cadaveric matter", while girls did not have this experience. He therefore issued an order requiring all medical students to wash their hands in a solution of chlorinated lime before treating women. The mortality from childbed fever promptly began to decrease, and for the year 1848 it fell to 1.27% in the First Division, even lower than the 1.33% in the Second.

It is remarkable that Semmelweis, after reaching what seemed to be an undeniable success, went in search of check-proofs, so performing a step which should be considered as a methodologically essential part of any abductive procedure. For instance, he infected the womb of rabbits with material

---

[2] See for instance [5, chapter 2.1].

derived from cases of endometritis, so obtaining the expected result. Also, by looking at the historical registers of the hospital, he found that no difference between the two clinics was registered before the practice of dissection. Notice that in this phase nobody was able to propose hypotheses about the causal mechanism which produced the effect. Only In 1879 Pasteur finally identified the haemolytic streptococcus responsible for childbed fever.

Semmelweis's case has been carefully studied, among others, by Peter Lipton[3]. As is well known, Lipton supported the idea of abduction as "inference to the best potential explanation" or, as he says, "inference to the potential *loveliest* explanation" where the loveliest explanation is the one that, if correct, provides the most understanding. "Through the use of judiciously chosen experiments, Semmelweis determined the loveliest explanation by a process of manipulation and elimination that left only a single explanation of the salient contrast" (p. 90). Lipton's IBE reconstruction of Semmelweis' enterprise is based on the idea of a contrastive explanation; the difference between two clinics was a central starting point for his research. In other words the central question had not the form "Why there is an $x\%$ of dead women in the first clinic?" but "Why there is an $x\%$ of dead women in the first clinic and $x - y\%$ in the second clinic?". According to this view the *explanandum* is not an event but a complex assertion whose form is $\exists x (m^1 - m^2 = x \wedge (x > \delta))$, where δ is a threshold value, and m^1 and m^2 are the recorded values of the mortality in the two clinics.

We may look at the episode in a slightly different way, going back to Hempel's analysis of the same episode.

Hempel remarked that Semmelweis reached his result by enumerating all causal hypotheses and eliminating all the hypothesized causal relations until remaining with the only one which resisted falsification. For instance, considering the priest's possible influence on the health of women, Hempel writes: "He [Semmelweis] asks himself: are there any readily observable effects that should occur if the hypothesis were true? And he reasons: If the hypothesis were true, then an appropriate change in the priest's procedure should be followed by a decline in fatalities. He checks this implication by a simple experiment and finds it false, and then rejects the hypothesis" (p. 20).

According to this reconstruction he considered various hypothetical causal relations of the form "H is causally related to E" and checked them by supposing $\neg H$ and seeing if $\neg E$ was true. This procedure suggests two remarks: 1) Semmelweis applied what has sometimes been called eliminatory induction, which is indeed what should be called (causal) selective abduction; 2) The causal relation is controlled not by seeing whether H makes E universally predictable, but seeing if $\neg H$ is lawfully followed by $\neg E$. This means that he intended causes essentially as necessary conditions and not as sufficient conditions for the effect (which, incidentally, seems at odds with Hempel's

---

[3] See [7].

reduction of causes to explanatory factors which is a salient feature of his epistemology).

2 Singular vs. General Causality

Nobody is willing to deny that Semmelweis' case offers a paradigm of a successful abductive procedure. However, there is the danger that at the present time it turns out to be misleading just in the realm of epidemiology, for the reasons which we are going to illustrate.

The first remark is that in studying populations we are not always we are in a position to manipulate the supposed causes. For instance, if the supposed cause is a genetic feature of a population or a meteorological phenomenon, we are not normally in a position to intervene on this feature of nature. What we can do is to limit ourselves to so called *observational studies*, which means having severe limitations in ascertaining correlations and in cross-checking.

In the second place we have to face a problem which is not simply terminological. Many people use to say such things as that the phenomenon "introduction of disinfection of hands and tools" is a cause of the phenomenon described as "reduction of puerperal fever", and Hempel's way of reconstructing Semmelweis' argument appears to suggest the legitimacy of such expressions.

This way of speaking is common, but incorrect if we assume that causal *relata* in primary sense are *token*-events, i.e. individual events endowed of a temporal index and having one or more subjects, such as for instance the event that Napoleon died on May 5th, 1821. The fact that no meaningful difference exists between the mortality of the two Departments in Semmelweis' story is not a *token*-event: it may have a temporal beginning, even if not exactly determined, but we are unable to identify subjects of the mentioned relations.

Even important epidemiologists, however, seem to have in mind *token*-events as causal *relata*. In a paper published in 1978, *Criteria for causation: an unified concept*, the epidemiologist A.S. Evans draws a parallelism between situations described by detectives and situations described by epidemiologists[4]

1. The criminal is present on the scene of the crime\The agent is found in the lesion.
2. Premeditation on behalf of the criminal\Temporal precedence of the cause.
3. Various instruments used in the crime\Multifunctioriality.
4. Seriousness of the wounds with respect to the conditions of the victim\Reaction to the touchiness of the guest.
5. Motivation to commit the crime\Biological plausibility.

---

[4] Now in [3].

6. No one else could have committed the crime\No other agent could have caused the disease in the given circumstances.
7. Guilt beyond any reasonable doubt\Role of the agent or factor established beyond any reasonable doubt in the specific case.

The parallelism outlined by Evans seems to imply that the event whose causes we are looking for is a singular event (see especially points 6. and 7., where reference is made to "the given circumstances" and to the "specific case"). However, when epidemiologists speak of their discoveries they do not mention singular events but diseases or macrophenomena which are their manifestations.

D.L. Weed writes for instance[5]: "Smoking is indeed a cause of lung cancer, laryngeal cancer, esophageal cancer, and bladder cancer... The list of chemical carcinogens – asbestos, arsenic, aniline dyes, diethylstilbestrol, and cadmium to cite a few examples – is long. Radiation of many types is responsible for – causes – skin cancer, breast cancer, and other diseases." (p. 7) ... "What epidemiologists do not do is to study disease causation in order to assign responsibility for harm caused to individuals; specific causation is not a traditional problem for epidemiologists". But the author adds "For judges, legal scholars, and others involved in toxic tort litigation, however, the problem of specific causation is paramount". So some epidemiologists have a clear perception of an intriguing problem. While epidemiologists are willing to speak of what is sometimes called *general causation*, juridical tradition both in Italy and in English-speaking countries is often oriented to endorse some variant of the *conditio sine qua non* theory of causality, which normally involves a conception of a cause as a relation among *token*-events.

The *conditio sine qua non* tradition has been recently revived by David Lewis' theory of causation. In such theory the causal relation takes this form:

$$e_1 \mathscr{C} e_2 =_{Df} Oe_1 \wedge Oe_2 \wedge \neg Oe_1 \mathbin{\Box\!\!\rightarrow} \neg Oe_2 \qquad (1)$$

where e_1, e_2 are *token* events, O is an operator forming propositions from event variables and $\Box\!\!\rightarrow$ is Lewis' conditional operator[6]. However, Lewis does not offer a formal analysis of *token*-events, which in his philosophical papers are seen as properties of spatio-temporal regions. For our purposes it is convenient, however, to give an exact treatment of *token*-events. We will treat them as generalized Kim-events. We may give a definition of a Kim-event as $[R^n, a_1, \ldots a_n, I]$, where R^n is a n-place relation, $a_1, \ldots, a_n$ are individuals, I is a time interval, which we define as an ordered couple of instants $\langle t_1, t_2 \rangle$ (intuitively, the first and last instant of the interval). The difference between original Kim-events and generalized Kim-events simply consists in introducing intervals in place of atomic instants.

---

[5] See [13].
[6] See [6].

However, there is a more important difference with Kim's philosophy. Kim speaks of *existence* of events, but we prefer to speak of existence of objects and *occurrence* of events[7].

Restricting attention to first-degree events, i.e. to events which are not compound of other events[8], we may define the notion of occurrence of events in this way:

a) $O[R^n, a_1, \ldots, a_n, I]$ is true at t if and only if

    1) at least one of the objects $a_1, \ldots, a_n$ has a real existence during I[9]
    2) $R^n a_1 \ldots a_n$ is true during the interval I
    3) t is the first instant of I
    4) $R^n a_1 \ldots a_n$ is false at some interval I' strictly preceding I.

As causal relations are concerned, we can then adopt Lewis' definition reported in (1) but with the restriction that $\square\!\!\rightarrow$ indicates a consequential relation among the clauses and the additional proviso that the event – cause is chronologically prior to the event-effect[10]. In order to stress the difference between these two kinds of conditionals, we will use the symbol $>$ instead of $\square\!\!\rightarrow$.

Let us then come to the main problem. When it is established beyond any reasonable doubt that a certain macrophenomenon F is a *conditio sine qua non* for a distinct macrophenomenon E, even if F and E were reconstructible as compound *token*-events, this does not mean that every case of E is caused by some case of F. Semmelweis' example is misleading since puerperal fever has a monocausal explanation: there is, in fact, only one determining cause of puerperal fever and Semmelweis was able to discover it. But, as a matter of fact, most diseases are multifactorial pathologies, and such disciplines as eco-epidemiology or socio-epidemiology are centered on this basic presupposition.

The case of the relation between smoking and lung cancer offers a paradigm case of an epidemiological discovery concerning a phenomenon which is very different from the case of puerperal fever. After many years of controversies it seems at present that there is a remarkable agreement on the fact that there is a non-spurious causal correlation between smoking and various kinds of cancer, among which lung cancer. But, as everyone knows, a wide number of non-smokers does not fall ill with lung cancer and a wide number of non-smokers does fall ill with lung cancer (among them we should record the anti-smoke guru Allen Carr, who however had been a chain-smoker for years).

---

[7] See [2].

[8] We cannot treat here this difficult question. Compound events may be conjunctive, disjunctive, negative events but also events consisting of causal or chronological relations among events.

[9] Real Existence is opposed to fictitious existence or "merely quantificational" existence. This distinction has no special importance in this context, while it is important in treating, for instance, omissive events.

[10] For more details see [11].

It is not surprising that this fact gave rise to a difficult puzzle in penal law. If someone having lung cancer asks for a compensation of damages to some cigarette industry, how can we know that he or she belongs to the class of the smokers who are victims of tobacco?[11].

We quote the words of an Italian epidemiologist, F. Berrino, with reference to a case which is not fictional: "when someone says that $x\%$ of the tumors are due to professional causes... there is no possibility of discriminating, among the exposed cases, who would not have fallen ill in absence of exposure and who, on the contrary, would have equally fallen ill (*note the use of counterfactuals*, n.R.). To make an example, in ten years of activity of the cancer-register in Lombardia, in the district of Varese we had knowledge of more than 3000 cases of lung cancer. Nearly 2000 of these patients during their activity have been in contact with one or more substances which are carcinogenic for the respiratory system.

We know beyond any reasonable doubt that nearly 1000 of such cases would not have occurred in absence of such specific professional factors. But we do not know which of them [...] Obviously all the names are at disposal of the judge. But what should he do?... Should he draw by lot?"[12]

As we will show in a more detailed way, on this question two opposite lines of thought have been developed. The first asserts that, in lack of the possibility of ascertaining correlations among *token*-events, we have not to do with causal connections. And since responsibility implies causal connection, we are not allowed to speak of responsibility. A second school of thought maintains that it is possible to speak of a causal correlation, even if not of the *token-token* kind which is postulated by the standard counterfactual theory of causation.

3 An Exercise of Formalization

Before going deeper into the analysis, it is fair to remark that counterfactuals are sometimes said to play a role in testing causal hypotheses which does not imply, or presuppose, a relation among *token*-events. The reference is mainly to Ronald Giere's theory of causation, which is based on ascertaining the truth of special counterfactual assertions: in the case of lung cancer, for instance, with reference to a given population we should ask for the truth of the counterfactual "if *nobody* smoked there would be less cases of cancer than if *everybody* smoked"[13]. This would establish the existence of what has

---

[11] A famous case is the one of Patricia Henley, a smoker ill with lung cancer, who obtained a reward from Philip Morris of $1,500,000 and a further reward of $50,000,000 for "punitive damages". It would interesting for the present paper to know the motivation of the sentence. The main fault imputed to Philip Morris was the omission of advices about danger of smoke on cigarette boxes.

[12] See [1, p. 167].

[13] See [4].

been named "general causality", i.e. a causal relation among generic events or event-types as "smoking" or "having lung cancer", with reference to some given population. But there are several difficulties in Giere's theory, which have received due evidence by several critics[14].

This does not mean, of course, that we cannot give some sense to causal relations between macrophenomena or event-types. We could stipulate, for instance, that saying that "smoke causes lung cancer" is simply an abridged way to say that *some token*-events of smoking are causes of *some token*-events of lung cancer.

What we will try to show is that, if the causal antecedent is a *token*-event, any alleged causal relation may be paraphrased in terms of *token-token* causal relations. As a matter of fact, the case in which the antecedent is a *token*-event is the only one which has penal relevance. In fact, when a person is supposed to be guilty of some crime, this can be said because he or she is responsible for some action or some specific event which may be represented as a Kim-event. For instance, establishing the beginning of the pollution at Porto Marghera near Venice, which was at the center of a famous trial, implies that some juridical subject (private person, firm and so on) promoted at a certain prior time a certain industrial project that was the origin of a causal chain which was supposed to lead to an impressive increase of sarcomas in that area. The first problem at stake is to identify exactly the logical form of the argument. For sake of simplicity, we will try to make a formal analysis of the real case exposed at page 280, with the only simplification that statistics are supposed to have provided not approximated but exact numbers.

To begin with, we have to decide what is included in the Background Knowledge BK. BK includes:

a) A set of accepted laws $L_1 \ldots L_n$

and among others the following true statements:

b) 3000 = the number of recorded cases of people who got lung cancer in Varese's district during the last ten years (Fact)

---

[14] See for instance [9]. From our viewpoint suffice it to remark, as a comment to the smoke-cancer example, that if nobody were to have an ashtray at home there would be less cases of lung cancer than if everybody had an ashtray at home, and that we may concede that there is a nomic correlation between the two classes of phenomena. So what appears is that Giere's counterfactuals may not be sufficient to detect spurious correlations. Things are different if we speak of *token-token* correlations. In fact, by our notion of occurrence we have to take into account the *real existence* of the involved objects, and to suppose counterfactually that such objects are unexisting in the given interval. So, supposing that no ashtray had been existing at home of the heavy smoker Mr.Smith, we conclude that he had anyway suffered of lung cancer. In order to establish this point we have to consider the class of smokers who do not use ashtray (a small class, surely!), for instance *clochard* or gypsies, and check the incidence of lung cancer in such a class.

c) 2000 = the number of people who normally get lung cancer and are not exposed to substance f, during any interval of ten years (Law statement)[15]

The preceding propositions jointly entail a proposition which is actually true, i.e.

d) 1000 = the number of people exposed to substance f who got lung cancer during the last ten years in Varese's district.

So, according to the Coveing Law schema of explanation, the fact described in d) turns out to be explained in terms of the *explanans* consisting of a)–c).

Furthermore, adding to the *explanans* the true statement

e) $1000 > \delta$ (δ being a fixed threshold value)

if m_2 is the number of people who got lung cancer without the exposition to f and m_1 the number of people who got lung cancer after the exposition to f we may derive also the existential statement quoted at page 277:

f) $\exists x (m_1 - m_2 = x \wedge (x > \delta))$.

namely the fact that there is a meaningful difference between the number of people who got lung cancer and were exposed to f and the number of those who got lung cancer but were not exposed to f.

It is doubtful, however, that the preceding argument gives a satisfactory causal explanation of the fact we are interested to explain (i.e. of the fact described in d)). Luckily, we have the elements to build also a counterfactual argument. Suppose in fact that no exposition to f had been made in the given circumstances. What is implied by the law expressed in c) is that in this condition the normal number of people ill with lung cancer would be 2000. Since $1000 = 3000 - 2000$, we conclude that under the supposition of the absence of f, 1000 of the 3000 recorded cases would not have taken place. So, if we agree that a conditional is true provided there is a consequential relation between the clauses, we are allowed to endorse the truth of the following counterfactual conditional:

(1) If f had not been introduced in the given circumstances, 1000 cases of the known 3000 people who fell ill with lung cancer would not have occurred.

The formal rendering of the preceding counterfactual seems to be not difficult, provided we perform some simplification of the involved numbers: for instance by dividing the number of cases by 1000, so to treat only with the numbers 1, 2, 3. In this simplified framework the counterfactual which we are trying to focus on is "if f had not been introduced in the given circumstances, exactly 1 person of the known 3 ones would not have developed lung cancer".

---

[15] Note that 3) follows from some nomic regularity, so it follows from the stock of laws $L_1 \ldots L_n$.

Let us call D the finite domain of people who lived in the area of Varese during the mentioned interval I and that got lung cancer, whose names are, suppose, a_1, a_2, a_3.

Let us write $E!^n Ax$ to say "exactly n objects in the domain exist who are A".

According to the well-known definition of this special quantifier we have the following equivalences, where M is supposed to stand for the predicate "getting ill with lung cancer".

(a) $E!^1 Mx \equiv \exists x(Mx \wedge \forall w(Mw \supset (w = x)))$
(b) $E!^2 Mx \equiv \exists x \exists y((x \neq y \wedge Mx \wedge My) \wedge \forall w(Mw \supset (w = x \vee w = y)))$
(c) $E!^3 Mx \equiv \exists x \exists y \exists z(x \neq y \neq z \wedge Mx \wedge My \wedge Mz \wedge \forall w(Mw \supset (w = x \vee w = y \vee w = z)))$

It should then be clear which form the long wffs, expressing the existence of exactly 1000, 2000 or 3000 people who got lung cancer, should have.

Within this simplified framework the relevant counterfactuals are:

(2) If no exposition to f had occurred in the given circumstances, there would have been two cases of cancer instead of three.

and also the following, which is a consequence of (2)

(3) If no exposition to f had occurred in the given circumstances, there would have been one and only person who would not have got lung cancer.

Which is the form of the second counterfactual?

(i) Looking at the surface form of (3), and considering that it is better to treat it as an explicit counterfactual (i.e. as a counterfactual who contextually states the falsity of its clauses), using f° as a symbol for "substance f has been introduced" one could suppose that the logical form of (3) is

(4) $f^\circ \wedge E!^3 xMx \wedge (\neg f^\circ > E!^1 x \neg Mx)$

(Note that from the fact that there are exactly three persons with lung cancer it follows that it is false that there is exactly one person with this kind of disease ($\neg E!^1 x \neg Mx$): so the consequent of the last conjunct is indeed contrary-to-fact).

Developing the equivalence (a) above we obtain the equivalence

(5) $\neg f^\circ > E!^1 \neg Mx \equiv \neg f^\circ > \exists x(\neg Mx \wedge \forall w(\neg Mw \supset w = x))$

Since $A > B$ jointly with the strict implication $B \dashv C$ implies $A > C$[16], we obtain, by (4), (5), and standard quantification theory, that $\neg f^\circ > E!^1 \neg Mx$ entails

---

[16] This logical law actually is not universally valid but holds for $>$ and $\dashv$ provided that the modal status of the clauses is the same (see for instance [10]). This restriction is satisfied in the present instantiation of the law.

(6) $(\neg f^\circ > \exists x \neg Mx) \wedge (\neg f^\circ > \exists x \forall w(\neg Mw \supset (w = x)))$

The first conjunct says that if f° were false there would be someone who is not ill with lung cancer: a conditional which is true by Lewis semantics but not in the framework of a consequentialist conception of conditionals (the consequent being true in the actual world independently from the antecedent).

The second conjunct is also false from a consequentialist viewpoint. It is equivalent in fact to

(7) $\neg f^\circ > \exists x \forall w(w \neq x \supset Mw)$,

whose consequent means that all people different from a certain x got lung cancer, which is intuitively false and anyway does not follows from the supposition that f has not been introduced. So the proposition represented as $\neg f^\circ > E!^1 \neg Mx$ is false since it entails a false proposition

(ii) If the formalization of (3) were given by moving the negation sign at the left of the quantifier in (4), i.e. by $f^\circ \wedge E!^3 xMx \wedge \neg f^\circ > \neg E!^1 xMx$ the result would be also mistaken, since the consequent would reduce itself to the negation of $\exists x(Mx \wedge \forall w(Mw \supset (w = x)))$, i.e. to the statement $\forall x(Mx \supset \exists w(Mw \wedge w \neq x))$, which is something trivially true in the actual world and cannot depend conditionally on $\neg f^\circ$.

We conclude then that the two preceding analyses were incorrect. To understand the mistake, we should remind that quantification in modal (conditional) contexts may be *sensu composito* or *sensu diviso*, and that the surface form of the sentences may be deceptive. In fact, the correct position of the quantifier should be represented as in the following formalization:

(8) $E!^1 x(f^\circ \wedge E!^3 xMx \wedge (\neg f^\circ > \neg Mx))$

What (8) says is that there is one and only one person such that, given the truth of $f^\circ \wedge E!^3 xMx$, we may assert that, in absence of f, he or she would not have developed lung cancer. Hence we are in front of a counterfactual which is itself existentially quantified, not of a counterfactual with a quantified consequent.

The form of (8) is developed, by definition of $E!^1$, into

(9) $\exists x(f^\circ \wedge E!^3 xMx \wedge (\neg f^\circ > \neg Mx) \wedge \forall x(f^\circ \wedge E!^3 xMx \wedge (w \neq x \supset \neg(\neg f^\circ > \neg Mx))^{17}$

---

[17] This formula could obviously be generalized to every numerical value different from 1 occurring in the special quantifier. Suppose for instance that the differential dead people were 3 instead of 1 and that the domain D of ill people had $n = 9$ elements. Then in place of Mx we would have in every occurrence $Mx \wedge My \wedge My \wedge Mz$ bounded by three quantifiers (see (c) at page 284). As regards the disjunction reported in (11), since the number of 3-elements subsets of a 9-elements set is calculated via by the formula of binomial coefficient m on n, the corresponding disjunction should have 84 disjuncts.

(9) says (i) that there is someone who, in the given circumstances, in absence of f would not have got lung cancer and (ii) that all other people would not have had this property. So (9) asserts the unicity of a subject who, in the given circumstances and in absence of f, would not have got lung cancer.

For our aims we may however introduce at this point a shortcut. It is enough to observe, neglecting the uniqueness clause, that (9) entails by quantificational logic the weaker

(10) $\exists x(f^\circ \wedge Mx \wedge (\neg f^\circ > \neg Mx))$

and, given that the matrix of this existentially quantified formula turns out to be true if the values of the variables are $a_1, a_2, a, 3$, (10) is equivalent to the finite disjunction which follows:

(11) $(Ma_1 \wedge f^\circ \wedge (\neg f^\circ > \neg Ma_1)) \vee (Ma_2 \wedge f^\circ \wedge (\neg f^\circ > \neg Ma_2)) \vee (Ma_3 \wedge f^\circ \wedge (\neg f^\circ > \neg Ma_3))$

But (11) has non trivial-implications. From any $\neg Ma_i$, given that Ma_i is supposed to have taken place during a certain interval I, thanks to the definition of occurrence formulated at page 280, one derives by Modus Tollens $\neg O[M, a_i, I]$.

Hence $\neg f^\circ > \neg Ma_i$ implies $\neg f^\circ > \neg O[M, a_i, I]$, for every a_i and a certain I, and the same holds for every disjunct of (11).

Let us go back to (11) and to every disjunct whose form is $Ma \wedge f^\circ \wedge (\neg f^\circ > \neg Ma)$ where a stands for a_1, a_2, a_3.

Let us recall that f° is by construction the occurrence of a *token*-event and that every Ma describes, in contexts having legal interest, something which takes places during some time-interval I – so that the truth of Ma implies and is implied by the truth of $O[M, a, I]$.

So every disjunct of (11) may be reformulated as something of the form $O[M, a, I] \wedge f^\circ \wedge (\neg f^\circ > \neg O[M, a, I])$. Since we know that every interval I is posterior to the interval of the occurrence of f°, the latter formula is equivalent by definition to $f^\circ \mathscr{C} O[M, a, I]$. Thus, by applying suitable transformations, (11) is actually equivalent to this disjunction:

(12) $f^\circ \mathscr{C} O[M, a_1, I] \vee f^\circ \mathscr{C} O[M, a_2, I'] \vee f^\circ \mathscr{C} O[M, a_3, I'']$

where a I, I', I'' are different intervals during which the mentioned *token*-events took place. By introducing quantification over intervals a consequence of (12) is

(13) $\exists i \exists x(f^\circ \mathscr{C}[M, x, i])$

where $[M, x, i]$ may be considered the event-type instantiated by each of the mentioned *token*-events.

This does not mean, of course, that f° is to be considered a *determining* cause of each of the mentioned effects. In order to state this we would have to prove that f° is also sufficient, jointly with other background conditions, to infer the consequent, and this may be matter of a specific demonstration which depends on variable context information.

4 The Juridical Problem

We have been able to deduce from a quantified explicit counterfactual statement, derived from an argument in which no causal notion is involved, a disjunction of causal statements having an unique causal antecedent f°, a proposition which has legal interest since by its construction contains the name of a possible responsible.

The proposition embodied in (12) could be read as saying that a determinate cause had a certain type of effect which actually occurred. But logic does not allow passing from (12) to any of its disjuncts, even in the presence of the uniqueness clause which asserts that one and only one subject has the stated property M: consequently we are unable to determine the specific effect of the given cause-event, even if the causal relation has been established beyond any reasonable doubt.

Let us observe that a specularly converse formula of (12) would be, for instance, the following:

(13) $f_1 \mathscr{C} e^\circ \vee f_2 \mathscr{C} e^\circ \vee f_3 \mathscr{C} e^\circ$

Here the effect is determined but the cause is not: and note that this is the situation which is basically at the ground of abductive reasoning. In fact we might describe abduction as a process of determination, which technically means the elimination of a disjunction.

The situation represented in (12) should not be considered as something anomalous. One could maintain indeed that, as a matter of fact, every description of every *token*- event is always placed at a certain level of determination. If we know that Mr. Smith dies of a heart attack in Oxford Street, it is not specified if he dies on the left part or on the right part of the street. So if the cause of the attack is f°, the causal relation could be also be described as a disjunction $f^\circ \mathscr{C} d' \vee f^\circ \mathscr{C} d''$, where d' and d'' are determinate *token*-events under the determinable event d, in place of the simple $f^\circ \mathscr{C} d$. By parity of argument, if Mrs. Smith died of lung cancer, it is equivalent to say that she died of cancer at the left lung or at the right lung, without reaching a deeper level of specification.

As a matter of fact, in this respect causal relations are not logically different from any other kinds of relations, as for instance the two-place relation of seeing. If I say a cat in the half light, I may correctly say

(14) I am seeing a female cat or I am seeing a male cat.

which is equivalent to saying that I am seeing a cat, even if I am unable to eliminate the disjunction and select one of the disjuncts.

In scientific inquiry indetermination may be not only qualitative indetermination but also metric indetermination. It is common to discover that the values of the ascertained effect which depends on a certain cause fall inside a certain interval, but existing information does not allow reducing the limits of the interval.

The juridical problem posed by the fact that what is established by the scientific inquiry is a disjunction like (12) but not any of its single disjuncts should be clear. Since in this case, according to a school of thought, we are we are unable to establish a specific *token-token* causal relation, (12) does not provide a genuine causal relation. As we are committed to the two principles *nullum crimen sine lege* and *in dubio pro reo*, there is no justification in condemning someone for being responsible of an effect which is not determined.

An opposite view has been taken by the theoreticians of so-called "alternative assessment", which has been cultivated especially by German jurisprudence.

Alternative assessment may be of two kinds. We may be certain about the facts which actually happened but uncertain about the species of crime which should be imputed to the guilty (*proper* alternative assessment). But it may also happen that we are uncertain about the facts but certain about the species of the crime (*improper* alternative assessment). The first case is not uncommon. It may happen, for instance, that we are unable to say if a certain subtraction was a theft or a receiving of stolen goods. In this case German laws states that we have to condemn the guilty for the less important crime. The second case is also recorded. Suppose that a guy gives two different names to a public officer. He surely has been lying in one the two circumstances, or perhaps in both. We are then in front of a disjunction of facts. So there is a underdetermination of facts, but what is clear is that the fellow has been guilty for at least one crime (*perjury to a public official*) and he should be condemned for this.

According to some scholars, epidemiological assessment, in case of multifactorial pathologies, is a species of the genus "improper alternative assessment"[18]. If two *token*-events e' and e'' are both damages of the same kind E, and we know that Smith caused e' or Smith caused e'', what follows is that Smith caused a damage of kind E, so he should be condemned according to what the law establishes for this kind of damage.

The only problem which can be a serious challenge for the judge rises if he has to assign a compensation for the victims of the injury. Since we are unable to specify the names of the victims, we cannot make a distinction between persons who deserve a reward and others who are not entitled to receive it. Furthermore, if the payment of the reward is the only punishment which law establishes for being guilty, the application of a sanction to the guilty may be a problem. But if other forms of punishment are established by law, as it normally happens in penal law, the guilty cannot escape punishment according to the criterion of alternative assessment.

---

[18] See [8].

References

1. Berrino, F.: Candido atteggiamento o denuncia di comportamernti inadeguati? La medicina del lavoro (1988)
2. Cresswell, M.J.: Why objects exist but events occur. Studia Logica 45(4), 371–376 (1986)
3. Evans, A.S.: Causation and Disease: A Chronological Journey. Plenum (1993)
4. Giere, R.: Understanding Scientific Reasoning. Holt, Rinehart and Winston (1979)
5. Hempel, C.G.: Philosophy of Natural Science. Prentice-Hall, Englewood Cliffs (1966)
6. Lewis, D.K.: Causation. Journal of Philosophy 7, 555–567 (1973)
7. Lipton, P.: Inference to the Best Explanation. Routledge & Kegan, London (2004)
8. Masera, L.: Accertamento alternativo ed evidenza epidemiologica nel diritto penale. Giuffrè, Milano (2007)
9. Miller, G.: Correlations and Giere's theory of causation. Philosophy of Science 52(4), 612–614 (1985)
10. Pizzi, C.: Aristotle's thesis between paraconsistency and modalization. Journal of Applied Logic 3, 119–131 (2004)
11. Pizzi, C.: Gestalt effects in abductive and counterfactual inference. Logic Journal of the IGPL 14(2), 257–269 (2006)
12. Sackett, D.L., Richardson, W.S., Rosenberg, W., Haynes, R.B.: Evidence-based Medicine. How to practice and teach EBM. Churchill-Livinsgtone (1997)
13. Weed, D.L.: Causation: An epidemiologic perspective (in 5 parts). Journal of Law and Policy 12, 43–53 (2003)

On a Theoretical Analysis of Deceiving: How to Resist a Bullshit Attack

Walter Carnielli

Abstract. This paper intends to open a discussion on how certain dangerous kinds of deceptive reasoning can be defined, in which way it is achieved in a discussion, and which would be the strategies for defense against such deceptive attacks on the light of some principles accepted as fundamental for rationality and logic.

1 How Can We Be so Easily Deceived?

Fallacies of weak induction are part of the arsenal of argumentative fallacies (for a traditional account see [16]) that we do commit sometimes, and that we agree we should not do: a logical analysis *a posteriori*, after having commit them, shows that we have fallen into a fallacy of weak induction when the premises are not strong enough to support the conclusion. A catalogue of weak induction fallacies includes the well-know cases of "appeal to hasty generalization", "weak analogy", "slippery slope" "appeal to unqualified authority" and "appeal to ignorance", among other possibilities - it seems that a complete catalogue of fallacies is just unfeasible. But how can we be deceived, if we know so well the roots of deception? Likewise, we apparently know equally well what a proof in mathematics or in logic is, and we make much less flaws in logic or mathematics than in common reasoning - some errors of the former are famous, but in the latter are just too numerous to even be counted. Which kind of forces may push us into jumping into conclusions, by assuming incorrect assumptions when we are not in possession of the whole knowledge about something?

Traditional fallacies, so argue Gabbay and Woods in [14], although conceived as "mistakes that are attractive, universal and incorrigible", may be

Walter Carnielli
GTAL/CLE and Department of Philosophy–IFCH, State University of Campinas
e-mail: `walter.carnielli@cle.unicamp.br`

L. Magnani et al. (Eds.): Model-Based Reasoning in Science & Technology, SCI 314, pp. 291–299.

subject to defensible strategies, and even be useful, as claimed in [12]. So some fallacies may have a higher degree of attractiveness (perhaps because of their pseudo-logical format), but the kind of deceptive reasoning we want to analyze here has a different status: it is almost logically inevitable (although not entirely fatidical).

We want to argue that falling into a specific deceptive reasoning which we call *bullshit attack* is not anything irrational from our side, but rather a rational response from an opponent maneuver, and that the entire episode can bee seen as a game, where logic and a certain principle of rational discussion play essential roles. Indeed, an opponent may act coercively into our reasoning process by using irrelevant facts or assertions, and by telling half truths in such a way that we feel forced to "complete" the story in a way that interest the opponent, perhaps contrary to our own interests.

Even to define what is "to deceive" is not easy. The act of deceiving would have to be intentional, and to involve causing a belief - but what about acting as to prevent a false belief to be revised by the other person? And to act as to make the other person to cease to have a true belief, or to prevent the person from acquiring a certain true belief? Of course one can deceive by gestures, by irony and also by just making questions. So there seems to be no universally accepted definition of "deceiving" yet; we assume currently a definition stated in [19]:

> To deceive $\overset{\text{Def}}{=}$ to intentionally cause another person to have or continue to have a false belief that is truly believed to be false by the person intentionally causing the false belief by bringing about evidence on the basis of which the other person has or continues to have that false belief.

Now, towards a concept of deceptive reasoning, we need some definitions. The intuitive idea is that, starting from a certain belief set Γ, we may not want a certain consequence α; so arriving at $\neg\alpha$ would be a safeguard against the unwanted α. Yet, if an opponent somehow "closes the door" to $\neg\alpha$ by subreptitiously imposing to us an extension Δ to our belief set, we may then turn to be susceptible to α. Moreover, as I argue below, we are as prone to the unwanted consequence α as much as we obey some principles accepted as fundamental for rationality and logic.

What I propose below is a model based on classical deduction $\vdash$ for understanding such movements– this does not mean of course that logic explains fallacies, or that there is any "logic of bullshit". The aims are rather to understand the forces behind errors of human intellect in the way Francis Bacon describes them, cf. [2], XLVI:

> The human understanding when it has once adopted an opinion (either as being the received opinion or as being agreeable to itself) draws all things else to support and agree with it. And though there be a greater number and weight of instances to be found on the other side, yet these it either neglects and despises, or else by some distinction sets aside and rejects, in

order that by this great and pernicious predetermination the authority of its former conclusions may remain inviolate.

Definition 1.1 *A set Δ is said to be* charged with α with respect to Γ *if:*

(i) $\Gamma \not\vdash \alpha$, *but*

(ii) $\Gamma \cup \Delta \not\vdash \neg \alpha$.

Example 1.2 *Consider $\Gamma = \{p \to \neg q, q \vee r\}$. Clearly, $\Gamma \not\vdash r$, and any consistent set Δ containing either $\neg q$, or p, or $q \to p$ will do the job of deriving r when adjoined to Γ that is, $\Gamma \cup \Delta \vdash r$ (and consequently $\Gamma \cup \Delta \not\vdash \neg r$, since $\Gamma \cup \Delta$ is consistent). Thus each such Δ is charged with r with respect to Γ.*

But also any consistent set Δ containing either $p \vee \neg q$, $p \vee r$, $\neg r \to p$ or $q \to r$ will satisfy $\Gamma \cup \Delta \not\vdash \neg r$, and thus will be charged with r with respect to Γ as well.

$\Gamma \vdash^\Delta \alpha$ denotes that the derivation of α from Γ is charged with Δ.

Now we can say that a reasoning from a player *Abelard* based upon a non-empty collection Δ of premises from a player *Eloise* is *deceptive* when[1]:

1. *Abelard* accepts a set of premises Δ from *Eloise* to be added to his beliefs Γ;
2. *Abelard* is forced to perform an inference α charged with Δ; and
3. *Abelard* does not accept α.

The intuitive idea is that α is "paradoxical", that is, it reveals to be pragmatically incoherent (contradictory or even undesirable) with respect to a knowledge basis Γ, though Δ by itself is not so. As we shall see, it is our rational capacity of forming maximal consistent sets combined with our inclination to follow the Principle of Charity or Rational Accommodation, more than our tolerance to bullshit, which will make us victims of bullshit attack.

By defining a deceptive attack as a strategic maneuver by an opponent to elicit a deceptive reply from our side, it seems clear that understanding the mechanism of deceptive attack contributes decisively for an argumentative self-defense.

2 Charity, or Rational Accommodation, and Its Dangers

The Principle of Charity, also known, specially as focused by D. Davidson, as the Principle of Rational Accommodation, is a very basic principle in argumentation and in critical thinking which governs our interpretation to other people statements, and supposedly also other people interpretations to our discourse. The Principle of Rational Accommodation thus functions as a warrant for the act of understanding a speaker's statement (or discourse) by

---

[1] *Abelard* and *Eloise* are usual labels for game players.

interpreting his or her statements to be in principle rational in its highest way and, in the case of any argument, by rendering the best, strongest possible interpretation of an argument. The principle forces us to find the most coherent or rational interpretation for the statements involved in an argument - in another words, the principle constrains us to interpret the assertions so as to maximize the truth or rationality of the opponent, but under certain conditions: it demands us to accommodate all statements in the best possible consistent way, if there is such a way. The principle has strong connections, besides rhetoric, to some philosophical problems concerning meaning, truth and belief, and how these are all connected, which led Donald Davidson to turn his interests to the question of how are apparently irrational beliefs and actions even possible: quoting J. Malpas from [20]:

> The basic problem that radical interpretation must address is that one cannot assign meanings to a speaker's utterances without knowing what the speaker believes, while one cannot identify beliefs without knowing what the speaker's utterances mean. It seems that we must provide both a theory of belief and a theory of meaning at one and the same time. Davidson claims that the way to achieve this is through the application of the so-called principle of charity (Davidson has also referred to it as the principle of rational accommodation) a version of which is also to be found in Quine. In Davidson's work this principle, which admits of various formulations and cannot be rendered in any completely precise form, often appears in terms of the injunction to optimise agreement between ourselves and those we interpret, that is, it counsels us to interpret speakers as holding true beliefs (true by our lights at least) wherever it is plausible to do (see Radical Interpretation [1973]).

But Davidson has more to say about this, besides recognizing the origin of the principle and the influence of Quine (cf. [9, p. 35]):

> Quine and I, following Neil Wilson, have called in the past the principle of charity. This policy calls on us to fit our own propositions (or our own sentences) in the other person's words and attitudes in such a way as to render their speech ad other behavior intelligible.

The reference is to the principle named by Neil L. Wilson in "Substances without substrata". *Review of Metaphysics* 12(4), page 521–539, 1959.

The realist versus anti-realist controversy, and the direct implications for this principle for the theories of truth implicited in this discussion do not concern us here, nor are we interested (at least at this moment) in its implications as a defense against scepticism. We are more interested in a theory of untruth, which could explain how misconceptions and false beliefs are imposed upon us. The key to such analysis, at least in our case, can be synthesized in the entry by Simon Blackburn on the Principle of Charity in the The Oxford Dictionary of Philosophy (cf. [3, p. 62]):

> it [the Principle of Charity] constrains the interpreter to maximize the truth or rationality in the subject's sayings.

Now, as I intend to show, the skilled the interpret is at fulfilling such constraint of maximizing the truth or rationality in the subject's sayings, more successful the deceptive attack can be. H. Frankfurt in [13, p. 20] qualifies bullshit as fundamentally distinct from falsehood, or from a lie. To utter a bullshit is to disregard truth or lack of truth at all; the bullshit statements just intend to produce an impression:

> It is impossible for someone to lie unless he thinks he knows the truth. Producing bullshit requires no such conviction. A person who lies is thereby responding to the truth, and he is to that extent respectful of it. When an honest man speaks, he says only what he believes to be true; and for the liar, it is correspondingly indispensable that he considers his statements to be false. For the bullshitter, however, all these bets are off: he is neither on the side of the true nor on the side of the false. His eye is not on the facts at all, as the eyes of the honest man and of the liar are, except insofar as they may be pertinent to his interest in getting away with what he says. He does not care whether the things he says describe reality correctly. He just picks them out, or makes them up, to suit his purpose.

As Frankfurt clearly puts it (cf. [13]), the bullshitter is faking things:

> The truth-values of his statements are of no central interest to him; what we are not to understand is that his intention is neither to report the truth nor to conceal it. (p. 20)

This subtle point does not seem to have been perceived in the traditional rationale on truth or lack of it. Augustine in *Contra Mendacium* (cf. [1], 23), for instance, maintained that hiding the truth is not (morally) equivalent to telling a lie. The latter implies the former, but not necessarily vice-versa. However, the bullshitter imposes a third attitude besides lying and hiding the truth, cf. [13, p. 22]:

> He does not reject the authority of the truth, as the liar does, and oppose himself to it. He pays no attention to it at all. By virtue of this, bullshit is a greater enemy of the truth than lies are.

So, the bullshitter is more dangerous than a lier, and what he or she does more often (see e.g. [19]) is to use facts or assertions that he or she knows to be irrelevant (ambiguity and equivocation), or telling truths that he or she knows to be just half-truths (concealment and exaggeration). An indication that this kind of communication is something to be taken with caution is the notion of "implicatures" of P. Grice which tries to characterize what can be suggested in an utterance without being positively informed (cf. [15]).

3   Illusions of Reasoning: A Defense against Deceptive Attacks

Bullshit is a very powerful source to elicit deceptive reasoning: suppose that the bullshitter act as making us to "buy" a collection Δ of statements (by adding it to our beliefs Γ) which he or she intends to be charged with a conclusion α to her benefit.

Now, since Δ is charged to α with respect to our beliefs Γ, we have:

1. $\Gamma \nvdash \alpha$, but
2. $\Gamma \cup \Delta \nvdash \neg\alpha$ as well.

However, by the Principle of Rational Accommodation, we are forced to maximize the truth or rationality in his or her sayings. In terms of logic, this implies that we have to extend the conjunction of our belief set Γ together with the bullshitter's set Δ to a *maximal consistent* set Γ^*. By the famous construction known as Lindenbaum Lemma concerning maximal consistent sets, we can always find Γ^* such that Γ^* is maximal consistent and $\Gamma^* \nvdash \neg\alpha$. However, this implies (by the maximality of Γ^*) that $\Gamma^* \vdash \alpha$ as our opponent wanted!

As an illustrative example, suppose that *Eloise*, acting as a bullshitter, tells *Abelard* that a common friend, *Carol*, is going to have birthday, and proposes to pay her a dinner.

Abelard accepts from *Eloise* the following agreement Δ: "Let's share a dinner in honor to *Carol*", and adds this to his beliefs Γ. Of course, *Abelard* did not even think into accepting α: to pay for the dinner without participating of it, that is, $\Gamma \nvdash \alpha$. However, *Eloise*'s maneuver makes it impossible for *Abelard* to totally avoid α (because he has agreed): $\Gamma \cup \Delta \nvdash \neg\alpha$.

Abelard, while following the Principle of Rational Accommodation, "maximizes" $\Gamma \cup \Delta$ to a consistent Γ^* (by expecting the best possible behavior from *Eloise*).

A few days later, however, *Eloise* gives *Abelard* the bill: "The dinner was fantastic, please pay half of it...". What went wrong with *Abelard*'s reasoning? He implicitly agreed to share a dinner, not to dine together!

It should be clear then that not only any conclusion charged with the bullshit set Δ risks to be deceptive, but deceptive conclusions can be strategically imposed. Of course there are several ways to extend $\Gamma \cup \Delta$ to a maximal consistent set. The more skillful we are into finding such extensions, the more conclusions (possibly of the interest of our opponent) we may derive. The opponent is by no means restricted to a single agent, but may be represented by a commercial company, or by an advertising campaign - lots of examples can be found in [17]. Our expertise plus our willingness to obey the Principle of Rational Accommodation as well as our tolerance to bullshit will maximize the force of the deceptive attack. However, the same capacity we have in finding maximal consistent extensions may be at our side, and this is our defense.

The article [4] investigates reasons to some empirical evidence implying that human behavior often acts in contradiction to expected utility theory. The authors proposes the *priority heuristic* as a way to explain the discrepancy between empirical data and expected utility theory. The paper shows that the priority heuristic, typically by consulting only one or a few reasons, really predicts choice behavior in several respects.

The aim of deriving a psychological process model able to predict choice behavior from empirical evidence of course does not exclude the logical side of it. The need for a deeper theory on why people simply abandon trade-off computation and embark into hasty conclusions (which seem to them to be correct) is explicitly recognized in [4, p. 429]:

> The heuristic provides an alternative to the assumption that cognitive processes always compute trade-offs in the sense of weighting and summing of information. We do not claim that people never make trade-offs in choices, judgments of facts, values, and morals; that would be as mistaken as assuming that they always do. Rather, the task ahead is to understand when people make trade-offs and when they do not.

As a connected problem, a question treated in several places is the possibility of self-deception: how can we deceive ourselves?

As the French (Russian born) chess Grandmaster Xavier Tartacover suggested, in some sense self-deception involves bullshitting against ourselves:

> A chess game is divided into three stages: the first, when you hope you have the advantage, the second when you believe you have an advantage, and the third... when you know you're going to lose!

The relevance of self-deception and several examples can be found in [10] and in [22]. There are also several aspects linking what I suggest here to abductive reasoning, but this is left for now. What is relevant for our purposes is to have it clear that the kind of deceptive reasoning analyzed here is not merely fallacious, but somehow philosophically imposed. If we agree with Davidson [8, p. xviii], and similarly defended by Scriven in [23], that

> Charity is forced upon us; whether we like it or not, if we want to understand others, we must count them right in most matters

then being aware of this philosophical imposition and of its effects when combined with the logic mechanism is a strategic advantage: the risks of combining well-accepted principles and ending up with something we may repudiate are explained and exemplified in [5] and [6]. Awareness of this hazard is our only safeguard against falling into bullshit attack.

4 Assessment and Critique

In writing $\Gamma \vdash^{\Delta} \alpha$ to denote that the derivation of α from Γ as charged with Δ, it should be clear that what I have in mind is the notion of classical deduction: I do not mean that there is any proof system that can express any distinctions between the notion of strong versus inductive arguments (as in [11]) – it is simply that the notion of classical deduction $\vdash$ is sufficient to demonstrate that there is a kind of limiting capacity of the idea of rationality if we take seriously certain presuppositions as the requirement to obey the Principle of Rational Accommodation (or Principle of Charity), and its logical counterpart, the notion of maximal consistency. This does not contradict the analysis of fallacies as done e.g. in Chapters IV and V of [11] (cf. also the appendix on rationality in the same book). The analysis I am attempting here points to that fact that there may be limits for sound reasoning (if for a sound reason we understand that one which does not push us into deceiving) as much as there are limiting results in logic or in rational choice theory. But such limits may not be like the well-known formidable limitative results of Tarski, Gödel, Turing, Church and Skolem, but perhaps like the more modest, but still burdensome, results on the impossibility of expressing irreflexitivy by a modal schema (see e.g. section 3.3 of [7]) or the fact that the transitive closure of a relation is not first-order definable (this and other weaknesses of first-order logic to treat some computational problems, as its lack of mechanisms for recursion and for counting, are discussed in [21]).

Kahneman and collaborators developed the notion of the *focusing illusion* or *affective forecasting* (see [18] and inside references to help explaining some mistakes people do when evaluating the effects of different scenarios on their future. This a typical illusion due to the fact that people tend to exaggerate the importance of a pretense positive factor, while overlooking other perhaps much more relevant negative factors. Another interesting and much studied problem is the *conjunction fallacy*, which happens when people assume that a number of specific events is more probable than a single general event (cf. [24]). Starting from the understanding that what I am proposing here is not to use methods of formal or informal logic to analyze fallacies, but to pay due attention to principles that also affect logic, discerning the reasons why we succumb under a bullshit attack may help us to understand why we commit other illusions of reasoning.

Acknowledgements. This research was supported by FAPESP Thematic Project ConsRel 2004/ 14107-2, by the CNPq research grant PQ 300702/2005-1 and by the Fonds National de la Recherche Luxembourg. The final version of this paper was prepared during a research stay at IRIT- Institut de Recherche en Informatique de Toulouse. I am indebted to Richard L. Epstein for discussions and criticisms, even if he strongly disagrees to the ideas contained in this paper.

References

1. Augustine: Contra Mendacium (Against Lying), 422. Ed. J-P. Migne. Patrologia Latina vol. 40. Paris (1887), English translation at,
   http://www.newadvent.org/fathers/1313.htm
2. Bacon, F.: The New Organon or True Directions Concerning The Interpretation of Nature (1620), http://www.constitution.org/bacon/nov_org.htm, Rendition based on James Spedding, Robert Leslie Ellis, and Douglas Denon Heath in The Works (Vol. VIII), published by Taggard and Thompson, Boston (1863)

3. Blackburn, S.: Charity, principle of. The Oxford Dictionary of Philosophy. Oxford University Press, Oxford (1994)
4. Brandstätter, E., Gigerenzer, G., Hertwig, R.: The priority heuristic: making choices without trade-offs. Psychological Review 113(2), 409–432 (2006)
5. Carnielli, W.A., Coniglio, M.E., Gabbay, D., Gouveia, P., Sernadas, C.: Analysis and Synthesis of Logics – How to Cut and Paste Reasoning Systems. Springer, Heidelberg (2008)
6. Carnielli, W.A., Coniglio, M.E.: Bridge principles and combined reasoning. In: Müller, T., Newen, A. (eds.) Logik, Begriffe, Prinzipien des Handelns (Logic, Concepts, Principles of Action), pp. 32–48. Mentis Verlag, Paderborn (2007)
7. Carnielli, W.A., Pizzi, C.: Modalities and Multimodalities. Springer, Heidelberg (2009)
8. Davidson, D.: Inquiries into Truth and Interpretation. Oxford University Press, Oxford (1984)
9. Davidson, D.: Problems of Rationality. Clarendon Press (2004)
10. da Fonseca, E.G.: Lies We Live by: The Art of Self-Deception. Bloomsbury (2000)
11. Epstein, R.L.: Five Ways of Saying "Therefore": Arguments, Proofs, Conditionals, Cause and Effect, Explanations. Wadsworth Pub. (2001)
12. Floridi, L.: Logical fallacies as informational shortcuts. Synthese 167(2), 317–325 (2009)
13. Frankfurt, H.G.: On Bullshit. Princeton University Press, Princeton (2005)
14. Gabbay, D.M., Woods, J.: Fallacies as cognitive virtues. In: Majer, O., Pietarinen, A.-V., Tulenheimo, T. (eds.) Games: Unifying Logic, Language, and Philosophy. Series Logic, Epistemology, and the Unity of Science, vol. 15, pp. 56–98. Springer, Heidelberg (2009)
15. Grice, P.: Studies in the Way of Words. Harvard University Press, Harvard (1989)
16. Hamblin, C.L.: Fallacies. Methuen (1970)
17. Harford, T.: The Logic of Life: The Rational Economics of an Irrational World. Random House, New York (2008)
18. Kahneman, D., Krueger, A., Schkade, D., Schwarz, N., Stone, A.: Would you be happier if you were richer? A focusing illusion. Science 312(5782), 1908–1910 (2006)
19. Mahon, J.E.: The Definition of Lying and Deception First. Stanford Encyclopedia of Philosophy,
http://www.science.uva.nl/seop/entries/lying-definition/
(Published February 21, 2008)
20. Malpas, J.: Donald Davidson. Stanford Encyclopedia of Philosophy,
http://plato.stanford.edu/entries/davidson/DonaldDavidson
(Published May 23, 2005)
21. Otto, M.: Bounded variable logics and counting. In: A study in finite models. Lecture Notes in Logic, vol. 9. Springer, Heidelberg (1997)
22. Pears, D.: Self-deceptive belief-formation. Synthese 89, 393–405 (1991)
23. Scriven, M.: Reasoning. McGraw-Hill, New York (1976)
24. Tversky, A., Kahneman, D.: Extension versus intuititve reasoning: The conjunction fallacy in probability judgment. Psychological Review 90, 293–315 (1983)

Using Analogical Representations for Mathematical Concept Formation

Alison Pease, Simon Colton, Ramin Ramezani, Alan Smaill, and Markus Guhe

Abstract. We argue that visual, analogical representations of mathematical concepts can be used by automated theory formation systems to develop further concepts and conjectures in mathematics. We consider the role of visual reasoning in human development of mathematics, and consider some aspects of the relationship between mathematics and the visual, including artists using mathematics as inspiration for their art (which may then feed back into mathematical development), the idea of using visual beauty to evaluate mathematics, mathematics which is visually pleasing, and ways of using the visual to develop mathematical concepts. We motivate an analogical representation of number types with examples of "visual" concepts and conjectures, and present an automated case study in which we enable an automated theory formation program to read this type of visual, analogical representation.

1 Introduction

1.1 The Problem of Finding Useful Representations

Antirepresentationalism aside, the problem of finding representations which are useful for a given task is well-known in the computational worlds of A.I. and cognitive modelling, as well as most domains in which humans work. The problem can be seen in a positive light: for instance, in "discovery",

Alison Pease · Alan Smaill · Markus Guhe
School of Informatics, University of Edinburgh, Informatics Forum,
10 Crichton Street, Edinburgh, EH8 9AB, United Kingdom
e-mail: A.Pease@ed.ac.uk

Simon Colton · Ramin Ramezani
Department of Computing, Imperial College London, 180 Queens Gate,
London, SW7 2RH, United Kingdom
e-mail: sgc@doc.ic.ac.uk

L. Magnani et al. (Eds.): Model-Based Reasoning in Science & Technology, SCI 314, pp. 301–314.
springerlink.com   © Springer-Verlag Berlin Heidelberg 2010

or "constructivist" teaching, a teacher is concerned with the mental representations which a student builds, tries to recognise the representations and provide experiences which will be useful for further development or revision of the representations [6]. As an example, Machtinger [22] (described in [6, pp. 94–95]) carried out a series of lessons in which groups of kindergarten children used the familiar situation of walking in pairs to represent basic number concepts: an even number was a group in which every child had a partner, and odd, a group in which one child had no partner. Machtinger encouraged the children to evolve these representations by physically moving around and combining different groups and partners, resulting in the children finding relatively interesting conjectures about even and odd numbers, such as:

even + even = even
even + odd = odd
odd + odd = even.

The problem of representation can also be seen in a negative light: for instance automated systems dealing in concepts, particularly within analogies, are sometimes subject to the criticism that they start with pre-ordained, human constructed, special purpose, frozen, apparently fine-tuned and otherwise arbitrary representations of concepts. Thus, a task may be largely achieved via the *natural* intelligence of the designer rather than any *artificial* intelligence within her automated system. These criticisms are discussed in [2] and [20].

1.2   *Fregean and Analogical Representations*

Sloman [29, 30] relates philosophical ideas on representation to A.I. and distinguishes two types of representation: Fregean representations, such as (most) sentences, referring phrases, and most logical and mathematical formulae; and representations which are analogous – structurally similar – to the entities which they represent, such as maps, diagrams, photographs, and family trees. Sloman elaborates the distinction:

> Fregean and analogical representations are complex, i.e. they have parts and relations between parts, and therefore a syntax. They may both be used to represent, refer to, or denote, things which are complex, i.e. have parts and relations between parts. The difference is that in the case of analogical representations *both* must be complex (i.e. representation and thing) and there *must* be some correspondence between their structure, whereas in the case of Fregean representations there need be no correspondence. [30, p. 2]

In this paper we discuss ways of representing mathematical concepts which can be used in automated theory formation (ATF) to develop further concepts and conjectures in mathematics. Our thesis is that an analogical representation can be used to construct interesting concepts in ATF. We believe that analogical, pre-Fregean representations will be important in modeling Lakoff and Núñez's theory on embodiment in mathematics [19], in which a situated, embodied being interacts with a physical environment to build up mathematical ideas[1]. In particular, we believe that there is a strong link between analogical representation and visual reasoning.

The paper is organized as follows: we first consider the role of vision in human development of mathematics, and consider some aspects of the relationship between mathematics and the visual, in section 2. These include artists using mathematics as inspiration for their art (which may then feed back into mathematical development), the idea of using visual beauty to evaluate mathematics, mathematics which is visually pleasing, and ways of using the visual to develop mathematical concepts. In section 3 we motivate an analogical representation of number types with examples of "visual" concepts and conjectures, and in section 4 we present an automated case study in which we enable an automated theory formation program to read this type of visual representation. Sections 5 and 6 contain our further work and conclusions. Note that while we focus on the role that the visual can play in mathematics, there are many examples of blind mathematicians. Jackson [16] describes Nicolas Saunderson who went blind while still a baby, but went on be Lucasian Professor of Mathematics at Cambridge University; Bernard Morin who went blind at six but was a very successful topologist, and Lev Semenovich Pontryagin who went blind at fourteen but was influential particularly in topology and homotopy. (Leonhard Euler was a particularly eminent mathematician who was blind for the last seventeen years of his life, during which he produced half of his total work. However, he would still have had a normal visual system.)

2   The Role of Visual Thinking in Mathematics

2.1   Concept Formation as a Mathematical Activity

Disciplines which investigate mathematical activity, such as mathematics education and philosophy of mathematics, usually focus on proof and problem solving aspects. Conjecture formulation, or problem posing, is sometimes addressed, but concept formation is somewhat neglected. For instance, in the philosophy of mathematical *practice*, which focuses on what mathematicians do (as opposed to proof and the status of mathematical knowledge), Giaquinto gives an initial list of some neglected philosophical aspects of mathematical activity as discovery, explanation, justification and application, where the goals are respectively knowledge, understanding, relative certainty and practical benefits [12, p. 75]. Even in research on visual reasoning in mathematics education, visualizers are actually *defined* as people who "prefer to use visual methods when attempting mathematical *problems* which may be

---

[1] One of the goals of a project we are involved in, the *Wheelbarrow Project*, is to produce a computational model of Lakoff and Núñez's theory.

solved by both visual and non-visual methods" [24, p. 298] (our emphasis; see also [25, 26, 34] for similar focus) and mathematical giftedness and ability are measured by the ability to *solve* problems [17].

2.2   Problems with Visual (Analogical) Representations in Mathematics

Debate about the role of visual reasoning in mathematics has tended to follow the focus on proof, and centers around the controversy about whether visual reasoning – and diagrams in particular – can be used to prove, or to merely illustrate, a theorem. In contrast, in this paper we focus on the often overlooked mathematical skills of forming and evaluating concepts and conjectures. We hold that visual reasoning does occur in this context. For instance, the sieve of Eratosthenes, the Mandelbrot set and other fractals, and symmetry are all inherently visual concepts, and the number line and Argand diagram (the complex plane diagram) are visual constructs which greatly aided the acceptance of negative (initially called fictitious) and imaginary numbers, respectively.

The principle objection to using visual reasoning in mathematics, and diagrams in particular, is that it is claimed that they cannot be as rigorous as algebraic representations: they are heuristics rather than proof-theoretic devices (for example, [31]). Whatever the importance of this objection, it does not apply to us since we are concerned with the formation of concepts, open conjectures and axioms, and evaluation and acceptance criteria in mathematics. The formation of these aspects of a mathematical theory is not a question of rigor as these are not the sort of things that can be provable (in the case of conjectures it is the *proof* which is rigorous, rather than the formation of a conjecture statement). However, other objections may be relevant. Winterstein [35, chap. 3] summarizes problems in the use of diagrammatic reasoning: impossible drawings or optical illusions, roughness of drawing, drawing mistakes, ambiguous drawing, handling quantifiers, disjunctions and generalization (he goes on to show that similar problems are found in sentential reasoning). Another problem suggested by Kulpa [18] is that we cannot visually represent a theorem without representing its proof, which is clearly a problem when forming open conjectures. Additionally, in concept formation, it is difficult to represent diagrammatically the lack, or non-existence of something. For instance, primes are defined as numbers for which *there do not exist* any divisors except for 1 and the number itself. This is hard to represent visually. However, we hold that these problems sometimes occur because a diagrammatic language has been insufficiently developed, rather than any inherent problem with diagrams, and are not sufficient to prevent useful concept formation.

2.3   Automated Theory Formation and Visual Reasoning

We support our argument that visual reasoning plays a role in mathematical theory formation with a case study of automated visual reasoning in concept formation, in the domain of number theory. Although the automated reasoning community has focused on theorem proving, it is also well aware of the need to identify processes which lead to new concepts and conjectures being formed, as opposed to solely proving ready made conjectures. Although such programs [8, 11, 21, 27] include reasoning in the domains of graph theory and plane geometry – prime candidates for visual reasoning – representation in theory formation programs has tended to be algebraic.

2.4   Relationships between Mathematics and Art

Artists who represent mathematical concepts visually develop a further relationship between mathematics and the visual. For instance, Escher [10] represented mathematical concepts such as regular division of a plane, superposition of a hyperbolic plane on a fixed two-dimensional plane, polyhedra such as spheres, columns, cubes, and the small stellated dodecahedron, and concepts from topology, in a mathematically interesting way. Other examples include Albrecht Dürer, a Renaissance printmaker and artist, who contributed to polyhedral literature [7]; sculptor John Robinson, who displayed highly complex mathematical knot theory in polished bronze [1]; and the artist John Ernest (a member of the British constructivist art movement), who produced art which illustrates mathematical ideas [9], such as his artistic representation of the equation for the sum of the first n natural numbers $1 + 2 + 3 + ... + n = n(n+1)/2$, a Möbius strip sculpture and various works based on group theory (Ernest's ideas fed back into mathematics as contributions to graph theory).

2.5   Using the Visual to Evaluate Mathematical Concepts

A visual representation of mathematical concepts and conjectures can also suggest new ways of evaluation. As opposed to the natural sciences, which can be evaluated based on how they describe a physical reality, there is no obvious way of evaluating mathematics beyond the criteria of rigor, consistency, etc. Aesthetic judgements provide a further criterion. [28] examines the roles of the aesthetic in mathematical inquiry, and describes views by Hadamard [13], von Neumann [32] and Penrose [23], who all argue that the motivations for doing mathematics are ultimately aesthetic ones. There are many quotes from mathematicians concerning the importance of beauty in mathematics, such as Hardy's oft-quoted "The mathematician's patterns, like the painter's or the poet's must be beautiful; the ideas, like the colors or the words must fit together in a harmonious way. Beauty is the first test: there is no permanent place in this world for ugly mathematics." [14, p. 85]. Clearly, beauty may not be visual (one idea of mathematical beauty is a "deep theorem", which establishes connections between previously unrelated domains, such as Euler's identity $e^{i\pi} + 1 = 0$). However, aesthetic judgements of beauty can often be visual: Penrose [23], for example, suggests many visual examples of aesthetics being used to guide theory formation in mathematics.

3   Visual Concepts and Theorems in Number Theory

3.1   Figured Number Concepts

Figured numbers, known to the Pythagoreans, are regular formulations of pebbles or dots, in linear, polygonal, plane and solid patterns. The polygonal numbers, for instance, constructed by drawing similar patterns with a larger number of sides, consist of dot configurations which can be arranged evenly in the shape of a polygon [15, 33]. In Figure 1 we show the first three triangular numbers, square numbers and pentagonal numbers. These provide a nice example of Sloman's analogical representations.

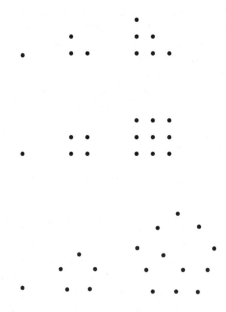

Fig. 1 Instances of the first three polygonal numbers: triangular numbers 1–3, square numbers 1–3 and pentagonal numbers 1–3.

The first six triangular numbers are $1, 3, 6, 10, 15, 21$; square numbers $1, 4, 9, 16, 25, 36$; and pentagonal numbers $1, 5, 12, 22, 35, 51$. Further polygonal numbers include hexagonal, heptagonal, octagonal numbers, and so on. These concepts can be combined, as some numbers can be arranged into multiple polygons. For example, the number 36 can be arranged both as a square and a triangle. Combined concepts thus include square triangular numbers, pentagonal square numbers, pentagonal square triangular numbers, and so on.

3.2 Conjectures and Theorems about Figured Concepts

If interesting statements can be made about a concept then this suggests that it is valuable. One way of evaluating a mathematical concept, thus, is by considering conjectures and theorems about it. We motivate the concepts of triangular and square numbers with a few examples of theorems about them. Firstly, the formula for the sum of consecutive numbers, $1 + 2 + 3 + ... + n = n(n + 1)/2$, can be expressed visually with triangular numbers, as shown in Figure 2.

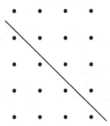

Fig. 2 Diagram showing that $1+2+3+4 = 4*5/2$. This generalizes to the theorem that the sum of the first n consecutive numbers is $n(n + 1)/2$.

Secondly, Figure 3 is a representation of the theorem that the sum of the odd numbers gives us the series of squared numbers:

$$1 = 1^2$$
$$1 + 3 = 2^2$$
$$1 + 3 + 5 = 3^2$$
$$1 + 3 + 5 + 7 = 4^2$$
$$...$$

Finally, Figure 4 expresses the theorem that any pair of adjacent triangular numbers add to a square number:

$$1 + 3 = 4$$
$$3 + 6 = 9$$
$$6 + 10 = 16$$
$$45 + 55 = 100$$
$$\ldots$$

Further theorems on polygonal numbers include the square of the nth triangular number equals the sum of the first n cubes, and all perfect numbers are triangular numbers.

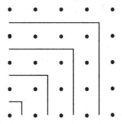

Fig. 3 Diagram showing that $1 + 3 + 5 + 7 + 9 = 5^2$; This generalizes to the theorem that the sums of the odd numbers gives us the series of squared numbers.

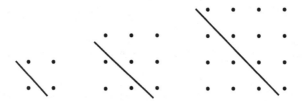

Fig. 4 Diagram showing that $S_2 = T_1 + T_2$; $S_3 = T_2 + T_3$; $S_3 = T_2 + T_3$; This generalizes to the theorem that each square number (except 1) is the sum of two successive triangular numbers.

4   An Automated Case Study

4.1   The HR Machine Learning System

The HR machine learning system[2] [3] takes in mathematical objects of interest, such as examples of groups, and core concepts such as the concept of being an element or the operator of a group. Concepts are supplied with a definition and examples. Its concept formation functionality embodies the constructivist philosophy that "new ideas come from old ideas", and works

---

[2] HR is named after mathematicians Godfrey Harold Hardy (1877–1947) and Srinivasa Aiyangar Ramanujan (1887–1920).

by applying one of seventeen production rules to a known concept to generate another concept. These production rules include:

- The *exists* rule: adds existential quantification to the new concept's definition
- The *negate* rule: negates predicates in the new definition
- The *match* rule: unifies variables in the new definition
- The *compose* rule: takes two old concepts and combines predicates from their definitions in the new concept's definition

For each concept, HR calculates the set of examples which have the property described by the concept definition. Using these examples, the definition and information about how the concept was constructed and how it compares to other concepts, HR estimates how interesting the concept is, [4], and this drives an agenda of concepts to develop. As it constructs concepts, it looks for empirical relationships between them, and formulates conjectures whenever such a relationship is found. In particular, HR forms equivalence conjectures whenever it finds two concepts with exactly the same examples, implication conjectures whenever it finds a concept with a proper subset of the examples of another, and non-existence conjectures whenever a new concept has an empty set of examples.

4.2 Enabling HR to Read Analogical Representations

When HR has previously worked in number theory [3], a Fregean representation has been used. For example the concept of integers, with examples $1 - 5$ would be represented as the predicates: integer(1), integer(2), integer(3), integer(4), integer(5), where the term "integer" is not defined elsewhere. In order to test our hypothesis that it is possible to perform automated concept formation in number theory using analogical representations as input, we built an interface between HR and the Painting Fool [5], calling the resulting system HR-V (where "V" stands for "visual"). The Painting Fool performs colour segmentation, thus enabling it to interpret paintings, and simulates the painting process, aiming to be autonomously creative. HR-V takes in diagrams as its object of interest, and outputs concepts which categorise these diagrams in interesting ways.

As an example, we used the ancient Greek analogical representation of figured numbers as dot patterns. We gave HR-V rectangular configurations of dots on a grid for the numbers 1–20, as shown in Figure 5 for 7, 8 and 9. Rearranging n dots as different rectangular patterns and categorizing the results can suggest concepts such as:

- *equality* (two separate collections of dots that can be arranged into exactly the same rectangular configurations),
- *evens* (numbers that can be arranged into two equal rows),
- *odds* (numbers that cannot be arranged into two equal rows),

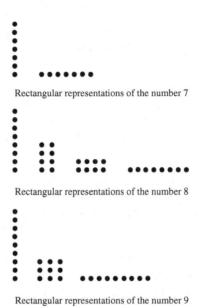

Rectangular representations of the number 7

Rectangular representations of the number 8

Rectangular representations of the number 9

Fig. 5 Rectangular configurations for the numbers 7, 8 and 9.

- *divisors* (the number of rows and columns of a rectangle),
- *primes* (a collection of dots that can only be arranged as a rectangle with either 1 column or 1 row),
- *composite numbers* (a collection of dots that can be arranged as a rectangle with more than 1 column and more than 1 row), and
- *squares* (a collection of dots that can be arranged as a rectangle with an equal number of rows and columns).

HR-V generated a table describing the rectangle concept, consisting of triples: [total number of dots in the pattern, number of rows in the rectangle, number of columns in the rectangle]. This is identical to the divisor concept (usually input by the user) and from here HR-V used its production rules to generate mathematical concepts. For instance, it used $match[X, Y, Y]$, which looked through the triples and kept only those whose number of columns was equal to the number of rows, and then $exists[X]$ on this subset, which then omitted the number of rows and number of columns and left only the first part of the triple: number of dots. This resulted in the concept of square numbers (defined as a number y such that there exists a number x where $x * x = y$) In another example, HR-V counted how many different rectangular configurations there were for each number – $size[X]$ – and then searched through this list to find those numbers which had exactly two rectangular configurations – $match[X, 2]$ – to find the concept of prime number. In this way, HR-V categorized the type or shape of configuration which can be made from different numbers into various mathematical concepts.

4.3 Results

After a run of 10,000 steps, HR-V discovered evens, odds, squares, primes, composites and refactorable numbers (integers which are divisible by the number of their divisors), as shown in Figure 6, with a few dull number types defined.

Even numbers can be represented as rectangles with two columns; odd numbers cannot

Square numbers can be represented as rectangles with the same number of rows as columns

Prime numbers have exactly two rectangular configurations; composite numbers have more than two

The number of different rectangular configurations of refactorable numbers corresponds to the number of rows in one of the rectangles

Fig. 6 The concepts which HR-V generated from rectangular configurations of numbers 1–20: evens, odds, squares, primes, composites and refactorables.

Thus, HR-V is able to invent standard number types using number rectangle definitions of them. This is a starting point, suggesting that automatic concept formation is possible using analogical representations. We would like to be able to show that analogical representations can lead to concepts which would either not be discovered at all, or would be difficult to discover, if represented in a Fregean manner.

5 Further Work

Although the representations can be seen as analogous to the idea they represent, they are still subject to many of the criticisms described in Section 1, such as being pre-ordained, human constructed, special purpose and frozen. In order to address this, we are currently enabling HR to produce its own dot patterns, via two new production rules. These production rules are different to the others in that they produce a new model or entity from an old one (or old ones), as opposed to the other seventeen production rules which produce a new concept from an old concept (or old concepts). In both production rules, an entity is a configuration of dots which is represented by a 100*100 grid with values of on/off where "on" depicts a dot.

The first new production rule will take in one entity and add dots to produce a new one. There are two parameters:

(i) the type of adding (where to put the dot);
(ii) the numerical function (how many dots to add).

The second new production rule will take in two entities and merge them to produce a new one. This is like the "compose" production rule but takes entities rather than concepts. Alternatively, it can take in one entity and perform symmetry operations on it.

While we have focused on concept formation in this paper, we believe that analogical representations, and visual reasoning in particular, can be fruitful in other aspects of mathematical activity; including conjecture and axiom formation, and evaluation and acceptance criteria. We hope to develop the automated case study in these aspects, as well as to extend concept formation to further mathematical domains.

6 Conclusion

We have argued that automatic concept formation is possible using analogous representations, and can lead to important and interesting mathematical concepts. We have demonstrated this via our automated case study. Finding a representation which enables either people or automated theory formation systems to fruitfully explore and develop a domain is an important challenge: one which we believe will be increasingly important, given the current focus in A.I. on embodied cognition.

References

1. Brown, R.: John Robinson's symbolic sculptures: Knots and mathematics. In: Emmer, M. (ed.) The visual mind II, pp. 125–139. MIT Press, Cambridge (2005)
2. Chalmers, D., French, R., Hofstadter, D.: High-level perception, representation, and analogy: A critique of artificial intelligence methodology. Journal of Experimental and Theoretical Artificial Intelligence 4, 185–211 (1992)

3. Colton, S.: Automated Theory Formation in Pure Mathematics. Springer, Heidelberg (2002)
4. Colton, S., Bundy, A., Walsh, T.: On the notion of interestingness in automated mathematical discovery. International Journal of Human Computer Studies 53(3), 351–375 (2000)
5. Colton, S., Valstar, M., Pantic, M.: Emotionally aware automated portrait painting. In: Proceedings of the 3rd International Conference on Digital Interactive Media in Entertainment and Arts, DIMEA (2008)
6. Davis, R.B., Maher, C.A.: How students think: The role of representations. In: English, L.D. (ed.) Mathematical Reasoning: Analogies, Metaphors, and Images, pp. 93–115. Lawrence Erlbaum, Mahwah (1997)
7. Dürer, A.: Underweysung der Messung (Four Books on Measurement) (1525)
8. Epstein, S.L.: Learning and discovery: One system's search for mathematical knowledge. Computational Intelligence 4(1), 42–53 (1988)
9. Ernest, P.: John Ernest, a mathematical artist. Philosophy of Mathematics Education Journal. Special Issue on Mathematics and Art 24 (December 2009)
10. Escher, M.C., Ernst, B.: The Magic Mirror of M.C. Escher. Taschen GmbH (2007)
11. Fajtlowicz, S.: On conjectures of Graffiti. Discrete Mathematics 72, 113–118 (1988)
12. Giaquinto, M.: Mathematical activity. In: Mancosu, P., Jørgensen, K.F., Pedersen, S.A. (eds.) Visualization, Explanation and Reasoning Styles in Mathematics, pp. 75–87. Springer, Heidelberg (2005)
13. Hadamard., J.: The Psychology of Invention in the Mathematical Field. Dover (1949)
14. Hardy, G.H.: A Mathematician's Apology. Cambridge University Press, Cambridge (1994)
15. Heath, T.L.: A History of Greek Mathematics: From Thales to Euclid, vol. 1. Dover Publications Inc. (1981)
16. Jackson, A.: The world of blind mathematicians. Notices of the AMS 49(10), 1246–1251 (2002)
17. Krutestskii, V.A.: The psychology of mathematical abilities in schoolchildren. University of Chicago Press, Chicago (1976)
18. Kulpa, Z.: Main problems of diagrammatic reasoning. part i: The generalization problem. In: Aberdein, A., Dove, I. (eds.) Foundations of Science, Special Issue on Mathematics and Argumentation, vol. 14(1-2), pp. 75–96. Springer, Heidelberg (2009)
19. Lakoff, G., Núñez, R.: Where Mathematics Comes From: How the Embodied Mind Brings Mathematics into Being. Basic Books Inc., U.S.A (2001)
20. Landy, D., Goldstone, R.L.: How we learn about things we don't already understand. Journal of Experimental and Theoretical Artificial Intelligence 17, 343–369 (2005)
21. Lenat, D.: AM: An Artificial Intelligence approach to discovery in mathematics. PhD thesis, Stanford University (1976)
22. Machtinger, D.D.: Experimental course report: Kindergarten. Technical Report 2, The Madison Project, Webster Groves, MO (July 1965)
23. Penrose, R.: The role of aesthetics in pure and applied mathematical research. In: Penrose, R. (ed.) Roger Penrose: Collected Works, vol. 2. Oxford University Press, Oxford (2009)

24. Presmeg, N.C.: Visualisation and mathematical giftedness. Journal of Educational Studies in Mathematics 17(3), 297–311 (1986)
25. Presmeg, N.C.: Prototypes, metaphors, metonymies and imaginative rationality in high school mathematics. Educational Studies in Mathematics 23(6), 595–610 (1992)
26. Presmeg, N.C.: Generalization using imagery in mathematics. In: English, L.D. (ed.) Mathematical Reasoning: Analogies, Metaphors, and Images, pp. 299–312. Lawrence Erlbaum, Mahwah (1997)
27. Sims, M.H., Bresina, J.L.: Discovering mathematical operator definitions. In: Proceedings of the Sixth International Workshop on Machine Learning, Morgan Kaufmann, San Francisco (1989)
28. Sinclair, N.: The roles of the aesthetic in mathematical inquiry. Mathematical Thinking and Learning 6(3), 261–284 (2004)
29. Sloman, A.: Interactions between philosophy and artificial intelligence: The role of intuition and non-logical reasoning in intelligence. Artificial Intelligence 2, 209–225 (1971)
30. Sloman, A.: Afterthoughts on analogical representation. In: Theoretical Issues in Natural Language Processing (TINLAP-1), pp. 431–439 (1975)
31. Tennant, N.: The withering away of formal semantics? Mind and Language 1, 302–318 (1986)
32. von Neumann, J.: The mathematician. In: Newman, J. (ed.) The world of mathematics, pp. 2053–2065. Simon and Schuster, New York (1956)
33. Wells, D.: The Penguin Dictionary of Curious and Interesting Numbers. Penguin Books Ltd., London (1997)
34. Wheatley, G.H.: Reasoning with images in mathematical activity. In: English, L.D. (ed.) Mathematical Reasoning: Analogies, Metaphors, and Images, pp. 281–297. Lawrence Erlbaum, Mahwah (1997)
35. Winterstein, D.: Using Diagrammatic Reasoning for Theorem Proving in a Continuous Domain. PhD thesis, University of Edinburgh (2004)

Good Experimental Methodologies and Simulation in Autonomous Mobile Robotics

Francesco Amigoni and Viola Schiaffonati

Abstract. Experiments have proved fundamental constituents for natural sciences and it is reasonable to expect that they can play a useful role also in engineering, for example when the behavior of an artifact and its performance are difficult to characterize analytically, as it is often the case in autonomous mobile robotics. Although their importance, experimental activities in this field are often carried out with low standards of methodological rigor. Along with some initial attempts to define good experimental methodologies, the role of simulation experiments has grown in the last years, as they are increasingly used instead of experiments with real robots and are now considered as a good tool to validate autonomous robotic systems. In this work, we aim at investigating simulations in autonomous mobile robotics and their role in experimental activities conducted in the field.

1 Introduction

The field of autonomous mobile robotics has recently started an effort to develop good experimental methodologies, after recognizing that experimentation has not yet reached a level of maturity comparable with that reached in other fields of engineering and in science. A number of initiatives have been put in place, including workshop series [4, 15], special issues [13], and funded projects [17, 19]. Along with the impulse toward the definition of good experimental methodologies, the role of simulation experiments has grown in the last years, as they are increasingly used instead of experiments with real robots and are now considered as a good tool to validate autonomous

Francesco Amigoni · Viola Schiaffonati
Artificial Intelligence and Robotics Laboratory, Dipartimento di Elettronica
e Informazione, Politecnico di Milano, Italy
e-mail: {amigoni,schiaffo}@elet.polimi.it

L. Magnani et al. (Eds.): Model-Based Reasoning in Science & Technology, SCI 314, pp. 315–332.
springerlink.com   © Springer-Verlag Berlin Heidelberg 2010

robotic systems [25, 16]. However, the very idea of simulation has been poorly analyzed so far.

In this paper, we discuss simulations in autonomous mobile robotics. In particular, we investigate two issues: the idea of simulation in autonomous mobile robotics and the role simulation can play in relation with experiments performed in this field. We explicitly note that this paper does not aim at providing definitive answers to the several questions it raises. Its main original contribution is in setting the ground for a further analysis of the issues it puts forward. Before starting the discussion, we briefly introduce autonomous mobile robotics.

Autonomous mobile robotics [20] is a discipline that aims at developing robots that operate in unpredictable environments without a continuous human control, like mobile robots for planetary exploration and service robots for performing housework. Intuitively, we can say that a robot is autonomous when its behavior is not fixed at design time, but is determined at execution time by the robot itself. The designer of an autonomous robot does not develop the robot by listing the sequence of all the operations the robot must perform (as it happens in many industrial robots used in manufacturing), but by equipping the robot with methods to autonomously decide, according to the history of past operations and states of the world perceived by sensors, the next operations to perform. Problems that are addressed in autonomous mobile robotics include locomotion (how a robot moves in an environment), perception (how a robot senses an environment and interprets the sensed data), localization (how a robot determines its position in an environment), navigation (how a robot moves safely from place to place), and mapping (how a robot builds a representation of an environment).

This paper is structured as follows. The next section presents a picture of the current main trends in experimental activity of autonomous mobile robotics. Simulations and their relation with experiments are discussed in Section 3, with a general perspective, and in Section 4, with specific reference to autonomous mobile robotics. Finally, Section 12 concludes the paper.

2 Experimental Trends in Autonomous Mobile Robotics

In this section, we discuss two main trends that are affecting the way experimental activities are performed in autonomous mobile robotics: the definition of good experimental methodologies and the use of simulations.

2.1 Good Experimental Methodologies

Experiments are essential ingredients of science, playing a role both to confirm/refuse a theory and to find out new theories. If a rigorous experimental approach has proved to be crucial for natural sciences, it is reasonable to expect that it can be also useful in engineering, for example when the behavior of an artifact and its performance are difficult to characterize analytically, as it is often the case in robotics.

The way the term 'experiment' is intended in robotics strongly varies from field to field. Some branches of robotics are based on rather standardized experimental procedures. For example, the performance of commercialized industrial manipulators (in terms of accuracy, repeatability, and spatial resolution) is typically assessed through rigorous experimental protocols. In other areas of robotic research (for example, in the field of autonomous robotics), the term 'experiment' seems to have a different meaning: it usually denotes a test made to show that a system works, or that it works better than other systems built for the same purpose. In this work, we concentrate on autonomous mobile robotics, in which experimental activities are often carried out with low standards of methodological rigor.

A critical analysis of the role of experiments in autonomous mobile robotics has recently taken place (see, for instance, the series of workshops on Good Experimental Methodology and Benchmarking [4] and the special issue of the *Autonomous Robots* journal [13]) and has widely recognized that experimentation in this field has not yet reached a level of maturity comparable with that reached in other scientific fields, like for example in physics, considered as the paradigm of mature, stable, and well-founded science. In the attempt to overcome this limitation, roboticists have proposed benchmarking as a way to foster scientific and technological advancement in autonomous mobile robotics research. The Rawseeds project [17], concerned with methodologies for comparing robotic systems' performance on the same data sets, and the Radish repository of experimental data sets [8], aptly illustrate this new trend. However, a wider analysis of the "principles" of experimentation has not yet been carried out and standard criteria for evaluating mobile robotic performance are still lacking. This is symptomatic of a sort of methodological negligence, which certainly hinders scientific and technological progress in autonomous robotics research, as outlined in [1]. Therefore, it is envisaged that, in the very next years, robotic engineers will exploit the valuable opportunity to come up with good experimental methodologies shaped after those of natural sciences, where they have been developed and optimized for centuries. From the analysis conducted in [2], it emerges that some works in autonomous mobile robotics are addressing in a more and more convincing way the general principles characterizing experimental activities: comparison, reproducibility and repeatability, and justification/explanation principles. A current limitation is that, although the *union* of experimental activities well covers the three above principles, their *intersection* is almost a null set. Put it in another way, the problem of experimental activity in autonomous mobile robotics is complex and involves multiple dimensions, from which each researcher selects a subset of dimensions, according to his/her current needs. This can be due to the relatively young discussion on good experimental practices that has not yet brought to a stable experimental methodology. However, current trends seem to go in the right direction.

If looking at the way experiments are assessed in natural sciences might be the first step to develop a good experimental methodology for autonomous mobile robotics, this requires reflecting on the nature of robotics, at the intersection between engineering and science. From the one hand, robotics appears more similar to engineering than to science, since differently from the objects of scientific investigation, robotic systems are artifacts built by humans and could not exist without human intervention [21]. According to this view, experiments in robotics have the goal of demonstrating that a given artifact is working or that it is better than another. From the other hand, the most advanced robotic systems are so complex that their behavior is hardly predictable, even by their own designers. In this perspective, experiments in robotics are somehow similar to experiments in natural sciences since, broadly speaking, both have the goal of understanding how complex systems work.

2.2 Simulations

The use of simulations for experimental purposes in autonomous mobile robotics has grown in the years. Today, several papers present only experiments performed in simulation to validate a proposed system. Simulations provide a convenient way to explore different robotic scenarios at lower cost with respect to using real robots. They reduce the effort needed for writing software programs, debugging them, and displaying results. They also often run much faster than their real counterparts. However, simulations have been often criticized for not being realistic enough, and for producing results that do not transfer easily to the real world. Such problems arise when simulation models are too naïve, because they embed insufficient knowledge on the real system, sensors and actuators models are not carefully calibrated, real world noise is not accounted for, and, most importantly, the correspondence between the simulated performance and that in the real world is not validated. Notwithstanding these problems, if until few years ago simulations were considered as "fake robotics" by most researchers in autonomous mobile robotics, the field now sees a growing attention both to the use of simulation and to the discussion on the evaluation criteria to compare simulation experiments with real ones [3].

In order to illustrate how they work, we describe two computational simulators that arguably are the most used in autonomous mobile robotics: Player/Stage [16] and USARSim [25]. The first one (Fig. 1) is composed of a program, Player, which provides a hardware abstraction layer, allowing the experimenter to control the hardware of robots without worrying about the actual components of the real robots. The other program of Player/Stage is Stage, that simulates the behavior of robots in environments. The researcher is required to write the code of the control subsystem of a simulated robot,

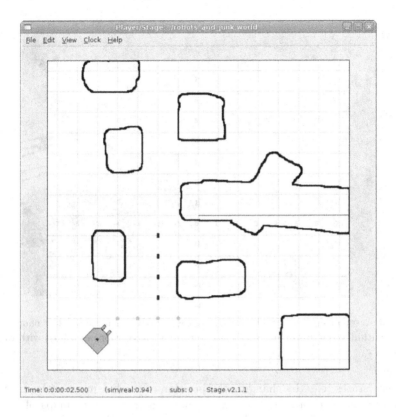

Fig. 1 A screenshot of Player/Stage (from the Player/Stage tutorial [16]): closed contours are obstacles, the area not included in them is free area, the green object at the bottom left is a robot, and small yellow dots and blue rectangles are objects that the robot is supposed to pick up.

namely to write the program that the robot runs to determine its next action according to the perceptions received and to the actions performed so far. At a very high level, a simulation with Player/Stage works as follows [16]:

- The control subsystem program sends commands to Player and receives sensor data from it.
- Player receives the commands and sends them to the robot. Moreover, it gets sensor data from the robot and sends them to the control subsystem.
- Stage interfaces with Player in the same way a robot's hardware would. It receives commands from Player, moves a simulated robot in a simulated two-dimensional world, gets sensor data from the robot in the simulation and sends them to Player.

The second simulator is USARSim [25]. It uses a client/server architecture in which the server maintains the states of all the objects in the simulation (e.g., robots), responds to the commands issued by the clients by changing the

Fig. 2 A screenshot of USARSim (from the USARSim manual [25]) showing a robot within an environment: green rays represent perceptions obtained with range sensors.

objects' states (e.g., by changing the position of a robot in an environment), and sends sensor data back to the clients. The server uses maps of three-dimensional environments and models of the robots (Fig. 2). A researcher is required to write a client that implements the control subsystem of a robot. A client is a program that receives sensor data from the server and sends commands to the server in order to make the robot behaving as desired. At an abstract level, both Player/Stage and USARSim have a similar architecture, in which control subsystems of the robots interface with the programs that simulate the interaction of the robots with environments.

Given this picture of experimental activities in autonomous mobile robotics, in the following we discuss simulations and their relation with experiments both from a general perspective and from the autonomous mobile robotics perspective.

3 Simulations and Experiments

In order to compare real and simulation experiments, in this section we start by proposing a very general definition of simulation, based on the concept of model but, at the same time, distinct from it. Then, we analyze why and how computer-based simulations can be used as experiments. Finally, we try to clarify the role of simulations used as experiments that, from the one side, allow making numerous and accelerated experiments but, from the other one, rise the problem of the validation of their results.

3.1   Simulation: Model + Execution

It is not easy to find out a general definition of what a simulation is. According to [7] a simulation imitates a process by another process, where the term 'process' refers to the evolution of some object or system whose state changes over time. In other words, a simulation reproduces the behavior of a system using another system, providing a *dynamic* representation of a portion of reality. Simulations are closely related to dynamic models that include assumptions about the time-evolution of a system; more precisely simulations are based on models that represent a dynamic portion of reality. We believe, however, that simulations represent reality in a different way than models do. In the following, we propose a distinction between models and simulations in the effort of clarifying their roles and relationships.

The exact meaning of the term 'model' remains open for discussion, notwithstanding the large effort devoted in the last years to its philosophical analysis [5]. For our purposes, a model is a *representation*: it can represent either a selected part of the world or a theory, in the sense that it interprets the laws of the theory. Let us consider a scale physical model of a bridge that is going to be constructed. This is a representation that replicates some features of the real object (the bridge that is going to be build) abstracting from the full details and concentrating on the aspects relevant to the purpose. For example, if the scale model has been constructed in order to show to its purchasers the final shape of the bridge, it would not be important its material, color, or dimension. What is represented in a model depends on the purposes for which the model has been conceived.

Now let us consider the case in which the same scale model is build to test the resistance of some materials, used in construction, to some atmospheric agents. In this case, the sole model is not enough for the purpose of testing the resistance of these materials; the model has to be put in a (controlled) physical environment where it can be subjected to the action of the atmospheric conditions. Here, we see something more than a simple description of a portion of reality; in a sense, the model is *executed* in the reality by means of the action performed by the environment: this is what we call a *simulation*. As the example shows, in order to have a simulation, it is necessary both to have a model and to execute it. Simulation involves more than a (static) model as it requires that the model is evolved to mimic the corresponding evolution that takes place in reality. The execution of a model to have a simulation is performed by an "agent" that can be the reality itself, as in the case of the bridge, or, much more frequently, a computer.

The difference between model and simulation, and between representation and execution considered above, is probably better evidenced in the case of computer simulations, which are based on *computational models*, namely formal mechanisms able to manipulate strings of symbols, that is to compute functions. The difference between model and simulation is tightly connected to that between a program and a process. Actually a completely specified computational model defines a program (as a sequence of operations), whereas the process resulting from the execution of a computational model, representing the behavior of a system, is a simulation process executed by a computer. It is important to notice that not every execution of a computational model is a simulation. To have a simulation, we need a computational model that represents a system behavior whose state changes in time. Moreover, it is worth stressing that without an underlying model (that represents the system) a simulation cannot occur and, at the same time, that a simulation is not just a representation, but an *executable* representation. This is in accordance with the concept of *core computer simulation* advanced in [16] that emphasizes both the representational aspect and the executional one[1].

With these examples in mind we can now discuss another element to characterize simulation, namely who or what is the *agent* that performs the execution required to have a simulation. In the case of a *computational* model, the execution is performed by a computational machine – a computer – and the model has been conceived as computational just in order to be executed by a machine. In the case of a *mathematical* model, for example when the time-varying behavior of a system is described by a set of interrelated differential equations, the model can be executed both by a machine and by a human being, even if nowadays it is much more common that a system of equations is solved by a computer[2]. In the case of a *physical* model, the execution is performed by the nature itself together with the human agent who sets up the scale model in the natural environment in order to be executed. For example in the scale model of the bridge built in order to test the resistance of its materials against some atmospheric conditions, we could say that the model is executed by the action of the atmospheric conditions themselves. In principle, it is possible that some models can be executed both by humans and by computers; however, a model for a simulation is usually designed having in mind its executor and so it would be utterly impractical to let it be executed by a different executor than that considered at the moment of its conception.

---

[1] "System S provides a core [computer] simulation of an object or process B just in case S is a concrete computational device that produces, via a temporal process, solutions to a computational model that correctly represents B." [16, p. 110]

[2] It is worth considering also the case in which no analytic solutions to the equations representing the evolution of a system exist. In this case, it is often possible to approximate the differential equations with difference equations and calculate numerically the values of these equations by a computer. The case in which analytical methods are unavailable and computer-implemented methods for exploring the properties of mathematical models are adopted is, in our opinion, a special case of computer simulation.

The conceptual framework just introduced allows for discussing different *degrees* of representation involved in the different types of simulations. Simulation based on a physical model (such as the case of the bridge) is inserted in the physical reality that works by itself. The representation of this physical reality is not needed for having a simulation; rather, the purpose is to isolate the relevant aspects of this reality in the underlying physical model. Computer simulations based on computational models (such as the simulation of a protein folding process) requires, instead, a higher degree of representation: a model of the phenomenon under investigation (the protein folding) is needed together with the model of the environment (e.g., the conditions under which proteins do fold into their biochemically functional forms) in which it takes place in order to evaluate their mutual interactions. As these examples show, when the agent executing a model in a simulation is not the reality, but an artifact, we need to model also (a part of) the reality. In this case, the purpose is not to isolate some aspects of interest, but to try to make this representation as precise as possible for the aim at hand. Referring to models at the basis of simulation, we could say that the degree of representation is proportional to the degree of control. When more representation is needed, as in the case of a computer simulation taking place in a represented reality, there is a higher degree of control due to the fact that this is a simplified representation not considering many elements present in the real environment. On the contrary, when the simulation takes places in the reality and no representation is required, there is less control due to the fact that reality is complex and rich of elements, most of which not even known.

3.2   Simulations Used as Experiments

In this section, we discuss the relation between simulations and experiments by starting from the evidence that, in the analysis of simulations, one of the core issues concerns their experimental capabilities. Nowadays simulations, and computer simulations in particular, are largely used in scientific research practice; they can even replace the conduction of real empirical experiments, and their outcomes are used to validate or demonstrate hypothesis. Within philosophy of science, it can be observed a variety of positions regarding the relation between simulations and experiments. Simulations have been seen just as techniques for conducting experiments on digital computers [14] or as substitutes for experiments impossible to make in reality [7]. Moreover, they have been considered as intermediate tools between theories and empirical methods [18], as novel experimental tools [16], or as special kinds of experiments [22].

We believe, however, that a better way to address the question of the relation between simulations and experiments is to say that (computer) simulations *can be used* as experiments, when the purpose for which a simulation is done coincides with the purpose for which an experiment is done. This is the case in which experiments/simulations are performed for discovering new explanatory hypotheses, for confirming or refusing theories, for choosing among competing hypotheses. Accordingly, we claim that it is misleading to identify simulations with experiments, as it is perfectly natural to have simulations that are not experiments, namely simulations that are made with no experimental purpose in mind. Think for example of the simulation of a protein folding process; it can be done for different purposes: either for a pedagogical purpose (to illustrate the simulated process to students) or for a scientific/experimental purpose (to inspire other experiments, to preselect possible configurations and set-ups, to analyze experiments, to find out suitable values for important parameters, and also to develop new hypotheses and models).

Therefore, we claim that simulations can be seen as *parts* of empirical experiments and that different degrees of extension of simulations within experiments are possible. In an extreme case, a simulation may be so extended to represent the whole experiment in a way that the experiment is "reduced" to a number of simulations and the two almost coincide. This is for example the case in which empirical experiments are impossible to make in reality and simulations are adopted (experimentally) as their substitutes. Let us think of the investigation of counterfactual situations or the analysis of the long term consequences of raising the income taxation by a given factor. In more common cases, simulations can be used as parts of experiments in a variety of ways. Simulations can be used as techniques in the cases in which it is impossible to derive analytical solutions to a system of differential equations formulated to describe certain phenomena. This is quite typical in physics, where theories are often expressed in mathematical terms and simulations can provide numerical solutions. Differently, in biology compact and elegant theories of the sort familiar in physics are rare. Explanations of phenomena are typically expressed by natural language narratives and are not always based on well-defined and complete paradigms. For this reason, simulations in biology often contribute in setting theoretical frameworks and in constructing theoretical knowledge. They play an important role in developing new hypotheses and models, as when the simulation results suggest new regularities that would not have been extracted from the model assumptions otherwise. This is the case, for example, when a model is incomplete and a simulation allows testing the hypothesis and participates in settling the model itself. These computer simulations can be seen as *explorative experiments* to get some hints that can become new knowledge to be further experimentally verified. In this case, the experiment is meant as explorative because it does not give the assurance of the correctness of a conjecture, but may anyway help to build it up.

Within this latest scenario, the validation of the results obtained via simulation represents a central problem. The reasons for trusting simulations are usually two-sided: either the models used in simulations are strongly grounded in well-founded theories or there are experimental data against which simulation results can be checked. This is not always possible when simulations are used as explorative experiments, where neither well-founded theories nor experimental data are present. In line with [27], the source of credibility for such simulations lies in the prior successes of the model building techniques adopted, in the production of outcomes fitting well with previously accepted data, observations, and intuitions, together with the capability of making successful predictions and of producing practical accomplishments. This strategy does not force to commit to any truthfulness claim: rather than on an inference from "success to truth" [11], it is based on an inference from "success to reliability". It is not a single strategy, but a pool of strategies that – as Ian Hacking [6] has pointed out many years ago for experiments in general – are fallible even if they offer good reasons to believe in the results of simulation procedures.

Taken for granted that computer simulations nowadays are used as experiments, the following question still remains open: *why* computer simulations can be used as experiments? We believe that two different kinds of reasons (theoretical and practical) can be identified. The theoretical reasons for using simulations as experiments are mainly rooted in the similarity between techniques of experimentation and techniques of simulation [26], as they both involve data analysis and constant concern with uncertainty and error. In our opinion, however, these reasons, while surely important to be considered, are not sufficient to epistemologically justify the use of simulations as experiments. Besides exploiting similar techniques, experiments and simulations share the ability and the necessity of controlling the features under investigation. Control deals with the idea that experiments consist in producing controlled circumstances, in which it can be assumed that a studied relationship does not depend on changes in non-controlled factors. Control, thus, is one of the key elements in the definition of experiment as *controlled experience*. It is not difficult to recognize that the same notion of control fits well with the very idea of simulation when used for experimental purposes, where experimental factors have to be chosen and controlled within an artificial setting. The set-up of a simulation, thus, is extremely similar to the set-up of an experiment: hypotheses to be tested, circumstances to be controlled, parameters to be set. In both cases (simulations and experiments) these activities are performed by humans who intentionally determine the setting up conditions. Moreover, also the execution of a simulation shares some similarities with the execution of an experiment. The execution of a simulation requires an agent: in the case of a computer simulation, the agent is the computer, whereas in the case of a physical simulation the agent is the environment in which the simulation takes place, together with the human who places the physical model required for the simulation in that environment. In both cases, it is not needed that the agent is constantly involved in the execution process. In a simulation, the agent acts as the *substratum* on which the execution can takes place. This is, again, similar to what happens with an experiment where, once the experimenter has put in place the execution, it follows its way without further inputs.

As said, there are also practical reasons for using simulations. First, simulations can be used to make several accelerated experiments, as the experiments can be repeated exactly, as many times as needed, with guarantee of a precision degree not always possible in empirical cases. Second, simulations can be used to perform experiments that are difficult to make in reality, being free from many of the practical limitations of real experiments, such as the possibility to change boundary and initial conditions. Third, simulations can be used to perform experiments that are impossible to make in reality, such as studying realms of reality which are not physically accessible (e.g., the processes involved in the formation of galaxies).

4 Simulation and Experiments in Autonomous Mobile Robotics

In this section, we cast the issues discussed in Section 3 in the context of autonomous mobile robotics. This analysis is particularly interesting given the ongoing process that aims at defining experimental methodologies for autonomous robotics (described in Section 2.1), in which simulations are expected to play a key role. We first discuss the nature of simulations in autonomous mobile robotics and then we discuss their role in experiments.

4.1 The Nature of Simulations

As we have seen, a simulation needs a dynamic model of the system it reproduces. In the case of autonomous mobile robotics, the system that is reproduced is a mobile robot that acts in an environment. The dynamic model must therefore include a representation of the robot and a representation of its interaction with the environment. Let us detail the elements comprised in these two representations. Roughly speaking, a mobile robot is modeled by representing its locomotion, sensing, and control subsystems (recall from Section 2.2 that the control subsystem is a program that determines the next action of the robot on the basis of the past history of perceptions and actions). The interaction of a mobile robot with an environment is a complex issue. For example, it involves a model describing the behavior of the robot in the environment after the control subsystem issued a command. If the command is 'go forward 50 cm', the actual movement of a wheeled robot in a real environment could be more or less than half a meter because of slipping wheels, of rough terrain, of errors in the motors moving the wheels, and of several other reasons. It is not easy to capture this variability in a computational model. Similar problems emerge in modeling the perception of the robot in the environment. Current robotic simulations model in different ways the uncertainties on the effects of actions and on the perceptions.

The first kind of uncertainty is usually managed by the physical engine (see below), while the second kind of uncertainty is artificially added to the data, according to different probability distributions.

Autonomy makes modeling a robot's interaction with the environment even more complicated, because interactions are hardly predictable. This is probably one of the reasons for the late adoption of simulation in autonomous mobile robotics. Until few years ago, the models of interaction between robots and the world were not enough accurate and using simulations based on these models was simply not convenient for the autonomous mobile robotic community. If a simulation is based on inaccurate models of the interaction with the world, it is not representative of the behavior of real autonomous robots and, as such, cannot be used to validate the behavior of the simulated robots and to generalize it to real robots. Nowadays, one of the most used simulators for autonomous mobile robots USARSim [25], models the interaction of robots with environment using a software, called Unreal engine, initially developed for a multiplayer combat-oriented first-person shooter computer game, Unreal Tournament 2004. Unreal engine contains a physics engine that simulates the interaction of three-dimensional physical objects and that allows to obtain highly realistic simulations. Resorting to components developed in the extremely competitive field of computer games is an interesting way to have state-of-the-art models of physical interaction between objects.

In discussing the nature of simulations in autonomous mobile robotics, it is interesting to mention a trend that is emerging in the last years: the use of publicly available data sets composed of data collected in the real world by some researchers [17, 8]. These data sets substitute sensor data in experimental tests of robotic systems. From the one hand, we can think of these data sets as models of the interaction between the robots and the real environment. Using these data sets appears very similar to perform a simulation, in which the underlying model is very precise, because it exactly records the interaction of real robots with real world. According to this view, the difficulty of building a model of the perception of a robot in an environment is addressed by letting the data collected during real operations be the model. From the other hand, using publicly available data sets can be considered as a real (not simulated) experiment, in which activities of collecting and processing data are performed in different places at different times. What is emerging here is a sort of continuum, ranging from performing completely simulated experiments, to using data sets like [17, 8], to performing real-world experiments.

4.2 The Role of Simulations in Experiments

We now turn to discussing the role of simulations in the experimental activity performed in autonomous mobile robotics. Simulations are common in some areas of robotics and less common in other areas. For instance, in the area of locomotion of legged robots, simulations are widely used to validate design choices. This might be due to the fact that there exist very good models (both cinematic and dynamic) of the behavior of robotic legs in a given situation (terrain, friction, forces, ...). On the other hand, in the area of robotic vision, simulations are almost absent, mainly because of the lack of good models of the behavior of cameras in a given situation (lights, shadows, ...). Autonomous mobile robotics is in an intermediate position, as good models of the behavior of a mobile robot in an environment are appearing (as discussed above) and, consequently, simulations are being increasingly employed. Using computer simulations has a potential impact on a number of issues that are relevant in the definition of good experimental methodologies for autonomous mobile robots. This impact is now even more relevant, because these methodologies are under development (see Section 2.1). We organized the following discussion around the three principles of experimentation individuated in [2] (comparison, reproducibility and repeatability, and justification/explanation), focusing in particular on the implications of using simulations as experiments. Before commenting on them, let us remark that several of the issues discussed below are highly interconnected and are inserted in a structured scheme only for presentation purposes.

- *Comparison.* Using simulations makes the creation of a common ground for comparing the performance of different systems easier. Simulations represent much more controlled settings than real world. For example, two systems for planning a path that brings a robot from a start pose (namely, from starting position and orientation, according to some reference frame) to a destination pose avoiding collisions with obstacles [12] should be compared in the same environments. To fairly compare the two systems, the shape, size, and position of obstacles in the environment are critical. The same environments can be reproduced in reality with much more difficulty than in simulation.

  Moreover, simulations allow to access a large number of measurable parameters (e.g., computational complexity, computational time, memory usage, precision, and accuracy) that can be used for comparison purposes. It is important to remark that, using a standard simulation tool, like Player/Stage [16], these parameters are measured in a uniform way over different runs of the simulation, which may be performed at different times and in different places. In this sense, simulations, being very controlled settings in which parameters can be accurately set and measured, allows to investigate more precisely the reasons for different performance of competing robotic systems.

  Generally speaking, as discussed in [10], there are three ways to compare performance of algorithms, in decreasing order of appeal:

  1. use the same code that was used in the previous experiments,
  2. develop a comparable implementation, starting from the description provided in papers and reports,
  3. compare the results with those obtained in other papers.

From the current state of the art in autonomous mobile robotics [2], most comparisons are presently conducted adopting the second way. The availability of standard simulation tools helps in moving toward the first way of comparison, because they ease the process of writing code and push researchers to make their code available, as in the OpenSLAM initiative [23].

The recent interest in developing benchmarks for autonomous mobile robotics, to assess the relative performance of robotic systems by running a number of standard tests, is an aspect of interest in the context of comparison. Proposed benchmarks [17] are based on publicly available data sets that, as discussed above, can be considered somewhere between simulations and real experiments. Benchmarks can be of two types [17]:

> Benchmark Problems (BPs), defined as the union of: (i) the detailed and unambiguous description of a task; (ii) a collection of raw multisensor data, gathered through experimental activity, to be used as the input for the execution of the task; (iii) a set of rating methodologies for the evaluation of the results of the task execution. The application of the given methodologies to the output of an algorithm or piece of software designed to solve a Benchmark Problem produces a set of scores that can be used to assess the performance of the algorithm or compare it with other algorithms.

> Benchmark Solutions (BSs), defined as the union of: (i) a BP; (ii) the detailed description of an algorithm for the solution of the BP (possibly including the source code of its implementation and/or executable code); (iii) the complete output of the algorithm when applied to the BP; (iv) the scores associated to this output, calculated with the methodology specified in the BP.

Sometimes, comparison of robotic systems is done in simulated competitions [24]. For example, in the RoboCup Rescue Simulation League, robotic systems for search and rescue in large scale disaster situations are simulated using USARSim. Beyond assessing performance in the specific applications, simulated competitions are also sought to foster research on advanced and interdisciplinary topics; for example, search and rescue involve engineering, medical, logistic, and social problems.

- *Reproducibility and repeatability.* Arguably, the major impact of simulations on experimental practice of autonomous mobile robotics is on reproducibility and repeatability. By *reproducibility* we mean the possibility to verify, in an independent way, the results of a given experiment. It refers to the fact that other experimenters, different from the one claiming for the validity of some results, are able to achieve the same results, by starting from the same initial conditions, using the same type of instruments, and adopting the same experimental techniques. By *repeatability* we refer to the fact that a single result is not sufficient to ensure the success of an experiment. A successful experiment must be the outcome of a number of trials, performed at different times and in different places. These requirements guarantee that the result has not been achieved by chance, but is systematic. In particular, being the setting up of simulations much more easier than the setting up of real robotic experiments (e.g., think of the hardware failures and of battery recharging), use of simulations is expected to facilitate both repeating the same experiment and reproducing the same conclusions. For example, several environments (indoor offices, indoor open spaces, outdoor crowded streets, outdoor parking lots, ...) can be considered without much effort during the testing of a simulated system. USARSim comes with a dozen of already available maps, in which autonomous mobile robotic systems can be tested.

- *Justification/explanation.* One of the techniques used to derive well-justified conclusions from experiments is to test a system in different settings (different environments, different parameter configuration). Simulation environments offer a way to easily change from a setting to another one and to provide robust results, that can be verified according to *ground truth*, namely to "real" results. For example, a robotic system for building maps of unknown environments can be simulated and the produced maps (representations of the obstacles and the free space build by the robot) can be compared with ground truth maps (representations of the obstacles and the free space available in the simulator) for evaluating their quality and, as a consequence, the quality of the mapping system. Ground truth is trivially available for simulated environments but it is seldom available for real environments (sometimes aerial images and Google Earth maps are considered as ground truth for real outdoor environments).

5 Conclusions

In this paper we have discussed the nature of robotic simulations and their role in experiments performed within autonomous mobile robotics. More precisely, according to the framework proposed in [2], we have investigated the role of simulations, intended as experimental tools, in autonomous mobile robotics, where simulations can be assessed both as a way to test the theoretical models at the basis of the building of robots and as a methodology for the development of robots.

We have raised a number of issues that are worth further investigating. For example, within the above theoretical framework, it emerges an asymmetry between experiments performed with simulated robots and with real robots: an element unverified in simulation will likely be unverified in reality, but an element verified in simulation can be or not be verified in reality. Generalizing a little, a fundamental issue that we have left open concerns the different validation strategies that can be adopted for assessing simulation results in autonomous mobile robotics. A promising way is to look for local solutions for validating simulations in specific cases, moving from a general discussion, as that conducted in this paper, to concrete proposals for single problems. More in general, future work will address the analysis of the several issues we raised in this paper about experimental use of simulations in autonomous mobile robotics.

References

1. Amigoni, F., Gasparini, S., Gini, M.: Good experimental methodologies for robotic mapping: A proposal. In: IEEE Int'l Conf. on Robotics and Automation, pp. 4176–4181 (2007)
2. Amigoni, F., Reggiani, M., Schiaffonati, V.: An insightful comparison between experiments in mobile robotics and in science. Autonomous Robots 27(4), 313–325 (2009)
3. Carpin, S., Lewis, M., Wang, J., Balakirsky, S., Scrapper, C.: Bridging the gap between simulation and reality in urban search and rescue. In: Lakemeyer, G., Sklar, E., Sorrenti, D.G., Takahashi, T. (eds.) RoboCup 2006: Robot Soccer World Cup X. LNCS (LNAI), vol. 4434, pp. 1–12. Springer, Heidelberg (2007)
4. EURON GEM Sig (2007), http://www.heronrobots.com/EuronGEMSig/
5. Frigg, R., Hartmann, S.: Models in science. In: Zalta, E.N. (ed.) The Stanford Encyclopedia of Philosophy (2009), http://plato.stanford.edu/archives/sum2009/entries/models-science/ (Summer 2009)
6. Hacking, I.: Representing and Intervening. Introductory Topics in the Philosophy of Natural Science. Cambridge University Press, Cambridge (1983)
7. Hartmann, S.: The world as a process: Simulations in the natural and social sciences. In: Hegselmann, R., et al. (eds.) Simulation and Modeling in the Social Sciences from the Philosophy of Science Point of View, pp. 77–100. Kluwer, Dordrecht (1996)
8. Howard, A., Roy, N.: The robotics data set repository, radish (2003), http://radish.sourceforge.net/
9. Humphreys, P.: Extending Ourselves. Computational Science, Empiricism, and Scientific Method. Oxford University Press, Oxford (2004)
10. Johnson, D.: A Theoretician's Guide to the Experimental Analysis of Algorithms. In: Goldwasser, M.H., Johnson, D.S., McGeoch, C.C. (eds.) Data Structures, Near Neighbor Searches, and Methodology: Fifth and Sixth DIMACS Implementation Challenges, pp. 215–250. American Mathematical Society, Providence (2002)
11. Kitcher, P.: Real realism: The Galileian strategy. Philosophical Review 110, 151–197 (2001)
12. LaValle, S.: Planning Algorithms. Cambridge University Press, Cambridge (2006)
13. Madhavan, R., Scrapper, C., Kleiner, A.: Special issue on characterizing mobile robot localization and mapping. Autonomous Robots 27(4), 309–481 (2009)
14. Naylor, T.: Computer Simulation Techniques. John Wiley, Chichester (1966)
15. Permis: Performance metrics for intelligent systems (2000), http://www.isd.mel.nist.gov/research_areas/research_engineering/Performance_Metrics/index.htm
16. Player Project (2009), http://playerstage.sourceforge.net/
17. Rawseeds (2006), http://rawseeds.elet.polimi.it/

18. Rohrlich, F.: Computer simulation in the physical sciences. In: Fine, A., Forbes, M., Wessels, L. (eds.) Proc. 1990 Biennal Meeting of the Philosophy of Science Association, pp. 145–163 (1991)
19. RoSta: Robot standards and reference architectures (2007), http://www.robot-standards.eu/
20. Siegwart, R., Nourbakhsh, I.: Autonomous Mobile Robotics. MIT Press, Cambridge (2004)
21. Simon, H.: The Sciences of the Artificial. MIT Press, Cambridge (1969)
22. Simpson, J.: Simulations are not models. In: Models and Simulations Conference, vol. 1 (2006)
23. Stachniss, C., Frese, U., Grisetti, G.: Openslam.org (2007), http://www.openslam.org/
24. The RoboCup Federation: RoboCup (1998), http://www.robocup.org/
25. USARSim (2010), http://usarsim.sourceforge.net
26. Winsberg, E.: Simulated experiments: Methodology for virtual world. Philosophy of Science 70, 105–125 (2003)
27. Winsberg, E.: Models of success vs. success of models: Reliability without truth. Synthese 152(1), 1–19 (2006)

The Logical Process of Model-Based Reasoning

Joseph E. Brenner

Abstract. Standard bivalent propositional and predicate logics are described as the theory of correct reasoning. However, the concept of model-based reasoning (MBR) developed by Magnani and Nersessian rejects the limitations of implicit or explicit dependence on abstract propositional, truth-functional logics or their modal variants. In support of this advance toward a coherent framework for reasoning, my paper suggests that complex reasoning processes, especially MBR, involve a novel logic of and in reality. At MBR04, I described a new kind of logical system, grounded in quantum mechanics (now designated as logic in reality; LIR), which postulates a foundational dynamic dualism inherent in energy and accordingly in causal relations throughout nature, including cognitive and social levels of reality. This logic of real phenomena provides a framework for analysis of physical interactions as well as theories, including the relations that constitute MBR, in which both models and reasoning are complex, partly non-linguistic processes. Here, I further delineate the logical aspects of MBR as a real process and the relation between it and its target domains. LIR describes 1) the relation between model theory – models and modeling – and scientific reasoning and theory; and 2) the dynamic, interactive aspects of reasoning, not captured in standard logics. MBR and its critical relations, e.g., between internal and external cognitive phenomena, are thus not "extra-logical" in the LIR interpretation. Several concepts of representations from an LIR standpoint are discussed and the position taken that the concept may be otiose for understanding of mental processes, including MBR. In LIR, one moves essentially from abduction as used by Magnani to explain processes such as scientific conceptual change to a form of inference implied by physical reality and applicable to it. Issues in reasoning involving computational and sociological models are discussed that illustrate the utility of the LIR logical approach.

Joseph E. Brenner
International Center for Transdisciplinary Research, Paris, France
e-mail: joe.brenner@bluewin.ch

L. Magnani et al. (Eds.): Model-Based Reasoning in Science & Technology, SCI 314, pp. 333–358.
springerlink.com

1   Introduction

Model-based reasoning (MBR) is a fascinating and complex concept. Despite the extensive work on the definition and explication of MBR by Lorenzo Magnani, Nancy Nersessian and others [26], the meaning of MBR has not become completely stabilized, at least for me. I have accordingly undertaken to explore the implications of the term MBR, the real process to which it refers and the relation between MBR and its targets.

From a pragmatic standpoint, building and using models, whether physical or mental, is an activity of adult human beings who possess substantial linguistic ability combined with a great deal of knowledge. Models are sophisticated tools for thought with well-recognized advantages in reasoning effectively about complex phenomena, scientific and other.

The term MBR is used to refer, in my opinion, to three different but related processes:

- reasoning, especially, in science, with the aid of models;
- reasoning about models, as process entities in their own right;
- reasoning that is model-determined.

In each case, models have different primary roles, in which the emphasis is on one or the other of the three terms in the concept:

- as *tools* for reasoning, which implies their prior construction and the reasoning necessary for that construction; (based)
- as the *targets* of reasoning; (model)
- as a unique *form* of reasoning that is applicable in any context, including the above. (reasoning)

The further questions one can ask, especially, what is the relation between a cognitive model and its target, seem to me to all refer to 1) aspects of MBR and of reasoning in general as processes; and 2) the substantial overlap or interaction between the elements of the processes, at least in the case of cognitive models and modeling.

The differences, if any, need to be established between model-based reasoning and other forms of high-level cognitive processes that correspond to the terms thinking, knowing and believing. Knowing, as a process, is usually simply, too simply in my view, contrasted with knowledge as a static product. Also germane to the utility of MBR is nature of the relation between the model, once "built", and the reality that it is supposed to represent. The reason is that the concept of a representation, which implies an intermediate structure or entity between the thing represented and an individual's awareness of it, are still widely used as having explanatory power for cognitive phenomena. Recent arguments against both the necessity and desirability of representations suggest that it might be important to examine further the relation between model and reality, to determine if a less problematical characterization of that relation might be found.

It is questions like these, to which I see only partial answers in the literature, which have prompted me to try to apply the principles of my novel Logic in Reality (LIR) to model-based reasoning. The core thesis and rationale of my paper is that by such application one can better see model-based reasoning (MBR) for what it is, namely, a complex process of and in the mind moving between real mental and physical structures, actual and potential.

I note that there is a degree of paradoxical self-reference involved in the definitions of MBR: model-based reasoning refers to how one uses models to reason *from* them *to* reality. But MBR also describes how one builds models, that is, goes *from* reality *to* models. These concepts are not new, but I wish to emphasize that both are themselves reasoning processes, involving the making of inferences and hence the use of logic, but what logic?

2 Logic and Reasoning

Logic is described as the theory of correct reasoning, where logic is understood as classical bivalent propositional and predicate logic or of its modern multivalent, modal, deontic or intuitionist versions [17]. Their machinery – symbolism, syntax and semantics – is essentially the same in the related disciplines of standard set and category theory. If model-based reasoning (MBR), reasoning using or with models, is included in that reasoning, these logics would constitute the theory of and/or be applicable to it.

However, models are used in reasoning to achieve practical results. The process of reasoning with models is thus neither topic-neutral nor context-independent, while standard logics are virtually required to be both. For this reason, and others, these logics are incapable of describing or capturing any but the simplest, abstract, syntactic and consistent aspects of reasoning, not of MBR in particular and reasoning as a real process in general.

2.1 Paraconsistent Logic

The first of two major advances in extending our logical grasp of reasoning has been the development of paraconsistent logic, first in Brazil by Newton da Costa and his students[1], and later by the Australian school around Graham Priest. Paraconsistent logics (PCL) are *defined* such that contradiction does not entail triviality and thus can mirror some of the contradictory aspects of real phenomena. In some paraconsistent logics [30] an ontological commitment is made and real contradictions are allowed. In others, such as the logics of formal inconsistency of Carnielli and Marcos [9], they are not.

---

[1] I would like to acknowledge here the help and encouragement I have received from Itala D'Ottaviano and Walter Carnielli of the State University of Campinas, Brazil, at which MBR09 took place.

The authors cited above have made extensions of their paraconsistent logical systems to explore aspects of reality that involve key issues in the foundations of science. These include inter-theoretic relationships, complementarity, the individuality of quantum entities and reasoning, among others, all of which are relevant to MBR. However, they tend to share the problem of the restrictions imposed by the concept of logic as a class of mathematical systems and their related formal tools, especially, standard set theory.

2.2 Abductive Logic and Abduction

The second development is the application by Lorenzo Magnani and Nancy Nersessian and their colleagues of the concept of abduction to reasoning, especially, model-based reasoning that goes well beyond standard conceptions of sentential abductive logic. The complexity of the interaction between abduction, considered as equivalent to abductive reasoning, models and reasoning as such is illustrated in Lorenzo Magnani's important paper "Inconsistencies and Creative Abduction in Science"[25]. Abductions of many kinds are model-based if they involve a reasoned problem solving process. The further definition of "model-based reasoning", due to Magnani and Nersessian, is "the construction and manipulation of various kinds of non-verbal representations, not necessarily sentential or formal". For the purposes of this paper, let me emphasize the attribution of a process character to models. It is primarily non-physical models that are, of course, most easily seen as processes, whereas most formal models have the mathematical structure that Batterman has called "applied mathematics" [1].

Formal models have inherent limitations in accounting for inconsistencies inherent in reasoning, and accordingly, in sentential abduction itself. There is a problem, however, with the expression "models of abductive reasoning", which are, presumably, cognitive models. Standard logical accounts of abduction fail to capture, in Magnani's words, "much of what is important in abductive reasoning", i.e. abduction, including, and we here exit a threatening circularity, the existence of model-based abductions, in which empirical, as well as theoretical inconsistencies should not be eliminated arbitrarily.

Magnani calls attention to the need, in creative scientific reasoning, to allow two rival theories not only to coexist but to compete, as epistemological and "non-logical" inconsistencies. I will claim that it is possible to describe all of these concepts of interaction in a more rigorous manner by extending the domain of logic to real phenomena, including theories and their "competition". As we will see, this means redefining the meaning of the phrase "simultaneous satisfaction of a set of positive and negative constraints, without the need for question-begging term of "spontaneous" in the connectionist model. I will return to the Magnani view of abduction later.

In their "Advice on Abductive Logic", Gabbay and Woods [11] outlined a schema of abduction in which its formal epistemological properties are clearly outlined. I agree in part with their differentiation between deduction, induction and abduction. Their proposition is that whereas deduction is truth-preserving and induction is probability-enhancing, abduction is ignorance preserving in the sense that "if you knew the answer, or the hypothesis you are making does not lack some degree of epistemic virtue, it wouldn't be abduction any more."

A simplification introduced is to restrict the abductive inferences considered to those that close a cognitive agenda, that is, a drive to execute a reasoning process. I would be reluctant to make this restriction; in the LIR conception of dynamics, processes go substantially, but not completely to "completion", potentialities remain for new actualizations, for new sequences of sequences, etc. Their open-ended existence is at least as ontologically important as closure. It is fascinating to note that even in their formal model of abduction, Gabbay and Woods find it necessary to introduce a technical rule that provides for backward propagation of the abductive process "against" the flow of deduction, in fact following the lines of non-formal natural reasoning described by LIR[2].

The difficulties for applying the Gabbay-Woods approach to model-based reasoning lie in its essential restriction of abduction to a place on a linguistic logical map, where inquiry involves premises searches and inference with premises projections, both following the rules of propositional, truth-functional logics. Both the degrees of complexity that the authors admit are not captured by their abductive logic, and some of the facts about real-life abduction that in most theories are suppressed, ignored or idealized are exactly those described by LIR, namely, the dialectical, non-linguistic properties of the abductive process.

At MBR04, I presented a paper entitled "A Transconsistent Logic for Model-Based Reasoning" [5]. I suggested that a logic extended to real systems and processes was the preferred logic for addressing issues in many areas in philosophy, epistemology and science, and had the potential for explicating outstanding problems in MBR.

My belief is, that this logic, which I now call Logic in Reality (LIR) does have such potential. The remainder of the paper is accordingly organized as follows: in Section 3, I describe LIR and propose the LIR perspective on reasoning as such; in Section 4 I look at problems with the concept of representations in theories of mind; in Section 5 I return to the description of abduction and MBR and analyze the relation between models, reasoning and reality based on the principles of LIR; and in Sections 6 and 7 I look at reasoning *with* models in different disciplines from the LIR perspective.

To summarize the thesis of this paper, logic is indeed the theory of complex reasoning, but it is not logic as it is generally understood. It is Logic in Reality, and it supports and explicates the Magnani approach.

---

[2] In LIR, it is the existence of two parallel chains of causality in complex, e.g. cognitive processes that provides the ontological basis for backward and forward movement at the same time in reality, cf. Brenner [6].

3   Logic in Reality (LIR)

LIR is a new kind of logical system, grounded in quantum mechanics, which postulates a foundational dynamic dualism inherent in energy and accordingly in causal relations throughout nature, including cognitive and social levels of reality [6]. This logic of real phenomena provides a framework for analysis of physical interactions as well as theory, including the relations that constitute MBR, in which both models and reasoning are complex, partly non-linguistic processes.

The term "Logic in Reality" (LIR) is intended to imply both 1) that the principle of change according to which reality operates is a *logical* principle embedded in it, *the* logic in reality; and 2) that what logic really *is*, or should be, involves this same real physical-metaphysical but also logical principle.

The four major components of this logic are: 1) the foundation of LIR in the dualities of nature; 2) its axioms and calculus that are intended to reflect real change; 3) the categorial structure of its related ontology; 4) a two-tier framework of relational analysis.

3.1   *Dualities and Fundamental Postulate*

LIR is based on Stéphane Lupasco's foundation of logic in the physical and metaphysical dualities in nature, attraction and repulsion (charge, spin, others), entropy and negentropy, actuality and potentiality, identity and diversity, continuity and discontinuity and so on.

LIR states that the characteristics of energy can be formalized as a structural logical principle of dynamic opposition, an antagonistic duality inherent in the nature of energy (or its effective quantum field equivalent) and accordingly of all real physical and non-physical phenomena and accordingly of all real physical and non-physical phenomena – processes, events, theories, etc. The overall theory is thus a metaphysics of energy and LIR is the logical part of that metaphysical theory. Lupasco basically combined Aristotelian potentiality with the essential aspects of a quantum picture of energy.

The fundamental postulate, as formulated by Lupasco [24], is that every real phenomenon, element or event **e** is always associated with an anti-phenomenon, anti-element or anti-event non-**e**, such that the actualization of **e** entails the potentialization of non-**e** and *vice versa*, alternatively, without either ever disappearing completely.

The point of equilibrium or semi-actualization and semi-potentialization is a point of maximum antagonism or "contradiction" from which, in the case of complex phenomena, a T-state (T for "*tiers inclus*", included third term) emerges, resolving the contradiction (or "counter-action") at a higher level of reality. The logic is a logic of an *included* middle, consisting of axioms and rules of inference for determining the state of the three dynamic elements involved in a phenomenon ("dynamic" in the physical sense, related to real rather than to formal change, e.g. of conclusions).

Based on this "antagonistic" worldview, axioms, a calculus, non-standard semantics and categorial ontology can be developed for LIR which have been presented elsewhere [6].

One has to follow, in this system, the different tendencies that principally characterize the different levels of reality, recognizing that processes at higher levels will instantiate, to a more or less actual or potential extant, the tendencies of the lower ones as summarized below:

The Primary Directions of Change

Level		Development Toward
Physical (inorganic)	$\rightarrow$	Non-contradiction of Identity
Biological	$\rightarrow$	Non-contradiction of Diversity Contradiction; T-state (emergence)
Cognitive	$\rightarrow$	Contradiction; T-state (emergence)

LIR is a non-arbitrary method for including contradictory elements in theories or models whose acceptance would otherwise be considered as invalidating them entirely. It is a way to "manage" contradiction, a task that also undertaken by paraconsistent, inconsistency-adaptive and ampliative-adaptive logics.

In the LIR calculus, the reciprocally determined "reality" values of the degree of actualization A, potentialization P and T-state T replace the truth values in standard truth tables. Its values can be handled as being similar to probabilities using a non-Kolmogorovian formalism. One can think of LIR as a kind of modal logic with a reality operator, different however from the existence operator of some free logics.

This logic contains that of the excluded middle as a limiting case, approached asymptotically but only instantiated in simple situations and abstract contexts, e.g., computational aspects of reasoning and mathematical complexity.

The fundamental postulate of LIR and its formalism can also be applied to logical operations, answering a potential objection that the operations themselves would imply or lead to rigorous non-contradiction. The LIR concept of real processes is that they are constituted by series of series of series, etc., of alternating actualizations and potentializations. However, these series are not finite, for by the Axiom of Asymptoticity they never stop totally. However, in reality, processes *do* stop, and they are thus not infinite, but *transfinite*.

Perhaps the most important point, which I am grateful to Walter Carnielli for helping me to formalize, is that the connectives of implication, conjunction and disjunction all correspond to real operators on the parameters of real elements.

In LIR one has something like an informational semantics, with information defined as energy, without, by definition, anything like classical truth conditions. The sense of truth that the semantics gives is the dynamic state of the event, phenomenon, judgment, etc, where the event is "on the way", more or less, as the case may be, between its actualization and the potentialization of its contradiction. There are no proofs in LIR in the semantic sense. The demonstrations are closer to those in science, in that they purport to describe in a coherent manner the processes and changes that are occurring or have occurred, by reference to a model of the elements involved and their relations.

3.2 Categories

The third major component of LIR is the categorial ontology that fits the above axioms. In this, the sole material category is Energy, and the most important formal category is Dynamic Opposition.

From the LIR metaphysical standpoint, for real systems or phenomena or processes in which real dualities are instantiated, their terms are *not* separated or separable! Real complex phenomena display a contradictional relation to or interaction between themselves and their opposites or contradictions. Of course, there are many phenomena in which such interactions are not present, and they, and the simple changes in which they are involved, can be described by classical logic or its modern versions. The most useful categorial division that can be made is exactly this: phenomena whose dualities show non-separability as an essential aspect of their existence, NSC, at their level of reality and those that instantiate separability, SC. Note that the requirements in classical category theory of exclusivity and exhaustivity do not apply: they are bivalent logic in another form.

LIR approaches in a new way the inevitable problems resulting from the classical philosophical dichotomies as well as such concepts as space and time, simultaneity and succession as categories with *separable categorial features*. Non-Separability underlies all other metaphysical and phenomenal dualities, such as cause and effect, determinism and indeterminism, subject and object, continuity and discontinuity, and so on. This is a "vital" concept: to consider elements that are contradictorially linked as separable is a form of category error. I thus claim that non-separability at the macroscopic level, like that being explored at the quantum level, provides a principle of organization or structure in macroscopic phenomena that has been neglected in science and philosophy.

3.3 A Two-Level Framework of Relational Analysis

The above development leads to a two-level framework of levels and meta-levels for analysis of the structure of reality and phenomena.

Based on the LIR axiom of *Conditional* Contradiction, aspects of phenomena involving levels of analysis that are generally considered independent can be understood as being in the dynamic relationship suggested, namely, as one is actualized, the other is potentialized.

One example is the relation between sets or classes of elements, events, etc. and the descriptions or explanations of those elements or events. Another example is the relation between object level and meta-level, as in metatheory, meta-philosophy and metalogic. It is easy to show here that not only can no clear distinction of, say, theory and metatheory be made, the levels are contradictorially connected.

Although in classical logic all consequence relations can depend only on the logical form of the premises and conclusion, in LIR, a complete disjunction between syntax and semantics does not exist. The functional relation between syntax and semantics is a further reflection of the dynamic relations underlying physical reality, also exemplified by the "interplay" between proof principles and principles of mathematical construction.

It is this principle of dynamic opposition, operating both *intra-* and *inter-*level, which I propose as logical, that is the basis for my core thesis (Brenner, 2008) that the extant domain (reality) is incorrectly described by current theories as following principles of classical logic. Reality is misrepresented by classical ontologies using these principles and the theories themselves embody classical logic and could be reconstructed according to LIR principles.

4 Models and Theories of Mind

4.1 The Use of Models

The concept that models are vehicles for learning about the world, allowing a new style of model-based reasoning seems straightforward. The question remains about how learning with a model is possible. In other words, what are the dynamics involved in the reasoning process and what are the characteristics of models that are relevant to their use.

In their article in the Stanford Encyclopedia of Philosophy, Hartmann and Frigg [13] point to a number of unresolved problems in understanding the use and value of models in science:

- The relationship between phenomenological models and complex theories is hazy and perhaps even contradictory.
- There is no systematic account of the different ways in which models can relate to reality and how these ways compare with one another.
- In modern logic and set theory, a model is a structure that makes all of the sentences of a theory true, but many models in science are not structures in this limited sense, but still represent target systems in the physical world.

The authors conclude by saying that there remain significant lacunae in our understanding of models and how they work.

In the LIR theory of the form or structure of reasoning with models, two sets of issues: the relation between models and what they model (their "targets") and between models and the reasoning based on them are addressed. In my "semantic", "non-statement view" [29], a scientific theory is not a class of statements or propositions but is itself a class of models, where a model is a non-mathematical structure consisting of some domains of entities and some relations defined over them that satisfy certain conditions. These are extra-linguistic, dynamic entities as opposed to the syntactic conception of a theory as a set of statements or formulas, especially mathematical formulas, inevitably governed by first-order predicate logic or its modal variants.

In the process view of MBR outlined in this paper, and the process view of human cognition I will develop that underlies it, I claim that these relations between models and the targets of those models on the one hand, and models and the reasoning based on them are best described by reference to the critical LIR ontological feature of non-separability, as follows. In LIR, real systems and their theoretical models are not totally independent entities. LIR provides for a dynamic, structural relation between them rather than a simple isomorphism. For complex cognitive phenomena, representations and the entity represented are not totally separated or separable.

4.2 Aspects of the LIR Epistemology

A view of complex cognitive models and model-based reasoning is only tenable if one can relate it to a sufficiently complex (in the Ashby sense of requisite variety) theory of mind, consciousness and knowledge. I will give only the briefest summary of the LIR epistemology that reflects the axioms and ontology outlined previously.

Reasoning is best described as an active process of knowing, a process of active knowing. In LIR, reasoner, reasoning and reasoned, like knower, knowing and knowledge, or known, are all processes, evolving dialectically in the following manner according to the schema I have outlined. Models of reasoning, both mental and physical, are accordingly also best seen as processes, the more or less temporarily stable resulting of the reasoning that went into their construction.

LIR views knowledge, as a complex of the knower, knowing and known as processes in which there is an opposition, conflict or contradiction between the knower and the known, resulting in the inability of the knower to see himself completely as knowing. The knower is not in the known, but the known is an element, entity or process that is contrary and contradictory to the knower.

In the LIR epistemology we as knowers are not totally external to what is known by us and not completely different from it. I must know, then, that if there are other knowers, as there are, they must be part of my known and *vice versa*. The source of human dignity *is* in ourselves as knowers, but if we avoid the error of solipsism, the origin of the sense of moral responsibility can only come from the relation to other knowers, in other words, all human beings, and by extension, other beings and perhaps even, as suggested by Magnani (2007), certain non-living entities. *A contrario*, one cannot find responsibility in oneself as an isolated agent. Since we are both a "not-other" *and* an "other" at the same time, a self-interest argument for morality holds, with self-interest losing its negative connotation in this case. Two or more human individuals and their relations constitute interactive systems in the LIR categorial sense of non-separable subjects and objects, sharing in part one another's characteristics. An individual is no more isolated logically, psychologically or morally than he or she is economically. It is thus because our will is *not* free that we, or at least some of us, must and will try to behave morally. Whether we have the capacity for doing so or not, however, will depend largely on our genetically determined capacities and propensities which I cannot go into further here.

4.3 Internal and External

A source of difficulty in understanding the dynamics of complex cognitive interactions has been the apparent absolute dichotomy between two individual human minds. While it is easy to see a cell in dynamic interaction with its environment or context, with change possible in both directions, it is difficult to understand how our cognitive context can be both internal and external without externality being determined by our consciousness. This would demand a full-blown anti-realist position.

As discussed above, however, in the LIR ontology, a strategy is available for describing not only a relation but a dynamic interaction between the familiar dualities of local and global, continuous and discontinuous and internal and external phenomena.

The LIR epistemological approach is to analyze the details of our acquisition of perceptions and effectuation of actions into actual and potential components. In the LIR theory of consciousness, afferent stimuli are "split" into conscious potentialities and unconscious actualities. Thus while your mind is physically external to mine, some of its perceptible potentialities can be internalized by me, perhaps by mirror neurons in the concept of Ramachandran. Individuals, as a "non-separable" part of a group, contribute in this way their individuality to it. But the group instantiates a group psychology and this becomes part of the individual. What is the "group part of the individual" is something instantiated at higher, more intuitive level, but not the less real for that. At all levels of reality, I will assume that there is a conflict or opposition between epistemological elements and the energetic processes to which they correspond. I may and in fact always will focus on one or the other aspect, but what I have called a contradictional relation is present, one aspect is actualized while the other is potentialized. I will return to this interpretation in my discussion of sociological models.

4.4 Representations

By now, we have become accustomed to talking about mental phenomena in terms of representations and the question of their existence as real entities is taken for granted. For example, model-based reasoning is used, by Magnani and others, to indicate the construction and manipulation of various kinds of representations, not primarily sentential or formal, but mental and/or related external processes. Bechtel [2] has attempted to capture the elusive characteristics of mental representations *qua* models in terms of mechanisms.

In representationalist theories dealing with cognition, internal entities of some sort stand for or correspond in some way to external processes and events. These mental representations explain or are explanatory devices for cognition in that they are, or correspond to (this vagueness is typical) intentional states, instances of intentionality considered as embodying the irreducible first-person properties that are alleged to characterize consciousness, reasoning and qualia.

As M.R. Bennett, a neuroscientist, and P.M.S. Hacker, a philosopher show in their recent massive document – *Philosophical Foundations of Neuroscience* [3] – representations are among a group of concepts for which no empirical evidence exists. These authors show how virtually *all* of the standard modern approaches to mental entities involve some form of confusion, to use their word. Worse, they give rise to a large amount of critical discussion that tends to obscure the real issues involved. For example, in the computational form of representationalism of Dretske and others, there is a symbolic entity between neurobiological and phenomenological data, and a host of secondary problems arise as to the properties and relations of the symbols involved.

Bennett and Hacker essentially deconstruct the concept of any mental entities including representations, qualia, models and concepts of self and free will – that are a substitute for, or an addition to, the mental processes themselves. A further difficulty with the standard picture of representations is the difference in treatment of mental states *vs.* intuition, which some people might consider a fiction. In the LIR view, intuitions, as diversities, and more permanent or salient mental states, as identities, are related contradictorially, as have shown above. The real existence of intuitive processes provides an argument against Fodor and against the introduction of what in my view is an unnecessary additional entity into the causal chain.

The unresolved difficulties in the relation between intuition and symbolic intentionality led Husserl to the postulation of still more relations and processes that I will not discuss here. I simply restate the major problem with the Fodorian picture, namely, how and where a symbol is to intervene and what, accordingly, are the properties of symbols.

The Bennett and Hacker approach is basically to focus on the human being as a psychophysical unity, avoiding the mistake of both neuroscientists and others in attributing perception, thought or knowing to the brain or its parts, such as its hemispheres. In the LIR view, such confusions are the consequence of the separations which have been the unavoidable consequence of applications of standard logics. Existing accounts of mental processes suffer from the need to introduce additional entities due to the lack of a principled categorial method of relating their critical concepts contradictorially. A mental phenomenon, which is not something other than the physical processes with emergent properties "displays" its contradictorial origins in appearing to have symbolic and non-symbolic aspects, and being closer or farther from the center of attention at a particular time.

In this view, it is a mistake to say that what we or some "mind" perceive is an image or representation of an object, or that perception involves *having* an image of the object. The so-called binding problem is a false problem, since the brain does not construct a perceived world, but enables an animal to see a visible scene. This view is echoed by Thompson [35] who states that "A phenomenal mental image is not a phenomenal picture in the 'mind's eye' (scare quotes mine), nor indeed is it any kind of static image; it is rather the mental activity of re-presenting an object by mentally evoking and subjectively stimulating a perceptual experience of that object".

Similarly, Damasio was mistaken in his distinction between having and feeling an emotion, as if emotions were some sort of somatic image or marker.

4.5 The Implications for MBR

Let us see now see what the implications of this approach might be to issues in MBR. First of all, we should agree that the representations referred to in standard views of sentential reasoning, and to which formal constructions are shown to apply, are theoretical constructs that are tautological with respect to those constructions.

A possible logic for an abductive approach to science (the only one possible), as shown in the paper by Itala D'Ottaviano and Carlos Hifume [10], is a logic of quasi-truth and partial structures, that harks back to Peirce's semiotic conception of knowledge[3], and looks forward toward applications to the theory of science, inconsistent beliefs, the realism-empiricism debate and so on. Current model theory, in regard to current scientific practice can be related, in these authors' views, to these concepts to handle incomplete, incoherent or contradictory information. Quasi-truth is proposed as mirroring the fact that representations in science are fallible and incomplete, therefore not true in a correspondence sense, but as partially true. However, the processes involved are not "approximately real" or containing some "real element", they are realities which can be captured by a logic of reality.

---

[3] The categories of experience of Peirce have in any case been criticized as overly dogmatic and restrictive, and the foundation of a theory of reasoning based on them is therefore open to question.

In substance, the LIR position is that da Costa, D'Ottaviano and their colleagues have taken linguistic truth as far as it can go in the direction of real processes in their formal – I give all the qualifiers – paraconsistent, modal, Jaskowski-type discussive logic.

LIR explicates the related process model of representation called "interactivism" by Bickhard and others[4], providing the necessary dynamic link between internally and externally related aspects of phenomena for which the term "representational content" is used, but to which I would object for the indicated reasons.

4.6 Anticipation

Anticipation is a property of conscious living systems that has been given a role in abstract modeling. Rather than as a self-representation at their level of reality, I see anticipation as simply another case of the projection of aspects of the real world, reality, into a model world consisting in a configuration space of lower dimensionality[4].

I have argued that potential states and processes, of which reasoning is an example, are causally effective and not epiphenomenal. If this is accepted, then the naturalization of anticipation follows logically, at least in my logic. I in fact assimilate anticipation at the cognitive level to particularly well-formed, homogeneous potential states that are opposed to the general fuzziness of the "stream of consciousness". Following Kolak (see below) and Bennett and Hacker, I believe it is important to focus on all high-level properties as properties of the whole human being, of whom the alleged parts are convenient abstractions for analysis.

What distinguishes anticipatory processes is a higher degree of potentiality, but anticipation does not define all processes. Anticipatory processes are a sub-class of a broader group of processes that constitute "consciousness".

4.7 Mind and Machine

Rapid advances in understanding, in many cases using extremely complex models, of the neurological bases of cognition or mental processes in general, including reasoning are being made. There seems to be, nevertheless, a strong tendency to retain either totally classical metaphysical concepts of individuals, a classic mind-body separation and emphasis on fully actualized aspects

---

[4] An example of this notion is to be found in the work of Gödel. The Gödel theorems and logic – as written – do not apply to physical or mental emergent phenomena, but LIR views the principle involved, the duality of consistency and completeness, axiomatically, as another instantiation of the fundamental duality of the universe. The logical and ontological development undertaken in LIR illuminates Gödelian dualism as another expression of the fundamental dynamic opposition at the heart of energy and phenomena.

of phenomena, all reflecting a strong bias in favor of the classic principle of logical bivalence. This underlies, in addition, standard set and category theory, which in my view are bivalent logic in modern dress. Arguing against what he considered to be the excessive concentration on computationalism in models of cognition (that minds can be explained as machines), Lucas said in 1990 that Turing's theorem might be applied to a computer that someone claimed to model a human mind, but it is not obvious that what the computer could not do, a mind could not. Gödel's theorem was a methodology, for Lucas, for proving that mechanistic explanations are false, and LIR supports Lucas' argument for the difference between mind and machine in terms of potentiality. Potentialities and actualities act as carriers of emergence between different levels of reality.

4.8 Models of Individual Identity

Logic in Reality suggests a new definition of what constitutes an individual, a group, and the relationship between them. It is in line with the proposal by Ladyman and Ross [19] of the elimination of the classic abstract a priori philosophic notion of an individual thing and its redefinition in terms of dynamic patterns. The consequent naturalized metaphysics is compatible, as is LIR, with fundamental physics.

In the LIR two-level framework for analysis, groups can no longer be considered, at the social level, as the equivalent of a set composed of individuals, equivalent to members of the set. Classical set theory requires that sets and their members be completely distinct; set theory is essentially bivalent logic in another form.

In Kolak's similar approach to personal identity [18], the critical move is to avoid a separation of the subject that is the bearer of personal identity from its psychological object identifications. LIR provides the rules for the relative, alternating dominance of the two perspectives: personal identity and the intuition of personal identity, the reality of subject-dependence and the appearance of subject-independence of experience are dynamically, dialectically related in the LIR logic. Logical, psychological and metaphysical perspectives intersect in this view. In LIR terms, Kolak's statement that one's essential subjectivity is obscured by the intuition of one's own existence and identity is that the former is potentialized by the latter. The conjoined personality experience by the subject from the inside as the identified self that expresses itself as *"I am I"*, not my brain, not my body not my body and not even my "self". Consciousness thus indeed constitutes personal identity, and is not a substance, etc., etc. LIR allows a principled minimal *metaphysical and ontological* process view of consciousness as constituted by systems of systems of processes following the indicated dynamics.

In the LIR sense of the non-separability of human beings, the additional questions that Kolak asks, "Am I You? Are you a model of me? Is your reasoning a model of or for my reasoning?" can receive logical answers.

5   Model-Based Reasoning in Reality

The question then becomes, of course, if the mental models of model-based reasoning are not representations, what are they and what are their characteristics? I have given the LIR answer to the first part of the question, which is that they are the processes of reasoning themselves. LIR is the logic for Thagard's view of representations consisting of patterns of activation (and passivation) of populations of neurons [34]. The discussion of the second part brings us to the Magnani concepts of abduction.

5.1   *Inferential Abduction and LIR*

Abduction is a form of reasoning that must be clearly differentiated from deduction and induction, as we have seen. However, as a partly truth-functional system, it remains limited in dealing with the actual cognitive processes involved in reasoning which, I believe, are required for dealing with the interaction in reasoning between models and their targets.

Following Magnani [27], I differentiate between selective, theoretical abduction or diagnostic reasoning and creative abduction[5]. The former selects from the library of existing theory with the objective of preserving consistency and making inferences to the best explanation, a linguistic entity. Its models are based on the concept of the epistemic agent. The latter generates or is the basis for the emergence of new hypotheses, ideas or concepts which I consider active, or more active processes. The former is the sentential view and corresponds roughly to analytical model building [31], the latter an inferential view that is like synthetic model-building. Logic in Reality is the logic appropriate to the inferences made about real systems in all their complexity and inconsistency. The agents in this approach are ontological. LIR thus applies to all high-level forms of conceptual change including, but not limited to scientific change. To emphasize this point, I claim that the process of creating new theories of sentential abduction, that follow standard logics, is an inferential, model-based one in the MBR sense, avoiding the use-mention trap.

5.2   *Manipulative Abduction and LIR*

The principles of Logic in Reality confer an additional rigorous character to Magnani's concept of manipulative abduction which occurs when we are thinking through doing and not only about doing. Manipulative abduction refers to extra-theoretical behavior that aims at creating communicable accounts of new experiments that can be integrated into an existing body of knowledge. Like all two level cognitive processes, theoretical and manipulative

---

[5] Lupasco pointed out that both deduction and induction should be viewed as active processes in a real antagonistic relationship [22].

abduction are not totally independent but interact in both the conscious and unconscious mind, with one or the other in the foreground at any one time. In Magnani's words, model-based abductive reasoning can extract and render explicit important information that is unexpressed at the level of data, transforming knowledge from tacit (potential) to explicit (actual) forms.

From my point of view, Logic in Reality *is* manipulative abduction in the sense that it sees as a logical structure the entire system composed of the mind-brain of a person *plus* any emergent external structures physical – tools, apparatus or machines and non-physical – rules, relations and constraints. A difference is, however, how one should view what Magnani calls internal representations that "consist in the knowledge and the structure in memory as propositions, productions, schemas, neural networks, models, prototypes and images". LIR sees all of these as *processes*, the stated entities themselves in potential form, not as separate representations of those entities.

In my opinion, this difference strengthens the Magnani concept of epistemic mediators, the result of the processes of manipulation "in reality". LIR has no difficulty in attributing theoretical aspects to embodied behavior, since no absolute separation is desirable or possible. LIR naturalizes, so to speak, the ontological features of manipulative abductive reasoning and epistemic mediators as regards the actions which they support. As in my discussion later of individual-group and group-group interactions in society, I completely agree with Magnani that "The cognitive process is *distributed* between a person (or a group of people) and external representations (JEB: or processes) and so is obviously *embedded* and *situated* in a society and in a historical culture". Finally, this view of abduction is the basis for Magnani's indication [28] of the need for a new *logic* of morality in a technological context, since simple deontic logic fails to include the relation up to and between internal intensions and external structures on which moral worth has been conferred.

Magnani develops the concepts of the disembodiment and externalization of minds to describe the emergence of higher-level cognitive processes from lower-level ones. Disembodiment of mind refers to what he calls the cognitive interplay between internal and external representations. In the second part of the paper, he analyzes the higher-level processes and their interaction with the lower ones in terms of model-based and manipulative abduction, that is, model-based (abductive) reasoning.

In view of the problematic character of representations, I prefer to recast the cognitive interplay in the LIR process dynamics where what is involved are the processes themselves, following the dialectical evolutionary rules outlined above.

Lupasco wrote [22] that instruments, apparatus and machines were the extension of the scientist such that the tacit inferences in procedures of manipulative reasoning that Magnani calls epistemic mediators, abductive movements, are logical in nature.

I will now turn to some specific ways in which the tools I have tried to develop may be used in reasoning with different types of models.

6   Reasoning with Computational Models

For reasoning with or about computational models, modified bivalent classical logics are generally adequate. Reasoning with computational models of complex, necessarily incomplete real-world situations is a process that requires a different concept of logic, as in the case of representational models. LIR can be used first to analyze the relation between objects and their models in general and computational models in particular, and I have proposed LIR as an addition to abductive logics for model-based reasoning in the real world, involving the emergence of new entities, in what Magnani has called the "iterative processes of creativity".

Symons [32] reasons that computational models are even more "screened off" from their targets than other types, although they are useful for explaining selected medium-scale phenomena. In focusing on MBR, I will not discuss this aspect further, except to point out that if the model is a limited one, it will screen off its target and satisfactory reasoning is likely to be difficult.

LIR in fact validates Symons' proposed modest role for computational models, by providing for *non-computational* models that have a high degree of generality for explanation (maximality), without being a "theory of everything".

6.1   *Computational Models and Simulations*

If a process structure even if involving recursion can be expressed by an algorithm, the applicable logic probably does not need to be LIR. However, as Gelfert argues [12], "a successful philosophical account of models and simulations must accommodate an account of mathematically rigorous results. Rigorous results are the "touchstone of success" are crucial to understanding the connections between different models. He also argues, however, for an epistemology of results as well as representation of reality, where there is a trade-off between quantitative exactness and explanatory value. In this sense, rigorous results are internal to a model, or class of models, and cannot be assimilated to either theory or data. As such, they illustrate the capacity of models to take on roles beyond both theory and performance in specific empirical and interventionist contexts. LIR basically proposes a new way of looking at the explanatory competence of a model without trying to justify its validity or usefulness in terms of the truth of statements about it.

6.2   *Computational Models of the Mind*

Let us now look at computational models of mind in relation to one of their critical domains, namely the characteristics of natural language. A standard gambit of skeptic philosophers is that one cannot give a determinate meaning to basic terms in natural language. More simply, it is problematic for them explain formally how someone can in fact understand someone else. Replies to the skeptic challenge involve disentangling the relation between language and the algorithms for language and hence the computability or non-computability of mental processes.

As suggested by Horst [16], a computational theory of mind combines a primarily representational theory of mind with a formal, that is, computational account of reasoning. In contrast, the most popular current models of psychological processes are neural network or connectionist models. Without going into this enormous debate, I can only point out that the goal of a naturalization of the psychological aspects of these theories – understanding them as physicalist – has not been achieved. In LIR, it has.

The computational theory of mind states that, regardless of whether the mind had a classical or connectionist architecture, it is a kind of computer and if that computer is responsible for our linguistic abilities, then the rules that govern those abilities must be computable.

The further computability assumption states that if natural language is computable, then it is computable by any universal computing machine and an algorithm can be defined for that machine. The argument then starts with the following: "If the human mind is a computer, then like any computer, it will have certain basic operations". This approach ignores the possibility of the reverse statement: "The human mind has certain basic operations, but they are not those of the computer". In this case, its basic terms may be similar to terms of natural language, and both are determinate.

The question then becomes: "What does it mean for the mind *not* to be a computer?" I claim that an adequate response to this question requires seeing mental processes as following the LIR theory of mind outlined above, that is, with both intra- and inter-level interactions operating according to the Axiom of Conditional Contradiction involving alternating partial actualization and potentialization of contradictory elements.

Theories of mind that are compatible with my model are available. I feel that something like my extension of logic to reality is implied in the influential book by Paul Thagard, *Computational Philosophy of Science* [33], as a more extreme but physically grounded form of practical inference. Especially in the domain of psychology, Thagard points out that empirical work "shows numerous systematic discrepancies between popular inferential practice and accepted logical norms", that is, classical or neo-classical logic. I believe LIR is a contribution to a methodology for mediating between the descriptive and the normative. Its principles are closer to how people actually reason and it has the capability of handling inconsistency, as Thagard seeks, without "Popperian oscillations".

I have tried to avoid strict localization of function as well as "absolute" holism, the unnecessary requirement that *all* of the brain is involved in *everything* it does. The principle of dynamic opposition in LIR provides a "domain-neutral" principle for characterizing the nature of and overlaps between cognitive domains. A key example is the action-sentence compatibility effect, the interaction between comprehension and motor control. Distribution or non-localization is thus necessary for functional integration of cognitive operations, but it is not sufficient without this additional principle.

I cannot go further here into applications of the LIR model of mind. What is essential is that it offers a solution to the outstanding difficulties of the "same-order" representation theory of consciousness, namely, that conscious mental states represent both the world and themselves. It answers Weisberg's complaint that there seem to be no good naturalistic candidate to connect higher-order and first order representation states, or, in the Magnani terminology, higher-level and lower-level cognitive processes.

7 Reasoning with Sociological Models

7.1 Castells' Network Model

Castells has discussed economic and political applications of new information and communication technologies in the emerging information society and knowledge-based economy. The major work of Castells, first published in 1993 [8], on the "Information Age: Economy, Society and Culture", has proven extremely prescient. He sees society as a complex system of networks that are a consequence of the new information and communication technologies. His views are of interest in the LIR context because of their reference to a "logic" of the network society and of its dynamics. Castells' network model of society as a "space of flows" can be analyzed from the LIR logical standpoint, as well as more standard sociological models, e.g. Leydesdorff's "triple helix" [20]. The LIR logical approach is applied to an analysis of the properties of the networks and their nodes, as well as to the segments of the society that are disfavored or excluded completely.

When we look at society, then, in Castells' model, as constituted by nodes and links, how should one think of oneself, or of the group of which one is a member – as a node, a link or both? LIR argues for "both", alternately and reciprocally, in process terms: when we are "node-ing" we are "link-ing" to a lesser extent, and *vice versa*.

Castells also describes the structure and dynamics of *resistance* to the hegemony of the network, and such resistance, like other oppositional elements, can be described in LIR terms, as follows: LIR is a methodology of reasoning, a way of doing reasoning. Here, reasoning about the evolution of the society using both aspects of the model, the network and the resistance to it, provides a better understanding of the forces at work and the relationships between the actors involved (inside and outside the network), given the ambiguities and inconsistencies in their approaches to society. The basis of an evolutionary theory of the information society [15] is suggested. The normative characteristics of LIR provide the basis for further discussion of ethics and points toward the development of an evolutionary ethical theory of the information society [7].

7.2   The Leydesdorff Approach and LIR Compared

The limitations of standard approaches are recognized elsewhere. As stated for example by the sociologist and information scientist Loet Leydesdorff, in inter-human communications, structured situations and double contingencies are involved that are structures by their latent, as well as current dimensions. The feedback loops involved in these non-Markovian processes cannot be handled without going to some form of non-Boolean logic and non-Kolmogorovian probability distribution of the variables involved.

Leydesdorff has developed a formal analytical approach [21] to the construction and operation of models in relation to cognitive agents and their assembly into social units by combining Luhmann's social systems theory and Giddens' structuration theory of action. His theory starts from several metaphysical-psychological (meta-psychological?) principles of which two are compared here:

- Double hermeneutics; Interactions. A double hermeneutics is possible in interhuman communication because one can understand someone else as a participant in the communication in addition to observing and interpreting her behavior.
- Structuration. Giddens proposed the concept of "structuration" and related this concept from its very origin to the double hermeneutics operating in intentional interactions among humans beings. Social systems are instantiated in observable networks of relations. The "duality of structure" is then proposed as the recursive operation which transforms aggregates of action into systems by invoking structuration as a "virtual" operation. Whereas structures can be analyzed as latent properties of communication systems (at each moment of time), structuration transforms both actions and structures (of social systems) over time by providing them with reconstructed meaning.

The starting points of Logic in Reality are formally similar to those of Leydesdorff:

- Double hermeneutics; Interactions. LIR is "designed" to study both the interactions between individuals in terms of the categorial non-separability of their cognitive systems. A double-hermeneutics is implied.
- Structuration. Structuration is a term also used by Lupasco, before Giddens, "*structuration*" in French, to emphasize the dynamic *process* aspects of complex structures, biological, cognitive or social. The answer he gave to his question "What is a structure?" [23] was that structures are also dynamisms, not to be objectified and reified. In the LIR perspective, structuration is a real operation on the relations between two individuals. Any

individual structure is never rigorously actual, that is, absolute in any sense, given the nature and logic of energy. It is a dynamic "structuring" that is always functionally associated with an antagonistic and contradictory potential structuring. Another way of saying this is that a structuring seen externally is a kind of form; looked at internally, it consists of the processes themselves.

Differences appear in the two approaches due to the basically epistemological orientation of the Leydesdorff construction and the ontological orientation of LIR. This results in the selection of different targets for application of the respective models.

From the point of view of model-based reasoning, this comparative "synthesis", that is the result of informal collaboration between Dr. Leydesdorff and myself, insures that two somewhat opposing but also complementary perspectives are available "as needed". In my LIR terms, they interact dialectically, with one or the other in the foreground, as the case may be. I can only discuss here the further points I consider most important.

In Leydesdorff's view, the construction of a social system begins with *action* involving Shannon-type information processing in an individual (first contingency) to *expectations* involving two or more individuals (second contingency) to *interaction* between expectations. Higher levels of interactions constitute the social system and its communication structure, most significantly in the way meaning is established and transferred.

Meaning development takes place through three levels of meaning processing, all involved in the dynamics of meaning. First, observable human actions can be considered as interactions in inter-human communication. Unlike agents, communications as events in the second contingency cannot directly be observed but have to be inferred from observable behavior. In other words, meaning is provided by a (mostly implicit) model. The model organizes the communications. When the meanings can additionally be exchanged, a next-order model of the components in the organization of meaning can be hypothesized. This next-order model structurates the structures, and thus can add to the reduction of uncertainty in the modeled system.

7.3 LIR Development

To model a social system from the LIR perspective in which interactions are present, despite the partial overlaps with Leydesdorff, a discussion of information, meaning and their relation in LIR is required for which there is no place here. With regard only to the above points, I can say the following:

- Expectations. I do not feel it necessary to limit the interactions between the two individuals in the double contingency to expectations as a separable categorial cognitive feature, but as anticipations that, as indicated, are both actual and potential. The dynamic interactions in the "first-contingency" are complex, structurated in Giddens' term, but include additional elements of dynamic structure that are both determined by lower-level biological and cognitive substrates and are determining for the higher, that is, the social level. In place of Leydesdorff's analytical distinction, the interaction is studied as existing between mutually dependent informational "variables", without separation between expectations and actions.

- Modeling. Cognitive models cannot be totally separated from their targets, provided the latter are non-physical. This is clearly seen in exactly the cases of interest, namely, where the target of the modeling process is reasoning itself, with or without models. Both share some of one another's properties.

- Empirical Applications. LIR is not intended to study patterns in data, and no direct comparison with the Leydesdorff application can be made. The major question that remains is to what extent this approach can be extended and used with information systems whose elements are not entities but complex biological or cognitive processes, e.g., of thought or individual-group interaction.

By using applicable aspects of both the Leydesdorff approach and LIR, one can partially model the operation of real interactions in dynamic terms. As one might expect, the disadvantage of this approach is that it does not yield the quantitative measure of the actual and potential cognitive states addressed by LIR that might be desired. One makes, in LIR, the inferences about their probability-like degree of actuality (presence) and potentiality (absence). However, the advantage in having a logical system that is potentially extendable to more complex group-group and individual-group interactions outweighs it.

8 Summary and Conclusions

In this paper, the critique of standard logic in relation to model-based reasoning, initiated by Magnani and Nersessian, is supported by a novel logic grounded in physics, logic in reality (LIR). LIR provides a process description for reasoning about complex context-dependent change at biological, cognitive and social levels of reality. The objective is to help insure that categorial errors and artificial theoretical constructs do not interfere with operation of MBR as a well-defined sub-domain of mental processing.

My arguments have covered issues in fundamental physics, mechanisms of perception and cognitive and social science. I am aware that the "transport dialectique", to use the term of Gilles Deleuze, may have been a difficult one. My vision of the world and theories of the world as related, consistent *and* inconsistent conflicts with much received wisdom. I ask, to begin with, that the reader renounce, for the sake of a "science of model-based reasoning", some standard (and cherished) notions not only of logic, but also set theory, category theory and a theory of models as mathematical objects and accept concepts from the latest quantum field views of the secondary ontological status of spacetime. The methodology of LIR means looking for structures in nature that are potential as well as actual, in a sense that is neither more, nor less than that a certain sequence of amino acids in an enzyme has the potential for binding with specific substrates under the appropriate conditions in the appropriate medium.

LIR describes 1) the relation between model theory – models and modeling – and scientific reasoning and theory (a confusing half-way situation according to Hodges [14]; and 2) the dynamic, interactive aspects of reasoning, not captured in standard logics. MBR and its critical relations, e.g., between internal and external models, are thus not "extra-logical" in the LIR interpretation. Essentially, one moves from abduction as used by Magnani (2006) to explain processes such as scientific conceptual change to a form of inference implied by physical reality and applicable to it, for example to issues in reasoning involving computational and sociological models.

My claim is that the LIR contradictorial picture of reasoning involves a *form* of identity theory of mind that avoids the difficulties of both standard identity and dualist theories by the introduction of the principle of dynamic opposition at all levels of perception, mental processing and action. The emphasis in the LIR realistic, austere but non-reductionist embodied theory of mind is on its enactive aspects, that is, of sensorimotor coordination in which afferent and efferent processes are tightly linked. No new, independent entities of the kind postulated in the various forms of representationalism are required, due to the availability, in LIR, of a dynamic relation between internal and external, actual and potential and identical and diverse aspects of phenomena. It is the alternating actualizations and potentializations derived from initial energetic inputs that *are* our ideas, images, beliefs, etc as well as models. Some further phenomenological classification of these process elements (such as that made by Husserl) is possible, but it does not change the overall structure of my proposed picture.

The instabilities, incoherency, incompleteness and inconsistency of human reasoning do not have to be relegated to some non-logical limbo, but are shown to be in a dialectical relation to their more commonly considered opposites. A further advantage of this approach is the elimination of the implied separation between reasoning and other mental functions.

LIR should be differentiated from logical frameworks that use models to analyze the formal properties of epistemic events [36]. The targets of LIR are the properties or characteristics of real instances of model-based reasoning, with or about models. The applications of the LIR logic to issues in reasoning with and about computational and sociological models illustrate the generality of this approach.

Models and their targets, and models and the reasoning based on them are above all viewed as processes. As in the interactivist approach of Bickhard, these are shown to be dialectically related, sharing in part one another's functional properties. LIR is thus, itself, a model for model-based reasoning.

Acknowledgements. Parts of this paper have appeared in Chapters 1 and 2 of my book *Logic in Reality*, published in 2008 by Springer Dordrecht. They are reproduced here with the kind permission of the copyright owner, Springer Science and Business Media B.V., Dordrecht, The Netherlands.

References

1. Batterman, R.W.: Idealization and modeling. Synthese 169, 427–446 (2009)
2. Bechtel, W.: Mental Mechanisms. Philosophical Perspectives on Cognitive Neuroscience. Taylor and Francis Group, Routledge (2008)
3. Bennett, M.R., Hacker, P.M.S.: Philosophical Foundations of Neuroscience. Blackwell Publishing, Massachusetts (2003)
4. Bickhard, M.H.: The interactivist model. Science 166, 547–591 (2009)
5. Brenner, J.E.: A transconsistent logic for model-based reasoning. In: Magnani, L. (ed.) Model Based Reasoning in Science and Engineering. Cognitive Science, Epistemology, Logic. Kings College Publications, London (2006)
6. Brenner, J.E.: Logic in Reality. Springer, Dordrecht (2008)
7. Brenner, J.E.: Prolegomenon to a logic for the information society. Triple-C 7(1), 38–73 (2009), http://www.triple-c.at
8. Castells, M.: The Information Age: Economy, Society and Culture. In: The Rise of the Network Society, 2nd edn., vol. I. Blackwell Publishing, Malden (2000)
9. Carnielli, W.: Logics of formal inconsistency. CLE e-Prints 5(1) (2005)
10. D'Ottaviano, I.M.L., Hifume, C.: Peircean pragmatic truth and da Costa's quasi-truth. Studies in Computational Intelligence 64, 383–398 (2007)
11. Gabbay, D.M., Woods, J.: Advice on abductive logic. Logic Journal of the IGPL 14(2), 189–219 (2006)
12. Gelfert, A.: Rigorous, results, cross-model justification, and the transfer of empirical warrant: the case of many-body models in physics. Synthese 169, 497–519 (2009)
13. Frigg, R., Hartmann, S.: Models in science. In: Zalta, E.N. (ed.) Stanford Encyclopedia of Philosophy (Summer 2009), http://plato.stanford.edu/archives/sum2009/entries/models-science/
14. Hodges, W.: Model Theory. In: Zalta, E.N. (ed.) The Stanford Encyclopedia of Philosophy (Fall 2008), http://plato.stanford.edu/archives/fall2008/entries/model-theory/
15. Hofkirchner, W., et al.: ICTs and Society: The Salzburg Approach. University of Salzburg Research Paper No. 3, December. ICT&S Center. Salzburg (2007)
16. Horst, S.: The Computational Theory of Mind. In: Zalta, E.N. (ed.) Stanford Encyclopedia of Philosophy (Winter 2009), http://plato.stanford.edu/archives/win2009/entries/computational-mind/
17. Jacquette, D.: Philosophy of Logic. In: Gabbay, D., Woods, J., Thagard, P. (eds.) Handbook of the Philosophy of Science Series. North-Holland Press (Elsevier), Amsterdam (2007)
18. Kolak, D.: Room for a view: On the metaphysical subject of personal identity. Synthese 162, 341–372 (2008)
19. Ladyman, J., Ross, D.: Every Thing Must Go. Metaphysics Naturalized. Oxford University Press, Oxford (2007)

20. Leydesdorff, L.: The Knowledge-Based Economy: Modeled, Measured, Simulated. Universal Publishers, Boca Raton (2006)
21. Leydesdorff, L.: Redundancy in Systems which Entertain a Model of Themselves: Interaction Information and the Self-organization of Anticipation. Preprint for ENTROPY (2009)
22. Lupasco, S.: Logique et Contradiction. Presses Universitaires de France, Paris (1947)
23. Lupasco, S.: Qu'est-ce qu'une structure? Christian Bourgois, Paris (1967)
24. Lupasco, S.: Le principe d'antagonisme et la logique de l'énergie. Editions du Rocher, Paris (1987); Originally published by Éditions Hermann, Paris (1951)
25. Magnani, L.: Inconsistencies and Creative Abduction in Science. In: AI and Scientific Creativity. Proceedings of the AISB99 Symposium on Scientific Creativity, University of Edinburgh, Edinburgh (1999)
26. Magnani, L., Nersessian, N.J. (eds.): Model-Based Reasoning. Kluwer Academic/Plenum Publishers, Dordrecht (2002)
27. Magnani, L.: Disembodying minds, externalizing minds. In: Magnani, L., Dossena, R. (eds.) Computing, Philosophy and Cognition. College Publications, London (2006)
28. Magnani, L.: Morality in a Technological World. Knowledge as Duty. Cambridge University Press, New York (2007)
29. Moulines, C.U.: Ontology, reduction, emergence: A general frame. Synthese 151, 313–323 (2006)
30. Priest, G.: In Contradiction. Martinus Nijhoff, Dordrecht (1987)
31. Schlimm, D.: Learning from the existence of models. Synthese 169, 521–538 (2009)
32. Symons, J.: Computational Models of Emergent Properties. Minds and Machines 18, 475–491 (2008)
33. Thagard, P.: Computational Philosophy of Science. MIT Press, Cambridge (1988)
34. Thagard, P.: How brains make mental models. In: The Conference on Model-Based Reasoning in Campinas, Brazil (2009)
35. Thompson, E.: Representationalism and the phenomenology of mental imagery. Synthese 160, 397–415 (2008)
36. Van Benthem, J., et al.: Merging frameworks for interaction. Journal of Philosophical Logic (2009)

Constructive Research and Info-computational Knowledge Generation

Gordana Dodig Crnkovic

Abstract. It is usual when writing on research methodology in dissertations and thesis work within Software Engineering to refer to Empirical Methods, Grounded Theory and Action Research. Analysis of Constructive Research Methods which are fundamental for all knowledge production and especially for concept formation, modeling and the use of artifacts is seldom given, so the relevant first-hand knowledge is missing. This article argues for introducing of the analysis of Constructive Research Methods, as crucial for understanding of research process and knowledge production. The paper provides characterization of the Constructive Research Method and its relations to Action Research and Grounded Theory. Illustrative examples from Software Engineering, Cognitive Science and Brain Simulation are presented. Finally, foundations of Constructive Research are analyzed within the framework of Info-Computationalism.

1 Understanding of Research Methodology in Computing on the Background of Philosophy of Engineering and Philosophy of Science

Information age is fundamentally dependent on engineering whose knowledge largely rests on the research within Computing/Informatics. Demonstrably, artefacts built on research results of Computing work remarkably well and in that sense the research within Computing yields a reliable and trustworthy knowledge about the behavior of computational (engineered) systems. One of the relevant methodological and epistemological questions is what sort of knowledge is it and how it relates to our understanding of the world,

Gordana Dodig Crnkovic
School of Innovation, Design and Engineering, Computer Science Laboratory, Mälardalen University, Sweden
e-mail: `gordana.dodig-crnkovic@mdh.se`

L. Magnani et al. (Eds.): Model-Based Reasoning in Science & Technology, SCI 314, pp. 359–380.
springerlink.com   © Springer-Verlag Berlin Heidelberg 2010

including our understanding of humans. It is of interest to know what aspects of Computing could be labeled "Science" in a rigorous classical sense, and how knowledge produced within Computing enriches both our understanding of the world and our potential for further learning about the world and acting in the world, (see [1] and [2]).

As we are building Knowledge Society, our understanding of knowledge production, both on the level of Cognitive Science and on the level of technological enhancements provided by ICT is becoming strategically important. In this context a very general view of Computing as Science and Computing as Engineering is needed wherefrom also necessity follows of understanding of Computing in both Philosophy of Science and Philosophy of Engineering contexts. In short, engineers have discovered the innovative potential of broad insights into the discipline, which gives tangible comparative advantage. Knowing about underlying mechanisms and structures for knowledge production helps us look out of the box and find novel approaches, which is especially important in constructive and design research.

2 Constructive Research as Based on Ontological Realism

The key idea of Constructive Research (or the Constructivist[1] knowledge production), is the construction, based on the existing knowledge used in novel ways, with possibly adding a few missing links. The construction proceeds trough design thinking that makes projection into the future envisaged solution (theory, artifact) and fills conceptual and other knowledge gaps by purposefully tailored building blocks to support the whole construction. Artifacts such as models, diagrams, plans, organization charts, system designs, algorithms and artificial languages and software development methods are typical constructs used in research and engineering. Constructivist solutions are designed and developed and not in the first place discovered, even though a lot of inspiration for many artifacts comes from nature, today especially evident in Natural and Organic Computing.

A construction, be it theoretical or a practical one, when it differs profoundly from anything previously existing, constitutes a new reality against which preexisting one can be examined and understood, so it has an undeniable epistemological value. E.g. constructing of a non-Euclidean geometry helps understand everyday Euclidean world as a special case. Many theories are constructivist in nature, and Einstein is known for his constructive approaches to Physics with the emphasis on the creative side of theory

---

[1] The meaning of terms Constructivist and Constructionist vary vastly, and one of often quoted distinctions (but far from the only one) is from Ackermann [4]. Also the content of Radical Constructivism is highly ambiguous such that for example Maturana is sometimes placed under Radical Realism, Critical Realism as well as Radical Constructivism.

building. Cognitive Science in its Info-Computational approach to human mind is one of the best examples of epistemological significance of constructive approaches in production of scientific knowledge.

The article will first describe Constructive Research Method, Constructivist Epistemology and Constructivist Learning Theory and make necessary distinction between Constructivism and Action Research. The awareness of the significance of Constructive Research appears to be lacking in both research education, Research Methodology and Philosophy of Computing.

To start with, let us first make some distinctions. The form of constructivism found in sciences and engineering is a *moderate* constructivism or *realist constructivism*, based on the realist ontological foundations. Realism in engineering consists in the conviction that the world exists independently of the observer and that it is knowable:

There is ample evidence that we still can adopt a critical realist outlook, even if every part of our world view is a construction. Cf. Saalmann [3].

Constructivism assumes creation of knowledge through interaction between the observer and the observed, and thus also recognizes the dependence of ontological constructs (all which exists and can exist) on the ways of interaction of the observer with the world. As a consequence, it also acknowledges social character of knowledge i.e. the fact that knowledge production occurs in networks of interacting agents[2]. Even though taking into account social aspects of construction of knowledge, Realist Constructivism does not go as far as some Radical Constructivisms as to deny the relevance of reality as the ultimate authority:

Appealing to reality as the ultimate arbiter of (scientific) disputes gives rise to the belief that there exists a mind-independent reality (MIR) which defines what is true and what is not. Cf. Riegler [5].

Realist Constructivism insists on the existence of stable world with which we interact and thus build increasingly rich knowledge about. The problem with the denial of the relevance of reality is that it conflates two claims, the ontological and the epistemological one. Reality *exists* independently of any mind, but it is mind-dependent in the epistemological sense and (re)constructed by cognizing agents through their interactions with the physical world and with other agents. The requirement for reproducibility of experiments under corresponding conditions (measurements, methods, interpretations) provides unambiguous and repeatable connections between an observer and the observed reality. The fundamental feature of reality of natural phenomena including humans as natural beings, is its remarkable stability, which is the basis of Realist Constructivism. That agrees with the following von Foerster's analysis:

---

[2] In this context the distinction must be made between Social Constructionism which concerns the phenomena related to social contexts (sociological constructs) and Social Constructivism with focus on individuals knowledge within a social context (psychological or cognitive construct).

The most we can say, therefore, is that the observer generates a description of the domain of reality through his or her interactions (including interactions with instruments and through instruments), and that the observer can describe a system of systems (a system of consensus) that leads to the emergence of systems that can describe: observers. As a consequence, because the domain of descriptions is closed, the observer can make the following ontological statement: The logic of the description is isomorphic to the logic of the operation of the describing system. Cf. von Foerster [6].

When it comes to Kant's [7] dichotomy between "phenomena" (things as they appear which constitute the world of common experience, which is illusion[3]) and "noumena" (things in themselves, constituting a transcendental world to which we have no empirical access, and which is reality) one can wonder: were X-rays in ancient Greece part of nuomena? Isn't the space of nuomena shrinking as we learn to observe the world in the domains to which our common experience does not have access?

Historically, we learned that reality is a resource much richer than what one human epoch can realize. Our knowledge about the world including ourselves is constantly growing. In the reality of ancient Greece there were no cell phones, no quantum mechanics, and microscopes, no quasars and DNA, no movies and TV, no process algebra, but the laws of mechanics and astronomy were the same and even human physiology and psychology, for all we know. If we look back and see how all those new discoveries and inventions came into being, we find that it is through efforts of generations of scientists and engineers who worked under assumptions of the mind-independent existence of knowable reality. Even though our interactions and new conceptualizations change our relationships with the world, all the constructive work of research is always constrained by the laws of the world itself through its elements that exist independently of mind and from which all the constructing work starts.

Devices and model systems are what socio-cultural studies of science refer to as the "material culture" of the community, but are also what cognitive studies of science refer to as "cognitive artifacts" participating in the representational, reasoning, and problem-solving processes of a distributed system. Our data lead to their interpretation as cognitive-cultural artifacts. They are representations and thus play a role in model-based reasoning and problem-solving; they are central artifacts around which social practices form; they are sites of learning; they connect one generation of researchers to another; they perform as cultural ratchets [...] in an epistemic community [...], enabling researchers to build upon the results of the previous generations, and thus move the problem solving forward [8].

In a similar way as physical reality presents fundamental constraints for knowledge construction, a community of practice (epistemic community) puts additional constraints through the process of interaction. Indeed, scientific work is in many ways defined by the interactions between members of a

---

[3] Later on I will give the example of the observation of a very distant quasar that appears to us as it was millions of years ago.

research community, but its role is often misunderstood by postmodern critics. Science is not just another narrative or just a different myth. It is not less stringent because it develops as a result of a collective effort, just on the contrary. It is a strictly regulated social system for knowledge production, and the result is, even though not absolute, still our best existing knowledge.

3   Characteristics of Constructive Research

The whole activity of manipulation is devoted to build various external epistemic mediators that function as an enormous new source of information and knowledge. Therefore, manipulative abduction represents a kind of redistribution of the epistemic and cognitive effort to manage objects and information that cannot be immediately represented or found internally (for example exploiting the resources of visual imagery). If we see scientific discovery like a kind of opportunistic ability of integrating information from many kinds of simultaneous constraints to produce explanatory hypotheses that account for them all, then manipulative abduction will play the role of eliciting possible hidden constraints by building external suitable experimental structures. Cf. Magnani [9].

Constructive research method implies building of an artifact (practical, theoretical or both) that solves a domain specific problem in order to create knowledge about how the problem can be solved (or understood, explained or modeled) in principle. Constructive research gives results which can have both practical and theoretical relevance. The research should solve several related knowledge problems, concerning *feasibility*, *improvement* and *novelty*. The emphasis should be on the theoretical relevance of the construct. What are the elements of the solution central to the benefits? How could they be presented in the most condensed form? [10, 11].

4   Research within Computing

In the beginning, Computing was accepted with skepticism among sciences but successively it is becoming increasingly important and presently in many respects it is seen as the prototype of an Ideal Science, thus replacing Physics [1]. The reason is the ability of Computing to provide a common language for not only sciences but also other forms of scholarship and arts. It tends to reflect the totality of human interests as information migrates from the physical world into the virtual world of computer networks.

No other field today presents such a lingua franca for all of our knowledge and agency as Computing. Concepts and paradigms from Computing are spreading "like a wild fire" (Cantwell Smith) through other fields. As Computing is a very wide area, including computer Science, Computer Engineering, Software Engineering and Information Systems, in what follows, we will concentrate on Software Engineering, SE. Classification of the issues of the SE discipline into Engineering and Scientific was proposed by [12] and [13] with two different kinds of objects of study: problems of building new software artefacts in Engineering, and the theoretical basis of the SE in Science.

Therefore, Engineering research differs greatly from traditional scientific research because while Sciences deal with the study of existing objects and phenomena, be it physically, metaphysically or conceptually, Engineering is based on how to do things, how to create new objects. Cf. Lázaro and Marcos [12].

In the above sense, constructive research is what suits both Engineering and Science, the former in the creation of new objects from the existing ones and the latter in the formation of new concepts and models.

5 Software Engineering Research

The received view is that methodology of engineering as a research field and Software Engineering in particular is fundamentally constructive. Most Software Engineering research groups work predominantly in 'construction mode', inventing new models and tools. Nevertheless, little effort has been invested into understanding of mechanisms of knowledge production in Constructive Research by means of distributed cognition and use of cognitive artifacts.

Constructive research takes off from the existing well understood ground and that is why research in Software Engineering often starts with empirical investigations where quantitative (Controlled experiment, Survey) or qualitative (Grounded Theory, Case studies) methods are used prior to the constructive work. Only when sufficient understanding of the research problem and the domain is obtained, one can start addressing a Software Engineering problem by Constructive Research method.

Both Grounded Theory and Action Research can be applied to the phase in engineering research/design research/economic research when artifact which is central for the first design or prototype phase is already done, when a new construct is employed in a social context (group of users applying a programming or theoretical tool or other artifact previously developed).

However the main knowledge production within engineering usually happens in the construction phase of the research. The major task of an engineer or a scientist who constructs an artifact is *the process of construction itself*, and it needs to be studied with the same rigor and should be given at least as much attention from the methodological and Philosophy of Science point of view as Grounded Theory and Action Research perspective of the same research project which address the implementation of the artifact (construct).

The idea of predominantly constructive nature of engineering exists but it is not clearly understood and the common view often identifies general methodology of Constructivism with radical post-modernist Social Constructivism which is not well received among scientists and engineers who take it to be hostile to sciences and engineering. Moreover, the role played by constructive research is an under-researched topic in the Theory of Science. New studies such as Nersessian [8] are important for recognition of research practices with constructive approaches and it is evident that a lot of work remains to be done. On the level of Philosophy and Methodology of Science, understanding is needed of what sort of knowledge can be produced by constructive methods and how it relates to knowledge produced by other research traditions, what the contribution of a constructive knowledge is and how it is justified.

6 Design Research as Constructive Research

Design research is often present as an important part of research within Engineering, Computer Science and Information Systems. It involves the analysis of the use and performance of designed artifacts (constructs) in order to understand, explain and improve designed systems. The outputs of Design Research are constructs, models, methods, theories, instantiations, algorithms, human-computer interfaces, system design methodologies, languages and other artifacts, [14].

Design Research as constructive research presents a bridge between natural and human spheres as it produces artifacts which are both natural and intentional. That implies understanding both of the workings of basic mechanisms (as found in sciences) and the role which a given construct may play in the broader context (societal aspects).

7 Constructive Research vs. Action Research

Constructivist epistemology emphasizes the fact that scientific knowledge is *constructed* by scientists with help of cognitive tools. It is the opposite of the *positivist epistemology* which sees scientific knowledge as *discovered* in the world. For a classical positivist, scientific facts are discovered and the connection between the world and the fact is unique. On the other hand constructivism entails that there is no single valid methodology for construction of scientific knowledge, so no unique prescription to establish "the facts" or provide the data, and no guarantee for a consensus. One can say that constructivism is more interested in the mechanisms of theory building [15] while positivism describes the steady state of theory where one dominant framework has been established among competing approaches.

Often mentioned is also the alleged opposition between *Constructivism* and *Realism*. Both come in two flavors – *ontological* and *epistemological*. The confusion among varieties of claims is vast. In short, constructivist approaches are primarily applied in the process of discovery where many possibilities are still open, in the sense of ontological choices, during concepts formation [16] and in the sense of epistemological approaches. In the new research, the reality appears as malleable and negotiable as an uncharted territory. Similar to the process of reinforcement learning in the brain, learning in the network of agents causes that those paths that are taken repeatedly get reinforced, and those that are not followed fade out – so finally in a long time perspective some approaches win and some get forgotten. That is a territory of steady state in which it can appear as if there is only one clear-cut approach to the phenomenon or a solution to the problem. As already mentioned, a received view in sciences and engineering is ontological realism in conjunction with epistemological constructivism.

8 Constructivism vs. Constructionism

Among numerous terminological confusions, the one around the distinctions between Constructivism vs. Constructionism is often found. We follow Ackermann who succinctly characterizes the two. Unlike some other authors she uses both terms when it comes to research practices as well as learning theories.

Piaget's *constructivism* offers a window into what children are interested in, and able to achieve, at different stages of their development. The theory describes how children's ways of doing and thinking evolve over time, and under which circumstance children are more likely to let go of – or hold onto – their currently held views. Piaget suggests that children have very good reasons not to abandon their worldviews just because someone else, be it an expert, tells them they're wrong. Papert's *constructionism*, in contrast, focuses more on the art of learning, or 'learning to learn', and on the significance of making things in learning. ... Integrating both perspectives illuminates the processes by which individuals come to make sense of their experience, gradually optimizing their interactions with the world. Cf. Ackermann [4] (emphasis mine).

9 Action Learning and Action Research

Action learning is a method where learners improve their knowledge and skills through actions, it is learning-by-doing and by means of examples and exercises. Often, action learning is performed in small groups (learning sets). In the same way as Constructive research is related to Constructive learning, Action research is related to Action learning.

Action-research, a comparative research on the conditions and effects of various forms of social action and research leading to social action that uses "a spiral of steps, each of which is composed of a circle of planning, action, and fact-finding about the result of the action. Cf. Lewin [17].

What distinguishes this type of research from general professional practices, consulting, or ordinary problem-solving is the emphasis on scientific approach which includes systematicity and interventions based on theoretical considerations. During the process, researcher is refining the methodological tools while collecting, analyzing, and presenting data on an ongoing basis simultaneously getting concerned people involved into research. The research takes place in real-world situations, and aims to solve real-world problems so it has a social dimension too. Action researchers do not try to remain objective, but recognize their bias to the other participants. Oquist [18] analyzes epistemological positions of Action Research and presents the reasons for its rejection by Empiricism, Logical Positivism and Structuralism and the reasons for its acceptance by Pragmatism.

Action Research may be one of the tasks of an engineer or scientist constructing an artifact. It may be used for validation of the results, and in the study of the research processes themselves. See for example Shaw [19] who exemplifies validation of a construct in Software Architecture Research. Benavides et al. [20] suggest using Action Research in Software Engineering as a method of resolving what they call "the triple schizophrenia of the Software Engineering researcher" – their split between researching, teaching and learning activities. This however represents a meta-level with respect to knowledge generation studies of constructive research mechanisms.

10 Grounded Theory

Borgatti defines Grounded Theory as a method of using empirical data by inductive structuring of observations without preconceived theories, in contrasts to theory derived deductively from the existing theories.

"Constant comparison is the heart of the process... Theory emerges quickly. When it has begun to emerge you compare data to theory"[4].

Notice that the above process of structuring of data is inherently constructive. In sciences and engineering this process is present in a construction of a model or an artifact, and is seldom done *ex nihilo*. Typically there is a starting point in the existing background knowledge, models and artefacts which present important constraints in the process of construction. In the research process of the design of an artifact of central interest is how this designed construction evolves and how the artifact relates to its users.

11 Computational Models and Simulations as Knowledge Production Tools

Mathematical modeling and numerical simulation are extensions of the traditional empirical and experimental approaches. They provide effective ways for virtual experimentation when real experiments are impractical or just too expensive. They enable study of the existing data, identify critical areas where data are missing, facilitate hypothesis generation, and similar.

---

[4] http://scu.edu.au/schools/gcm/ar/arp/grounded.html

Simulations can be used to predict the behavior of the system for specific choices of parameters and/or initial conditions [21].

An experiment with an artifact model provides an extension of the scientific practice of thought experimenting (. . .) – more complete and complex and potentially less subject to individual bias and error. It affords a more detailed representation, a wider range of manipulations, and more control than is possible for a thought experiment. In a manner similar to how microscopes and telescopes extend scientists' capacity to see, artifact simulation models extend their biological capacity to perform simulative reasoning. As with thought experimenting, the model system simulation contributes to predictions about how the in vivo system might perform under specified conditions. Cf. Nersessian [8].

Best computational methods are results of an exhaustive theoretical analysis. At the same time, the analysis of results is similar to the analysis of experimental data. Computational methods/simulations in science are very useful when the problem is too difficult to handle analytically, an approximate theoretical result might not be reliable or an experiment is expensive or not feasible to perform. Ultimately it is of course experiment which decides on the nature of the real-world phenomena, but computational methods present very effective cognitive tools.

The problem-solving practices of engineering scientists provide an excellent resource for advancing the agenda of environmental perspectives because their simulation technologies are simultaneously cognitive and cultural artifacts. Within the confines of this paper attention is directed primarily to one aspect – the function of physical models as cognitive artifacts in distributed cognitive processes – but in fact the socio-cultural dimensions are always present if not attended to directly in this brief analysis. The epigraph quote from Daniel Dennett provides a pithy summary of a major premise of range of research under headings such as "distributed cognition", "the extended mind thesis", "situated cognition", and "activity theory", (. . .) Within this framing, to understand how problem solving is achieved requires examining the generation, manipulation, and propagation of salient representations within the system; that is, examining how representational states flow across media and accomplish cognitive work. Cf. Nersessian [8].

The central claim is that reasoning in this distributed cognition framework involves *co-processing of information in human memory and in the environment*, where artifacts are integrated into cognitive processing.

12   Research as Learning. Knowledge Generation as Information Processing

Furthermore, all researchers are learners. As we have found, the researchers and their technologies have intersecting developmental trajectories. Cf. Nersessian [8].

The Constructive Research Paradigm we suggest as closest to actual practices in problem solving and research within natural sciences and engineering is Realist Constructivism, or Grounded Constructivism, grounded in a sense that it ontologically presupposes the mind-independent *existence* of reality. (N.B. that is not to say that Grounded Constructivism presupposes existence of mind-independent reality which is a different and wrong claim.) In sum, Realist Constructivism is based on ontic *realism and epistemic constructionism*.

Following are characteristics of Realist Constructivism:

- The world exists and is knowable.
- All knowledge is embodied in agents.
- Knowledge is constructed via computational processing of information.
- Reality is informational in nature (Informational Structural Realism, Floridi)
- Dynamics of information is computation. (Pancomputationalism)

13 Info-Computationalist View of Knowledge Generation and Constructive Research Paradigm

Within Info-computational naturalism [22] knowledge is seen as a result of successive structuring of data, where data are understood as simplest information units, signals acquired by a cognizing agent through the senses/sensors/ instruments [23]. Information is meaningful data, which can be turned into knowledge by an interactive computational process going on in the agent. Potential information exists in the world even if no cognizing agents are present. The world (reality) for an agent presents potential information, both outside and within an agent. Knowledge, on the other hand, always resides in a cognitive agent. The interplay of an agent with the environment which leads to adaptation, clearly presents developmental advantage which increases agent's ability to cope with the changing environment.

An agent is a physical system, living organism or a robot possessing cognitive structures capable of adaptive behaviors. Living agents evolved successively starting with prebiotic evolution of increasingly complex molecules able to adapt to its environment and reproduce. Semantics of information is relative to an agent. Meaning that proto-information will take for a biological organism is in the use the agent (organism) has for it. Information is always embodied in a physical signal, molecule, particle or event which will induce change of structure or behavior in an agent. Semantics develops as data /information/knowledge structuring process, in which complex structures are self-organized by the computational processing from simpler ones. Meaning of information is thus defined for an agent and a group of agents in a network (ecology) and it is given by the use information has for them. Knowledge generation as information processing in biological agents presupposes natural computation, defined by MacLennan [24] as computation occurring in nature or inspired by that in nature, which is the most general current computing paradigm.

Taking information and computation as basic principles in a dual-aspect Info-Computational model, a common framework is formulated in [22, 25], providing the argument that knowledge generation should be understood as natural computation. Knowledge is a characteristic of intelligent agents. Intelligence in a living organism as well as its cognitive capability is a matter of degree. Artificial intelligent agents can be ascribed knowledge as long as they function as agents, exchanging information with the world, in analogy with an organism. In the same way as we do not ascribe knowledge to a dead living organism, we do not ascribe knowledge to a non-functional robot. Information, on the other hand can be stored in non-living systems, but for information to be knowledge, process of interaction between different pieces of information is necessary.

When we talk about computation in living organisms, models of computation beyond Turing machine are needed to handle complexity of phenomena such as creation, maintenance and reproduction of cognizing agency, perception, learning, intelligence, consciousness, social phenomena, etc. Nowadays, the source of inspiration for new models of computing is found in natural processes, and especially in organic computing seen as evolutionary development of intelligent self-organized processes and structures necessary for living organisms to survive in the complex and dynamical world.

All of our knowledge of the world is based on information we gain from the world, be it from direct experience (interaction with the world) or by learning (getting directly or indirectly information) from other people. Physicist Zeilinger [26] suggests possibility of seeing information and reality as one. This is in accord with Informational Structural Realism which says that reality is made of informational structures [27, 28] as well as with info-computational ontology [25] based on Informational Structural Realism (ISR). What Floridi assumes to be mind-independent data corresponds to proto-information (information in the world, world as information) of Info-computationalism. Reality is informational and mind-independent and consists of structural objects, which brings together metaphysical views of Wiener ("information is information, not matter or energy") and Wheeler ("it from bit").

14 The Computing Universe

The complementary basic concept of information is computation, which is the dynamics of informational structure. Floridi's Informational Structural Realism with information as the fabric of the universe, has as a logical consequence that the process of dynamical changes of the universe makes the universe a huge computational network. Here is how Chaitin describes the computing universe:

And how about the entire universe, can it be considered to be a computer? Yes, it certainly can, it is constantly computing its future state from its current state, it's constantly computing its own time-evolution! And as I believe Tom Toffoli pointed out, actual computers like your PC just hitch a ride on this universal computation! Cf. Chaitin [29].

Within Info-computationalism, computation is both symbolic and sub-symbolic[5] information processing, such as natural computation or organic computation. As it corresponds to the dynamic of processes that exist in the universe, it is necessarily both discrete and continuous. In the Info-computational naturalist framework, information and computation are two fundamental and inseparable elements necessary for naturalizing meaning, cognition and mind, cf. Dodig Crnkovic [22].

In sum: information is the structure, the fabric of reality. The world exists independently from us (realist position of structural realism) in the form of proto-information or ur-(in)formation, the richest potential form of existence corresponding to Kant's Ding an sich. That proto-information becomes information ("a difference that makes a difference" according to Bateson) for a cognizing agent in a process of interaction through which specific aspects of the world get uncovered. In science this process of successive transformation of manifold potential proto-information in the world into particular actual information in an agent becomes evident when for example observing the same physical object in different wavelengths[6]. The world as it appears to an agent is dependent on the type of interaction through which the agent acquires information. Potential information in the world is obviously much richer than what we observe, containing invisible worlds of molecules, atoms and sub-atomic phenomena, distant cosmological objects and similar. Our knowledge about this proto-information which reveals with help of scientific instruments will surely continue to increase with the development of new devices i.e. the new ways of interaction with the world [2].

Computation and information are the backbone of sciences. They are inseparable – there can be no structure without a process (a structure actualizes through a process) and the vice versa, there is no process without a structure to act on, [22, 23]. Formalization of info-computational approach within category theory may be found in Burgin [30].

15   Information and Computation in Biological and Intelligent Artificial Systems

Recent studies in biology, ethology and neuroscience, which have increased our knowledge of biological cognitive functions, have led to the insight that

---

[5] Sub-symbolic computations go on in neural networks.

[6] Results of observations of the same physical object (celestial body) in different wavelengths (radio, microwave, infrared, visible, ultraviolet and X-ray), see http://bb.nightskylive.net/asterisk/viewtopic.php?f=8{\&}t=14561

the most important feature of cognition is its ability to deal efficiently with complexity [31]. Insights into natural intelligence, together with the increase in power of electronic computing bring us closer to the modeling of intelligent behavior. Artificial intelligence (AI) is based on the belief that intelligent behavior can be understood in such a way that a machine can be constructed able to simulate it [23]. In the same paper it is argued that Info-computationalism is the most appropriate theoretical framework for understanding of the phenomena we call intelligence. From the computationalist point of view intelligence may be seen as based on several levels of data processing [32, 33] in a cognizing agent. As AI attempts to construct non-living intelligent agents, we include even AI agents in the discussion. Information (produced from sensory data processed by an agent) can be understood as an interface between the data (world) and an agent's perception of the world, [34]. Patterns of information should thus be attributed both to the world and to the functions and structures of the brain. In an analogous way, knowledge can be understood as an interface between perception and cognition. Structures of knowledge are related both to percepts (sensory information) and to the brains cognitive functions and organization. Meaning and interpretation are the results of the processes of temporal development of information, and its refinement by relating to already existing memorized information. The meaning of an object in the external world is recognized through the process of perception of sensory signals, their processing through nervous system, comparison with memorized objects, anticipation from memorized experiences of possible scenarios in which the object played different roles, etc.

Data, information, perceptual images and knowledge are organized in a multiresolutional (multigranular, multiscale) model of the brain and nervous system [32]. Multiresolutional representation has proven to be a good way of dealing with complexity in biological systems, and they are also being implemented in AI [33].

Cognitive robotics research presents us with a sort of laboratory where our understanding of cognition can be tested in a rigorous manner. From cognitive robotics it is becoming evident that intelligence is closely related to agency. Anticipation, planning and control are essential features of intelligent agency. A similarity has been found between the generation of behavior in living organisms and the formation of control sequences in artificial systems.

Self-organization, self-reproduction and self-description (or self-representation) are fundamental intrinsic properties of natural intelligent systems. More and more of intelligent robotics builds on similar basic principles called *organic computing*. Learning is an essential part of each of the above three capabilities and it requires among others the development of a symbolic system which is easy to maintain and use. It is possible to build intelligent control systems that can collect and process information, as well as generate and control behavior in real time, and cope with situations that evolve among the complexities of the real world.

In intelligent biological systems based upon a hierarchy of functional loops, each of these loops can be treated as a control system per se [32]. Generation of structures resulting from sensory processes (data), and information organized into knowledge to be used in decision making are built in a multiresolutional way, with many pattern recognition and control mechanisms hardwired.

Along with the above constructive approach to AI with roots in engineering, logic and programming, there is a different (also constructive) one which tries to implement in silico info-computational biological functions translating biological structures starting on molecular level – notably the Blue Brain Project of ETF which will be discussed later on. Common to both approaches is the belief that intelligent behavior can be implemented in a machine (non-biologically). Up to now there are only specific aspects of intelligence (weak AI) that machines can possess, while general AI seems still a far-reaching goal, after more than fifty years of research.

16   Knowledge Generation as Natural Computation

Present day's computers perform syntactic mechanical symbol manipulation which in the future has to be extended to include information processing with semantic aspects. Burgin [35] identifies three distinct components of information processing systems: hardware (physical devices), software (programs that govern its functioning), and infoware which represents information processed by the system. Infoware is a shell built around the software-hardware core which is the traditional domain of automata and algorithm theory. Semantic Web is an example of infoware. Bio-computing and especially Organic computing are expected to substantially contribute to the infoware by adopting strategies for handling of complexity found in the organic world.

Classical theoretical models of computers are mathematical objects equivalent to abstract automata (Turing Machines) which are equivalent to algorithms/effective procedures, recursive functions, or formal languages. Turing Machine is a logical device, a model for execution of algorithms. However, if we want adequately to model biological structures and processes understood as embodied physical information processing, highly interactive and networked computing models beyond Turing Machines are needed. In order to develop general theory of the networked physical information processing, we must also generalize the ideas of what computation is and what it might be. For new computing paradigms, see [36, 37, 38, 22]. Compared with new computing paradigms, Turing machines form the proper subset of the set of information processing devices.

Naturalized epistemology according to [39, 40] is an idea that knowledge should be studied as a natural phenomenon. The subject matter is not our concept of knowledge, but the knowledge itself as it appears in the world[7]. An evolving population incorporates information about its environment through changes by natural selection, as the state of the population matches the state of the environment. The variants which adapt and survive in the environment increase in frequency in the population. Harms [39] proves a theorem showing that natural selection will always lead a population to accumulate information, and so to "learn" about its environment. Okasha [41] points out that any evolving population "learns" about its environment, in Harms' sense, even if the population is composed of organisms that lack minds entirely, hence lack the ability to have representations of the external world at all.

That is correct, and may be seen not as a drawback but as strength because of the generality of naturalistic approach. It shows how mind is a matter of degree and how it slowly and successively develops with evolution. There is fresh evidence that even simple "lifeless" prion molecules are capable of evolutionary change and adaptation [43].

However, this understanding of basic evolutionary mechanisms of accumulating information at the same time increasing information processing capacities of organisms (such as memory, anticipation, computational efficiency) is only the first step towards a full-fledged evolutionary epistemology, but the most difficult and significant one. From bio-computing we learn that in living organism biological structure (hardware) is at the same time a program (software) which controls the behavior of that hardware.

By autopoetic process [42, 43], biological system changes its structures and thus the information processing patterns in a self-reflective, recursive manner. Natural selection of organisms, responsible for nearly all information that living systems have built up in their genotypes and phenotypes, is a simple but costly method to develop knowledge capacities. Higher organisms (which are "more expensive" to evolve) have grown learning and reasoning capability as a more efficient way to accumulate knowledge. The step from "genetic learning" (typical of more primitive life forms) to acquisition of cognitive skills on higher levels of organization of the nervous system (such as found in vertebrata) will be the next step to explore in the project of naturalized epistemology.

17 Why Our Perception of the World Is an Illusion

Of all information processing going on in our bodies, perception is only a tiny fraction. Our perception of the world depends on the relative slowness of conscious perception. Time longer than one second is needed to synthesize conscious experience. At time scales shorter than one second, the fragmentary

---

[7] Maturana was the first to suggest that *knowledge is a biological phenomenon*. He argued that life should be understood as a process of cognition which enables an organism to adapt and survive in the changing environment.

nature of perception reveals. The brain creates a picture of reality that we experience as (and mistake for) "the actual thing" (Ballard [45]).

As already mentioned, Kant argued that "phenomena" or things as they appear and which constitute the world of common experience are an illusion. Kaneko and Tsuda discuss why.

Hence the brain does not directly map the external world. From this proposition follows the notion of "interpreting brain", i.e. the notion that the brain must interpret symbols generated by itself even at the lowest level of information processing. It seems that many problems related to information processing and meaning in the brain are rooted in the problems of the mechanisms of symbol generation and meaning. Cf. Kaneko and Tsuda [44].

Consciousness provides only a rough sense of what is going on in and around us, in the first place what we take to be essential for us. The world as it appears for our consciousness is a sketchy simulation. Belief that we ever can experience the world 'directly' is the biggest illusion [46].

What would that mean anyway to experience the world "directly as it is", without ourselves being part of the process? Who would experience that? It is important to understand that, as Kaneko and Tsuda [44] emphasize, the brain maps the information about the (part of the) world into itself, but the mapped information is always affected by the activity of the brain itself. This seems to be the view of Maturana [47] as well. The question of what is reality "an sich" in the sense of proto-information and understanding of our interactions with the world outside through the (conscious and sub-conscious) exchange of information is fundamental, and we have to keep searching for a deepened understanding of that reality in relation to ourselves. To that end, the awareness of the presence of models (simulations) in understanding (and perception) of the world is essential.

18 Science. The World and a Model. Real and Virtual

If what we perceive of the world is a simulation our brain plays for us in order to manage complexity and enable us to act efficiently, then our understanding of the world must also be mediated by this modeling nature of cognition. Not even the most reliable knowledge about the physical world as it appears in sciences is independent of the modeling frameworks which indirectly impact what can be known. The common positivist optimism about observations independent of the observer proved problematic in many fields of physics such as quantum mechanics (wave function collapse after interaction), relativity (speed dependent length contraction and time dilatation) and chaos (a minor perturbation sufficient to switch to a different attractor). In general, observer and the systems observed are related and by understanding their relationship we can gain insights into limitations and power of models and simulations as knowledge generators [6].

Models are simplified representations, made for a purpose and they ignore aspects of the system which are irrelevant to its purpose. The properties of a system itself must be clearly distinguished from the properties of its models. All our knowledge of systems is mediated by models. Engineers using different models often get so familiar with the model and its functions that they frequently act as if the model was the actual reality itself [48].

Awareness of the modeling character of knowledge and the active role of the cognizing agent in the process of generation of knowledge is specifically addressed by second order cybernetics. Cybernetic epistemology is constructivist in recognizing that knowledge cannot be passively transferred from the environment, but must be actively constructed by the cognizing agent based on the elements found in the environment in combination with information stored in the agent. The environment eliminates inadequate models, which are getting into conflict with it. Model construction thus proceeds through variation and selection. This agrees with von Glasersfeld's [50] two basic principles of constructivism:

Knowledge is not passively received either through the senses or by way of communication, but is actively built up by the cognizing subject.

The function of cognition is adaptive and serves the subject's organization of the experiential world, not the discovery of an "objective ontological reality".

Finally in words by Kaneko and Tsuda:

> The key question that arises in this 'constructive' approach lies in the relationship between the virtual world (the model) and reality. The virtual world should not just be an imitation of reality, but a sort of abstraction from reality, and be constructed from our side by utilizing some abstracted essential features of reality. Understanding the relationship between the virtual world and reality is a fundamental issue in the study of complex systems with a constructive approach. [44]

19   The Blue Brain Project as an Example of Constructive Research in Fundamental Science

The Blue Brain Project[8] studies functions of the brain modeled as consisting of neocortical columns, NCC which are structural units in the cerebral cortex. Millions of NCCs compose the brain. At present stage Blue Brain Project is attempting to accurately model a single NCC of a young rat brain. Project has succeeded to create an artificial column that responds to electrical impulses in the same way as the biological one. The model of an NCC is built down to the molecular level and simulates the real-world electrical activity of rat brain tissue. The Blue Brain Project group hopes eventually to use the same simulation approach just scaled up to model the whole rat brain and

---

[8] http://bluebrain.epfl.ch/

ultimately even the human brain. Currently the Blue Brain runs on IBM's Blue Gene supercomputer. The present computing power is just enough to simulate a single cortical column, while modeling millions of columns is far beyond the scope of computational capacity of a Blue Gene, so new solutions must be found to increase the computational power.

An interesting question is what phenomena besides simultaneous firing of neurons may be expected to emerge in the simulation. It is speculated that among emergent properties of the whole brain simulation consciousness will emerge when rat brain simulation is connected in sensorimotor loops with a robotic rat. This presents an eminently constructivist program for understanding consciousness and the brain itself by re-constructing it in silico.

The Blue Brain project is in many respects an example of constructive science at its best with classical elements such as reproducibility, third person perspective and meticulously performed reconstruction of biological neurons computationally simulated in a great detail on the Blue Gene supercomputer.[9] The big triumph of the Blue Brain project was a spontaneous firing of simulated neurons when given electrical stimuli.

The way knowledge is generated in Blue Brain Project is typically info-computational, with simulation representing epistemological laboratory of a kind in which computational models are tested against biological "reality". The prominent role of computerized visualization is also characteristics of Info-computationalism as well as a mix of knowledge from many different disciplines – from Biology, Neuroscience to Computer Science implemented in info-computational tools. Among others an advanced patch clamp robot technique is used contributing to huge effectivization by automatization of experimental procedure, through combining functions of a microscope with computer-assisted reconstruction of neuron structure.

20   Conclusions

When writing chapter on methodology in dissertations and thesis work within Software Engineering it is common practice to refer to Empirical Methods, Grounded Theory and Action Research. Attention is seldom paid to the analysis of Constructive Research methods which are fundamental for all engineering and sciences when it comes to concept formation, modeling and the use of artifacts. This article argues for the necessity of changing of this practice and focusing on the analysis of constructive research practices, as they lie at the core of the knowledge production.

Justification for the claims about necessary constructive character of all knowledge is found in Info-Computational approach which provides models of knowledge generation on a fundamental level of information processing in a cognizing agent.

---

[9] The same constructive strategy in developing scientific understanding is described by Kaneko and Tsuda in the case of study of chaotic systems [44].

References

1. Dodig-Crnkovic, G.: Shifting the paradigm of the philosophy of science: the philosophy of information and a new renaissance. Minds and Machines 13, 521–536 (2003), http://www.springerlink.com/content/g14t483510156726
2. Dodig-Crnkovic, G., Müller, V.: A dialogue concerning two world systems: Info-computational vs. mechanistic. In: Dodig-Crnkovic, G., Burgin, M. (eds.) Information and Computation. Series in Information Studies. World Scientific Publishing Co., Singapore (2010) (forthcoming)
3. Saalmann, G.: Arguments opposing the radicalism of radical constructivism. Constructivist Foundations 3, 16–18 (2007)
4. Ackermann, E.: Piaget's constructivism, Papert's constructionism: What's the difference? In: Constructivism: uses and perspectives in education, vol. 1, 2, pp. 85–94, Geneva, Research Center in Education (2001)
5. Riegler, A.: Towards a radical constructivist understanding of science. Foundations of Science 6(1), 1–30 (2001)
6. von Foerster, H.: Understanding Understanding: Essays on Cybernetics and Cognition. Springer, Heidelberg (2003)
7. Kant, I.: Critique of pure reason. St Martin's Press, New York (1965)
8. Nersessian, N.J.: How do engineering scientists think? Model-based simulation in biomedical engineering laboratories. Topics in Cognitive Science 1, 730–757 (2009)
9. Magnani, L.: Model-based and manipulative abduction in science. Foundations of Science 9, 219–247 (2004)
10. Lukka, K.: The constructive research approach. In: Ojala, L., Hilmola, O.-P. (eds.) Case study research in logistics. Publications of the Turku School of Economics and Business Administration. Series B 1, pp. 83–101 (2003)
11. Kasanen, E., Lukka, K., Sintonen, A.: The constructive approach in management accounting research. Journal of Management Accounting Research 5(1), 243–263 (1993)
12. Lázaro, M., Marcos, E.: Research in software engineering: Paradigms and methods. In: J. Castro, E. Teniente (eds.) Advanced Information Systems Engineering, 17th International Conference, pp. 517–522. CAiSE 2005, Porto, Portugal, Proceedings of the CAiSE 05 Workshops, 2 (2005)
13. Lavrishcheva, E.M.: Software engineering as a scientific and engineering discipline. Cybernetics and Systems Analysis 44 (2008)
14. Vaishnavi, V., Kuechler, W.: Design Research in Information Systems (2004), http://ais.affiniscape.com/displaycommon.cfm?an=1&subarticlenbr=279
15. Magnani, L., Nersessian, N., Thagard, P. (eds.): Model-Based Reasoning in Scientific Discovery. Plenum, New York (2000), Chinese translation published by China, Science and Technology Press
16. Thagard, P.: Conceptual Revolutions. Princeton University Press, Princeton (1992)
17. Lewin, K.: Resolving Social Conflicts. In: Lewin, G.W. (ed.) Selected Papers on Group Dynamics, Harper & Row, New York (1948)
18. Oquist, P.: The epistemology of action research. Acta Sociologica 21, 143–163 (1978)
19. Shaw, M.: The coming-of-age of software architecture research. In: Proceedings of ICSE-2001, pp. 657–664. IEEE Computer Society Press, Los Alamitos (2001)

20. Benavides, D., et al.: The triple schizophrenia of the software engineering researcher. In: Castro, J., Teniente, E. (eds.) Advanced Information Systems Engineering, 17th International Conference, Proceedings of the CAiSE 2005 Workshops, vol. 2, pp. 529–534 (2005)
21. Stauffer, D., de Oliveira, S.M., de Oliveira, P.M.C., Sa Martins, J.S.: Biology, Sociology, Geology by Computational Physicists. Monograph Series on Nonlinear Science and Complexity. Elsevier, Amsterdam (2006)
22. Dodig-Crnkovic, G.: Information and Computation Nets. VDM Verlag (2009)
23. Dodig-Crnkovic, G.: Knowledge generation as natural computation. Journal of Systemics, Cybernetics and Informatics 6 (2008)
24. MacLennan, B.: Natural computation and non-Turing models of computation. Theoretical Computer Science 317, 115–145 (2004)
25. Dodig-Crnkovic, G.: Investigations into Information Semantics and Ethics of Computing. Mälardalen University Press (2006)
26. Zeilinger, A.: The message of the quantum. Nature 438, 743 (2005)
27. Floridi, L.: Against digital ontology. Synthese 168, 151–178 (2009)
28. Floridi, L.: A defence of informational structural realism. Synthese 161, 219–253 (2008)
29. Chaitin, G.: Epistemology as information Theory. In: Dodig-Crnkovic, G., Stuart, S. (eds.) Computation, Information, Cognition – The Nexus and The Liminal, pp. 2–18. Cambridge Scholars Publishing (2007),
http://www.cs.auckland.ac.nz/~chaitin/ecap.html
30. Burgin, M.: Information dynamics in a categorical setting. In: Dodig-Crnkovic, G., Burgin, M. (eds.) Information and Computation. Series in Information Studies, World Scientific Publishing Co., Singapore (2010) (forthcoming)
31. Gell-Mann, M.: The Quark and the Jaguar: Adventures in the Simple and the Complex. Owl Books (1995)
32. Minsky, M.: Interior grounding, reflection, and self-consciousness. In: Dodig-Crnkovic, G., Burgin, M. (eds.) Information and Computation. Series in Information Studies. World Scientific Publishing Co., Singapore (2010) (forthcoming)
33. Goertzel, B.: The Evolving Mind. Gordon and Breach (1993)
34. Hoffman, D.: The interface theory of perception: Natural selection drives true perception to swift extinction. In: Object Categorization: Computer and Human Perspectives. Cambridge University Press, Cambridge (2009)
35. Burgin, M.: Super-Recursive Algorithms. Springer Monographs in Computer Science (2005)
36. Wegner, P.: Interactive foundations of computing. Theoretical Computer Science 192, 315–351 (1998)
37. Kampis, G.: Self-Modifying Systems in Biology and Cognitive Science: A New Framework for Dynamics, Information and Complexity. Pergamon Press, Oxford (1991)
38. Kurzweil, R.: The Singularity is Near. Viking, New York (2005)
39. Harms, W.F.: Information and Meaning in Evolutionary Processes. Cambridge University Press, Cambridge (2004)
40. Harms, W.F.: Naturalizing epistemology: Prospectus 2006. Biological Theory 1, 23–24 (2006)
41. Okasha, S.: Review of William F. Harms, information and meaning in evolutionary processes. Notre Dame Philosophical Reviews 12 (2005)

42. Maturana, H., Varela, F.: Autopoiesis and cognition: The realization of the living. In: Cohen, R.S., Wartofsky, M.W. (eds.) Boston Studies in the Philosophy of Science, vol. 42. D. Reidel Publishing, Dordecht (1980)
43. Maturana, H., Varela, F.: The Tree of Knowledge. Shambala (1992)
44. Kaneko, K., Tsuda, I.: Complex Systems: Chaos and Beyond. A Constructive Approach With Applications in Life Sciences. Springer, Heidelberg (2001)
45. Ballard, D.: Our perception of the world has to be an illusion. Journal of Consciousness Studies 9, 54–71 (2002)
46. Nørretranders, T.: The User Illusion: Cutting Consciousness Down to Size, translated by Jonathan Sydenham. Viking, New York (1999)
47. Maturana, H.: Systemic versus genetic determination. Constructivist Foundations 3, 21–26 (2007)
48. Heylighen, F., Joslyn, C.: Cybernetics and second order cybernetics. In: Meyers, R.A. (ed.) Encyclopedia of Physical Science & Technology, 3rd edn., vol. 4, pp. 155–170. Academic Press, New York (2001)
49. Scripps Research Institute.: Lifeless prions capable of evolutionary change and adaptation. ScienceDaily (January 3, 2010),
http://www.sciencedaily.com/releases/2009/12/091231164747.htm
(Retrieved January 12, 2010)
50. von Glasersfeld, E.: Radical constructivism: A way of knowing and learning. Falmer Press, London (1995)

Emergent Semiotics in Genetic Programming and the Self-Adaptive Semantic Crossover

Rafael Inhasz and Julio Michael Stern

Abstract. We present SASC, Self-Adaptive Semantic Crossover, a new class of crossover operators for genetic programming. SASC operators are designed to induce the emergence and then preserve good building-blocks, using meta-control techniques based on semantic compatibility measures. SASC performance is tested in a case study concerning the replication of investment funds.

1 Introduction

Genetic Programming (GP) are evolutionary algorithms that work on populations, whose individuals represent possible (viable) solutions to the optimization problem, see [2] and [8, 9]. The solution functions, code or programs defining an individual are its *genotype*, while the image, graph or output of these functions are the individual's *phenotype*. An *adaptation*, *cost* or *fitness* function, computed from an individual's phenotype, represents the objective function of the optimization problem.

GP are meta-heuristics based on some key functions and operators inspired on evolution theories for biological species. *Reproduction* operators generate new individuals, the *children*, from existing ones, their *parent(s)*, hence expanding the population. *Mutation* operators act on single individuals, for asexual reproduction, while *crossover* operators act on pairs of individuals, for sexual reproduction. A mutation operation generates a random change in the parent's code. This change is usually small, but may have important consequences for the individual fitness, often bad, but sometimes good. A

Rafael Inhasz
Institute of Mathematics and Statistics, University of São Paulo, Brazil
e-mail: `rafael.inhasz@yahoo.com.br`

Julio Michael Stern
Institute of Mathematics and Statistics, University of São Paulo, Brazil
e-mail: `jstern@ime.usp.br`

L. Magnani et al. (Eds.): Model-Based Reasoning in Science & Technology, SCI 314, pp. 381–392.
springerlink.com   © Springer-Verlag Berlin Heidelberg 2010

crossover operation generates new children by swapping portions of their parents' codes at randomly selected *recombination points*.

Reproduction operators are random operators. However, they only introduce a limited amount of entropy (noise or disorder) in the process, making it possible for children to *inherited* many characteristics coded by their parents' genotype. GP starts from an initial population, that may be randomly generated. The population then evolves according to the random reproduction and selection stochastic processes. The entropy introduced at reproduction allows for creative innovation, while the selection processes induce learning constraints. Under appropriate conditions, after many generations (near) optimal individuals are likely to emerge in the population.

The *schemata theorem*, arguably the most characteristic result of GP theory, shows that, under appropriate conditions, the emerging optimal solutions naturally exhibit a hierarchical modular organization. Such modules are known as *genes, schemata* or *building blocks*, see [6, 11, 17, 19, 23]. In light of the Schemata theorem, it is easy to understand that efficient crossover operators must be compatible with, preserve, favor, or even induce the emerging modular structure. More efficient operators are less likely to break down existing building blocks during reproduction, an unfortunate event known in the literature as *destructive crossover*.

This paper presents a new crossover operator, named SASC or *Self-Adaptive Semantic Crossover*. SASC is based on *meta-control* techniques designed to guide the random selection of recombination points by a measure of *semantic compatibility* between the portions of code being swapped. It is important to realize that SASC's meta-control system is not hard-wired or pre-defined. On the contrary, it is an emerging feature, co-evolving with the population. The meta-control system is based on the history of each individual in the population. However, the required historical information, accumulated during the individual's evolutionary line, is very limited. Hence, its implementation only generates a minor computational overhead.

Section 2 gives a short review of genetic programming. Section 3 explains some meta-control concepts and defines SASC – the self-adaptive semantic crossover operator. Section 4 presents some ideas of semiotics and cognitive constructivism that inspired this line of research. Section 5 gives some implementation details and section 6 compares the performance of SASC and standard crossover operators at a case study concerning the replication of financial investment portfolios. Section 7 presents our conclusions and final remarks.

2 Genetic Programming in Functional Trees

In this and the following sections, we deal with GP in the context of functional trees. In this setting, the objective is to find the correct specification, the best functional form, or just a good emulation of a complex *target* function. The

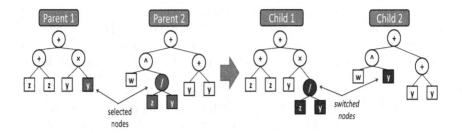

Fig. 1 Example of destructive crossover.

only information available about the target function is an input-output data-bank. An individual in the population is represented as a tree, with atoms at the leaves representing constants or input variables, and primitive operators at internal nodes. The root node output, at the top of the tree, expresses the individual's phenotype. Atoms and primitive operators are taken from finite sets, $A = \{a_1, a_2, \ldots\}$ and $OP = \{op_1, op_2, \ldots op_p\}$. Each operator, op_k, takes a specific number of arguments, $r(k)$, known as the arity of op_k.

Figure 1 shows four individuals in the population of a GP trying to emulate the target function, $f(w, y, z) = y^2 + w^{z/y}$, from the primitive set of expanded arithmetic operators, $OP = \{+, -, \times, /, \wedge\}$. Inputs at the leaves are represented in a square, and operators at internal nodes or at the root are represented in a circle. Figure 1 also shows a crossover, having the first two individuals as parents and the last two as children. The recombination points in the parent trees are highlighted. Notice that the first parent contains the component, partial solution or building block for the first term in the target function, y^2, while the second parent contains the building block for the second term, $w^{z/y}$. Since none of these interesting building blocks are preserved in the children, we call this a destructive crossover. A child inherits its root node, and hence usually most of its code, from the parent we call its *mother*, while from its *father* the child receives a, usually smaller, sub-tree. Hence, in this example, parent 1 and 2 are, respectively, mother and father of child 1, and father and mother of child 2.

Angeline, [1], proposed the SSAC – *Selective Self-Adaptive Crossover* – in order to make destructive crossovers less likely. Standard crossover selects recombination points in a parent tree with uniform distribution.

In SSAC like crossovers, each node, $n(i)$, stores a meta-control variable, ρ_i, a real number bounded to the normalization constraint: $0 \leq \rho_{min} \leq \rho_i \leq \rho_{max}$. The probability of selecting node $n(i)$ for recombination is proportional to ρ_i. That is, the probability of choosing node $n(i)$ as the recombination point in that tree is $p_i = \rho_i / \sum_j \rho_j$.

After a crossover, nodes at the children, carry along the meta-control variables they had at the parents, and afterwards suffer the effect of random noise. For example, the meta-control variable in node $n(i)$ can be updated as $\rho'_i = (1 + \mu_i + \sigma_i \epsilon)\rho_i$, where ϵ is the standard Normal random variable, μ_i is a zero or positive drift, and σ_i is a positive scale factor. All ρ_i are initialized at the minimum value, ρ_{min}, and allowed to move inside the normalization bounds. For details on Angeline's original implementation, see [1]. Several interesting variations can be found in the proceedings in the reference list.

The intuition behind SSAC is that survivors in the GP competition process are well adapted individuals, containing good building blocks. Moreover, successful breeders must be able to give these building blocks intact to their children. At these breeders, large meta-control variables should mark plausible building blocks, indicating good recombination points to be used (again) in the future. Genotype codes and meta-control variables should both co-evolve, facilitating the emergence, marking, and preservation of good building blocks.

Angeline [1] also presents an alternative method, SAMC – *Self-Adaptive Multi-Crossover*, where the meta-control variables can be interpreted as absolute probabilities, that is, $\rho_{max} = 1$. SAMC selects a recombination point in a two step process: First, all nodes in the tree receive a Boolean mark, 1 with probability ρ_i, and 0 otherwise. At the second step, the recombination point is selected from the nodes marked 1 with uniform distribution.

Before ending this section we make some additional comments about the schemata theorem. As already mentioned in the introduction, it is in the light of the schemata theorem that we can understand why efficient crossover operators must be compatible with, preserve, favor, or even induce the emerging modular structure. However, Holland's original theorem was stated for a very particular case, namely, genetic algorithms using string coded programs. *Schemata theories* extend this fundamental result to genetic programming using functional trees, see [15, 16, 18]. Hence, we must rely on Rosca, Poli and Langdon's results to keep our work on well founded theoretical ground.

3   The Self-Adaptive Semantic Crossover

SASC descends from Angeline's SSAC and SAMC operators, but it also incorporates information concerning the sub-trees rooted at the nodes in possible recombination points. The first information used for this purpose is captured through the notion of similarity. (Sub)Trees A and B are phenotypically similar if their output, computed at the records available on the data bank, agree within a specified tolerance.

We assume that two parents, father A and mother B, have been selected for crossover according to the mating distributions used at the GP. SASC starts by using a first heuristic procedure to define new meta-control variables, δ_i, at the nodes, $n(i)$, of the father, A. Let $A(i)$ be the sub-tree of A rooted at $n(i)$. For each sub-tree, $A(i)$, the procedure searches the mother, B, for sub-trees, $B(j)$, that are similar to and also either the same size or shorter than $A(i)$. If such a short similar sub-tree is found, $\delta_i = \rho_{min}$. Otherwise, $\delta_i = \rho_i$. Finally, the recombination point at the father is randomly selected with probabilities $p_i = \delta_i / \sum_j \delta_j$. The intuition behind the first heuristic procedure is to stimulate innovation, that is, to only chose recombination points at the father that, by the crossover operation, are able to contribute with an innovative component, $A(i)$, that is not already present in the mother or, at least, to contribute with a similar component that is more efficiently coded.

After the recombination point at the father, $n(i)$ – root of sub-tree $A(i)$, has been chosen, a second heuristic procedure selects the recombination point at the mother, $m(j)$ - root of sub-tree $B(j)$. Again, new meta-control variables, λ_j are defined for the nodes $m(j)$, followed by a random selection with probabilities $p_j = \lambda_i / \sum_j \lambda_j$. The idea behind this second heuristic procedure is to stimulate the crossover to exchange sub-trees, $A(i)$ and $B(j)$, with analogous meanings, compatible semantics, similar interpretations, etc. This heuristic procedure draws inspiration from biology, where analogy is defined as compatibility in function but not necessarily in structure or evolutionary origin.

The formal expression used to evaluate the meta-control variables at the second heuristic procedure is:

$$\lambda_j = w_0 + \left[\sum_{d=1}^{D} w_d C_k \Big(A(i), B(j) \Big) \right]$$

The index d spans D semantic dimensions or factors. The positive weights, w_d, add to one, and the semantic compatibility measures, C_k, are normalized in the interval $[0, 1]$.

The functional form of the compatibility measures, $C_k(\ )$, are completely dependent on insights and interpretations for the actual problem being solved. In the case of the arithmetic functional tree presented at this section, the analogy between two sub-trees could be established, for example, simply by the fraction of input variables they share in common. In this case, blocks coding y^2 e $2y$ would have compatibility measure equal to 1, while the blocks coding y^2 and $w^{z/y}$ would have compatibility measure equal to $1/3$.

After a SASC crossover, the children's nodes carry along the meta-control variables, ρ_i, they had at the parents, and are afterwards updated by a random perturbation. We used a standard Normal multiplicative noise with drift μ_i and scale factor σ_i, that is, $\rho'_i = (1 + \mu_i + \sigma_i \epsilon) \rho_i$. At practical implementations we always used a positive drift at the recombination points, and a null drifts elsewhere. Sometimes we also used scale factors, σ_i, that decrease with the height of node $n(i)$. For instance, take σ_i inversely proportional to the depth of sub-tree $A(i)$. Using larger scale factors at lower nodes can help to induce the emergence of smaller building-blocks, that are more efficiently coded, and less prone to destructive crossover.

4   Emerging Building Blocks and Semiotics

In the following sections we explain our implementation of SASC methods, present an application case, and gauge its performance. However, before proceeding to finer details, this section tries to provide a larger picture, presenting the general framework that lead us to this line of research and some of the intuitions that inspired the name and definition of the SASC operator. Those readers interested mainly in the algorithmic aspects of the SASC operator can skip this section without prejudice. Nevertheless, the ideas presented in this section may provide a conceptual framework to examine similar algorithms and, in so doing, encourage or influence future research.

As stated in the introduction, under appropriate conditions, evolutionary systems naturally exhibit a hierarchical modular organization. The spontaneous emergence of hierarchical modular structures in natural or artificial evolving organisms is further studied in [19] and [23–25].

At the same time, the need or willingness to understand such systems leads to the attribution of meanings or interpretations corresponding to the systems' constituent parts. Specially in the case of inferential systems, this attribution of meaning leads, in turn, to consider the semiotic character of such parts. In this condition, we consider the system's modules as symbols representing concrete referents, that is, signs pointing to real things existing in the world or truthful relations occurring in the systems' environment. In this setting, rules of composition or coherent organization for the system's modular structures can be considered as articulation rules for terms in a systemic language. Hence, in this perspective, the emergence of a hierarchical modular organization in an evolutive inferencial system corresponds to a process of linguistic ontogenesis.

In the epistemological framework of cognitive constructivism, the semiotic character and the consequent semantic interpretation of the system's modular components arises from the inherent complementarity of a dual perspective:

(1) In the *autopoietc system* perspective, these components are seen as building blocks of an autopoietic unit, structured in a hierarchical and modular organization.

(2) In the *reasoning model* perspective, the same components are seen as constituent parts of a reasoning system, like sub-routines of a complex code, functions of a large program, etc.

The semiotic character and semantic interpretations of building blocks correspond to coherent and consistent (although possibly multimodal) forms in which these modules are used as operational tools, instrumental agents, partial production units, etc. used to implement solutions for the problems faced by the autopoietic system interacting with its environment.

The motivating and validating argument for the superposition of these two distinct and complementary views of the emerging structures, namely, that of an autopoietic system and that of a reasoning model, is given by Humberto

Maturana and Francisco Varela celebrated principle stating that - every autopoietic system is an inferential system, and its domain of interactions a cognitive domain. This principle is stated in [13, p. 10], and it is further explored, within the epistemological framework of cognitive constructivism, in [4] and [20-23].

The ideas briefly discussed in this section are far to general and abstract to directly generate specific algorithms or formulate explicit modeling solutions. Nevertheless, we hope that these ideas can be useful to investigate new techniques bearing some similarity to the specific methods presented in this article. If so, these ideas could be helpful in the "downward" direction leading to the development of new heuristic techniques used to accelerate the convergence or to induce the emergence of meaningful components in the underlying evolutive process.

At the same time, it is also true that the precise meaning of concepts used in a given epistemological framework can only be fully accessed analyzing specific theories, well defined hypothesis, concrete models or existing embodied system. Hence, we hope that the ideas discussed in this section can also be useful in the "upward" direction, fostering further research in the fields of applied semiotics, cognitive constructivism, and epistemological aspects of inferential systems.

5 Implementation

Our implementation of SASC methods is based on ECJ, an open-source evolutionary computing system written in Java. ECJ is developed at George Mason University's ECLab Evolutionary Computation Laboratory. ECJ maintains a well organized object-oriented design. Its powerful classes and methods proved to be very flexible, and could be easily extended to our purposes. The SASC package, developed by the first author, extends some ECJ classes in order to easily implement the methods under discussion. Most of the new code is concentrated at the class *SASCNode*, used to represent functional trees evolving by SASC GP. This class also includes abstract methods that facilitate the implementation of semantic compatibility measures, specified at sub-classes implemented for each specific problem.

Finally we should mention that ECJ supports distributed computing, specifying the desired number of parallel threads as a parameter to be set according to the available resources offered by the hardware and operating system. This feature was especially useful for multi-population scenarios, to be described in the next section, where SASC GP had an excellent performance.

6 Case Study

SASC operator was compared to standard crossover operators at a test case problem concerning the replication of an hypothetical investment fund.

Although hypothetical, this problem has strong similarities with real problems regarding the construction of synthetic portfolios faced by the first author in his professional activities. Portfolios of this kind are typical of correlation trade, since its return statistics are sensitive to the correlation matrix for the returns of various components in a basket. Such portfolios can be easily synthesized using readily available exotic derivatives like rainbow options, that is, calls or puts on the best or worst of several underlying assets.

Lemon, the hypothetic fund, is based on stocks negotiated at *BM&F-Bovespa - São Paulo Securities, Commodities and Futures Exchange*. Lemon's daily log-return, r_t, is given by the log-return average of four components, r_t^k, corresponding to key economic sectors. These are, using standard *BM&F-Bovespa* equity codes:

$$r^1 = \min(BBDC4, PETR4, BBAS3),$$
$$r^2 = \min(LAME4, LREN3, NETC4),$$
$$r^3 = \max(TNLP4, TCLS4, VIVO4) \text{ and}$$
$$r^4 = \max(CYRE3, ALLL11, GFSA4).$$

These components represent four key economic sectors: Telecommunications, construction and transports, finance and cyclic consumption.

An asset manager wants to synthesize a second fund, Lime, with the objective of tracking fund Lemon. However, only the daily share values of fund Lemon are available, not its operational rules. Of course, GP was the method chosen to find the best specification of the synthetic portfolio Lime. The atoms for this problem are the log-returns of 63 of the most liquid stocks negotiated at *BM&F-Bovespa*, that include all the stocks used to specify fund Lemon. The primitive operators are {max, min, mean}, the maximum, minimum and mean value of two real numbers. The training data bank consists of the daily log-returns of fund Lemon and all 63 stocks, computed from 04-Nov-2008 to 01-Apr-2009.

The fitness function for this problem is the mean squared error between the synthetic and the target log-returns, plus a regularization term adding, for each node, $n(i)$, a penalty $\pi(i)$. For the application at hand, we used $\pi(i) = c_{h(i)} 2^{h(i)-1}$, where $h(i)$ is the height of node $n(i)$. For the example at hand, we used $c_{h(i)} = 1$ at the root node and zero otherwise. The purpose of regularization term is to avoid needless complexity and over-fitting in the final model, see [5].

In the GP experiments, we used two distinct population scenarios. Scenario 1: One population of 300 individuals evolving over 700 generations, Scenario 2: 8 populations of 300 individuals each, that first evolve in isolation over 400 generations and are then allowed to merge and evolve for 100 generations more. In both scenarios the GP is allowed to warm-up using the standard crossover, 200 generations for scenario 1 and 100 for scenario 2, and then switch to (or not) to SASC crossover. SASC's semantic compatibility function is the Boolean indicator of having at least one atom in common.

The actual GP implementation uses a dual tree representation for each individual in the population, as suggested in Angeline original paper, [1]. The first tree only stores the genotype used to code the function expressed by the individual's phenotype. Meanwhile, the second tree only stores meta-control variables.

GP meta-parameters were set as follows: mutation rate was set at 5%, using a 3-round tournament selection process. Crossover rate was set at 95%, using a 7-round high pressure / 3-round low pressure combination of father / mother selection, see [23]. $\rho_{min} = 0.001$, $\rho_{max} = 0.999$, $w_0 = 0.01$, $w_1 = 0.99$, $\sigma_i = 0.4$ for $h(i) = 2$ and approximately inversely proportional to the node height for $h(i) > 2$. Further details about the algorithm fine tuning can be seen at the source code documentation, available from the first author.

Figure 2 compares the GP results using standard and SASC crossover operators. The use of Angeline's original SSAC instead of the standard crossover operator had only a minor impact in GP performance, and is not shown in the figure. This figure displays 95% confidence intervals for the mean square error of the best solution found over 50 independent GP runs.

Figure 3 shows the best empirical solution found by SASC GP. The figure also highlights the building blocks encapsulated by meta-control variables larger than a critical threshold. This solution replicates very well the target fund. Notice that each of the highlighted building blocks corresponds to one of the key economic sectors used to define the operation rules of fund Lemon.

Each best solution found at a batch of 50 SASC GP experiments under scenarios 1 and 2 was categorized according to the number of key economic sectors represented by a constituent building block. Table 1 displays the average mean square error of each category. This table shows that better adjusted

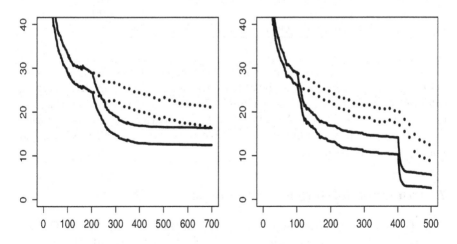

Fig. 2 Confidence interval for best solution MSE by generation. Crossovers' comparative performance: Standard ($\cdots$) and SASC (—).

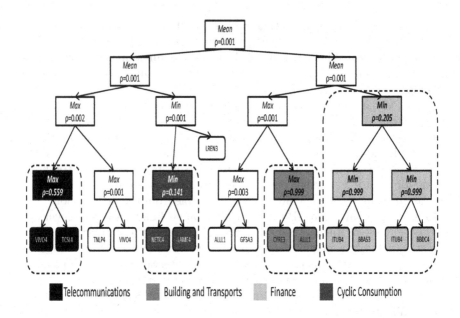

Fig. 3 Emerging building-blocks in near-optimal solution.

Table 1 Number of key economic sectors represented by building blocks

Category	Scenario 1	MSE	Scenario 2	MSE
One key sector	14%	12.3	10%	8.9
Two key sectors	16%	8.1	30%	1.9
Three key sectors	8%	9.3	38%	1.4
Four key sectors	0%	-	4%	0.1
Other (spurious) blocks	62%	21.7	18%	10.2

functional trees have more of the four key economic sectors present as a building block. This conclusion may be obvious to someone knowing the operating rules of Lemon, the original target fund. However, it is remarkable that the best solutions offered by SASC GP for the replication fund Lime, synthesized only from input-output data, are able to capture so well the logic and semantics of fund Lemon.

7   Conclusions and Final Remarks

From Figure 2, we can conclude that, at least for the test case at hand, GP has a much better performance when using SASC than the standard crossover operator. At scenario 2 the best empirical solution, shown at Figure 3, is found repeatedly. At scenario 1, SASC not only achieves better results, but also seems to greatly accelerate the finding of good solutions. These effects are even stronger at scenario 2, where a second acceleration effect is clear just after the populations merge. At this final stage, one can observe that the best solution are formed purging spurious building blocks and combining good building blocks that had emerged at the previously isolated populations. It is as if SASC were able to isolate, identify, and collect good building blocks.

The explanatory power of the emergent building blocks, that is, on one hand, how well they capture the semantics of the system under study and, on the other hand, how much they contribute to its prediction accuracy, is made even clearer by Table 1. Accordingly, Figure 3 suggests that SASC GP can also provide an implicit method of semantic analysis. That is, at least in our case study, the internal operational logic and the semantics of the target system is adequately represented by the building blocks of the best solutions synthesized by SASC GP. Nevertheless, it is important to keep in mind that these logical and semantic relations were not externally imposed or driven, but are truly emergent properties co-evolving with the GP solutions.

Future Research

In future research we plan to investigate techniques of self-adaptive meta-control using abstract type node labels as auxiliary control variables. Transformation rules for label mutation and label compatibility rules for permissible recombination points should be able to induce building block formation and encapsulation, and also be able to foster emergent semantic interpretations, even in problems lacking natural heuristics for explicit semantic compatibility measures.

Acknowledgements. The authors are grateful for the support of *IME-USP*, The Institute of Mathematics and Statistics of the University of São Paulo, *FAPESP*, Fundo de Amparo à Pesquisa do Estado de São Paulo, *CNPq*, The Brazilian National Research Council, and *BM&F-Bovespa*, The São Paulo Securities, Commodities and Futures Exchange. The authors are also grateful for the helpful comments of Marcelo Lauretto and an anonymous referee.

References

1. Angeline, P.: Two self-adaptive crossover operators for genetic programming. In: Angeline, P.J., Kinnear, K.E. (eds.) Advances in Genetic Programming. Complex Adaptive Systems, vol. 2, ch.5, pp. 89–110. MIT Press, Cambridge (1996)
2. Banzhaf, W., Francone, E.D., Keller, R.E., Nordin, P.: Genetic Programming, an Introduction: On the Automatic Evolution of Computer Programs and its Applications. Morgan Kaufmann, San Francisco (1998)
3. Banzhaf, W., Poli, R., Schoenauer, M., Fogarty, T.C. (eds.): EuroGP 1998. LNCS, vol. 1391. Springer, Heidelberg (1998)

4. Borges, W., Stern, J.M.: The rules of logic composition for the bayesian epistemic e-values. Logic Journal of the IGPL 15(5-6), 401–420 (2007)
5. Cherkaasky, V., Mulier, F.: Learning from Data. Wiley, NY (1998)
6. Holland, J.H.: Adaptation in Natural and Artificial Systems. University of Michigan Press, Ann Arbor (1975)
7. Iba, H., Sato, T.: Meta-level strategy for genetic algorithms based on structured representation. In: Proc. of the Second Pacific Rim International Conference on Artificial Intelligence, pp. 548–554 (1992)
8. Koza, J.R.: Genetic programming: On the programming of computers by means of natural selection. MIT Press, Cambridge (1992)
9. Koza, J.R.: Genetic programming II: automatic discovery of reusable programs. MIT Press, Cambridge (1994)
10. Koza, J.R., Deb, K., Dorigo, M., Fogel, D.B., Garzon, M., Iba, H., Riolo, R.L. (eds.): Genetic Programming 1997: Proceedings of the Second Annual Conference. Morgan Kaufmann, San Francisco (1998)
11. Langdon, W.B., Poli, R.: Foundations of Genetic Programming. Springer, Heidelberg (2002)
12. Lauretto, M., Nakano, F., Pereira, C.A.B., Stern, J.M.: Hierarchical forecasting with polynomial nets. In: [14], p. 305–315 (2009)
13. Maturana, H.R., Varela, F.J.: Autopoiesis and Cognition. The Realization of the Living. Reidel, Dordrecht (1980)
14. Nakamatsu, K., Phillips-Wren, G., Jain, L.C., Howlett, R.J. (eds.): New Advances in Intelligent Decision Technologies. Springer, Heidelberg (2009)
15. Poli, R., Langdon, W.B.: A new schema theory for genetic programming with one-point crossover and point mutation. In: [10], pp. 278–285 (1997)
16. Poli, R., Langdon, W.B.: A review of theoretical and experimental results on schemata in genetic programming. In: [3], pp. 1–15 (1998)
17. Reeves, C.R.: Modern Heuristics for Combinatorial Problems. Blackwell Scientific, Malden (1993)
18. Rosca, J.P.: Analysis of complexity drift in genetic programming. In: [10], pp. 286–294 (1997)
19. Simon, H.A.: The Sciences of the Artificial. MIT Press, Cambridge (1996)
20. Stern, J.M.: Cognitive constructivism, eigen-solutions, and Sharp statistical hypotheses. Cybernetics and Human Knowing 14(1), 9–36 (2007a)
21. Stern, J.M.: Language and the self-reference paradox. Cybernetics and Human Knowing 14(4), 71–92 (2007b)
22. Stern, J.M.: Decoupling, sparsity, randomization, and objective bayesian inference. Cybernetics and Human Knowing 15(2), 49–68 (2008a)
23. Stern, J.M.: Cognitive Constructivism and the Epistemic Significance of Sharp Statistical Hypotheses. In: Tutorial book for MaxEnt 2008, The 28th International Workshop on Bayesian Inference and Maximum Entropy Methods in Science and Engineering, Boracéia, São Paulo, Brazil, July 6-11 (2008b)
24. Stern, J.M., Colla, E.C.: Factorization of sparse bayesian networks. In: [14], pp. 275–294 (2009)
25. Stern, J.M.: The Living and Intelligent Universe. Tech.Rep. MAP-IME-USP-2009-04. Presented at MBR-09, Campinas, Brazil (2009)

An Episodic Memory Implementation for a Virtual Creature

Elisa Calhau de Castro and Ricardo Ribeiro Gudwin

Abstract. This work deals with the research on intelligent virtual creatures and cognitive architectures to control them. Particularly, we are interested in studying how the use of episodic memory could be useful to improve a cognitive architecture in such a task. Episodic memory is a neurocognitive mechanism for accessing past experiences that naturally makes part of human process of decision making, which usually enhances the chances of a successful behavior. Even though there are already some initiatives in such a path, we are still very far from this being a well known technology to be widely embedded in our intelligent agents. In this work we report on our ongoing efforts to bring up such technology by building up a cognitive architecture where episodic memory is a central capability.

1 Introduction

The research agenda on intelligent virtual creatures [4, 1, 2, 10] is a very intense one, both in terms of philosophy and in computer science. In a general sense, a virtual creature is an intelligent agent [7] which is embodied in a virtual world, capturing data through its sensors and autonomously acting on its environment, in order to meet some internal purpose or goal. An artificial creature is a special kind of autonomous agent, which is particularly embodied in a given environment (there may be autonomous agents which are not embodied). A virtual creature, for its turn, is a special kind of artificial creature, where the environment is a virtual environment, so the creature's

Elisa Calhau
DCA-FEEC-UNICAMP
e-mail: ecalhau@dca.fee.unicamp.br

Ricardo Ribeiro Gudwin
DCA-FEEC-UNICAMP
e-mail: gudwin@dca.fee.unicamp.br

L. Magnani et al. (Eds.): Model-Based Reasoning in Science & Technology, SCI 314, pp. 393–406.
springerlink.com © Springer-Verlag Berlin Heidelberg 2010

body is not a concrete one, like in a robot, but just an avatar in a virtual environment.

In the literature, there are multiple reports on many different strategies for building up artificial minds to control such virtual creatures. Among more traditional artificial intelligence techniques [10], some proposals of cognitive architectures were presented as possible implementations for such artificial minds.

Cognitive architectures [13, 8] are mainly inspired by human neuro-cognitive and psychological abilities, where typical human cognitive tasks as perception, learning, memory, emotions, reasoning, decision-making, behavior, language, consciousness, etc. are in some way modeled and used as a source of inspiration in order to enhance the capabilities of artificial creatures. Many of such cognitive abilities were successfully reported as very useful in making smarter creatures. Among others, abilities as emotions, learning, language evolution, action selection and either consciousness brought the performance of such virtual creatures to an amazing level.

Nevertheless, there seems to be at least one of such cognitive abilities which was not so widely explored so far. This ability is what we may refer from now on as *episodic memory* [18]. The first virtual creatures used to live only in the present, sensoring its surroundings and choosing its action based only on the current situation. Next generations of creatures enhanced that by living not only on the present, but also with an eye on the future, being able to making plans and creating expectations, which clearly sophisticated its behavior. But few of them were able to refer to its past, just like we do as humans.

We (humans) are able to remember what we did by this morning, some issues we lived last week, 2 months ago or even years ago. And more than this, we are able to build up a chronological time line, and order such events and locate them in this time line. We use this memory in order to learn things and to help us in performing our daily behavior. This is currently a missing gap in cognitive systems research. It will be an important improvement if our creatures were able to remember that they already were in such and such location, where they met such and such objects and creatures, and where such and such episodes were testified by them.

This is the main motivation of this work. Even though some related initiatives already started to appear in the literature [8, 6, 14, 3, 5, 12, 15, 16, 9], we are still very far from having this as a well known technology to be widely used in intelligent agents. In this work we report on our ongoing efforts to bring up such technology by building up a cognitive architecture where episodic memory is a central capability.

2 Human Memory System and Episodic Memory

The term "memory" can be used in many different contexts, addressing different kinds of things. It can be used, e.g. in the context of dynamical systems, to designate a specific state variable, which maintains its value through time, and is able to make an influence on the overall system state. We can also use the term "memory" in the context of a computer architecture, and so a memory will be an addressable array of flip-flop circuits, carrying on some value, during many cycles of machine clock. But the term "memory" can also be used in the context of human memory. Human memory, opposite to a dynamical system or a computer memory, is a very sophisticated system, with many different behaviors, which comprise, in a deeper analysis, an inter-related complex of different kinds of memory systems. We will see, next, that episodic memory is a specific subsystem which is a part of the whole human memory system.

2.1 Human Memory System

The human memory system has received a special attention from the scientific community in general, therefore several research areas have focused their efforts in better understanding this complex system. Although the research on memory, in different areas, vary in aims and perspectives, they usually consider the memory system divided in the following basic aspects [17]:

- Working Memory
- Short Term Memory
- Long Term Memory

    - Non-declarative Memory
        · Perceptual Memory
        · Procedural Memory
    - Declarative Memory
        · Semantic Memory
        · Episodic Memory

In a first glance, the human memory system is divided between Working Memory, Short Term Memory and Long Term Memory.

The *Working Memory* is used to store transient information during per-ception, reasoning, planning or other cognitive functions. Its capacity in time and space is very short, ranging from a few dozen itens, and periods ranging from a few seconds up to a few minutes.

The *Short Term Memory* is an intermediary kind of transient repository, which accomodates conscious information (information which reached con-sciousness) in a buffer ranging from 3 to 6 hours, during a process of consol-idation when this information is permanently stored in long term memory.

Finally, the *Long Term Memory* is a very complex memory subsystem, where different kinds of information are stored for long term retrieval. It can be decomposed into many different subsystems. The division described here, though, is not a consensus among memory experts. Some experts may say that the same kind of subdivisions employed here for long term memory, may apply also to short term memory and working memory. Following [17], we will be dividing long term memory into non-declarative and declarative memory.

Declarative Memories are memories that refers to *facts* that can be explicitly declared, like e.g. a proposition given by a phrase in a particular language. *Non-declarative memories*, on the other side, constitute the many different parts involved in a declarative memory, like e.g., the many different words used in a phrase. In this sense, non-declarative memories are used to record perceptions and actions, given rise to a further sub-division of non-declarative memories into Perceptual and Procedural memories.

The *Perceptual Memory* is the memory of categories of things which can be perceived by a Perceptual System. It includes different things attributes and patterns which can be categorized by a perceptual system. Each instance of a perceptual memory is a representation of a category used during perception.

The *Procedural Memory* is the memory of actions and behaviors of a system. It is a non-declarative memory which refers to a "how to" kind of information, usually consisting of a record of possible motor and behavioral skills.

Declarative memories, on the other side, are used to describe *knowledge*, as it appears in complete sentences in a natural language. They can be used to store both atemporal, general common-sense knowledge, like e.g. "Dogs are a specific kind of animal", or "My name is Paul", or used to store specific temporal event knowledge, like e.g. "Yesterday, from 23:00 to midnight I was sleeping in my bed". So, declarative memory can be divided into two different subsystems: Semantic Memory and Episodic Memory.

The *Semantic Memory* is used to record facts of a general kind, not contextualized in time and space. The *Episodic Memory*, on the other hand, is used to store facts particularly contextualized in time and space, forming "episodes" which refers to information specific to a particular location and timeframe.

Cognitive studies with humans which had some kind of impairment on their memory system, due to brain damage, show that it is possible to have damage in some kinds of memories while still retaining other kinds of memories. In cognitive systems research, we can also address the same observation. There may be cognitive systems which are responsible for providing memory capabilities of one kind, while not providing memory capabilities of other kinds.

In this work, we are particularly interested in the Episodic Memory, so we will focus our attention and detain ourselves to more deeply explore its inherent cognitive capabilities.

3   The Episodic Memory

Episodic Memory is a neurocognitive mechanism for accessing timing contextualized information that naturally makes part of the human process of

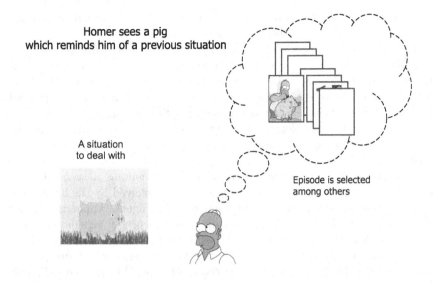

Homer sees a pig which reminds him of a previous situation

A situation to deal with

Episode is selected among others

Fig. 1 Episodic Memory.

decision making, usually enhancing the chances of a successful behavior. This assertion is supported by human psychological research which indicate that the knowledge of his/her personal history enhances one's person ability to accomplish several cognitive capabilities in the context of sensing, reasoning and learning.

Take as an example what is happening in figure 1. The character in the figure is dealing with a particular situation, and needs to decide what to do. Using objects which are perceived in the current situation as a hint, the episodic memory system is triggered, and a past situation where this object appeared is recovered as an episode. The character is now able to use this information in order to decide what to do in the current situation.

As in the case of the character in figure 1, in an artificial system, we would like to include an episodic memory subsystem, whose purpose is to assist the process of learning and ultimately providing a mechanism for better performance of intelligent autonomous agents in dynamic and possibly complex environments.

The main unit of information in an Episodic Memory system is called an *episode*. An episode is a record defined within a period of time and formed from information regarding the agent's task or other specific data observed in the environment. It also contains a measurement of "how successful" or relevant that information was for accomplishing a task in a past situation. In other words, the episode links particular data to a particular time and place in the environment and provides an indication of how to use that information to successfully perform a task. Therefore, along the time, it builds a repository of previous gathered experiences. Then whenever the creature faces certain situations and has to decide how to proceed next, it makes use of its repository of episodes to evaluate the best decision to make.

Episodes may be *State-based Episodes* or *Scene-based Episodes*. State-based episodes store the episode as sequences of an agent's states (including environmental sensed states). State-based episodes are easier to store, but more difficult to be used by higher-level cognitive functions. In artificial cognitive systems they are the most popular option for implementation, due to the easiness of its implementation, but they can be used only on specific kinds of applications, as its use on more sophisticated applications will be difficult.

Scene-based Episodes encode a time-space segment as a scene. In this scene, there are objects which were consciously perceived by the agent, and an action, performed by the agent itself or other agents appearing in the scene. Scene-based Episodes, can be viewed as interpreted versions of state-based episodes. They are easier to be used by high-level cognitive functions, as they already segment the scene into discrete elements, which are playing its own role in the scene dynamics. At the same time, they are more difficult to be implemented in artificial systems, because they require a process of interpretation of sensorial information in order to discover the objects and actions being performed in the environment.

Episodes can also be *autobiographic* and *non-autobiographic*. Autobiographic episodes are those episodes where the agent itself is performing the action being described in the episode. On the contrary, on non-autobiographic episodes, the subject of the action is another agent. In this case, these actions are being observed by the current agent and memorized as something seen, but not done by the agent itself.

An episodic memory system do require three major subsystems [14]:

- Encoding Subsystem
- Storage Subsystem
- Retrieval Subsystem

The *Encoding* subsystem is responsible for capturing the episode from the perception system (and maybe the behavior system, in the case of autobiographical episodes), and setting up the way the episodes are captured and stored. This subsystem addresses issues concerning the proper time to record the episodes and what information is to be stored within the episodes.

The *Storage* subsystem is responsible for getting the episode from the encoding subsystem and recording it in a permanent storage. This subsystem is responsible to define how the episodes are maintained, addressing issues such as memory decay and possibly merging of episodes to compact storage.

The *Retrieval* subsystem is responsible for providing episodes for being used by other cognitive functions. In other words, it defines how memory retrieval is triggered. This subsystem addresses issues related to the cue determination (which key data is used to trigger an episode) and how to use the retrieved episode.

Finally, to summarize this brief accounting of episodic memory, it is important to point out some possible uses of episodic memory in a cognitive architecture. The main use of episodic memory is to implement a cognitive capability called "mental time travel". Mental time travel is the capacity of "going back in time and space" and retrieving episodes related to a present situation. This capacity can be used to improve and enhance other cognitive capabilities, like e.g. perception, learning, planning, decision-making and action selection and execution.

For example, in *perception*, episodes may help in the process of detecting repetition and relevant input. Besides that, the mechanism provides the retrieval of features outside current perception which are relevant to the current task. Episodic memory also assists the mechanisms of action modeling and environment modeling.

In *learning*, episodic memory aids the learning processes providing an efficient mechanism of reviewing experiences and learning from them. Comparing multiple events simultaneously, the learning system is able to generalize knowledge. In addition, provides a way of recording previous failures and successes, which can be useful later for planning and decision-making.

It also aids in *planning* and *decision making* processes through predicting the outcome of possible courses of actions. Basically, the episodic memory allows the person/agent to review its own past action or of another one. Decisions which were useful in the past may be employed to solve current situations. Decisions which did not succeed may be avoided.

In *action selection and execution*, episodic memory may be used to keep track of progress and manage long-term goals. Using episodic memory, the system may be able to know that specific parts of a plan have already been executed, so the action-selection algorithm may define the next steps of a plan to be executed.

Besides that, episodic memory allows the person/agent to develop a sense of identity, as the episodes creates what could be accounted as the personal history of an individual. This personal history encompass information of events which were consciously perceived and performed by the person (or agent).

4   Episodic Memory in Cognitive Systems Research

Most of the research concerning Episodic Memory within Computer systems was published in the last five years. Though being still an incipient area of research, the computational study of Episodic Memory has provided interesting insights and these first works exploring its capabilities have presented very stimulating and promising results. The following research have been the main references for our work.

- Nuxoll and Laird's Episodic Memory for Soar
- Dodd's Episodic Memory for ISAC (Intelligent Soft Arm Control)

- Brom's virtual RPG actor with Episodic Memory
- Kim's virtual creature Rity's and its Episodic Memory
- Ho's Autobiographic Memory Control Architecture
- Tecuci's Generic Episodic Memory Module

Soar (originally known as SOAR: State Operator And Result) is a general purpose cognitive architecture being developed since a long time by the team of Prof. John Laird at University of Michigan, which was recently enhanced with an Episodic memory module [14], developed by Andrew Nuxoll. They performed several different experiments where different approaches for episode were tested and results extensively analyzed. For instance, they have analyzed effectiveness of partial versus complete matching algorithms during the retrieval phase, providing insights and alerting to trade-offs to be considered when dealing with cue and feature selection. The work presented very promising results and concepts were explored within a computer game environment.

The team of Prof. Kazuhiko Kawamura from the Cognitive Robotics Lab at Vanderbilt University developed ISAC (Intelligent Soft Arm Control), a cognitive robotic system - more specifically - a humanoid robot equipped with airpowered actuators designed to work safely with humans and used as a research platform for human-humanoid robotic interaction and robotic embodied cognitive systems. Will Dodd, a member of Prof. Kawamura team presented interesting results [6] when enhanced the ISAC with an Episodic Memory module. They have analyzed the impact of the use of Episodic Memory in terms of the robot performance and computational resources.

Prof. Cyril Brom, from Charles University in Prague, Czech Republic, developed a project to enhance an RPG (i.e. a role-playing game) actor, a non-player character, with a Memory module that allows it to reconstruct its personal story [3]. Although the focus of the project is regarded with linguistics, it explores and analyzes the basis of an Episodic Memory architecture, where episode structure, feature relevance and computational resources demanded are special issues to consider and which are crucial to the architecture efficiency. They show that in their game scenario, actors with Episodic Memory present a better performance than those without it, but only in low dynamic worlds and that the memory consumption is acceptable.

Prof. Jong-Hwan Kim and his team from the Korea Advanced Institute of Science and Technology (KAIST), developed Rity, a dog-like virtual creature that is the "software robot" unit of the *Ubibot*: the ubiquitous robot system project at KAIST, which has largely evolved during the last ten years. The Episodic Memory architecture was mainly developed by researchers N. S. Kuppuswami and Se-Hyoung Cho, from Kim's team, in the middle of the decade in order to provide a cognitive task selection mechanism for Rity. The creature's architecture explores the advantages of a reactive architecture with the higher level planning offered by the Episodic Memory, in addition to provide a learning mechanism that evolves with time, since Rity's decision making process is more efficient as the creature's experience grows [12].

In the Adaptive Systems Research Group at the University of Hertford-shire, UK, coordinated by Profs. Kerstin Dautenhahn and Chrystopher Ne-haniv, the researcher Wan Ching Ho developed an autobiographic memory control architecture (a kind of Episodic Memory) for virtual creatures [9]. The architecture is mainly focused on navigation problems, but its results are very promising confirming the effectiveness of the use of Episodic Mem-ory in decision-making problems. In the architecture, whenever certain inter-nal states of the creature are lower than a threshold, the creature searches through all the records in memory and reconstructs an event using a "mean-ingful search key" to recognize the possible sequence of how an event should be organized (event reconstruction mechanism). The records that match the key then provide the target resources to satisfy the current internal needs.

Dan Tecuci, from the University of Texas, designed a generic Episodic Memory module that can be attached to a variety of applications. He proposes that each generic episode presents thee dimensions that will be used during the retrieval phase and according to the type of application: *context*: general setting in which an episode occurs, for example, it could be the initial state and the goal of the episode, *contents*: ordered set of events that make up the episode and *outcome*: the evaluation of the episode's effect. The kind of task (planning oriented, goal recognition or classification) to be executed defines a scope focusing its procedures on one dimension of the episodes. For example, a classification-like task mainly recognizes whether a goal is solvable according to a state of the world. This corresponds to retrieval based on *episode context* and using the *outcome* of the retrieved episodes (i.e. their success) for classification. The generic module provides an API with two basic functions: store and retrieve. *Store* takes a new Episode represented as a triple [context, contents, outcome] and stores it in memory, indexing it along the three dimensions and *retrieve* takes a cue (i.e. a partially specified Episode) and a dimension and retrieves the most similar prior episodes along that dimension [15, 16]. This work provides interesting insights in how to efficiently establish the features that an episode must present in order to be actually useful after being retrieved and interpreted and those features that a *cue* must address to allow the retrieval of the most promising episodes.

5 The CACE Project - Cognitive Artificial Creatures Environment

5.1 General Characteristics and Motivation

The CACE project – Cognitive Artificial Creatures Environment, being de-veloped by our group at the University of Campinas, Brazil, consists of a computer game where robots (virtual creatures), controlled by a cognitive architecture working as a mind, try to accomplish a given task. The task is a "leaflet" containing a sequence of specific objects that must be collected in the environment and delivered in a specific place. The performance is basically measured in how fast the robots correctly accomplish their tasks along the game time. Figure 2 presents a screen shot of the scenario of the game.

The environment is essentially dynamic, since the robots can change the position of the objects by hiding them under the ground or simply moving them to other positions in the game space. Figures 3 and 4 show the robots and other entities of the game: food (nuts and apples), obstacles (in pink) and objects (bricks with different colors).

In the current version of the game, competition among the robots is encouraged and they never help each other or form teams. Consequently, simply moving an object that does not belong to its private *leaflet*, but that may belong to others, may be an interesting move to interfere in the other robots' performance. In addition, homeostatic internal states must be observed: the robots spend energy along the time, which has to be reestablished by food consuming. However, the food may be perishable or not. Consequently, along the time, it is expected that the robots develop a strategy where perishable food is consumed preferably and within their validity period and the best place to store the non-perishable food for future consumption and precaution.

In this work, our main purpose is the development of an "episodic memory" module for CACE, mainly consisted in storing and using the agents' previous actions and other specific data, while exploring the game environment. This module could be interpreted as a metaphor of a simplified "declarative memory" of each agent. The agent could access this information whenever a similar situation emerges and then decide how to proceed. Ultimately, the project aims in verifying if the use of the "episodic memories" actually enhances the agents' performance in the game.

5.2   The Use of Episodic Memory

In our work, the Episodic Memory is mainly used in decision planning. More specifically, it must aid in handling and analyzing three issues that are described in the following sections.

5.2.1   Path Planning

When the environment is large, it is not feasible to plan using all known obstacles and objects. So, the information within the episodes is used to evaluate feasible paths (e.g. without obstacles) during the navigation mechanism. Figure 5 illustrates the idea. In other words, the information within the episodes are used during the generation of a path plan and, consequently, may also anticipate problems while the creature is navigating. Since the environment is dynamic, there is no certainty regarding the path, but the episode provides certain level of probability once refers to a path previously observed.

After the path planning module evaluate all candidate paths, the *way points* that form each path are analyzed based on the information present in

Fig. 2 Screenshot of the game.

Fig. 3 Robot avoiding obstacles and moving.

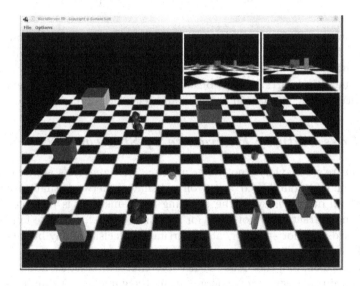

Fig. 4 Two robots looking for colored objects and food towards a non-perishable food (nut).

the Episodic Memory. If there is no obstacle along the path it is considered feasible and the shortest path among those evaluated as feasible is chosen.

5.3 Episodes Instead of World Map

Information regarding obstacles, food and other creatures perceived by the visual system are recorded within an episode. Instead of storing this information in a "world map", they are maintained in episodes within the Episodic Memory. During planning process, episodes are recollected in the Working Memory, and only "remembered" things are considered during the decision

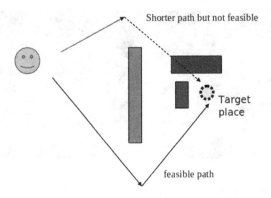

Fig. 5 Use of Episodic Memory in path-planning.

making process. This "remembered" information comes from episodes that matched the current situation in a certain level of similarity. A partial matching algorithm using different approaches have been considered when comparing the cue of the current situation with the episode in Memory: number of same features, key features in common and relevance of features in common.

5.4 Possibility of Emergence of Strategies

The creature's behavior is not deterministic. The creature's action decision mechanism is accomplished based on a behavior network that provides a certain level of flexibility to the planning mechanism. It is possible thanks to inherent characteristics of behavior networks. Therefore, while following a plan towards a "short-term" goal, opportunistic decisions may be taken that satisfies "long-term" goals.

In order to explore this behavior network characteristic, we intend to analyze if certain strategies may emerge during the game. One example is based on what we define in our work as "non-autobiographical episodes": those in which the creature is a mere observer and not the subject of the action.

For example, observing the opponents behaviors (perceiving episodes where the opponent is the agent performing actions) a creature may infer the opponents' goals. Then, an agent may try to hide the objects that the opponents need in order to decrease their performance.

6 Conclusion

Despite being still a young research area, in the context of computational systems, the study of Episodic Memory in cognitive systems has provided very interesting insights. The works that have explored its computational capabilities have presented very promising results which have consequently increased the scientific community curiosity. On the other hand, exactly for being such an incipient area, there is still too much to be explored, analyzed and experienced. Recent research have shown that trade-offs must be taken into account and not all scenarios may get much benefit from the use of Episodic Memory. However, as cognitive systems become more and more complex and have to handle more and more information, mechanisms with certain cognitive capabilities, such as Episodic Memory, will be a prerogative.

The current work is still in progress and the final results and analysis will be published in future papers.

References

1. Aylett, R., Cavazza, M.: Intelligent virtual environments: A state-of-the-art report. In: Eurographics 2001, STAR Reports, vol. 2001, pp. 87–109 (2001)
2. Balkenius, C.: Natural Intelligence in Artificial Creatures. Lund University Cognitive Studies (1995)
3. Brom, C., Peskova, K., Lukavsky, J.: What does your actor remember – towards characters with a full episodic memory. In: Cavazza, M., Donikian, S. (eds.) ICVS-VirtStory 2007. LNCS, vol. 4871, pp. 89–101. Springer, Heidelberg (2007)
4. Dean, J.: Animats and what they can tell us. Trends in Cognitive Sciences 2(2) (1998)
5. Deutsch, T., Gruber, A., Lang, R., Velik, V.: Episodic memory for autonomous agents. In: Proceedings of IEEE HSI Human System Interactions Conference, Krakow, Poland (2008)
6. Dodd, W.: The Design of Procedural, Semantic and Episodic Memory Systems for a Cognitive Robot. Master's thesis, Vanderbilt University (2005)
7. Franklin, S., Graesser, A.: Is it an agent, or just a program? A taxonomy for autonomous agents. In: Tambe, M., Müller, J., Wooldridge, M.J. (eds.) IJCAI-WS 1995 and ATAL 1995. LNCS, vol. 1037, Springer, Heidelberg (1996)
8. Franklin, S., Kelemen, A., McCauley, L.: Ida: A cognitive agent architecture. In: IEEE Conf. on Systems, Man and Cybernetic. IEEE Press, Los Alamitos (1998)
9. Ho, W.C., Dautenhahn, K., Nehaniv, C.L.: Autobiographic agents in dynamic virtual environments - performance comparison for different memory control architectures. In: Proceedings of IEEE Congress on Evolutionary Computation, pp. 573–580 (2005)
10. Isla, D., Blumberg, D.: New challenges for character-based ai for games. In: Proceedings of the AAAI Spring Symposium on AI and Interactive Entertainment, Palo Alto, CA (2002)
11. Kim, J.H., Lee, K.H., Kim, Y.D.: The origin of artificial species: Genetic robot. International Journal of Control, Automation and Systems 3(4), 564–570 (2005)
12. Kuppuswami, N.S., Se-Hyoung, C., Jong-Hwan, K.: A cognitive control architecture for an artificial creature using episodic memory. In: Proc. SICE-ICASE Int. Joint Conf., Busan, Korea, pp. 3104–3110 (2006)
13. Langley, P., Laird, J.: Cognitive architectures: Research issues and challenges. Cognitive Systems Research 10(2), 141–160 (2009)
14. Nuxoll, A.M.: Enhancing Intelligent Agents with Episodic Memory. Ph.D. thesis, University of Michigan (2007)

15. Tecuci, D.: Generic Episodic Memory Module. Tech. rep., University of Texas in Austin (2005)
16. Tecuci, D.: A Generic Memory Module for Events. Ph.D. thesis, University of Texas in Austin (2007)
17. Tulving, E.: Concepts of human memory. In: Squire, L., Lynch, G., Weinberger, N.M., McGaugh, J.L. (eds.) Memory: Organization and locus of change, pp. 3–32. Oxford Univ. press, Oxford (1991)
18. Tulving, E.: Episodic memory: From mind to brain. Annual Review of Psychology 53, 1–25 (2002)

Abduction and Meaning in Evolutionary Soundscapes

Mariana Shellard, Luis Felipe Oliveira,
Jose E. Fornari, and Jonatas Manzolli

Abstract. The creation of an artwork named RePartitura is discussed here under principles of Evolutionary Computation (EC) and the triadic model of thought: Abduction, Induction and Deduction, as conceived by Charles S. Peirce. RePartitura uses a custom-designed algorithm to map image features from a collection of drawings and an Evolutionary Sound Synthesis (ES-Synth) computational model that dynamically creates sound objects. The output of this process is an immersive computer generated sonic landscape, i.e. a synthesized Soundscape. The computer generative paradigm used here comes from the EC methodology where the drawings are interpreted as a population of individuals as they all have in common the characteristic of being similar but never identical. The set of specific features of each drawing is named as genotype. Interaction between different genotypes and sound features produces a population of evolving sounds. The evolutionary behavior of this sonic process entails the self-organization of a Soundscape, made of a population of complex, never-repeating sound objects, in constant transformation, but always maintaining an overall perceptual self-similarity in order

Mariana Shellard
Instituto de Artes (IA) – UNICAMP
e-mail: marianashellard@gmail.com

Luis Felipe Oliveira
Departamento de Comunicao e Artes. Univ. Federal de Mato Grosso do Sul – UFMS
e-mail: oliveira.lf@gmail.com

Jose E. Fornari
Núcleo Interdisciplinar de Comunicação Sonora (NICS) - UNICAMP
e-mail: tutifornari@gmail.com

Jonatas Manzolli
Instituto de Artes (IA) – UNICAMP and Núcleo Interdisciplinar de Comunicação Sonora (NICS) - UNICAMP
e-mail: jonatas@nics.unicamp.br

L. Magnani et al. (Eds.): Model-Based Reasoning in Science & Technology, SCI 314, pp. 407–427.
springerlink.com

to keep its cognitive identity that can be recognize for any listener. In this article we present this generative and evolutionary system and describe the topics that permeates from its conceptual creation to its computational implementation. We underline the concept of self-organization in the generation of soundscapes and its relationship with computer evolutionary creation, abductive reasoning and musical meaning for the computational modeling of synthesized soundscapes.

1   Introduction

One of the foremost philosophical problems is to rationally explain how we interact with the external world (outside of the mind), in order to understand reality. We take the assumption that human mind understands, recognizes and rapport with reality through a constant and dynamic process of mentalmodeling. The process is here seen as divided in three states: 1) **Perception**, where the mind receives sensory information from outside, throughout its bodily senses. This information comes from distinct mediums, such as mechanical (e.g. hearing and touch), chemical (e.g. olfaction and taste) and electromagnetic (e.g. vision). According to evolutionary premises, these stimuli are non-linearly translated into electrochemical information to the nervous system. 2) **Cognition**, the state that creates, stores and compares models with the gathered information, or from previously reasoned models. This is the information processing stage. 3) **Affection**, where emotions are aroused, as an evolutionary strategy to motivate the individual to act, to be placed in-motion, in order to ratify, refute or redefine the cognitive modeling of a perceived phenomenon. Here we introduce RePartitura; a case study in which we correlate these three stages with a pragmatic approach that combines logic principles and synthetic simulation of creativity using computer models.

RePartitura is here analyzed based on the assumption of mental model re-construction and re-building. This cycle of model recreation has insofar proved to be an eternal process in all fields of human culture; as well as in Arts and Science. As described by G. Chaitin[1], the search for a definite certainty along of the history of mathematics has always led to models that are: incomplete, uncomputable and random [7]. Inspired by Umberto Eco's book "The Search for the Perfect Language", Chaitin describes herculean efforts of great minds of science to find completeness in mathematics, such as Georg Cantor's unresting (and unfinished) pursuit of defining infinity, Kurt Godel's proves that "any mathematical model is incomplete". Following, Alan Turing's realization of uncomputability in computational models, and lastly, Chaitin's own Algorithmic Information Theory, that leads to randomness. In conclusion, "any formal axiomatic theory is fated to be incomplete". In another hand, he also recognizes that, "viewed from the perspective of Middle

---

[1] Chaitin, G. "The search for the perfect language."
http://www.cs.umaine.edu/~chaitin/hu.html

Ages, programming languages give us the God-like power to breathe life into (some) inanimate matter". So, computer modeling can be used to create artworks that resembles life evolution in a never-ending march for completeness, in an unreaching process of eternal self-recreation.

RePartitura is a multimodal installation that uses the ESSynth [9] method for the creation of a synthetic soundscape[2] where formant sound objects are initially built from hand-made drawings used to retrieve artistic gesture. ESSynth is a sound synthesis that uses Evolutionary Computation (EC) methodology, that was initially inspired in the Darwinian theory of evolution. ESSynth was originally constituted by a *Population* of digital audio segments, that were defined as the population *Individuals*. This population evolved in time, in generation steps, by the interaction of two processes: 1) *Reproduction*, that creates new individuals based on the ones from the previous generation; and 2) *Selection*, that eliminates poorly-fit individuals for the environmental conditions and select the best-fit individual, that creates (through the process of Reproduction) the next generation of its population [3]. In this way, ESSynth is an adaptive model of non-deterministic sound synthesis that present complex sonic results, at the same time that these sounds were bounded by a variant similarity, given the overall generated sound, somehow similar to the perceptual quality of a soundscape.

In section two we introduce the conceptual artistic perspective of RePartitura. We describe the process of creating the drawing collection and mapping its graphic features, inserted by the hand-made gesture that created the drawings, into genotypes used by the ESSynth that creates the soundscapes. We also describe the abduction process that emerges the sonic meaning of a soundscape. In section three, we discuss the possibility of self-organization in the computer-model sonic output, which is here claimed to describe an immersive self-similar perceptual environment; a soundscape. In section four we discuss the capacity of this evolutionary artistic system in emulating a creative process of abduction by expressing an algorithmic (computational) behavior here described as artificial abduction. In section five, it is discussed the aesthetic meaning for the dynamic creation of soundscapes where this is compared with musical meaning, in terms of its cognitive process, emotional arousal (through a "prosody" of expectations). Finally, we end this article with a conclusion, reassessing the ideas and concepts from previous sections and offer further perspectives into the designing of artificial creative systems.

2   Conceptual Perspective

In this section we elucidate the interaction between concepts that were in the genesis of RePartitura. Firstly, we relate the concept of abduction reasoning,

---

[2] Soundscape refers to both the natural and human acoustic environment, consisting of a complex and immersive landscapes of sounds that is self-similar but always new.

as presented by Charles S. Peirce, to the computational adaptive methodologies, such as EC. Secondly, we create RePartitura in line with the concept of Generative Art and the idea that iterative processes can be related to the Peircean concept of *habits*.

2.1   Abduction and Computational Adaptive Methods

The pragmatism of Peirce, points out to the conceptualization of three categories of logic reasoning as: 1) *Deduction*, 2) *Induction* and 3) *Abduction*. Abduction is the process of hypothesis building, by the generation of an initial model, as an attempt of understanding or explaining a perceived phenomenon. Induction tests this model against other factual data and makes the necessary adjustments. Deduction applies the established model of the observed phenomenon. This model will be used for deductive reasoning insofar as the advent of further information that may jeopardize its model trustworthy, or require its tackling to a reality change (which is always), where the whole process of Abduction, Induction and Deduction creates a new model of reasoning.

In this article our goal is to present a computer methodology related to the Peircean pragmatic reasoning. In computational terms, it is usual to refer to an observed phenomenon as a problem. In the concept expressed by this article, we consider Peircean triadic logical process as related to the following methodological taxonomy: a) **Deduction** corresponds to **Deterministic Methods** , as they can present predictable solutions to a problem; b) **Induction** is related to **Statistic Methods** once that they present not a single but a range of possible solutions to the same problem; c) **Abduction** is then related to Adaptive Methods that can redefine and recreate themselves, based on the further understanding of a problem, or its dynamic change.

Among computational adaptive methods, Evolutionary Computation (EC) is the one inspired into the biological strategy of adapting populations of individuals, as initially described by Charles Darwin. EC is normally used to find the best possible solution to problems when there is not enough information to solve it through formal (deterministic) methods. An EC algorithm usually seeks out for the best solution of a complex problem, into an evolving landscape of possible solutions. In our research group at NICS, we have studied adaptive methodologies in line with the creation of artworks, such as the system: 1) VoxPopuli to generate complex and harmonic profiles using genetic algorithms [21], 2) the RoBoser system, created in collaboration with the SPECS group from UPF, Barcelona, uses the Distributed Adaptive Control (DAC) to develop a correlation between robotic adaptive behavior and algorithmic composition [20] and 3) the Evolutionary Sound Synthesis (ESSynth) [9] a method to generate sound segments with spectral dynamic changes using genetic algorithms in the reproduction process and Euclidean distance between individuals as fitness function for the selection process.

ESSynth showed the ability of generating a queue of waveforms that were perceptually similar but never identical, which is a fundamental condition of a soundscape. This system was later developed further to also manipulate the spacial sound location of individuals in order to create the dynamical spreading acoustic landscape, so typical of a soundscape [11].

In all of these studies, we considered that adaptive methods, such as EC, could be used in artistic endeavours. Particularly, in this paper we will describe the RePartitura research, that relates multimodal installation and the ESSynth method. Furthermore, the discussion presented here is also related to the works of [23] where is discussed the process of musical meaning and logical inference from the perspective of Peircean pragmatism. This idea is discussed in the section 5 "Soundscape Meaning" where we focus our discussion on how listeners deduce some general patterns of musical structures that are inductively applied to new listening situations such as computer generated soundscapes.

2.2 Habits, Drawings and Evolution

The collection of drawings that proceeded RePartitura (see example in Figure 1) was based in the concept of defining a generative process as artwork. Particularly, the process analyzed here was defined as a daily habit of repetitive actions, which lasted ten months and generated almost three hundred drawings. This action was done by the artist's right arm in repetitive movements, from down-up and semi-circular. The movement pattern, along time, evolved from thick and short curves to long and narrows ones. This evolutionary characteristic of a gestural habit reflected an adaptation of the arm's movement to the area within the paper sheet.

Our first assumption here was to consider this long process of adaptation producing a visual invariance as a creation of a visual habit. Initially, different kinds of paper sheets were tested, such as: newsprint, rice, and a type of coffee filter paper. The filter paper was better suited for the characteristics of the movement, it was resistant, absorbent and with a nice tone of slightly yellowish white. The Indian ink was appropriate to the dynamics of gesture and, as black color is neutral, it did not cause visual noise. The paper size was established when the movement was stable, after a period of training. Japanese brushes and bamboo pen were tested. The second one produced a better result, by allowing a greater number of movement repetitions without loss of sharpness. Once that was defined, the material (filter paper, black ink pen and bamboo) remained the same throughout the entire process. The standardization of the material restrained the action and helped to create the habit of the arm's movement. As the gesture became a habit, the drawings stretched and the repetition was concentrated in a reduced area, showing a narrow and long curve (Figure 1), compared to initial ones (Figure 2). During the process new experiments occurred resulting in new patterns, such

Fig. 1 Sequence of Original Drawings that preceded RePartitura.

Fig. 2 Sequence of initial drawings created during the experimentation period.

as pouring ink on the paper to avoid the gesture interruption due to the necessity of loading the pen with ink. But, in doing so, the paper was softened by the ink tearing easily and this new method was discharged.

The gradual and progressive adaptation of the gesture and stabilization of drawing is consider here as a way of generating a habit, which can be associated, according to Peirce, with the *removal of stimuli* [25, p. 261]. At the same time, each drawing was influenced by the environment (physical and emotional) which led to the disruption of *habit*. Considering Peirce's affirmation that *the breaking up of habit and renewed fortuitous spontaneity will, according to the law of mind, be accompanied by an intensification of feeling* [25, p. 262], the emotional and physical conditions involved in the moment of the action, interfered in the individual gestures and resulted in accidental variations (e.g. outflow of ink or paper ripping), causing changes and triggering new possible repetitions.

The collection of drawings shown in Figure 1 was presented as an installation named *Mo(vi)mento*. After that, an analysis of visual features and perceived graphical invariance led us to create a reassignment of this process in the sonic domain. This was the genesis of RePartitura. The first idea was to represent similar behaviors in different mediums. After identifying invariant patterns in all drawings, they were parameterized and used in the creation of sound objects. ESSynth was chosen because of its similarity with the artistic process that created the collection of drawings, described above, which was also characterized by an evolutionary process.

2.3 Repetition, Fragments and Accumulation Mapped into Sound Features

We developed an analytical approach in order to identify visual invariance in the original drawings to represent them into the sound domain. Our idea was to describe the habits embedded in the drawings, in parametrical terms, to further use them to control the computer model of an evolutionary sound generation process. We found out three categories of visual similarity in each drawing of the collection. They were named as: 1) Repetitions; thin quasi-parallel lines that compose the drawing main body, 2) Fragments; spots of ink smeared outside the drawing main body, and 3) Accumulation; the largest concentration of ink at the bottom of the drawing (where the movement started). These three aspects are shown in Figure 3.

The identity of each drawing was related to the characteristics of these three categories. It was developed an algorithm to automatically map these ones from the drawings digital image and attribute to them specific parametric values. These categories were related to the evolution of the gesture and the conditions of each drawing moment. Their evolution was characterized by the habit of the movement to create the drawings. The values of the parameters of the drawings created within the same day tended to be similar. However, at times when emotional inference and external intervention were higher, the drawings underwent a break in the gesture habit, which could be detected by the changes in the parametric values of the three categories. From this visual perspective, we developed a translation into the sonic features of the next stage.

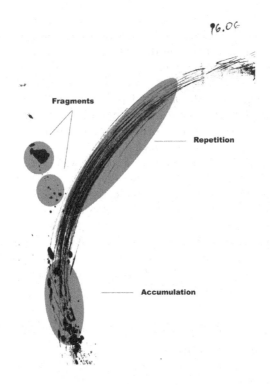

Fig. 3 The three categories of graphic objects found in all drawings.

Table 1 Mapping of formal aspects of the drawings into their sonic equivalents.

Invariance	Drawing Aspects	Sonic Aspects
Accumulation	Concentration of ink in the lower area of the drawing, characterized by ink stains.	Constant,long-term duration and low frequency noisy sounds.
Repetition	Number of repetition curve.	Cycles of sinusoidal waves with average duration.
Fragments	Drips of ink.	Very short sounds, varying from noisy to sinusoidal waveforms.

   Initially, we established: long-term, middle and very short duration sounds. The first ones were associated to the Accumulation parameter and were represented by low frequency noisy sounds. Repetition parameter was associated with cycles of sinusoidal waves. Fragments were related to sharp sounds varying from noisy to sinusoidal ones. This mapping is presented in Table 1.

   The duration of each element of the mapping was also related to the idea that Perception, Cognition and Affection can be expressed in different time scales of the sonic ambient. In this domain, the perceptive level can be related to the sensorial activation of auditory aspects, such as intensity, frequency, and phase of sounds, which is studied by psychoacoustics. Cognition is related to the sonic characteristics that can be learned and recognized by the listener. Its time scale was initially studied by the psychologist William James, who developed this concept, which refers by "specious present" the seemingly present time of awareness for a sonic or musical event. It can be argued that the 'special present' is related to short-term memory, which can vary from individual to individual and acording to the direction of the mode or range in which the musical information is perceived as a whole, such as a language sentence, a sound signal or a musical phrase [27]. Some experiments have shown that, in music, their identification is approximately the order of one to three seconds of duration [17]. The emotional aspects are those that evoke emotion in the listener. Affective characteristics are associated with a longer period of time (up to thirty seconds) and may be processed with long-term memory, of which it is possible to recognize the genre of a music or soundscape. The recognition of the whole sonic environment and its association with listeners expectations is further explored in this article, when we discuss the research of [16] and [19].

2.4   Drawings, Adaption and Abduction

In RePartitura, the gestures that engendered drawings were mapped to sonic objects, and became individuals within an evolutionary population that compounded the soundscape. This infers an analogy with the evolution of habits of gestures throughout time. The sound objects are like a mirror for the striking differences expressed by the visual invariances of the drawing categories. The application of EC methodology can be seen as a way of representing the drawing habits in the sonic domain and the trajectories of these individuals (sound objects) are correlated to the evolution of the initial drawing gestures. The unique aspects of each drawing, influenced by several conditions, such as the artist variations of affection and mood, and by the environmental conditions, such as the external interruptions of any sort, characterizes the hidden organizing force that make possible the adapting evolution of habits in this system, which is a paramount characteristic of abduction.

As postulated by Peirce: "... diversification is the vestige of chance-spontaneity; and wherever diversity is increasing, there chance must be operative. On the other hand, wherever uniformity is increasing, habit must be operative. But wherever actions take place under an established uniformity, there so much feeling as there may be takes the mode of a sense of reaction" [15]. The difference between drawings gestures, that generated the seed of chance for the change of habits on the sound system, is a representation of the spontaneity embedded in the process of making each drawing unique, yet similar. In our work we are inferring a correlation of this idea to the notion of Abduction, when Peirce defines that: "method of forming a general prediction without any positive assurance that it will succeed either in the special case or usually, its justification being that it is the only possible hope of regulating our future conduct rationally, and that Induction from past experience gives us strong encouragement to hope that it will be successful in the future" [30].

In another paragraph, Peirce correlates habits to the listening of a piece of music: "... whole function of thought is to produce habits of action; and that whatever there is connected with a thought, but irrelevant to its purpose, is an accretion to it, but no part of it. If there be a unity among our sensations which has no reference to how we shall act on a given occasion, as when we listen to a piece of music, why we do not call that thinking. To develop its meaning, we have, therefore, simply to determine what habits it produces, for what a thing means is simply what habits it involves. Now, the identity of a habit depends on how it might lead us to act, not merely under such circumstances as are likely to arise, but under such as might possibly occur, no matter how improbable they may be. What the habit is depends on when and how it causes us to act. As for the when, every stimulus to action is derived from perception; as for the how, every purpose of action is to produce some sensible result. Thus, we come down to what is tangible and conceivably practical, as the root of every real distinction of thought, no matter how subtle it may be; and there is no distinction of meaning so fine as to consist in anything but a possible difference of practice" [29, 5.400].

Thus, meaning is pragmatically connected to habit, and habit is a necessary condition for the occurrence of action. Meaning is at the heart of actions of inquiry and of predicting consequences of future actions. For each inquiry there is an action that occurs in a very specific way. At the core of such process, there is a very special category of reasoning (or action); the Abduction.

Abductive reasoning can be considered as a valuable analytical tool for the expansion of knowledge, helping with the understanding of the logical process of formulating new hypotheses. In regular and coherent situations, the mind operates deductively and inductively upon stable habits. When an anomalous situation occurs, abduction comes into play, helping with the reconstruction of articulated models (the generation of explanatory hypotheses) so that the mind can be free of doubts. We elucidate this point of view by presenting here the artwork RePartitura, a computer model that uses a pragmatic approach paradigm to describe the creative process in sound domain. Here we used processual gestures and adaptive computation in order to digitally generate soundscapes. Our focus in this article is to examine the theoretical implications of that methodology towards a synthetic approach for the logic of creativity in the sound domain involving interactive installations. Logic of discovery is a theory that attempts to establish a logical system for the process of creativity. Peirce argued that in order to have creativity manifesting, new habits must firstly emerge as signs in the mental domain; taking that any semiotic system is primarily a logical system.

2.5 Computer Modeling

The computer design and implementation of RePartitura is further discussed in [11, 12]. In the next paragraphs we present a brief overview on that. The collection of drawings were mapped by an algorithm written in Matlab, where the features, classified in three categories, where processed in different sonic time-scaling. **Accumulation** were mapped into long time scale, representing affective aspects. **Repetitions** went into middle-time scale, related to the specious present, as defined by James Williams, and thus representing the cognitive aspects of sounds. **Fragments** were mapped into short time scales, corresponding to the perceptual aspects. The first feature retrieved was given by a simple metric defined by the equation below:

$$m = \frac{4 \cdot \pi \cdot \text{Area}}{\text{Perimeter}^2}$$

to describe the roundness of each object. For $m = 1$, the object is a circle. For $m = 0$, the object is a line. The second feature retrieved was the object Area, in pixels, where the object with the biggest value of Area is the Accumulation. The third feature was the object distance to the image origin, given by two numbers of their coordinate (x, y) into the image plan. We set apart Fragments and Repetitions using the value of m. The roundest objects ($m < 0.5$) were classified as Fragments. The stretched objects ($m < 0.5$) were classified as Repetitions. Each of these objects features were mapped into Sound Object genotype.

The genotypes were transferred to an implementation of ESSynth written in PD (PureData) language. The individuals (sound objects) were also designed in PD, as PD patches (in PD jargon). Our model of individual is created by the main system, as a meta-programming strategy, where "code writes code", at certain extent. The individuals would "born", live within the population, as sound objects, and, once their life-time was over, they would dye, to never be repeated again. The initial individuals received their genotypes from the drawing mapping. After that, by the reproduction of individuals, new genotypes would be created and eliminated, as the individuals died. Each genotype is described by the acoustic descriptors of a sound object. In this work, the sound object features used are divided into two categories: deterministic (melodic or tonal) and stochastic (percussive or noisy). For each category, there was: intensity, frequency and distortion, which would bridge this two sonic worlds (deterministic to stochastic) as a metaphor to the reasoning processes of, respectively: deduction and induction. For that, the abduction would be represented by the evolutionary process per se; the soundscape. These ones are given by the self-organization of the population of sound objects whose overall sound output is the output of the system.

3   Self-Organizing Soundscapes

After presenting the conceptual framework related to the creation and analysis of RePartitura, we will now discuss the sonic aspects of this work. Our attention is focused on the idea that a computer generative process can synthesize a sonic process that resembles a soundscape. Thus, firstly we present a formal definition of soundscape and correlate that to the computer model that implements the evolutionary process used here to produce RePartitura dynamic sonification.

Soundscape is a term coined by Murray Schafer that refers to the immersive sonic environment perceived by listeners that can recognize it and even be part of its composition [28]. Thus, a soundscape is initially a fruit of the listener's acoustic perception. As such, a soundscape can be recognized by its cognitive aspects, such as foreground, background, contour, rhythm, space, density, volume and silence. According to Schafer, soundscapes can be formed by five distinct categories of analytical sonic concepts, derived from their cognitive units (or aspects). They are: *Keynotes, Signals, Soundmark, Sound* Objects, and Sound Symbols. Keynote is formed by the resilient, omnipresent sounds, usually in the background of listeners' perception. It corresponds to the musical concept of tonality or key. Signals are the foreground sounds that grasp listener's conscious attention as they may convey important information. Soundmarks are the unique sounds only found in a specific soundscape. Sound Objects are the atomic components of a soundscape. As defined by Pierre Schaeffer, who coined the term, a Sound Object is formed by sounds that deliver a particular and unique sonic perception to the listener. Sound symbols are the sounds which evoke cognitive and affective responses based on the listener's individual and sociocultural context. The taxonomy used by Schafer to categorize soundscapes based on its cognitive units, serves us well to describe them from the perspective of its macro-structure, as it is easily noticed by the listener. These cognitive units are actually emergent features self-organized by the complex sonic system that forms a soundscape. As such, these units can be retrieved and analyzed by acoustic descriptors, but they are not enough to define a process of truly generating soundscapes. In order to do that, it is necessary to define not merely the acoustic representation of sound objects but their intrinsic features that can be used as a recipe to synthesize a set of similar-bound but always original sound objects.

In terms of its generation, as part of an environmental behavior, soundscapes can be seen as self-organized complex open systems, formed by sound objects acting as dynamic agents. Together, they orchestrate a sonic environment that is always acoustically original but, perceptually speaking, this one withholds enough self-similarity to enable any listener to easily recognize (cognitive similarity) and discriminate it. This variant similarity or invariance is a trace found in any soundscape. As such, in order to synthesize a soundscape using a computer model it is necessary to have an algorithm able to generate sound objects with perceptual sound invariance. Our investigation is to associate this perceptual need to a class of computer methods that are related to adaptive systems. Among them, we studied the EC methodology. Next section, we are going to correlate EC systems with the concept of Artificial Abduction. With the next considerations, we aim to link the computer generative process and the conceptual perspective presented in Section 2.

4 Artificial Abduction

Abduction is initially described as an essentially human mental reasoning process. However, its concept has a strong relation with Darwinian natural selection, as both may be seen as "blind" methods of guessing the right solution for not-well defined problems. In such, EC methodology, that is inspired in the Darwinian theory, may be able to emulate, to some extent, abductive reasoning. This is what is named here as Artificial Abduction, and is explained below. Most of the ideas in this section were discussed in [22]. Here, we point out the main topics that are linked to RePartitura creative process.

4.1 Abduction and Evolution

As already mentioned, abduction is related to the production of more convincing hypotheses to explain a given phenomenon through relative evaluation of several candidate hypotheses, as also discussed in [8]. In short, the general scheme of Abductive arguments consists in the proposition of alternative hypothesis to explain specific evidence (a fact or set of facts), and the availability of an appreciation (or recognition) mechanism, capable of attributing a relative value to each explanation. The best one is probably true if, besides comparatively superior to the others, it is good in some absolute sense. In opposition to the deductive arguments, the conclusion in abductive inference does not follow logically from the premises, and does not depend on their contents. In opposition to the inductive arguments, the conclusion not necessarily consists of the uniform extension of the evidence.

Our main concern here is simply the existence and specificity of abductive inference, and its spread application to perform customary reasoning. As mentioned above, this article examines the theoretical implications of a model for the logic of creativity in the sound domain. Our aim is to relate the construction of an alternative hypothesis in the search for the best explanation for a phenomenon, with the possibility of simulating an artificial evolution using evolutionary algorithms. EC simulates an artificial evolution categorized by hierarchical levels: the gene, the chromosome, the individual, the specie, the ecosystem. The result of such modeling is a series of optimization algorithms that result from very simple operations and procedures (crossover, mutation, evaluation, selection, reproduction) applied to a computer represented genetic code (genotype). These procedures are implemented in a search algorithm, in this case, a *population-based search*. The *revolutionary idea* behind *evolutionary algorithms* is that they work with a *population of solutions subject to a cumulative process of evolutionary steps*. Classic problem-solving methods usually rely on a single solution as the basis for future exploration, attempting to improve that solution. But there is an additional component that can make population-based algorithms essentially different from other problem-solving methods: the concept of competition and/or cooperation among solutions in a population [3]. Essentially, the degree of adaptation of each candidate solution will be determined in consonance with the effective influence of the remainder candidates. As a competitive aspect, each candidate has to fight for a place in the next generation. On the other hand, symbiotic relationships may improve the adaptation degree of the population individuals. Moreover, random variation is applied to search for new solutions in a manner similar to natural evolution [3] This adaptive behavior produced by EC is also related here with the notion of Abuductive reasoning.

4.2   Evolution and Musical Creativity

Probably, the most famous enquiry about the music creative capacity of computers was formulated by Ada Lovelace. She realized that Charles Babbage's "Analytical Engine" – in essence, a design for a digital computer – could "compose and elaborate scientific pieces of music of any degree of complexity or extent". But she insisted that the creativity involved in any elaborated pieces of music, emanating from the Analytical Engine, would have to be attributed not by the engine but by the engineer [5]. She said: "The Analytical Engine has no pretensions whatsoever to originate anything. It can do [only] *whatever we know how to order it to perform*". That Analytical Engine have never been built, but Babbage supposes that, in principle, his machine could be able of playing games such as checkers and chess by looking forward to possible alternative outcomes, based on current potential moves.

Since that, for many years artworks have been emerged from computer models for many years. The main goal is to understand, either for theoretical or practical purposes, how representational structures can generate behavior, and how intelligent behavior can emerge out of unintelligent (machinery) behavior [5]. The usage of EC presented here can be seen as an effective way to produce art based on an efficient manipulation of information. A proper use of computational creativity is devoted to incrementally increase the fitness of candidate solutions without neglecting their aesthetic aspects. A new generation of computer researchers is applying EC and looking for some kind of artistic creativity simulation in computers with some surprising results. The ideas discussed here suggest an effective way of producing art, based on an dynamic manipulation of information and a proper use of a computational model resembling the Abductive processes, through EC with an interactive interface. EC seems to be a good paradigm for computational creativity, because the process of upgrading hypotheses is implemented as an interactive and iterative population-based search.

5 Soundscape Meaning

The concept of musical meaning is controversial and has led to a myriad of different perspectives in the philosophy of western music, and the problems of musical meaning are conceptually even more daring when considering the pure music, without words, a.k.a. instrumental music. This very distinct essence that music has and its non-conceptual nature gives to that subject a distinct consideration in modern aesthetics. It is from the rising of Modern Age that these kind of problem emerges, when music looses its connection with the old cosmologies that assured its proper role in the human knowledge and culture. Roughly, since the music of Modern Age was understood in terms of language and rhetoric analysis; a sort of special language, or the language of the emotions, as in the philosophy of 19th century.

Notwithstanding, also in the 19th century, Edward Hanslick initiated a formalist perspective of musical aesthetics that takes music as music, without any necessary connection with emotions or natural language, for its meaningfulness. Apart from the common-sense understanding of music, the formalist approach dominated musicology and related fields in 20th century. Regarding the problem of meaning, the formalist approach led to the question of how music is understood by the human mind and the result of affection reactions and emotions in the listener[3]. In the last century, music psychologists, still in a very formalist perspective, furnished some hypothesis on how the mind engages with musical form in (meaningful and affective) listening. Mainly, it is assumed that the mind operates *logically* in listening to music actively, and the models so far proposed in psychology are instantiations of a deductive-inductive perspective [16, 19].

Those models claim that by exposition to a cultural environment the listener deduces some general patterns of music structures that are inductively applied to new listening situations, assuming the general inductive belief that the future should conform to the past. Thus, a key concept of meaning in music is expectation; a meaningful music is the one in which the listener can engage structurally with it and predict consequent relations. Emotions arise in the struggle of the expected patterns and that actual patterns the music display; when they are similar there is a limbic reward for the efficient prediction, made when the prediction is false, there is a contrastive valence that results in the surprise effect (see Huron [16]).

---

[3] Hanslick never denied that music induce emotions in the listener but considered that a secondary effect a secondary one and claimed that the meaning of music is not by the mimesis of emotions, as usually said, but by the perception of its structures.

The process of acquisition of knowledge, or inquiry, as Peirce usually points out, is not sufficiently accounted with a deductive-inductive model for the very reason that before any deduction could be made, a hypothesis should be presented to the mind. Abduction is the logical process by witch hypotheses are generated. This threefold logical model of inquiry offers another viewpoint to consider musical meaning and affect, not opposed to the models of music psychology but complementary to them. In fact, through the perspective of the Logic of Discovery, creativity turns out to be a logical process, instead of a mysterious and obscure one, beyond understanding. The abductive creation of hypothesis is the very basis of inquiry and, by extension, of knowledge itself. In Peirce's philosophy, this threefold logicality is involved in any process of signification, assuming the possibility of different distributions of the three kinds of reasonings in each particular case. The maxim of pragmatism, as formulated by Peirce, claims that the whole meaning of an idea is the sum of all the practical consequences of such idea. In this sense, the concept of meaning is a matter of: habits and believes, that, consequently, govern our actions. Habbits and beliefs are firstly and priorly design by abduction. There is, thus, a connection between logic, habit and action, in the pragmatic conception of meaning.

Musical (structural) listening is an action (as much as thought is an action for Peirce). As such, it is active rather than passive. This action, as any action, is guided by beliefs[4] and habits, that form a conceptual space which is the interface between the listener and his cultural ambient [5]. It is in the coupling interaction between habits and structures that music becomes meaningful and affective. Habits are created by the logical process of Abductive reasoning. In ordinary music listening, when the audience is familiar with the stimuli, i.e., it is culturally embedded and have habits embodied that respond properly to that music genre, listening might be a more deductive-inductive logical process. The more predictable is the music, the more inductive is its thinking action. In listening situations with unfamiliar music or when a music piece presents non-culturally-standards structures, habitual action might not conform to that structures and expectations could not be derived properly. This music requires a process of habit reformulation by the active listener, i.e., Abduction.

The conceptual space is altered every time a new habit is called into existence, shifting the listening experiences from that moment. That is why one could have a lifelong listening experience with one piece of music and it is absolutely not the repetition of such experience over and over again. Even if that daily appreciation is made with the same recording of the piece, the conceptual space is not the same because it is dynamically altered by abduction processes. Signification is an emergent property of such conceptual space, i.e., the dynamic coupling of a listener (with his audition history embodied as habits and beliefs) and musical works (culturally embedded).

---

[4] For the relevance of belief in aesthetic appreciation see, for instance, Aiken [2, 1].

Similarly, in the case of soundscapes, the conceptual space is also created and recreated by the Abductive reasoning of listeners, when they recognize and even contribute to it, as parts of this environment (such as in a crowded audience). Soundscapes are formed anywhere as long as there is at least one listener to abduct it. As asked by the old riddle; "If a tree falls in the forest and no one is around to hear it, does it make a sound?". If there is no listener to abduct the meaning of the sound waves generated by this natural process, there is no soundscape, as its meaning depends upon its reasoning.

In the case of RePartitura, the EC computer model that synthesizes soundscapes, attempts to create a doorway to pass through the signification emerged from the habits acquired by the artist during the drawing collection production, into a population of sound objects whose genotype is given by the drawings features mappings. The conceptual space of the synthesized soundscape is dynamically recreated in a self-similar fashion, which guarantees that a listener, although not (yet) able to participate of its recreation, can easily abduct its perpetuated meaning.

6 Discussion

RePartitura, is a computational model-based that attempts to create artificial abduction; thus emulating the reasoning process that an artist has when creating an artwork. The artist abducts done since the first insight, when this one has its initial idea of creating a piece of artwork, and afterwards, during the process of its confection, when habits are developed while the artwork is being shaped and reshaped according to the bounding conditions imposed by the environment, being they external (e.g. material, ambient, etc.) or internal (e.g. subjective, affective, mood, willingness, inspiration, etc.). To model that in a computational system, we used an evolutionary sound synthesis system, the ESSynth, based on EC methodology, that was inspired on the natural evolution of species, as described by Darwin. EC is sometimes defined as a non-supervised method of seeking solution, mostly used for problems not-well defined (non-deterministic). The idea of a non-supervised method that is able of finding complex solutions, such as the creation of living beings, without the supervenience of an even more complex and sophisticated system, such that would be an "intelligent designer", is the core of Darwinism and is being increasingly used in a broad range of fields in order to try to explain the natural law that allow systems to be self-organized and/or becoming autopoietic. For that perspective, a complex system can emerge as habits of its compounding agents, under the influence of permeating laws that regulate their environment and their mutual interactions. Similarly, abduction can be seen as a mental process that allow us to naturally identify the self-similarity of a self-organized system. Peirce himself acknowledges that abduction must be a product of natural evolution, when he points out that: "... if the universe conforms, with any approach to accuracy, to certain highly pervasive laws, and if man's mind has been developed under the influence of these laws, it is to be expected that he should have a natural light, or light of nature, or instinctive insight, or genius, tending to make him guess those laws aright, or nearly aright" [25]. As an adaptive model that generates self-organized soundscapes, considered here as embodying aesthetic value, RePartitura seemed to fulfill the pre-requisites of being a system that presents a form of Artificial Abduction.

As the sound objects population of RePartitura evolves in time, so does its soundscape. Thus, new sound events can emerge during this process. In the computational implementation presented here, we didn't set an interaction of the system with the external world. This can be further done using common sensors such as the ones for audio (microphone) and/or image (webcam). Nevertheless, the soundscape will present ripples in its cognitive surface of self-similarity, which is welcome. We had RePartitura exhibit for several days in an art gallery (Sesc – São Paulo, 2009) and it was interesting to realize that, despite the long hours of exposition in this sonic ambient, it did not tire the audience as much as it should if it where given by the same acoustic information, although its overall sound was always very similar. This feature is found in natural soundscapes, such as the sonic ambient nearby waterfalls, forests, or by the sea. These seemingly constant sonic information have a soothing affective response for most of people. Maybe it is done by the fact that our abduction reasoning is always activated to keep track of the continuity of sameness. Expectations will, however, be minimal as, cognitively speaking, this information doesn't bring novelty to uprise limbic reactions, as the ones related to: fight, flight or freeze. This prosody is smooth, as being similar, yet enticing, as it brings a constant flux of perceptual change. We might say, in poetic terms, that the prosody of a soundscape is Epic, as it describes a thread of perceptual change; a cognitive never-ending sonic story, instead of Dramatic, as it normally doesn't startle emotive reactions in the listeners by drastic changes in their expectations [16].

If aesthetic appreciation were governed only by the subjective opinion, there would not be means to obtain automatic forms of artistic production, with some aesthetic value, without a total integration human(artist)-machine. On the other hand, if the rules and laws that conduct art creation did not allow the maintenance of a set of degrees of free expression, then the automation would be complete, despite the apparent complexity of the artwork. Since both extremes do not properly reflect the process of artistic production, the general conclusion is that there is room for automation either in the exploration of degrees of free expression, through a human-machine interactive search procedure, or in the application of mathematical models capable of incorporating general rules during the computer-assisted creation. In few words, the degrees of freedom can be modeled, in the form of optimization problems, and the general rules can be mathematically formalized and inserted in computational models, as restrictions or directions to be followed by the algorithm. The single trait of each creation will be understood as the result of a specific exploration of the search space, by the best blend of free attributes among all possibilities.

7 Conclusion

We started this article describing that the drawings used in RePartitura explored the development of a gesture over a period of time. The drawings showed pattern changes according to the day of its execution. The pattern variation was associated with physical and physiological influences. The analysis of pattern variation led us to associate the formation gesture to acquisition of habit and it's breaking up. The acquisition of habit was associated with gradual and progressive aspect of the drawings (elongated, narrow curve aspect). The breaking up of habit was associated with the influence of chance (resulting in drawings with overflow of ink). The first was characterized by drawings with less visual information and the second more visual information. In turn, all these ideas were associated with Peircean perspective on the formation of habits.

In RePartitura we used the ESSynth for the creation of computer-generated soundscapes where the formant sound objects are generated from patterns and invariance's of the drawings. The image invariances were identified and parameterized to create genotypes of sonic objects, which became individuals within a sonic evolutionary ambient. The sound objects orchestrate a sonic environment that is always acoustically original but, perceptually this one withholds enough self-similarity to enable any listener to easily recognize and discriminate it.

The soundscape meaning is different from the musical meaning due to its absence of a prior and paradigmatic syntax. Soundscapes have a discourse less affective, but rather more perceptual and cognitive, thus differing from the traditional aesthetic of Western music. However, some relations can be observed if one compares the components of soundscapes with traditional concepts employed in music analysis. For instance, a soundmark or a signal may have the rule a theme or motive usually has; motivic developments are made on similarities and differences in the spectro-morphology of sound objects and the relations on these sound objects are unique for each composition, as the thematic development of a symphony that has no other one similar to it. But besides this similarities, the absence of a priori syntactical rules makes the listening less directional and opened to other alternative ways of understanding it. However, the signification over this less directional listening occurs by the very same logical processes: a deductive-inductive bases updated and adapted by abductive inferences. But soundscape meaning is more abductive because it has not a priori syntactical rules of development that can be presumed by the listener and incorporated in his listening habits and aesthetical beliefs. Thus, each soundscape is an unique aesthetic experience that calls for the logic of guessing more often to be understood. We may say that evolutionary soundscapes are twice abductive, as adaptation and abduction occurs together in such sonic environment, by its algorithmic generation, as well in its listener's meaningful and affective appreciation, as a piece of art.

References

1. Aiken, H.D.: The aesthetic relevance of belief. Journal of Aesthetics and Art Criticism 9(4), 310–315 (1951)
2. Aiken, H.D.: The concept of relevance in aesthetics. Journal of Aesthetics and Art Criticism 6(2), 152–161 (1947)
3. Bäck, T., Fogel, D.B., Michalewicz, Z. (eds.): Evolutionary Computation 2: Advanced Algorithms and Operators. Institute of Physics Publishing (2000)
4. Boden, M.: What is creativity? In: Boden, M. (ed.) Dimensions of creativity, pp. 75–117. MIT Press, London (1996)
5. Boden, M.: Creativity and artificial intelligence. Artificial Intelligence 103(1-2), 347–356 (1998)
6. Csikszentmihalyi, M.: Creativity: Flow and the Psychology of Discovery and Invention. HarperPerennial, New York (1996)
7. Chaitin, G.J.: Information Randomness and Incompleteness. World Scientific, Singapore (1990)
8. Chibeni, S.S.: Cadernos de História e Filosofia da Ciência, Series 3, 6(1), 45–73 (1996); Center from Epstimology and Logic, Unicamp (1996)
9. Fornari, J., Manzolli, J., Maia Jr., A., Damiani, F.: The evolutionary sound synthesis method. In: Proceedings of ACM Multimedia, Toronto (2001)
10. Fornari, J., Maia Jr., A., Manzolli, J.: Soundscape design through evolutionary engines. Special Issue Music at the Leading of Computer Science. JBCS – Journal of the Brazilian Computer Society (2000)
11. Fornari, J., Shellard, M., Manzolli, J.: Creating evolutionary soundscapes with gestural data. Article and presentation. SBCM - Simpósio Brasileiro de Computação Musical (2009)
12. Fornari, J., Shellard, M.: Breeding patches, evolving soundscapes. Article presentation. In: 3rd PureData International Convention – PDCon09, São Paulo (2009)
13. Harman, G.: The inference to the best explanation. Philosophical Review 74(1), 88–95 (1965)
14. Holland, J.H.: Emergence: From Chaos to Order. Helix Books. Addison-Wesley, Reading (1999)
15. Hoopes, J.: Peirce on Signs: Writing on Semiotic. The University of North Caroline Press, USA (1991)
16. Huron, D.: Sweet Anticipation: Music and the Psychology of Expectation. The MIT Press, Cambridge (2006)
17. Leman, M.: An auditory model of the role of short-term memory in probe-tone ratings. Music Perception 17(4), 481–509 (2000)
18. Manzolli, J.: Auto-organização um paradigma composicional. In: Debrun, M., Gonzales, M.E.Q., Pessoa Jr., O. (eds.) Auto-organização: Estudos Interdisciplinares, Campinas, CLE/Unicamp, pp. 417–435 (1996)
19. Meyer, L.B.: Emotion and Meaning in Music. Chicago University Press, Chicago (1956)

20. Manzolli, J., Verschure, P.: Roboser: A real-world composition system. Computer Music Journal 29(3), 55–74 (2005)
21. Moroni, A., Manzolli, J., Von Zuben, F., Gudwin, R.: Vox populi: an interactive evolutionary system for algorithmic music composition. Leonardo Music Journal 10, 49–54 (2000)
22. Moroni, A., Manzolli, J., Von Zuben, F.: Artificial abduction: A cumulative evolutionary process. Semiotica 153(1/4), 343–362 (2005)
23. Oliveira, L.F., Haselager, W.F.G., Manzolli, J., Gonzalez, M.E.Q.: Musical meaning and logical inference from the perspective of peircean pragmatism. Journal of Interdisciplinary Music Studies 4(1), 45–70 (2010)
24. Peirce, C.S.: The Collected Papers of Charles S. Peirce. Harvard University Press, Cambridge (1931-1965)
25. Peirce, C.S.: Essays in the Philosophy of Science. In: Tomas, V. (ed.). Bobbs-Merrill, New York (1957)
26. Peirce, C.: Pragmatism and pragmaticism. In: Hartshorne, C., Weiss, P. (eds.) The Collected Papers of Charles Sanders Peirce, vol. V-VI. Harvard University Press, Cambridge (1933)
27. Poidevin, R.L.: The perception of time. In: Zalta, E. (ed.) The Stanford Online Encyclopedia of Philosophy (2000), http://plato.stanford.edu/
28. Murray Schafer, R.: The Soundscape (1957)
29. Truax, B.: Handbook for Acoustic Ecology (1979)
30. Peirce Edition Project (ed.): The Essential Peirce. Selected Philosophical Writings, vol. 2, pp. 1893–1913. Indiana University Press, Bloomington (1998)

Consequences of a Diagrammatic Representation of Paul Cohen's Forcing Technique Based on C.S. Peirce's Existential Graphs

Gianluca Caterina and Rocco Gangle

Abstract. This article examines the forcing technique developed by Paul Cohen in his proof of the independence of the Generalized Continuum Hypothesis from the ZFC axioms of set theory in light of the theory of abductive inference and the diagrammatic system of Existential Graphs elaborated by Peirce. The history of the development of Cohen's method is summarized, and the key steps of his technique for defining the extended model $M[G]$ from within the ground model M are outlined. The relations between statements in M and their correspondent reference values in $M[G]$ are modeled in Peirce's Existential Graphs as the construction of a modal covering over the sheet of assertion. This formalization clarifies the relationship between Peirce's EG-β and EG-γ and lays the foundation for theorizing the abductive emergence of the latter out of the former.

1 Introduction

C.S. Peirce's concept of abduction represents one of the very few fundamental and substantial innovations in the history of logic since Aristotle and the Stoics. In addition to the two broadest categories of logical inference – deduction and induction – Peirce proposes a third, abduction, which would share certain characteristics with both. The term abduction is meant to capture the specific form of inference involved in hypothesis-formation. As such, it intrinsically links logic, experimental method and intellectual creativity in a single process. To give some sense of the scope Peirce accorded to the concept,

Gianluca Caterina · Rocco Gangle
Endicott College, 376 Hale Street, Beverly, Massachusetts 01915, U.S.A.
e-mail: {gcaterin,rgangle}@endicott.edu

L. Magnani et al. (Eds.): Model-Based Reasoning in Science & Technology, SCI 314, pp. 429–443.
springerlink.com © Springer-Verlag Berlin Heidelberg 2010

it may be sufficient to state that Peirce considered all forms of perceptual judgment to be species of abductive inference.

One question that remains essential to the debates over the status of abduction is whether abduction itself is susceptible to formalization and, more generally, as Hoffman inquires, "Is there a logic of abduction" at all [5]? Hoffmann's reflections on Peirce emphasize the context-dependent character of all logic and representation, and he argues that the mechanism of abduction (if, indeed, mechanism is a fair attribution) may be best understood as a re-arranging of contexts of knowledge, description and practice. Practice takes priority here – as Hoffman writes:

> A central condition for taking new perspectives is activity. Peirce emphasizes this element of activity in particular in respect to discoveries in mathematics: Proofs and deductive reasoning are essential characteristics of mathematics, but if we want to prove that the sum of angles in a triangle is exactly 180°, we need a form of reasoning which Peirce called "diagrammatic"... The essence of diagrammatic thinking is to create new representations out of a given one. The point is that *one* representation in a continuum of possible representations "compels us" to perceive new relations or a new organizing structure of a set of data.

In this way, the eminently practical character of abduction intersects with diagrammatic method, which will prove crucial in what follows.

We may identify two particular issues that arise in addressing the question whether there is a "logic of abduction". The first is the specifically creative character of hypothesis-formation. Peirce himself emphasizes that one of the marks distinguishing abduction from both deduction and induction is that a new term appears in the conclusion of the abductive inference that cannot be found in either the major or minor premise. In this way, abduction introduces genuine novelty into thought, and it would seem that this precludes any formalization of its logic.

The second issue is the relation of abduction to Peirce's synechistic conception of logic and ontology. In his later work, Peirce stresses that the nature of thought cannot be adequately grasped without reference to continua [7]. In particular, he took tentative steps towards a topological redefinition of logic. It is not clear if and how this development in Peirce's mature thought recontextualizes the problem of abduction.

Both of these issues lead us to pose the question of a possible logic of abduction in more specific terms, namely with respect to the system of Existential Graphs which Peirce himself identified as his most important contribution to logic.

The purpose of this work is not definitively to answer the question of the formalization of abduction. Rather, we provide an iconic representation of the forcing relation – or, more precisely, the relationship between this relation itself and its reference values in the extended model – in Peirce's EG. In particular, the *forcing* relation as defined by Cohen between *conditions* and statements S, denoted by

$$\pi \models S$$

which may be expressed in Peirce's EG at the β–level, serves as the basis for the indication of modal statements in EG-γ which apply to "possibilia" in an extended universe of reference. This representation of forcing in EG, in conjunction with Badiou's extended argument that Cohen's mathematics is the formal context within which hypothesis formation and generic truth must be framed, lays the foundation for future work in the formalization of abductive inference within EG[1].

2 The Continuum Hypothesis

Peirce's contemporary G. Cantor formulated what has come to be called the Continuum Hypothesis. The technique that we will be using as the *trait d'union* between abduction and the diagrammatic representation of the emergence of EG-γ from EG-β was developed by Paul Cohen in his work to prove the independence of this hypothesis from the Zermelo-Frænkel axioms of set theory. What follows is a brief summary.

Let us denote by ω the set $\omega = \{0, 1, 2, \ldots\}$, by $|S|$ the cardinality of a set S, by $\aleph_0$ the cardinality of ω and by $\mathcal{P}(S)$ the power set of S. Then the Continuum Hypothesis (abbreviated CH) is the conjecture, formulated by Cantor, that, if S is an infinite subset of $\mathcal{P}(\omega)$, then either $|S| = \aleph_0$ or $|S| = \mathcal{P}(\omega)$. In particular, CH asserts that there is no set whose cardinality is strictly between that of the integers and that of the real numbers.

This conjecture was the first on the list of problems presented by David Hilbert in the year 1900. About thirty years later, Kurt Gödel [4] proved that CH cannot be disproved from the Zermelo-Frænkel set of axioms (abbreviated ZF) and that therefore, if ZF is consistent, ZF+CH is still consistent.

More precisely, Gödel defined the class of *constructible* sets and showed that the statement (which, interestingly, he denoted simply with the letter "A"):

Every set is constructible

is consistent with ZF and, moreover, that CH can be proved from ZF as restricted to constructible sets and therefore, that ZF+CH is consistent.

The intrinsic difficulties in constructing alternate models of the Zermelo-Frænkel axioms was a serious obstruction towards any substantial progress about CH. It was not until the early 1960s that Paul Cohen found a method to construct models of ZF in which neither proposition A, nor CH holds. Combined with Gödel's results, Cohen showed that CH is *independent* from ZF.

---

[1] The authors would like to thank Ahti-Veikko Pietarinen, Jaakko Hintikka and Priscila Borges for their comments and suggestions at the MBR 2009 Symposium on Peirce's Existential Graphs.

Cohen's work relies on a novel technique that he invented *ex novo*, called *forcing*. In what follows we will outline the fundamental, simple ideas behind forcing and discuss how this method constitutes a natural framework for the emergence of both of Peirce's "Logic of Continuity" as expressed in EG$-\gamma$ and abduction.

3   The Generic Extension of Standard Models of ZF

The following section is a brief overview of the background preparation necessary to define forcing. Our presentation follows the structure that Badiou [1] uses in his own summary of Cohen.

Cohen's technique of forcing makes use of two models of ZF, one of which is taken to be the "ground model", which we will denote by M, and its *extension* $M[G]$ obtained by adjoining what will be called a *generic set* G to M. G is *generic* in the sense that it remains indiscernible from the point of view of what is codable within M, so that M and $M[G]$, while both models of ZF, function at distinct levels. Forcing will make use of this difference. The precise character of this difference is elaborated in the following sequence of steps.

3.1   *The Generic Set*

A model for the axioms of set theory can be simply thought as a "realization" of those axioms, exactly in the same way as a group, say $(\mathbb{Z}, +)$ (the group of the integers with the operation of addition), is a "realization" for the group axioms.

The goal of this subsection is to define a notion of *generic set*: intuitively, a set G is generic with respect to a model M, if "from inside" M, G can be defined but yet cannot be discerned. In order to unfold this only apparent contradictory definition, we need to outline precisely the steps of the construction. In what follows we only assume familiarity with the basic notation of set theory.

3.1.1   Standard Models

We start with a *standard* model M for ZF, where by *standard* we mean that:

- every element x of M is a well-founded set: that is, x is constructed inductively from the empty set using operations such as taking unions, subsets, powersets, etc.;
- M is transitive: every member of an element of M is also an element of M;
- M is countable.

3.1.2 Set of Conditions

Within M, we discern a set of *conditions*, that we denote by $\mathcal{C}$, such that:

- $\mathcal{C} \in M$;
- there is a partial order on $\mathcal{C}$, denoted by $\subset$.

We say that π_1 *dominates* π_2 if $\pi_2 \subset \pi_1$ and that π_1 is compatible with π_2 if there is π_3 such that $\pi_1 \subset \pi_3$ and $\pi_2 \subset \pi_3$. If we represent the order with the branches of a tree as below

then this can be expressed graphically by the following diagram:

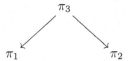

Then the last property that identifies $\mathcal{C}$ is that

- every condition is dominated by two conditions which are incompatible among themselves:

$$\forall \pi \, \exists \pi_1, \pi_2 \text{ such that } (\pi \subset \pi_1) \ \& \ (\pi \subset \pi_2) \text{ with } \pi_1 \text{ and } \pi_2 \text{ incompatible.}$$

Graphically, that means that, if π_1 and π_2 are incompatible, then a picture like the following is impossible within $\mathcal{C}$:

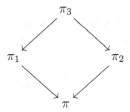

Example 1. Let $\mathcal{C}$ be the set of all the finite binary sequences. That is

$$\mathcal{C} = \bigcup_{n \in \mathbb{Z}^+} <a_1, a_2, \ldots, a_n>$$

where $a_i \in \{0, 1\} \ \forall i \in \{1, 2, \ldots, n\}$. Let the order be defined by

$$< a_1, a_2, \ldots, a_n > \; \subset \; < b_1, b_2, \ldots, b_k >$$

if and only if the first n entries of $< b_1, b_2, \ldots, b_k >$ coincide with $< a_1, a_2, \ldots, a_n >$; that is if and only if

1. $n < k$
2. $a_i = b_i \; \forall i \in \{1, 2, \ldots, n\}$

In this case we have, for instance, that $< 0, 1, 1 > \; \subset \; < 0, 1, 1, 1 >$. Graphically:

$$< 0, 1, 1, 1 >$$

$$\downarrow$$

$$< 0, 1, 1 >$$

By the same token, it easy to see that, for instance, the sequences $< 0, 1, 1 >$ and $< 0, 1, 0 >$ are incompatible because there does not exist a sequence which dominates both:

3.1.3   Correct Subsets of $\mathcal{C}$

Within $\mathcal{C}$, we further discern a set δ, called a *correct* set of conditions, such that:

1. $(\pi_2 \in \delta, \pi_1 \subset \pi_2) \Rightarrow \pi_1 \in \delta$;
2. $(\pi_1 \in \delta, \pi_2 \in \delta) \Rightarrow \exists \, \pi_3 \in \delta$ such that $\pi_1 \subset \pi_3, \pi_2 \subset \pi_3$.

This definition is reminiscent of that of an ideal in ring theory – in our case an element of a correct set δ is such that it forces every object that is below it in the order to belong to δ.

Example 2. Following up on the previous example, let $\mathcal{C}$ be again the set of all the finite binary sequences with the order defined above. It can be checked that the subset of $\mathcal{C}$ whose elements are the finite sequences having only $1's$ as entries is a correct set (let us call such a subset δ_1). Indeed, only finite sequences of all $1's$ can be dominated by sequences of the same kind and two such sequences are clearly compatible (it is enough to take a third sequence of all $1's$ longer than the two chosen ones).

Once we realize that a correct set can be discerned in an unambiguous way by its defining property (i.e. sequence of all 1's), it is remarkable to notice that the interplay between the structure of $\mathcal{C}$ and that of δ generates a natural boundary between the discernible and its complement. Indeed, since any element, say π_1 in $\mathcal{C}$ must be dominated by two incompatible conditions, say π_2 and π_3, by the property 2 above for δ at least one of them, say π_2, cannot be an element of δ. Therefore π_2 does not possess the property which discerns δ!

Badiou cleverly notes that "the concept of correct set is perfectly clear for an inhabitant of $M\ldots$". What he means by this is that the correct set can be discerned from within M: a suitable formal language that is codable in M is powerful enough to "see" the crisp boundaries of δ within $\mathcal{C}$. Badiou continues, "What is not yet known is how to describe a correct set which would be an indiscernible part of $\mathcal{C}$, and so of the model M" [1]. This is our next task.

3.1.4   Dominations

We start with an example which reconnects with those given above ($\mathcal{C}$ is again the set of all the finite binary sequences and the order is the usual one).

Example 3. The property discerning the complement of δ_1 in $\mathcal{C}$, say δ_1^c is given by: "binary sequences containing at least one 0". For any element in δ_1 there will be an element in δ_1^c which dominates it. But this offers a beautiful way to let the discernibility of δ_1 be characterized in a structural way, with no reference to the language: for any element of δ_1, for instance $< 1, 1, 1 >$ there is at least one element in the complement, say $< 1, 1, 1, 0 >$ which dominates it. This only apparently innocent observation is at very heart of the foundation of the generic.

To formalize what we were just discussing in this example, consider a correct set δ. δ is *discernible* from an inhabitant of M if there is an explicit property λ (expressible within the model M) that distinguishes δ unambiguously:

$$\alpha \in \delta \iff \lambda(\alpha)$$

Since every condition $\pi_1 \in \delta$ is dominated by two incompatible conditions π_2 and π_3, by the second property defining δ we have that either π_2 or π_3 has to live outside δ. The importance of this remark cannot be overestimated. It is indeed upon this simple observation that we can define what a *domination* is. Let's define a *domination* D as a set of conditions such that any condition outside the domination is dominated by at least one condition inside the domination. In symbols:

$$\sim (\pi_1 \in D) \Rightarrow (\exists \pi_2)[(\pi_2 \in D)\&(\pi_1 \subset \pi_2)]$$

(here $\sim$ denotes negation).

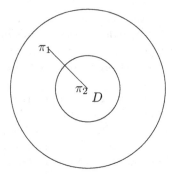

3.1.5   The Generic Set

We may now define a generic set. A correct set G will be said generic for M if, for any domination D which belongs to S, we have $D \cap G \neq \emptyset$. In other words, G is a correct set that has the property of having at least one element in common with all the dominations. From the point of view of M, it should be clear why G is indiscernible: "...otherwise it would not intersect the domination which corresponds to the negation of the discerning property" [1].

3.2   From M to $M[G]$

Let us remember that our goal is to find another model of ZF by *adjoining* G to M. In order to accomplish this task, we need two more intermediate steps.

3.2.1   Names

In spite of its indiscernibility, elements of G must be given names from within M. This is done by using a technique very close, in spirit, to transfinite induction. We begin by defining names μ of *rank 0*:

$$rank(\mu) = 0 \iff [(\gamma \in \mu \Rightarrow \gamma = < \emptyset, \pi >)]$$

where $\pi \in \mathcal{C}$ is any condition. Essentially, names of rank 0 are thus (any possible) sets of ordered pairs, of which each ordered pair simply indexes a particular condition (an element of $\mathcal{C}$) to the empty set.

Inductively we can then define names of rank greater than 0:

$$rank(\mu) = \alpha \iff [(\gamma \in \mu \Rightarrow \gamma = < \mu_1, \pi >)] \ \& \ rank(\mu_1) < \alpha.$$

In effect, this defines an inductive hierarchy of names in which some name at a given ordinal rank greater than 0 is itself constituted as a set of ordered pairs, of which each ordered pair indexes a particular condition to some name as determined at a lower rank. These names, whose ordinal induction may be entirely defined from within M but which cannot in fact all be themselves elements of M, will be used to produce the "excess" of $M[G]$ over and above M by assigning a specific value to each name according to a second inductive procedure.

3.2.2   Reference Values

Once we can name elements of G, we want to assign *referential values* to these names which will in effect determine the essential features of the adjunction of G to M through the action of G (or rather, the strictly formal definition of G from within M as outlined above in section 3.1) upon the ordinal hierarchy of names. In this way, the names will be used to indicate the generic set G sufficiently without thereby discerning or fully determining it. We start by defining referential values for names of rank 0: For names of rank 0 (which are composed of pairs $< \emptyset, \pi >$) we posit:

- $R_G(\mu) = \{\emptyset\} \iff \exists < \emptyset, \pi > \in \mu \mid \pi \in G$
- $R_G(\mu) = \emptyset$ otherwise.

Inductively, let us suppose that the referential value of the names has been defined for any rank less than α. Then, if μ_1 is a name such that $rank(\mu_1) = \alpha$, we have that:

$$R_G(\mu_1) = \{R_G(\mu_2) \mid (\exists \pi)(< \mu_2, \pi > \in \mu_1 \ \& \ \pi \in G)\}$$

Each of the names is thus formally assigned a unique value that depends solely upon which conditions are indeed elements of G in any given case, although G itself remains relatively undetermined (save for its genericity).

3.2.3   Adjoining the Generic Set to the Ground Model

At this point we are in a position to define $M[G]$:

$$M[G] = \{R_G(\mu) \mid \mu \in M\}$$

In plain words, $M[G]$ is the set of all the G-referential values of names that are themselves elements of M. We will not go into further technical details, but it is worthwhile, for the sake of clarity, to mention that the following, fundamental facts can be proven:

- $M \subset M[G]$, that is, $M[G]$ is a non-trivial extension of M
- $G \in M[G]$ and, moreover, G itself can be named from the point of view of M.

3.3 Diagrammatic Recapitulation

The diagram below summarizes the order and connections between the various steps of the construction of $M[G]$. The two triangles represent determinations involving more than one of the previous results.

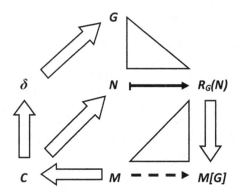

4 Modeling the Forcing Relation in EG

Cohen's mathematical technique of forcing produces a correspondence between a well defined forcing relation in M and possible truth values of statements in $M[G]$. We represent this correspondence iconically in Peirce EG as a lifting of EG-β statements on a single sheet of assertion into a modal sheet of EG-γ. In our diagrammatic construction, discrete modal statements at the level of the EG-β sheet of assertion take as their referents the emergent continuous domain of $M[G]$ as expressed in EG-γ.

4.1 The Quasi-implicational Structure of Forcing and Peirce's Existential Graphs

On the one hand Cohen, in his exposition of forcing, highlights its modal nature, noting that

> Clearly there are some properties of S which no reasonable procedure could interpret as being true or false for a 'generic' set G... Given π we will then ask, if under some procedure to be given, it is reasonable to expect that π forces a statement S about G to hold or forces $\sim S$ to hold, or whether the condition π does not force S one way or the other. *Although forcing will be related to the notion of implication it will differ from it in that given that π forces S it will not be true that any G that satisfies π will also satisfy S. What will be true is that any generic G satisfying π will also satisfy S* [notation altered to conform with the presentation above and emphasis added].

On the other hand, in his system of Existential Graphs, Peirce distinguishes between the single sheet of assertion at the level of α and β graphs and a "book of separate sheets, tacked together at points" at the γ-level, where modality emerges from relations of continuity and discontinuity between the sheets.

> In the gamma part of the subject all the old kinds of signs take new forms... Thus in place of a sheet of assertion, we have a book of separate sheets, tacked together at points, if not otherwise connected. For our alpha sheet, as a whole, represents simply a universe of existent individuals, and the different parts of the sheet represent facts or true assertion concerning that universe. At the cuts we pass into other areas, areas of conceived propositions which are not realized. In these areas there may be cuts where we pass into worlds which, in the imaginary worlds of the outer cuts, are themselves represented to be imaginary and false, but which may, for all that, be true, and therefore continuous with the sheet of assertion itself, although this is uncertain (cited in [9]).

Forcing is intrinsically relational – it distributes a determination or "control" across two heterogeneous domains. $M[G]$ is built up solely from resources available in M, but the very genericity of G guarantees that the extended model $M[G]$ is indiscernable from the standpoint of M and thus radically different or "Other". This is why the definition of G on the basis of the dominations is so important – dominations effectively formalize an intuitively epistemological or semiotic concept, namely the discernability of a "property" of some given correct set. It is the power of the generic to name the indiscernible [1] that corresponds to the creative moment that Peirce emphasizes is central in the logical operation of abduction.

The second crucial element of the heterogeneity of M and $M[G]$ is the pair of transfinite inductions that determine the hierarchy of names on the one hand and the reference-values on the other. Notice that the names are defined independently of G while the reference-values are determined precisely by coordinating names with elements of G. If the names function first of all as indices of conditions and then, through the succession of ordinal ranks, indices of possible collections of these indices as paired with conditions, and so on, the reference-values function as the well-defined link that connects the names in M to elements of $M[G]$.

Forcing is, quite simply, a carefully-defined relation between conditions and statements in M. What matters of course is precisely how this relation is defined: the effort of building up the extended model $M[G]$ in the way outlined above becomes worthwhile solely because it enables this definition itself. Forcing is a relation in M between a given condition and a statement about the names that holds if and only if the corresponding statement about the reference-values of those names is verifiable in $M[G]$ and the given condition is an element of G.

Thus, a relation that is wholly determined within M is able to signify the verifiability of statements about the extended model $M[G]$, depending upon whether or not particular conditions in M are also elements of G. Because of this, it becomes entirely determinable within M given a statement S whether or not the condition of the empty set forces S, whether or not some but not all conditions force S, and whether or not no conditions whatsoever force S. Indeed it is easily shown that these three cases or possibilities are both mutually exclusive and exhaustive. To each of the three cases corresponds a specific kind of knowledge about the statement S which takes the reference-values of the names as its argument, and hence is a statement about $M[G]$. The following table lays out the relevant correspondences:

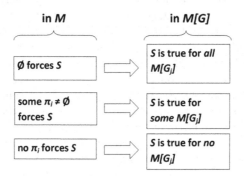

As this table shows, forcing is thus both an explicit relation in M (between conditions and statements) and an implicit correlation between M and $M[G]$ (determining at least some knowledge of the latter from the standpoint of the former, namely whether the corresponding statement about $M[G]$ is always, sometimes but not always or never verifiable).

4.2 Diagramming Forcing as the Abductive Emergence of EG–γ from EG–β

Within the newly constituted horizon of statements about $M[G]$ from the standpoint of M, the three "cases" may therefore be interpreted in terms of the three logical modalities expressible in Peirce's EG-gamma: necessity, contingency and impossibility. Peirce's EG-γ notation is as follows:

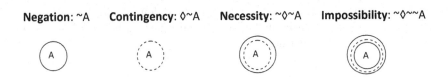

Negation: ~A   **Contingency:** ◊~A   **Necessity:** ~◊~A   **Impossibility:** ~◊~~A

An essential aspect of abduction is that its conclusions, like those of induction, are at best merely probable. Yet even more strongly, Peirce insists that abduction is not even determinately probable – this is due directly to its "creative" element. It is exactly this character of abduction that has eluded formalization.

Yet the modalities expressible in EG-γ reflect the bare, minimal knowledge that is both necessary and sufficient for abduction. Thus if we are able to express formally the relation between non-modal EG-β statements in M and modal EG-γ statements in $M[G]$, then we will have within Peirce's graphical system a diagrammatic representation of the specific correlational structure of forcing insofar as it defines modal truth-values in $M[G]$ from within M.

The following diagram represents, in EG, the lifting of the forcing relations in the ground model M into the modal truths of $M[G]$ (for the sake of concise representation, n_k here represents the sequence $n_1, n_2, \ldots, n_k$):

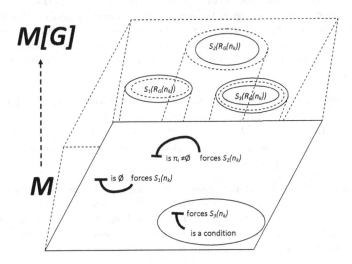

This representation shows how the set-theoretical statements and relations of forcing may be translated into the formal-iconic language of EG. This form itself is productive of new content: the EG-γ representation of forcing is itself a diagram whose internal relations carry additional implications and consequences. In particular, it illustrates the abductive character of forcing from an essentially topological perspective. In this way, we produce a rigorous framework for the further investigation of the problem of a logic of abduction in terms of the problematic of the topological relations between the continuum and its possible coverings.

5 Conclusions

Our formalization of the Peircean abductive structure of forcing extends the philosophical interpretation of Cohen's work offered by Badiou in [1] while remaining fully within the formal language of EG. Badiou understands the universe of Gödel's constructible sets to correspond to Being qua Being whereas the models of ZFC that adjoin generic sets in accordance with Cohen's technique indicate a domain in which infinite procedures of truth may exceed Being as such. As Badiou himself points out, such "truth procedures" rely upon the formation of hypotheses that are always necessarily in excess of what has already been given, but are expressed in terms of the "finite fragment made up of the present state of the enquiries":

> ...the subject solely controls – because it is such – the finite fragment made up of the present state of the enquiries. All the rest is a matter of confidence, or of knowing belief. Is this sufficient for the legitimate formulation of a hypothesis of connection between what a *truth* presents and the *veracity* of a statement that bears upon the names of a subject-language? Doesn't the infinite incompletion of a truth prevent any possible evaluation, *inside* the situation, of the veracity to-come of a statement whose referential universe is suspended from the chance, itself to-come, of encounters, and thus of enquiries?

Cohen's proof demonstrates mathematically that the answer to this question is no. The forcing relationship in M allows us to make definite, but limited, assertions about truths in $M[G]$. For Badiou, this relationship between the discretely statable and the generically indiscernible instantiates what he calls the *fundamental law of the subject*:

> if a statement of the subject-language is such that it will have been veridical for a situation in which a truth has occurred, this is because *a* term of the situation exists which both belongs to that truth (belongs to the generic part which *is* that truth) and maintains a particular relation with the names at stake in the statement.

Our diagram formalizes the relationship in question as the emergence in Peirce's EG-β of a continuous sheet in excess of the sheet of assertion which corresponds to the reference-domain of $M[G]$.

In his seminal paper, Louis Kauffman [6] establishes deep connections between Peirce's EG and infinitesimals from the topological perspective that naturally emerges from Peirce's ideas. We hope that our work can contribute to opening new ground for the construction and investigation of a formal topological theory of continuous modalities and abductive processes.

References

1. Badiou, A.: Being and Event. O. Feltham(trans.) Contiuum, London, New York (2005)
2. Chow, T.: A beginner's guide to forcing. Contemporary Mathematics 479, 35–40 (2009)
3. Cohen, P.: Set Theory and the Continuum Hypothesis. Dover Publications, Mineola (1966)
4. Gödel, K.: The Consistency of the Continuum Hypothesis. Princeton University Press, Princeton (1940)
5. Hoffmann, M.: Is There a 'Logic' of Abduction?, http://user.uni-frankfurt.de/~wirth/texte/hoffmannabdu.htm
6. Kauffman, L.: The mathematics of Charles Sanders Peirce. Cybernetics & Human Knowing 8(1-2), 79–110 (2001)
7. Peirce, C.S.: Reasoning and the Logic of Things. In: Ketner, K.L. (ed.), Harvard University Press, Cambridge (1992)
8. Roberts, D.D.: The Existential Graphs of Charles S. Peirce. Mouton and Co., The Hague (1973)
9. Zalamea, F.: Peirce's Logic of continuity: Existential graphs and non-cantorian continuum. Review of Modern Logic 9(29), 115–162 (2003)

Part III
Models, Mental Models, Representations

How Brains Make Mental Models

Paul Thagard

Abstract. Many psychologists, philosophers, and computer scientist have written about mental models, but have remained vague about the nature of such models. Do they consist of propositions, concepts, rules, images, or some other kind of mental representation? This paper will argue that a unified account can be achieved by understanding mental models as representations consisting of patterns of activation in populations of neurons. The fertility of this account will be illustrated by showing its applicability to causal reasoning and the generation of novel concepts in scientific discovery and technological innovation. I will also discuss the implications of this view of mental models for evaluating claims that cognition is embodied.

1 Introduction

Mental models are psychological representations that have the same relational structure as what they represent. They have been invoked to explain many important aspects of human reasoning, including deduction, induction, problem solving, language understanding, and human-machine interaction. But the nature of mental models and the processes that operate on them has not always been clear from the psychological discussions. The main aim of this paper is to provide a neural account of mental models by describing some of the brain mechanisms that produce them.

The neural representations required to understand mental models are also valuable for providing new understanding of how minds perform abduction, a kind of inference that generates and/or evaluates explanatory hypotheses. Considering the neural mechanisms that support abductive inference makes it possible to address several aspects of abduction, some first

Paul Thagard
Department of Philosophy, University of Waterloo, Waterloo, Canada
e-mail: pthagard@uwaterloo.ca

L. Magnani et al. (Eds.): Model-Based Reasoning in Science & Technology, SCI 314, pp. 447–461.
springerlink.com   © Springer-Verlag Berlin Heidelberg 2010

proposed by Charles Peirce, that have largely been neglected in subsequent research. These aspects include the generation of new ideas, the role of emotions such as surprise, the use of multimodal representations to produce "embodied abduction", and the nature of the causal relations that are required for explanations.

The suggestion that abductive inference is embodied raises issues that have been very controversial in recent discussions in psychology, philosophy, and artificial intelligence. This paper argues that the role of emotions and multimodal representations in abduction supports a moderate thesis about the role of embodiment in human thinking, but not an extreme thesis that proposes embodied action as an alternative to the computational-representational understanding of mind.

2 Mental Models

How do you solve the following reasoning problem? Adam is taller than Bob, and Bob is taller than Dan; so what do you know about Adam and Dan? Readers proficient in formal logic may translate the given information into predicate calculus and use their encoding of the transitivity of "taller than" to infer that Adam is taller than Dan, via applications of the logical rules of universal instantiation, and-introduction, and modus ponens. Most people, however, report using a kind of image or model of the world in which they visualize Adam as taller than Bob and Bob as taller than Dan, from which they can simply read off the fact that Adam is taller than Dan.

The first modern statement of the hypothesis that minds use mechanical processes to model the world was by Kenneth Craik, who in 1943 proposed that human thought provides a convenient small-scale model of a process such as designing a bridge [3, p. 59]. The current popularity of the idea of mental models in cognitive science is largely due to Philip Johnson-Laird, who has used it extensively in explanations of deductive and other kinds of inference as well as many aspects of language understanding (e.g. [16, 18, 19]. In his history of mental models, Johnson-Laird cites as an important precursor the ideas of Charles Peirce about the class of signs he called "likenesses" or "icons", which stand for things by virtue of a relation of similarity [17]. Earlier precursors may have been Locke and Hume with their idea that ideas are copies of images. Many recent researchers have used mental models to explain aspects of thinking including problem solving [13], inductive learning [15], and human-machine interaction (e.g. [39]). Hundreds of psychological articles have been published on mental models[1].

Nevertheless, the nature of mental models has remained rather fuzzy. Nersessian [26, p. 93] describes a mental model as a "structural, behavioral, or functional analog representation of a real-world or imaginary situation, event or process. It is analog in that it preserves constraints inherent in what is

---

[1] http://www.tcd.ie/Psychology/other/Ruth_Byrne/mental_models/

represented." But what is the exact nature of the psychological representations that can preserve constraints in the required way? One critic of mental model explanations of deduction dismisses them as "mental muddles" [34].

This paper takes a new approach to developing the vague but fertile notion of mental models by characterizing them in terms of neural processes. A neural approach runs counter to the assumption of mental modelers such as Johnson-Laird and Craik that psychological explanation can proceed at an abstract functional and computational level, but I will try to display the advantages of operating at the neural as well as the psychological level. One advantage of a neural account of mental models is that it can shed new light on aspects of abductive inference.

3 Abduction

Magnani [22] made the explicit connection between model-based reasoning and abduction, arguing that purely sentential accounts of the generation and evaluation of explanatory hypotheses are inadequate (see also [23, 5, 45]). A broader account of abduction, more in keeping with the expansive ideas of Peirce [29, 30], can be achieved by considering how mental models such as ones involving visual representations can contribute to explanatory reasoning. Sententially, abduction might be taken to be just "If p then q; why q? Maybe p". But much can be gained by allowing the p and q in the abductive schema to exceed the limitations of verbal information and include visual, olfactory, tactile, auditory, gustatory, and even kinesthetic representations. To take an extreme example, abduction can be prompted by a cry of "What's that awful smell?" that generates an explanation that combines verbal, visual, auditory, and motor representations into the answer that "Joe was trying to grate cheese onto the omelet but he slipped, cursed, and got some cheese onto the burner".

Moreover, there are aspects of Peirce's original descriptions of abduction that cannot be accommodated without taking a broader representational perspective. Peirce said that abduction is prompted by surprise, which is an emotion, but how can surprise be fitted into a sentential framework? Similarly, Peirce said that abduction introduces new ideas, but how could that happen in sentential schemas? Such ideas can generate flashes of insight, but both insight and their flashes seem indescribable in a sentential framework. Another problem concerns the nature of the "if-then" relation in the sentential abductive schema. Presumably it must be more than material implication, but what more is required? Logic-based approaches to abduction tend to assume that explanation is a matter of deduction, but philosophical discussions show that deduction is neither necessary nor sufficient for explanation (e.g. [35]). I think that good explanations exploit causal mechanisms, but what constitutes the causal relation between what is explained and what gets explained? I aim to show that all of these difficult aspects of abduction

– the role of surprise and insight, the generation of new ideas, and the nature of causality – can be illuminated by consideration of neural mechanisms.

Terminological note: Magnani [24] writes of "non-explanatory abduction", which strikes me as self-contradictory. Perhaps there is a need for a new term describing a kind of generalization of abduction to cover other kinds of backward or inverse reasoning such as generating axioms from desired theorems, but let me propose to call this generalized abduction *gabduction* and retain abduction for Peirce's idea of the generation and evaluation of explanatory hypotheses.

4 Neural Representation and Processing

A full and rigorous description of current understanding of the nature of neural representation and processing is beyond the scope of this paper, but I will provide an introductory sketch (for fuller accounts, see such sources as [2, 5, 10, 27, 42].

The human brain contains around 100,000,000,000 neurons, each of which has many thousands of connections with other neurons. These connections are either excitatory (the firing of one neuron increases the firing of the one it is connected to) or inhibitory (the firing of one neuron decreases the firing of the one it is connected). A collection of neurons that are richly interconnected is called a neural population (or group, or ensemble). A neuron fires when it has accumulated sufficient voltage as the result of the firing of the neurons that have excitatory connections to it. Typical neurons fire around 100 times per second, making them vastly slower than current computers that operate at speeds of billions of times per second, but the massive parallel processing of the intricately connected brain enables it to perform feats of inference that are still far beyond the capabilities of computers.

A neural representation is not a static object like a word on paper or a street sign, but is rather a dynamic process involving ongoing change in many neurons and their interconnections. A population of neurons represents something by its pattern of firing. The brain is capable of a vast number of patterns: assuming that each neuron can fire 100 times per second, then the number of firing patterns of that duration is $(2^{(100)})^{100000000000}$, a number far larger that the number of elementary particles in the universe, which is only about 10^{80}. I call this "Dickenson's theorem", after Emily Dickenson's beautiful poem "The brain is wider than the sky". A pattern of activation in the brain constitutes a representation of something when there is a stable causal correlation between the firing of neurons in a population and the thing that is represented, such as an object or group of objects in the world [9, 28]. The claim that mental representations are patterns of firing in neural populations is a radical departure from everyday concepts and even from cognitive psychology until recently, but is increasingly supported by data acquired through experimental techniques such as brain scans and by rapidly developing theories about how brains work (e.g. [1, 37, 42]).

5   Neural Mental Models

Demonstrating that neural representations can constitute mental models requires showing how they can have the same relational structure as what they represent, both statically and dynamically. Static mental models have spatial structure similar to what they represent, whereas dynamic mental models have similar temporal structure. Combined mental models capture both spatial and temporal structure, as when a person runs a mental movie that represents what happens in some complex visual situation such as two cars colliding.

The most straightforward kind of neural mental models are topographical sensory maps, for which Knudsen, du Lac, and Esterly [21, p. 61] provide the following summary:

> The nervous system performs computations to process information that is biologically important. Some of these computations occur in maps – arrays of neurons in which the tuning of neighboring neurons for a particular parameter value varies systematically. Computational maps transform the representation into a place-coded probability distribution that represents the computed values of parameters by sites of maximum relative activity. Numerous computational maps have been discovered, including visual maps of line orientation and direction of motion, auditory maps of amplitude spectrum and time interval, and motor maps of orienting movements.

The simplest example is the primary visual cortex, in which neighboring columns of neurons process information from neighboring small regions of visual space [21, 20]. In this case, the spatial organization of the neurons corresponds systematically to the spatial organization of the world, in the same way that the location of major cities on a map of Brazil corresponds to the actual location of those cities.

Such topographic neural models are useful for basic perception, but they are not rich enough to support high level kinds of reasoning such as my "taller than" example. How populations of neurons can support such reasoning is still unknown, as brain scanning technologies do not have sufficient resolution to pin down neural activity in enough detail to inspire theoretical models of how high-level mental modeling can work. But let me try to extrapolate from current views on neural representation, particularly those of Eliasmith and Anderson [10], to suggest how the brain might be able to make extra-topographic models of the world (see also [9]).

Neural populations can acquire the ability to encode features of the world as their firing activity becomes causally correlated with those features. (*A* and *B* are causally correlated if they are statistically correlated as the result of causal interactions between *A* and *B*). Neural populations are also capable of encoding the activity of other neural populations, as the firing patterns of one population becomes causally correlated with the firing patterns of another population that feeds into it. If the input population is a topographic map, then the output population can become a more abstract representation of the features of the world, in two ways. The most basic retains some of the topographic structure of the input population, so that the output population is still a mental model of the world in that it shares some (but not all) relational structure with it. An even more abstract encoding is performed by an output neural population that captures key aspects of the encoding performed by the input population, but does so in a manner analogous to the way that language produces arbitrary, non-iconic representations. Just as there is no similarity between the word "cat" and cats, so the output neural population may have lost the similarity with the original stimulus: not all thinking uses mental models. Nevertheless, in some cases the output population provides sufficient information to enable decoding that generates an inference fairly directly, as in the "taller-than" example. The encodings of Adam, Bob, and Dan that include their heights makes it possible to just "see" that Adam is taller than Dan.

A further level of representation is required for consciousness, such as the experienced awareness that Adam is taller than Dan. Many philosophers and scientists have suggested that consciousness requires representation of representation (for references see [43]), but mental models seem to require several layers: representation of representation of representation. The conscious experience of an answer to a problem comes about because of activity in top-level neural populations that encode activity of medium-level modeling populations, that encode activity of low-level populations, that topographically represent features of the world. To put it another way, conscious models represent mid-level models that represent low-level topographic models that represent features of the world. The relation *representation* need not be transitive, but in this case it carries through, so that the conscious mental model represents the world and generates inferences about it.

So far, I have been focusing on mental models where the similar relation-structure is spatial, but temporal relations are just as important. When you imagine your national anthem sung by Michael Jackson, you are creating a mental model not only of the individual notes and tones but also of their sequence in time. Similarly, a mental model of a working device such as a windmill requires both visual/spatial representations of the blades and base of the windmill and also temporal representations of the blades of the mill. Not a lot of research has been done on how neurons can encode temporal relations, but I will explore two possibilities.

Elman [11] and other researchers have shown how simple recurrent networks can encode temporal relationships needed for understanding language. A recurrent network is one in which output neurons feed back to provide input to the input neurons, producing a kind of temporal cycle that can retain information. Much more complex neural structures, however, would be needed to encode a song or running machine, perhaps something like the neural representation of a rule-based inference system being developed by Terry Stewart using Chris Eliasmith's neural engineering framework [38]. On this approach, a pattern of neural activation encodes a state of affairs that can be matched by a collection of rules capturing if-then relations. Then running a temporal pattern is a matter of firing off a sequence of rules, not by the usual verbal matching employed by rule-based cognitive architectures such as Anderson's [1] ACT, but by purely neural network operations. If the neural populations representing the states of affairs are mental models of either the direct, topographic kinds or the abstracted, structure-preserving kinds, then the running of the rule-based system would constitute a temporal *and* spatial mental model of the world.

6 Generating New Ideas

Peirce claimed that abduction could generate new ideas, but he did not specify how this could occur. If abduction is analyzed as a logical schema, then it is utterly mysterious how any new ideas could arise. The schema might be something like: "q is puzzling, p explains q, so maybe p." But this schema already includes the proposition p, so nothing new is generated. Hence logic-based approaches to abduction seem impotent to address what Peirce took to be a major feature of this kind of inference (cf. [45, 4]). Thagard [4] gave an account of how new concepts can be generated in the context of explanatory reasoning, but this account only applied to verbal concepts represented as frames with slots and values.

In contrast, the view of representations as patterns of activity in neural populations can be used to describe the generation of new multimodal concepts. Here I give only a quick sketch, as full details including mathematical analysis and computer simulations are provided in [46].

Assuming that two concepts are represented by patterns of activity in neural populations, which may be disjoint or overlapping, then a new combined concept can be represented by a new pattern of activity in a neural population which may also be disjoint or overlapping. A mathematical operation that combines patterns of neural activity is called convolution, which was originally a method for combining waves in signal processing theory. Tony Plate [31] adapted convolution to apply to vectors that stand for high-level symbolic representations, and Chris Eliasmith [8] developed a method for using biologically realistic neural networks to perform convolution. Thagard and Stewart [46] describe how many kinds of creativity and innovation, including scientific discovery, technological invention, social innovation, and artistic imagination, can be understood in terms of mechanisms of representation combination.

The convolution model of creative conceptual combination is fully multimodal, applying to whatever neural populations can represent, including information that is visual, auditory, olfactory, gustatory, tactile, kinesthetic, or pain-related. Moreover, the Thagard and Stewart [46] account of creativity also applies to emotions, which can also be understood as patterns of activity in neural populations involving multiple brain areas involved in both cognitive appraisal and physiological perception [43]. In particular, the wonderful AHA! experience that attends creative breakthroughs can be understood as a neural process that involves a triple convolution:

1. Two representations are convolved to produce a novel one.
2. An emotional reaction to the new representation requires a convolution of cognitive appraisal and physiological perception.
3. The AHA or EUREKA reaction is a convolution of the new representation and the emotional reaction to it.

Thus the mechanism of convolution in neural networks is capable of modeling not only the combination of representations but also the emotional reaction that successful combinations generates.

Creative conceptual combination does not occur randomly, but rather in the directed context of attempts to solve problems, including ones that require generation of new explanations. Let us now consider how abductive inference can operate with neural populations.

7 Neural Abduction and Causality

Following ideas suggested in [41], Thagard and Litt [44] presented a neural network model of abductive reasoning based on the account of reasoning with conditionals developed by Eliasmith [8]. At one level, our neural model of abduction is very simple, using thousands of neurons to model a transition from q and p causes q to p. The advantage in taking a neural approach to modeling, as I have already described, is that p and q need not be linguistic representations, but can operate in any modality. To take a novel example, q could be a neural encoding of pain that I feel in my finger, and p could be neural encoding of a picture of splinter in my finger. Then my abductive inference goes from the experience of pain to the adoption of the visual representation that there is a splinter.

Moreover, the neural model of abduction tracks the relevant emotions. Initially, the puzzling q is associated with motivating emotions such as surprise and irritation. But as the hypothesis p is abductively adopted, the emotional reaction changes to relief and pleasure. Thus neural modeling can capture emotional aspect of abductive reasoning.

But how can we understand the causal relation in "p causes q"? Thagard and Litt ([44], see also [41]) argue that causality should not be construed formally as a deductive or probabilistic relation, but as a schema that derives from patterns of visual-motor experience. For example, when a baby discovers that moving its hand can move a rattle, it is forming an association that combines an initial visual state with a subsequent motor and tactile state (pushing the rattle and feeling it) with a subsequent visual-auditory state (seeing the rattle move and make noise). I do not know whether such sensory-motor-sensory schemas are innate, having been acquired by natural selection in the form of neural connections that everyone is born with; alternatively they may be acquired very quickly by infants thanks to innate learning mechanisms. But on the basis of perceptual experiments in both adults and children, there is evidence that understanding of causality is tied to such multimodal representations (e.g. [25]). Moreover, the concept of force that figures centrally in many accounts of physical causality has its cognitive roots in body-based experiences of pushes and pulls. Hence it seems appropriate to speak of "embodied abduction", since both the causal relation itself and the multimodal representations of many hypotheses and facts to be explained are tied to sensory operations of the human body. However, the topic of embodiment is highly controversial, so I now discuss how I think the embodiment of abduction and mental models needs to be construed.

8   Embodiment: Moderate and Extreme

I emphatically reject the extreme embodiment thesis that thinking is just embodied action and therefore incompatible with computational-representational approaches to how brains work [6]. I argue below that even motor control requires a high degree of representation and computation. Much more plausible is the moderate embodiment thesis that language and thought are inextricably shaped by embodied action, a view that is maintained by Gibbs [14], Magnani [24] and others. On this view, thinking still requires representations and computations, but the particular nature of these depends in part on the kind of bodies that people have, including their sensory and motor capabilities. My remarks about multimodal representations and the sensory-motor-sensory schemas that underlie causal reasoning provide support for the moderate embodiment thesis.

However, there are two main reasons for not endorsing the extreme embodiment thesis. First, many kinds of thinking including causal reasoning, emotion, and scientific theorizing take us well beyond sensorimotor processes, so explaining our cognitive capacities requires recognizing representational/computational abilities that outstrip embodied action. Second, even the central case of embodied action – motor control – requires substantial representational/computational capabilities.

I owe to Lloyd Elliott the following summary of why motor control is much harder than you might think. Merely reaching to pick up a book requires solutions to many difficult problems for the brain to direct an arm and hand to reach out and pick up the book. First, the signals that pass between the brain and its sensors and muscles are very noisy. Information about the size, shape, and location of the book is transmitted to be brain via the eyes, but the process of translating retinal signals into judgments about the book involved multiple stages of neural transformations [37, ch. 2]. Moreover, when the brain directs muscles to move the arm and hand in order to grasp the book, the signals sent involve noisy activity in millions of nerve cells.

Second, motor control is also made difficult by the fact that the context is constantly changing. You may need to pick up a book despite the fact that there are numerous changes taking place, not only in the orientation of your body, but also in visual information such as light intensity and the presence of other objects in the area. A person can pick up a book even though another person has reached across to pick up another book. Third, there are unavoidable time delays as the brain plans and attempts to move the arm to pick up the book.

Fourth, motor control is not an automatic process that occurs instantly to people, but usually requires large amounts of learning. It takes years for babies to become adept at handling physical objects, and even adults require months or years to become proficient at difficult motor tasks such as playing sports. Fifth, motor control is not a simple linear process of the brain just telling a muscle what to do, but requires non-linear integrations of the movements of multiple muscles and joints, which operate with many degrees of freedom. Picking up a book requires the coordination of all the muscles that move different parts of fingers, wrists, elbows, and shoulders.

Hence grasping and moving objects is a highly complex task that has been found highly challenging by people attempting to build robots. Fortunately for humans, millions of years of animal evolution have provided humans with the capacity to learn how to manipulate objects. Recent theoretical explanations of this capacity understand motor control as representational and computational, requiring mental models (see e.g. [4, 47]). What follows is a concise, simplified, synthesis of their accounts.

The brain is able to manipulate objects because its learning mechanisms, both supervised and unsupervised, enable it to build powerful internal models of connections among sensors, brain, and world. A brain needs a *forward model* from movements to sensory results, which enables it to predict what will be perceived as the result of particular movements. It also needs an *inverse model* from sensory results to movements, which enables it to predict what movement will produce the desired perceived result. Forward and inverse models are both dynamic mental models in the sense I discussed earlier: the relational structure they share with what they represent is both spatial and temporal, concerning the location and movement of limbs to produce changes in the world. Motor control in general requires a high-level control process in which the brain enables the body to interact productively with the world through a combination of representations of situations and goals, forward and inverse models, perceptual filters, and muscle control processes. The overall process is highly complex and not all like the kinds of manipulations of verbal symbols that some philosophers still take as the hallmark of representation and computation. But the brain's neural populations still stand for muscle movement and visual changes, with which their activity is causally correlated, so it is legitimate to describe the activities of such populations as representational. Moreover, the mental modeling, both forward and inverse, is carried out by systematic changes in the neural populations, which hence qualifies as "principled manipulation of representations" [7, p. 29].

Let me summarize the argument in this section. Embodied action requires motor control. Motor control requires mental models, both forward and inverse, to identify dynamic relations among sensory information and muscle activity. Mental models are representational and computational. Hence embodied action requires representation and computation, so that it cannot provide an alternative to the representational/computational view of mind. Therefore considerations of multimodal representations, embodied abduction, and sensory-motor conceptions of causality only support the moderate embodiment thesis, and in fact require rejection of the extreme version.

Proponents of representation-free intelligence like to say that "the world is its own best model". As an advisory that a robot or other intelligent system should not need to represent everything to solve problems, this remark is useful; but literally it is clearly false. For imagining, planning, explaining, and many other important cognitive activities, the world is a very inadequate model of itself: far too complex and limited in its manipulability. In contrast, mental models operating at various degrees of abstraction are invaluable for high-level reasoning. The world might be its own best model if you're a cockroach, with very limited modeling abilities. But if you have the representational power of a human or powerful robot, then you can build simplified but immensely useful models of past and future events, as well as of events that your senses do not enable you observe. Hence science uses abductive inference and conceptual combination to generate representations of theoretical (i.e. non-observable) entities such as electrons, viruses, genes, and mental representations.

Cockroaches and many other animals are as embodied, embedded, and situated in the world as human beings, but they are far less effective than people at building science, technology, and other cultural developments. One of the many advantages that people gain from our much larger brains is the ability to work with mental models, including ones used for abduction and the generation of new ideas.

9 Conclusion

I finish with a reassessment of Peirce's ideas about abduction from the neural perspective that I have been developing. Peirce did most of his work on inference in the nineteenth century, well before the emergence of ideas about computation and neural processes. He was a scientist as well as a philosopher of science, and undoubtedly would have revised his views on the nature of inference in line with subsequent scientific developments.

On the positive side, Peirce was undoubtedly right about the importance of abduction as a kind of inference. The evaluative aspect of abduction is recognized in philosophy of science under the headings of inference to the best explanation and explanatory coherence, and the creative aspect is recognized in philosophy and artificial intelligence through work on how hypotheses are formed. Second, Peirce was prescient in noticing the emotional instigation of abduction as the result of surprise, although I do not know if he also noticed that achieving abduction generates the emotional response of relief. Third, Peirce was right in suggesting that the creation of new ideas often occurs in the context of abductive inference, even if abduction itself is not the generating process.

On the other hand, there are several suggestions that Peirce made about abduction that do not fit well with current psychological and neural understanding of abduction. I do not think that emotion is well described as a kind of abduction, as it involves an extremely complex process that combines cognitive appraisal of a situation with respect to ones goals and perception of bodily states [43, 42]. At best, abductive inference is only a part of the broader parallel process of emotional reactions. Similarly, perception is not a kind of abduction, as it involves many more basic neuropsychological processes that are not well described as generation and evaluation of explanatory hypotheses (see e.g. [37, ch. 2]).

Finally, Peirce's suggestion that abduction requires a special instinct for guessing right is not well supported by current neuropsychological findings. Perhaps evolutionary psychologists would want to propose that there is an innate module for generating good hypotheses, but there is a dearth of evidence that would support this proposal. Rather, I prefer the suggestion of Quartz and Sejnowski [32, 33] that what the brain is adapted for is adaptability, through powerful learning mechanisms that humans can apply in many contexts. One of these learning mechanisms is abductive inference, which leads people to respond to surprising observations with a search for hypotheses that can explain them. Like all cognitive processes, this search must be constrained by contextual factors such as triggering conditions that cut down the number of new conceptual combinations that are performed [4]. Abduction and concept formation occur as part of the operations of a more general cognitive architecture.

I see no reason to claim that the constraints on these operations include preferences for particular kinds of hypotheses, which is how I interpret Peirce's instinct suggestion. Indeed, scientific abduction has led to the generation of many hypotheses that scientists now think are wrong (e.g. humoral causes of disease, phlogiston, caloric, and the luminiferous ether) and to many hypotheses that go against popular inclinations (e.g. Newton's force at a distance, Darwin's evolution by natural selection, Einstein's relativistic account of space-time, quantum mechanics, and the mind-brain identity theory). Although it is reasonable to suggest that the battery of innate human learning mechanisms includes ones for generating hypotheses to explain surprising events, there is no support for Peirce's contention that people must have an instinct for guessing *right*. Evolutionary psychologists like to compare the brain to a Swiss army knife that has many specific built-in capacities; but a more fertile comparison is the human hand, which evolved to be capable of many different operations from grasping to signaling to thumb typing on smartphones. Peirce's view of abduction as requiring innate insight is thus as unsupported by current research as the view of Fodor [12] that cognitive science cannot possibly explain abduction: many effective techniques have been developed by philosophers and AI researchers to explain complex causal reasoning.

I have tried to show in this paper how Peirce's abduction is, from a neural perspective, highly consonant with psychological theories of mental models, which can also productively be construed as neural processes. Brains make mental models through complex patterns of neural firing and use them in many kinds of inference, from planning actions to the most creative kinds of abductive reasoning. I have endorsed a moderate thesis about the importance of embodiment for the kinds of representations that go into mental modeling, but critiqued the extreme view that sees embodiment as antithetical to mental models and other theories of representation. Further developments of neural theories of mental models should further clarify their roles in many important psychological phenomena.

Acknowledgements. For valuable ideas and useful discussions, I am grateful to Chris Eliasmith, Terry Stewart, Lloyd Elliott, and Phil Johnson-Laird. This research has been funded by the Natural Sciences and Engineering Research Council of Canada.

References

1. Anderson, J.R.: How Can the Mind Occur in the Physical Universe? Oxford University Press, Oxford (2007)
2. Churchland, P.S., Sejnowski, T.: The Computational Brain. MIT Press, Cambridge (1992)
3. Craik, K.: The Nature of Explanation. Cambridge University Press, Cambridge (1943)
4. Davidson, P.R., Wolpert, D.M.: Widespread access to predictive models in the motor system: A short review. Journal of Neural Engineering 2, S313–S319 (2005)
5. Dayan, P., Abbott, L.F.: Theoretical Neuroscience: Computational and Mathematical Modeling of Neural Systems. MIT Press, Cambridge (2001)
6. Dreyfus, H.L.: Why Heideggerian AI failed and how fixing it would require making it more Heideggerian. Philosophical Psychology 20, 247–268 (2007)
7. Edelman, S.: Computing the Mind: How the Mind Really Works. Oxford University Press, Oxford (2008)

8. Eliasmith, C.: Cognition with neurons: A large-scale, biologically realistic model of the Wason task. In: Bara, B., Barasalou, L., Bucciarelli, M. (eds.) Proceedings of the XXVII Annual Conference of the Cognitive Science Society, pp. 624–629. Lawrence Erlbaum Associates, Mahwah (2005)

9. Eliasmith, C.: Neurosemantics and categories. In: Cohen, H., Lefebvre, C. (eds.) Handbook of Categorization in Cognitive Science, pp. 1035–1054. Elsevier, Amsterdam (2005)

10. Eliasmith, C., Anderson, C.H.: Neural Engineering: Computation, Representation and Dynamics in Neurobiological Systems. MIT Press, Cambridge (2003)

11. Elman, J.L.: Finding structure in time. Cognitive Science 14, 179–211 (1990)

12. Fodor, J.: The Mind Doesn't Work that Way. MIT Press, Cambridge (2000)

13. Gentner, D., Stevens, A.L. (eds.): Mental Models. Lawrence Erlbaum, Hillsdale (1983)

14. Gibbs, R.W.: Embodiment and Cognitive Science. Cambridge University Press, Cambridge (2006)

15. Holland, J.H., Holyoak, K.J., Nisbett, R.E., Thagard, P.R.: Induction: Processes of Inference, Learning, and Discovery. MIT Press/Bradford Books, Cambridge (1986)

16. Johnson-Laird, P.N.: Mental Models. Harvard University Press, Cambridge (1983)

17. Johnson-Laird, P.N.: The history of mental models. In: Manktelow, K., Chung, M.C. (eds.) Psychology of Reasoning: Theoretical and Historical Perspectives, pp. 179–212. Psychology Press, New York (2004)

18. Johnson-Laird, P.N.: How We Reason. Oxford University Press, Oxford (2006)

19. Johnson-Laird, P.N., Byrne, R.M.: Deduction. Lawrence Erlbaum Associates, Hillsdale (1991)

20. Kaas, J.H.: Topographic maps are fundamental to sensory processing. Brain research bulletin 44, 107–112 (1997)

21. Knudsen, E.I., du Lac, S., Esterly, S.D.: Computational maps in the brain. Annual Review of Neuroscience 10, 41–65 (1987)

22. Magnani, L.: Model-based creative abduction. In: Magnani, L., Nersessian, N.J., Thagard, P. (eds.) Model-Based Reasoning in Scientific Discovery, pp. 219–238. Kluwer/Plenum, New York (1999)

23. Magnani, L.: Abduction, Reason, and Science: Processes of Discovery and Explanation. Kluwer/Plenum, New York (2001)

24. Magnani, L.: Abductive Cognition. The Epistemological and Eco-Cognitive Dimensions of Hypothetical Reasoning. Springer, Berlin (2009)

25. Michotte, A.: The Perception of Causality. Miles, T.R., Miles, E. (trans.), Methuen, London (1963)

26. Nersessian, N.J.: Creating Scientific Concepts. MIT Press, Cambridge (2008)

27. O'Reilly, R.C., Munakata, Y.: Computational Explorations in Cognitive Neuroscience. MIT Press, Cambridge (2000)

28. Parisien, C., Thagard, P.: Robosemantics: How Stanley the Volkswagen represents the world. Minds and Machines 18, 169–178 (2008)

29. Peirce, C.S.: Collected Papers. Harvard University Press, Cambridge (1931-1958)

30. Peirce, C.S.: Reasoning and the Logic of Things. Harvard University Press, Cambridge (1992)

31. Plate, T.: Holographic Reduced Representations. CSLI, Stanford (2003)

32. Quartz, S.R., Sejnowski, T.J.: The neural basis of cognitive development: A constructivist manifesto. Behavioral and Brain Sciences 20, 537–556 (1997)
33. Quartz, S.R., Sejnowski, T.J.: Liars, Lovers, and Heroes: What the New Brain Science Reveals about How We Become Who We Are. Morrow, William, New York (2002)
34. Rips, L.J.: Mental muddles. In: Brand, M., Harnish, R.M. (eds.) The representation of knowledge and belief, pp. 258–286. University of Arizona Press, Tucson (1986)
35. Salmon, W.C.: Four decades of scientific explanation. In: Kitcher, P., Salmon, W.C. (eds.) Scientific Explanation. Minnesota Studies in the Philosophy of Science, vol. XIII, pp. 3–219. University of Minnesota Press, Minneapolis (1989)
36. Shelley, C.P.: Visual abductive reasoning in archaeology. Philosophy of Science 63, 278–301 (1996)
37. Smith, E.E., Kosslyn, S.M.: Cognitive Psychology: Mind and Brain. Pearson Prentice Hall, Upper Saddle River (2007)
38. Stewart, T.C., Eliasmith, C.: Spiking neurons and central executive control: The origin of the 50-millisecond cognitive cycle. In: Howes, A., Peebles, D., Cooper, R. (eds.) 9th International Conference on Cognitive Modeling ICCM 2009, Manchester, UK (2009)
39. Tauber, M.J., Ackerman, D. (eds.): Mental Models and Human-Computer Interaction, vol. 2. North-Holland, Amsterdam (1990)
40. Thagard, P.: Computational Philosophy of Science. MIT Press, Cambridge (1988)
41. Thagard, P.: Abductive inference: From philosophical analysis to neural mechanisms. In: Feeney, A., Heit, E. (eds.) Inductive Reasoning: Experimental, Developmental, and Computational Approaches, pp. 226–247. Cambridge University Press, Cambridge (2007)
42. Thagard, P.: The Brain and the Meaning of Life. Princeton University Press, Princeton (2010)
43. Thagard, P., Aubie, B.: Emotional consciousness: A neural model of how cognitive appraisal and somatic perception interact to produce qualitative experience. Consciousness and Cognition 17, 811–834 (2008)
44. Thagard, P., Litt, A.: Models of scientific explanation. In: Sun, R. (ed.) The Cambridge Handbook of Computational sychology, pp. 549–564. Cambridge University Press, Cambridge (2008)
45. Thagard, P., Shelley, C.P.: Abductive reasoning: Logic, visual thinking, and coherence. In: Dalla Chiara, M.L., Doets, K., Mundici, D., van Benthem, J. (eds.) Logic and Scientific Methods, pp. 413–427. Kluwer, Dordrecht (1997)
46. Thagard, P., Stewart, T.C.: The Aha! experience: Creativity through emergent binding in neural networks. Cognitive Science (forthcoming)
47. Wolpert, D.M., Ghahramani, Z.: Computational principles of movement neuroscience. Nature Neuroscience 3, 1212–1217 (2000)

Applications of an Implementation Story for Non-sentential Models

Jonathan Waskan

Abstract. The viability of the proposal that human cognition involves the utilization of non-sentential models is seriously undercut by the fact that no one has yet given a satisfactory account of how neurophysiological circuitry might realize representations of the right sort. Such an account is offered up here, the general idea behind which is that high-level models can be realized by lower-level computations and, in turn, by neural machinations. It is shown that this account can be usefully applied to deal with problems in fields ranging from artificial intelligence to the philosophy of science.

1 Introduction

Here I offer an elaboration and defense of the cognitive models hypothesis (CMH), which is the proposal that human cognition is sometimes constituted by the utilization of non-sentential representations that are like scale models in crucial respects. Scale models are representations that can be used to effect truth-preserving inferences regarding modeled systems in virtue of their instantiation of the very same properties as those systems. They are also sometimes termed *physically isomorphic* representations [1]. On this broad construal, even simple spatial matrix representations of spatial properties and relations count as scale models.

There is, of course, ample historical precedent for the CMH, having been advanced by Aristotle, Berkeley, Locke, and many others. Many contemporary cognitive scientists also favor the CMH, an hypothesis to which they often refer (or at least allude) with terms like "mental image", "depictive" or "non-propositional representation", "mental model", or "analog representation". The CHM has in recent decades been fruitfully applied in the context

Jonathan Waskan
Department of Philosophy, University of Illinois at Urbana-Champaign
e-mail: waskan@illinois.edu

L. Magnani et al. (Eds.): Model-Based Reasoning in Science & Technology, SCI 314, pp. 463–476.
springerlink.com

of theories of perception, comprehension, grammar, classification, deductive reasoning, planning, abduction, naïve physics, and mind reading. Despite its many useful applications, it has been clear for some time that the CMH faces a major realization crisis. Here I will describe the nature of this crisis and chart out what, at present, appears to be the only way past it. I will then show that this way of resolving the crisis can be used to address a set of issues that has recurred in various guises across a number of fields, including artificial intelligence, logic, psychology, and the philosophy of science.

2   The Realization Crisis

By way of introducing the realization crisis facing the CMH, consider first another hypothesis in whose favor the realization crisis has been resolved – namely, the computational theory of cognitition (CTC). The CTC is just the idea that cognition is constituted (some think entirely) by representation-transforming processes that involve the application of syntax-sensitive rules to sentential representations. McCulloch and Pitts provided one of the earliest indications of how a bridge might be built from neural machinations to strict computations with their description of how collections of neuron-like process-ing units might implement a set of logic gates and, ultimately, a universal Turing machine [2]. More recently, it has been shown that recurrent neural networks are (infinite memory notwithstanding) universal-Turing equivalent [3]. This sort of research has left few doubts that the brain is the sort of sys-tem that can, in principle, realize the sorts of strict (i.e., syntax-crunching) computations posited by proponents of the CTC.

In contrast, a satisfactory demonstration that neural machinations might realize the sorts of representations posited by the CMH has proven far more elusive. The CMH is, recall, the proposal that humans utilize non-sentential representations that are like scale models in crucial respects. Unfortunately, when attempts have been made to specify in precisely in what respects these putative cognitive models are like scale models, the resulting proposals have come out looking either too weak to adequately distinguish cognitive models from other sorts of representations or too strong to be compatible with basic brain facts.

A proposal of the latter sort is that the cognitive representations in ques-tion are like scale models in that they too are physically isomorphic with what they represent. Kosslyn, for example, has claimed, based upon the fact that certain areas of visual cortex exhibit retinotopy, that "[t]hese areas represent depictively in the most literal sense [. . .]" [4]. Unfortunately, any physical isomorphisms exhibited by these areas are highly distorted due, for instance, to the disproportionate amount of area devoted to the fovea. Moreover, such areas exhibit at best 2-D isomorphisms, but in order for cognitive repre-sentations to play the role for which they are slated by most theories that invoke the CMH they generally need to exhibit isomorphisms with regard to

3-D spatial properties, not to mention kinematic and dynamic ones. In other words, in many cases there would, *per impossibile*, need to be literal buckets, balls, doors, and so forth, in head.

Other ways of fleshing out the notion of a cognitive model have proven too weak to adequately distinguish cognitive models from other sorts of representations. Some say, for instance, that cognitive models are representations that are *merely* isomorphic (i.e., isomorphic without any further restrictions) with what they represent. This seems to be what Craik had in mind when he claimed, "By a model we thus mean any physical or chemical system which has a similar relation-structure to that of the process it imitates" [5]. However, a very wide range of representations will count as models on this view, including those created using the notations of formal logic and, relatedly, those harbored by production systems. Craik's own example of a representational system with a similar relation structure to what it represents – namely, Kelvin's Tide Predictor[1] – does little to narrow down the relevant field. If anything it helps prove the point that in order to support truth-preserving inferences regarding a represented domain a representational system *of any sort* must exhibit isomorphisms with that domain.

Also widely regarded as too weak to distinguish cognitive models from other sorts of representations is the proposal that cognitive models are *functionally* isomorphic with what they represent – that is, that they function in the ways that physically isomorphic representations such as scale models function. Unfortunately, it was recognized some time ago that one may always constrain a syntax-crunching system so that it functions like a scale model [1, 6, 7, 8]. Case in point, computational matrix representations have been constructed that function in the ways that 2-D and 3-D spatial matrix representations function (e.g., in terms of how changes in relative location are updated). However, accepted wisdom has it that these are computational representations in the strict sense described above and, accordingly, that they should be considered sentential representations in good standing. As Block puts it, "[o]nce we see what the computer does, we realize that the representation of the line is descriptional" [9]. In other words, the received view, which has gone nearly unchallenged, is that if a representation of spatial, kinematic, or dynamic properties (let us call these *corporeal* properties) is implemented using a high-level computer program, then it must be sentential in character [10, 11, 12].

In sum, proponents of the CMH have, as yet, failed to adequately articulate what makes cognitive models distinct from other sorts of representations in a way that is compatible with basic brain facts. This concern, which has been forcefully articulated by Pylyshyn and other proponents of the CTC, by no means entails that research in the CMH tradition should grind to a halt. However, it does detract from the relative plausibility of theories that invoke the CMH (e.g., as compared to those that invoke the CTC). As a proponent

---

[1] See http://www.math.sunysb.edu/~{}tony/tides/machines.html#kp1 (last accessed January 9, 2010).

of the CMH, I found it important to search for a favorable way of resolving
this realization crisis facing the CMH.

3 The Realization Story

What soon became apparent was that, with their insistence that computational implementation entails sentential representation, proponents of the
CTC were turning their backs on the principle of property independence
(POPI), a principle that more than any other helped pave the way for a
favorable resolution of the realization crisis facing their own position.

POPI: Properties found to characterize a system when it is studied at a
relatively low level of abstraction are often absent when it is studied at a
higher level, and vice versa.

This principle is in large part what justifies one in saying that at level n
a certain system contains a set of electronic switches, relays, and so forth,
that at level $n + 1$ it is best described in terms of the storing of bits of
information in numerically addressable memory registers, and that at level
$n+2$ it is best understood in terms of the application of syntax-sensitive rules
to syntactically structured representations. However, nothing about POPI
entails that at the *highest* level these systems must be characterized in terms
of sentences and inference rules. Indeed, POPI opens up at least logical space
for systems that engage in syntax crunching at one level and that harbor
and manipulate non-sentential models at a higher level. As it turns out, in
this logical space reside actual systems, such as those that implement finite
element models (FEMs). These computationally realized representations of
actual and possible physical conditions were first developed in the physical
(e.g., civil and mechanical) engineering disciplines, but now they are used in a
variety of fields for purposes of testing designs, exploring the ramifications of
theories, generating novel predictions, and facilitating understanding. What
is important about FEMs for our purposes is that they constitute an existence
proof that computational processes can realize non-sentential representations
that are like scale models and unlike sentential representations in crucial
respects.

To see why, notice first that there are at least two levels of abstraction at
which a given FEM may be understood. Just as with scale models, there is,
to start with, the relatively low level of the modeling medium. In the case of
FEMs, at this level what one finds are sentential specifications of coordinates
(e.g., polygon vertices) along with rules constraining how they may change
(e.g., due to collisions and loads) (see Figure 1). This, clearly, is the level
upon which enemies of the idea of computationally realized non-sentential
models are fixating when they suggest that computational realization entails
sentential representation. It is, however, not obvious that at this level what
we are even dealing with are representations (i.e., of worldly objects and properties), any more than we are when, for instance, we fixate on the constraints

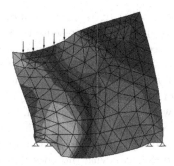

Fig. 1 Polymesh representation of a blunt impact to a semi-rigid sheet of material.

governing the behavior of individual Lego blocks. Rather, the representations are to be found when we turn our attention to the higher level of the models that are realized, *and multiply realizable*, by such modeling media. Moreover, when we take a closer look at the properties of the high-level FEMs realized through massive amounts of low-level number crunching, we find that they share several characteristics that are – most notably, by those who suggest that computational implementation entails sentential representation – taken to distinguish sentential representations from scale models.

To start with, scale models are often said to be distinct from sentential representations in that, *taken by themselves*, the former are all but incapable of representing either non-concrete properties (e.g., the property of being a war criminal), genera (e.g., triangularity), or the singling out of specific properties of specific objects (e.g., the mere fact that Joe's house is blue)[2]. Taken by themselves FEMs suffer from these exact same limitations. Like a scale model, an FEM is always a representation of a specific, concrete object, though it can be used as a proxy for many different objects of the same type. In addition, it always represents many properties of a given object rather than singling out any specific one. Thus, by these reasonable standards, FEMs ought to be considered computationally-*realized* non-sentential models that are at least the close kin of scale models.

What is far more important for present purposes is that representational genera are also often distinguished in terms of whether they constitute extrinsic or intrinsic representations. This distinction, which was introduced by Palmer [1], has been picked up on by opponents of the idea that computers realize non-sentential models. On Palmer's view, extrinsic representations

---

[2] In Waskan [17] I argue, without undermining the present point, that these representational limitations might be overcome in the human case through the use of extra-representational cognitive resources, such as those implicated in emotional, attentional, and analogical processes. On this view, the sentences with which we give voice to mental states are, somewhat as Dennett and Churchland would have it, compressed representations of a far more complex underlying cognitive reality.

are those that need to be arbitrarily constrained in order to respect the non-arbitrary constraints governing their represented systems, whereas intrinsic representations do not need to be so constrained. Sentence-and-rule-based representations (e.g., formal-logic and production-system representations) are thought to best exemplify the former [11, 13]. Scale models are thought to best exemplify the latter. While this distinction does get at an important difference between the two sorts of representation, it has unfortunately been drawn in a way that leans far too heavily upon the unclear notion of an arbitrary constraint. However, it is possible to preserve the key intuitions behind it in a way that does away with the questionable appeal to arbitrary vs. non-arbitrary constraints.

A better way to draw the distinction is to take as extrinsic those representations that only support predictions concerning particular types of alterations to a represented systems on the basis of distinct data structures. Extrinsic representations are those wherein the consequences of different types of alteration must be spelt out and built in – that is, by hand, learning, or evolution – antecedently and explicitly. For instance, in order to predict the consequences of alterations to even a simple system, such as one containing a doorway, a bucket, and a ball, a production system must incorporate distinct statements or operators that represent the consequences of those alterations. With intrinsic representations, on the other hand, the consequences of different types of alterations can instead be determined on demand and as needed simply by manipulating the representation in the relevant ways and reading off the consequences. For instance, a scale model of the system containing a door, bucket, and a ball can be manipulated in countless ways in order to predict how the consequences of many distinct alterations might play out. In order to predict what happens when the bucket is placed over the ball and slid through the door, one simply carries out the corresponding alteration to the model. Thus, one need not incorporate information about the consequences of this, and countless other alterations antecedently and explicitly.

Now the received view is that FEMs and their brethren are extrinsic representations because the constraints governing how the coordinates of primitive modeling elements may change must be explicitly, antecedently, and arguably even arbitrarily imposed [13]. Indeed, at the level of coordinates and transformation rules nothing is gotten for free in the case of FEMs, for both the coordinate system and the constraints governing changes to vertex coordinates must be antecedently and explicitly imposed. However, once a modeling medium has been used to construct a suitable FEM of a collection of objects, it can be altered in any of countless ways in order to determine the (at least possible) consequences of the corresponding alterations to the represented objects. One can, for instance, use an FEM of the door, bucket, ball system to infer, among countless other things, what would happen were we to place the bucket over the ball and slide the bucket through the door, what would happen were the bucket used to throw the ball at the open door-way, what would happen were the air pressure dramatically decreased, and so on indefinitely [14]. The consequences of these alterations need not be anticipated and explicitly incorporated into the system. Indeed, the very point of constructing FEMs is to *find out* how a system will behave in light of whichever alterations an engineer or scientist can dream up.

Those who would contend that FEMs are, *qua* computational, necessarily extrinsic representations once again overlook the fact that there are multiple levels of abstraction at which a given FEM model can be understood. There is, to be sure, the relatively low level of the modeling medium, and, insofar as there are representations at this level at all, the representations in question are unquestionably extrinsic. It is clear that the goings-on at this level inspire the above-mentioned contention, but what is once again being overlooked is that there is also a higher level, a level at which one finds models of collections of objects. These models are, every bit as much as the scale models they were created to replace, unquestionably intrinsic representations. Thus, once again, by the very standards employed by critics of the idea that some computers realize non-sentential models, FEMs are like scale models and unlike paradigmatic sentential representations.

All of this is bears directly on the longstanding concern that there is no way to bridge the gap between neural machinations and the non-sentential models hypothesized by proponents of the CMH. What the foregoing makes clear is that computational systems can realize non-sentential models that share with scale models the main characteristics that have long been used, even by opponents of the CMH, to distinguish scale models from sentential representations. Insofar as one already thinks that the brain is capable, at least in principle, of realizing computational processes, then one must also agree that brains can realize non-sentential models. This, I submit, is not just the most promising, but also (as yet) the only satisfactory account whatsoever of how a set of electrochemical circuits might realize non-sentential models of the sort posited by proponents of the CMH[3]. These considerations, in turn, give a real boost to the credibility of the CMH and, by extension, to its many specific applications. Indeed, FEMs are generally intrinsic representations, not just of spatial properties, but of kinematic and dynamic properties as well, and so their hypothesized cognitive counterparts ought to be fully capable of playing the roles for which they are slated by the proponents of the CMH.

4   Applications

The foregoing realization story turns out to have ramifications for work in a number of different fields, ranging from artificial intelligence (A.I.) to the philosophy of science. Of particular concern here is the notorious, albeit somewhat ephemeral, frame problem.

---

[3] Elsewhere I claim that the underlying recipe that is typically followed when constructing FEMs suggests that there may also be a kind of Northwest Passage, one that takes us directly from neural goings-on to non-sentential models [17].

4.1 Artificial Intelligence

Though it first came to light as a consequence of early work in logic-inspired, sentence-and-rule based A.I., we shall see that there are good reasons for understanding the frame problem in a more generic way, as that of determining how, through finite means, a creature or device can come to have human-like knowledge of the consequences of alterations to world. This sort of knowledge often enables us to choose beneficial and avoid harmful courses of action, and it also often enables us to formulate creative solutions to the many challenges that we face.

The frame problem can be broken up into at least two component problems. One, the prediction problem, has to do with fact that we humans have ability to predict the (at least possible) consequences of countless alterations to the world [15]. For instance, with regard to the ball, bucket, door scenario discussed earlier, we all know (to some admittedly fallible degree) that were the bucket placed over the ball and moved through the doorway the ball would also move through the doorway. We also know what would happen were the bucket used to throw the ball through the doorway, and so on indefinitely. While our knowledge of such alterations is immense, we are here still only dealing with a quite limited, 'toy' world.

Another component of the frame problem – namely, the qualification problem [16] – has to do with the fact that we humans are also able to envision countless possible defeaters of specific predictions. For instance, what we actually know about placing the bucket over the ball and moving it through the doorway is actually far more complex than was described above, for what we really know is something like this: If the bucket is placed over the ball and moved through the doorway *and it is not the case that either* there is a hole in the floor, or there is a hole in the side of the bucket, or the ball is affixed to the floor, or what have you, then the ball will move through the doorway.

The prediction problem, as concerns sentence-and-rule-based A.I., is that when we try to endow a system with knowledge of the consequences of countless alterations to a given situation using, at the highest level, sentence-and-rule-based representations of objects, we find that that we must incorporate countless, separate data structures (statements or rules) for each alteration-consequence pair. This problem is compounded by the qualification problem, for in order to truly match what we know, each such statement or rule would also have to incorporate countless distinct qualifications. Put formally, a sentence-and-rule-based system would have to contain countless distinct, endlessly qualified statements or rules of the following form (S's represent starting conditions, A's alteration conditions, Q's qualifiers, and C's consequences)[4]:

---

[4] There are also possible problems having to do with the sets of S's, A's, and C's.

$$[(S_1 \& S_2 \& \ldots \& S_n) \& (A_1 \& A_2 \& \ldots \& A_n) \& $$
$$(\sim Q_1 \& \sim Q_2 \& \ldots \& \sim Q_n)] \rightarrow (C_1 \& C_2 \& \ldots \& C_n)$$

Restricting ourselves to systems that employ, at the highest level, sentential representations of the world, these problems look to be insoluble. However, they also look insoluble for any approach on which high-level extrinsic representations carry the inferential load, whether it be sentences and rules or activation and weight vectors. There is simply too much knowledge for it to be explicitly encoded in any form.

These problems do, nevertheless, admit of a determinate computational solution [17]. The solution is to constrain syntax-crunching operations so that they realize modeling media from which can be built *intrinsic*, non-sentential models (e.g., FEMs) of mechanisms. A device that can construct such models of its environment and wield them as its core inference engine will be endowed with what might be termed *inferential productivity*, the capacity for boundless inferences through finite means. Such models are, we saw, like scale models in that they can be manipulated in any of countless ways in order to make inferences about how alterations to the world might play out and, by the same token, about the ways in which those consequences might be defeated. Admittedly, this solution does engender problems all its own, but they are far more tractable by comparison [18].

4.2 Psychology and Logic

The frame problem is actually even more generic than a mere problem facing A.I., for it must also be dealt with by any theory of how humans (and perhaps other creatures) are able engage in this sort of reasoning. Even in the human case, there is just too much knowledge for it all to be encoded explicitly. Indeed, centuries ago, European rationalists were so impressed by this 'universal' reasoning ability as to conclude that no mere machine (biological or otherwise) could possibly account for it. The human mind, they thought, had to be of non-corporeal origin.

Perhaps, however, the vast bulk of what we know about the consequences of worldly alterations is only *tacit*, which is to say that it is not stored explicitly anywhere in memory but is rather produced on demand and as needed. Take, for instance, the knowledge that virtually all of us possess about how an airship can (i.e., unless it is transparent, there is an elaborate set of mirrors, a cloaking device, etc.) prevent a flagpole from casting a shadow. We all possess this knowledge, though few of us have ever had occasion to encode it explicitly. One way to account for how we come to possess knowledge about this and countless other scenarios through finite biological means is to say that what we have is an ability to construct non-sentential, intrinsic models of objects and to manipulate them in relevant ways on demand and as needed. Indeed, this may be the first full-blown mechanical explanation for our 'universal' reasoning ability.

This suggests, in turn, that we humans possess another mode of monotonic reasoning apart from deduction. Deduction, it is well-known, is monotonic in that if we validly deduce some conclusion which turns out to be incorrect, some of the information from which the conclusion was derived must also be incorrect. But deduction is also formal in that representations of abstract logical particles and principles are what bear the inferential load; the specific contents consistently quantified over and connected drop out as largely irrelevant. Deduction is, of course, sometimes effected externally through the use of truth tables and formal logical notations. Many psychologists believe that deduction is also effected internally through cognitive counterparts to these external methods – namely, through so-called *mental models* [19] or a *mental logic* [20].

Consider, however, the sorts of spatial inferences we are able to make using external, intrinsic representations. For instance, suppose we know that Linus is about a 1/4th taller than Prior and Prior is about 1/4th taller than Mabel [7]. Using an intrinsic representation of their relative heights (e.g., broken matchsticks), we can effect some simple monotonic inferences, such as that Linus is taller than Mabel or that arranging them side-by-side with Linus in the middle would form a kind of pyramidal shape. We also use more sophisticated intrinsic models to make monotonic inferences about kinematic and dynamic happenings. Importantly, in all of these cases, insofar as our representations are accurate, our conclusions must be as well. Conversely, insofar as the conclusions reached on the basis of these models are inaccurate, so too must be the representations from which they were derived. This form of monotonic reasoning is, however, clearly not deductive in nature. It is not effected by abstracting away from specific contents and allowing representations of logical particles and principles to bear the inferential load. Instead, it is the representations of specific contents that bear this load. As yet, however, no name has been assigned to this non-formal mode of monotonic reasoning. To assign it one, let us call it *exduction* (*ex-* out + *duce-* lead). Exduction is obviously effected externally using scale models and, more recently, FEMs and the like. The above proposal is just that we also sometimes engage in exduction internally through the use of non-sentential, intrinsic cognitive models. Indeed, we have seen that there are good reasons for thinking that we do engage in this non-formal mode of monotonic reasoning internally.

If all of this is correct, then exduction must be added to our taxonomy of reasoning processes alongside deduction, both of which are to be classified as monotonic. Inductive generalization, analogical reasoning, and abduction, on the other hand, count as non-monotonic. It also bears mentioning that abduction (by which I mean inference to the best explanation) is, though non-monotonic, also unique in that it may be partly constituted by *any* of the other forms of reasoning. Indeed, matters are complicated further by the fact that explanations lie at the core of all abductive reasoning, and explanations may themselves involve reasoning of a certain sort.

4.3 Explanation

The idea that monotonic reasoning lies at the core of all explanations is not new, as it formed the basis for what for a long time was, and in some quarters still is, the dominant model of explanation – namely, the deductive-nomological (D-N) model[5]. This model fell out of favor in mainstream philosophy of science as problems with it began to accrue. Two of the best known were its seeming inability to account for statistical explanations and its failure to distinguish explanations from non-explanatory deductions. Even more germane to the present discussion, however, are the surplus meaning problem and the problem of provisos.

The first of these has to do with the fact that explanations have countless implications beyond the happenings that they explain. To take a non-scientific example, consider that a mechanic may explain why an automobile engine exhibits a loss of power in terms of its possessing faulty rings. On the D-N model, this explanation involves a deduction of the happening to be explained from information about laws and boundary conditions in something like the following manner:

- If an engine's cylinder has faulty rings, then the engine will exhibit a loss of power.
- One of the engine's cylinder has faulty rings.
- Therefore, the engine exhibits a loss of power.

However, even where one is able to provide a plausible-sounding D-N reconstruction of an explanation such as this one, such reconstructions seldom do justice to the full complexity of the explanations they represent. Consider, for instance, that what the mechanic knows is not only that the faulty rings will result in a loss of power, but the many other implications of his explanation being correct, such as that oil will leak into the combustion chamber, the exhaust will look smoky, the end of the tailpipe will become oily, the spark-plugs will turn dark, replacing the rings will restore power, replacing the filter will not restore power, and so on indefinitely. Any suitable reconstruction of the explanation must thus imply not only the *explanandum*, but countless other things as well. The problem with the D-N model is that it relies on an extrinsic representational scheme, and so no D-N reconstruction can embody all of an explanation's *surplus meaning* [21]. The problem here looms especially large given that these additional explanatory implications are not idle justificatory bystanders. They are what we largely rely upon when assessing the adequacy of explanations.

---

[5] Admittedly, the D-N model was not meant to be in any way psychological, though elsewhere this view has been contested [17].

Making matters worse, one who possesses an explanation such as this one also knows of the countless ways in which each of its countless implications is qualified. The mechanic, for instance, knows that bad rings will lead to a loss of power, but only if the engine is not augmented with an NO2 supply, the other cylinders are not bored out to a higher displacement, and so on. The D-N model is, however, no more able to account for this kind of knowledge than is any other theory that relies upon extrinsic representational apparatus. What makes this problem of *provisos* [22] especially troubling is that, as Quine famously noted, the knowledge at issue here is what enables us to hang on to our explanations in the face of unruly evidence.

Though no alternative theory of explanation has yet proven capable of filling the substantial void left by the D-N model's demise, the mechanistic approach to explanation is increasingly viewed as a promising contender [23, 24]. This approach was pioneered in large part by Salmon, who claimed that explanations are to be identified with the objective mechanisms at work in the world. On his view, an explanatory mechanism is roughly just an arrangement of parts that act and interact so as to collectively yield the happening in question. One limitation of this *ontic* version of the mechanistic approach is that it fails to allow for the possibility of explanations that are either right or wrong, good or bad. Nor, therefore, does it leave room for a process of inference to the best explanation. This limitation is overcome by adopting a psychologistic version of the approach [25]. Broadly speaking, according to the psycho-mechanistic approach, to have an explanation is to have the belief that a certain mechanism is, or may be, responsible for producing some happening, where such beliefs are constituted by mental representations of those mechanisms. It is largely in virtue of our awareness of the information conveyed by these representations that events and physical regularities are rendered intelligible.

The specific variant of the psycho-mechanistic approach suggested by the foregoing is that the mental representations in question are intrinsic cognitive models. It should now be clear that this *model model* of explanation can overcome such limitations of its deductive counterpart as the surplus meaning problem and the problem of provisos. Though these problems were discovered on quite independent grounds by philosophers of science, they are just variants on the prediction and qualification problems uncovered through work in deductive-logic-inspired A.I. Accordingly, the same solution seems to apply – namely, to eschew the appeal to formal-deductive reasoning processes in favor of an appeal to exductive reasoning effected through the manipulation of non-sentential, intrinsic cognitive models. On this view, our exductive inferences give us explicit knowledge of the mechanisms by which a happening may have been produced, but, constituted as they are by intrinsic models, they also endow us with boundless tacit knowledge of an explanatory mechanism's further implications and of the countless ways in which those implications are qualified. This knowledge is, once again, what enables us to determine the testable implications of our explanations and to hang onto those explanations come what may, all of which is essential to forward progress in science [26].

5 Conclusion

The CMH has long been plagued by the concerns about the in-principle neurological plausibility of appeals to non-sentential cognitive models. I have proposed here a particular way of addressing these concerns according to which models that share central and important characteristics with scale models piggy-back atop a thick layer of strict computational processing and, in turn, upon a neurophysiological bedrock. Further research is needed in order to show that, and precisely how, the human brain implements cognitive models, and it is to be expected this research will reveal other ways of bridging the gap between brain and model, and to refine our views about the differences between cognitive models and scale models. I would caution, however, that in order to be considered true extensions of the CMH the central characteristics of non-sentential models discussed here will need to be preserved.

References

1. Pamer, S.: Fundamental aspects of cognitive representation. In: Rosch, E., Lloyd, B. (eds.) Cognition and Categorization, pp. 259–303. Lawrence Erlbaum Associates, Hillsdale (1978)
2. McCulloch, W., Pitts, W.: A logical calculus of the ideas immanent in nervous activity. Bulletin of Mathematical Biophysics 5, 113–115 (1943)
3. Franklin, S., Garzon, M.: Computation by discrete neural nets. In: Smolensky, P., Mozer, M., Rumelhart, D. (eds.) Mathematical Perspectives on Neural Networks, pp. 41–84. Lawrence Earlbaum Associates, Mahwah (1996)
4. Kosslyn, S.M.: Image and Brain: The Resolution of the Imagery Debate. The MIT Press, Cambridge (1994)
5. Craik, K.J.W.: The Nature of Explanation. Cambridge University Press, Cambridge (1952)
6. Shepard, R.N., Chipman, S.: Second-order isomorphism of internal representations: Shapes of states. Cognitive Psychology 1, 1–17 (1970)
7. Huttenlocher, J., Higgins, E.T., Clark, H.: Adjectives, comparatives, and syllogisms. Psychological Review 78, 487–514 (1971)
8. Anderson, J.R.: Arguments concerning representations for mental imagery. Psychological Review 85, 249–277 (1978)
9. Block, N.: Mental pictures and cognitive science. In: Lycan, W.G. (ed.) Mind and Cognition, pp. 577–606. Basil Blackwell, Cambridge (1990)
10. Pylyshyn, Z.W.: Computation and Cognition: Toward a Foundation for Cognitive Science. The MIT Press, Cambridge (1984)
11. Sterelny, K.: The imagery debate. In: Lycan, W.G. (ed.) Mind and Cognition, pp. 607–626. Basil Blackwell, Cambridge (1990)
12. Fodor, J.A.: The Mind Doesn't Work That Way. The MIT Press, Cambridge (2000)

13. Pylyshyn, Z.W.: Mental imagery: In search of a theory. Behavioral and Brain Sciences 25, 157–182 (2002)
14. Waskan, J.A.: Intrinsic cognitive models. Cognitive Science 27, 259–283 (2003)
15. Janlert, L.: The frame problem: Freedom or stability? With pictures we can have both. In: Ford, K.M., Pylyshyn, Z.W. (eds.) The Robot's Dilemma Revisited: The Frame Problem in Artificial Intelligence, pp. 35–48. Ablex Publishing, Norwood (1996)
16. McCarthy, J.: Applications of circumscription to formalizing common-sense knowledge. Artificial Intelligence 28, 86–116 (1986)
17. Waskan, J.: Models and Cognition. The MIT Press, Cambridge (2006)
18. Waskan, J.: A virtual solution to the frame problem. In: Proceedings of the First IEEE-RAS International Conference on Humanoid Robots. Electronic only (2000), https://netfiles.uiuc.edu/waskan/www/77.pdf
19. Johnson-Laird, P.N., Byrne, R.M.J.: Deduction. Lawrence Erlbaum Associates, Hillsdale (1991)
20. Rips, L.J.: Cognitive processes in propositional reasoning. Psychological Review 90, 38–71 (1983)
21. MacCorquodale, K., Meehl, P.E.: On a distinction between hypothetical constructs and intervening variables. Psychological Review 55, 95–107 (1948)
22. Hempel, C.G.: Provisoes: A problem concerning the inferential function of scientific theories. Erkenninis 28, 147–164 (1988)
23. Salmon, W.: Scientific Explanation and the Causal Structure of the World. Princeton University Press, Princeton (1984)
24. Bechtel, W., Richardson, R.C.: Discovering Complexity: Decomposition and Localization as Strategies in Scientific Research. Princeton University Press, Princeton (1993)
25. Waskan, J.: Knowledge of counterfactual interventions through cognitive models of mechanisms. International Studies in Philosophy of Science 22, 259–275 (2008)
26. Lakatos, I.: Falsification and the methodology scientific research programmes. In: Lakatos, I., Musgrave, A. (eds.) Criticism and the Growth of Knowledge, pp. 91–195. Cambridge University Press, Cambridge (1970)

Does Everyone Think, or Is It Just Me?

A Retrospective on Turing and the Other-Minds Problem

Cameron Shelley

Abstract. It has been roughly 60 years since Turing wrote his famous article on the question, "Can machines think?" His answer was that the ability to converse would be a good indication of a thinking computer. This procedure can be understood as an abductive inference: That a computer could converse like a human being would be explained if it had a mind. Thus, Turing's solution can be viewed as a solution to the other-minds problem, the problem of knowing that minds exist other than your own, applied to the special case of digital computers. In his response, Turing assumed that thinking is a matter of running a given program, not having a special kind of body, and that the development of a thinking program could be achieved in a simulated environment. Both assumptions have been undermined by recent developments in Cognitive Science, such as neuroscience and robotics. The physical details of human brains and bodies are indivisible from the details of human minds. Furthermore, the ability and the need of human beings to interact with their physical and social environment are crucial to the nature of the human mind. I argue that a more plausible solution to Turing's question is an analogical abduction: An attribution of minds to computers that have bodies and ecological adaptations akin to those of human beings. Any account of human minds must take these factors into consideration. Any account of non-human minds should take human beings as a model, if only because we are best informed about the human case.

1 Introduction

Roughly 60 years ago, Turing wrote his famous article beginning with the question, "Can machines think?" [30]. The answer is obviously "yes": We

Cameron Shelley
Centre for Society, Technology, & Values, University of Waterloo, Waterloo, Canada
e-mail: `cam_shelley@yahoo.ca`

L. Magnani et al. (Eds.): Model-Based Reasoning in Science & Technology, SCI 314, pp. 477–494.
springerlink.com   © Springer-Verlag Berlin Heidelberg 2010

ourselves are machines that think. What Turing was really after, of course, was the issue of whether or not fundamentally non-human machines, e.g., digital computers, could ever rightly be said to think. He was optimistic and urged that a program of empirical research be undertaken to determine the truth of the matter.

Sixty years of research in Artificial Intelligence and Cognitive Science has not confirmed Turing's optimism. Developments in the last 20 or so years in particular have undermined the rationale for Turing's response to his question, that is, the famous *Turing Test*. Briefly put, Turing proposed that we would be justified in inferring that a computer has a mind if it can fool people into accepting it as a human being through conversational ability alone. The computer that Turing had in mind for this test would be a kind of chip-in-a-vat, housing a program that would live in a purely simulated environment.

The Turing Test represents an attempt to address the *other-minds problem*. That is, how do we know that anyone has a mind other than ourselves? Does everyone think, or is it just me? The problem arises because we do not possess the same information about other minds as we do about our own minds. We can sense our own minds through introspection. We cannot sense other minds in this way. In that case, we face the question, "What evidence should convince us that other people do (or can) think?" Even more perplexing is the question, "What evidence should convince us that exotic machines do (or can) think?"

Turing held that the solution to the other-minds problem should be the same for both humans and machines. That is, the same principles – whatever they are – that we use to determine that other human beings have minds should be the principles used to determine that machines have (or do not have) minds.

Turing's argument for this position rests on (at least) two assumptions:

1. Thinking ability is a matter of software and not hardware, and
2. Thinking ability could be achieved through simulation alone.

Recent developments in Cognitive Science have tended to undermine both assumptions. Research in robotics, embodiment, and neural networks suggests that details about the physical implementation of a thinking being are crucial to its nature. Investigations of situatedness and extended cognition suggest that thinking ability results from the occupation of a special ecological niche in response to the demands of the real world.

In this article, I will take a look back at the other-minds problem and its solution as Turing saw it, and indicate what I see as its shortcomings. To state the matter briefly, Turing saw the solution to the problem as an abductive inference based on general principles about minds and thinking. I will argue that work done since the publication of Turing's article favors a somewhat different view, that the solution is an analogical abductive inference using knowledge of human beings and their minds as a model. The difference may seem slight at first glance, but it ties the other-minds problem to a difficulty that faces anyone using model-based reasoning, that models are a potential source of bias as well as of helpful information.

2 The Other-Minds Problem

Does everyone think, or is it just me? This question is not one that would normally detain anybody. However, it became a question of philosophical interest when philosophers such as Descartes began to wonder how it could be rigorously answered. Although we are normally confident that other people have minds, our confidence seems hard to justify when their thoughts remain beyond the reach of our senses. Descartes argued that the existence of the thoughts and minds of others could be deduced through rigorous argument.

In other words, the problem for Descartes arose because he accepted the following two claims:

1. I have access to my own mind only, and
2. Minds and bodies are distinct kinds of things (substances).

The first claim captures Descartes' view that knowledge of one's own mind is immediate and private. The briefest consideration will assure me that I have a mind; otherwise, how could I consider the problem? This claim is the famous *cogito* argument: I think, therefore I am. However, I have no such access to the minds of others. In the absence of telepathy, no amount of introspection can reveal the minds of other people to me. Therefore, the existence of other minds must be inferred.

Descartes [9] provides a lengthy (and problematic) deduction of the existence of other minds in his exploration of the cogito argument. Since a discussion of this detailed argument lies outside the scope of this article, I will instead present a much simpler argument given elsewhere by Descartes [8]. Descartes argues that the existence of other minds may be deduced from the presence of fluency in language. In brief, whatever has a mind, also speaks a language and vice versa. The argument could be represented in the following form:

Something speaks language fluently if, and only if, it is a being with a mind

You speak language fluently

Therefore, you are a being with a mind

This argument is plainly a *deduction* in the sense that if the premises are true, then the conclusion must be true also. So, the argument would be perfectly convincing if the premises were unassailable. Unfortunately, as Descartes' contemporaries pointed out, the first premise is false: There are counterexamples to it. For instance, infants seem to have minds (as any parent will attest) even though they lack the ability to speak fluently [11]. So, having a mind and being able to speak a language are not the same thing.

A very different approach to the other-minds problem was taken by Wittgenstein. He rejected the need for any sort of elaborate inference to prove the existence of other minds. Instead, Wittgenstein argued that the other-minds problem could be solved only if we could become clear about what we mean when we say that someone has a mind. Here, I will summarize Wittgenstein's account as well as I understand it, while acknowledging that his account developed over his career [27] and is open to interpretation [5].

First of all, Wittgenstein argued that the practice of attributing mental states comes in two varieties:

1. The attribution of mental states to oneself, and
2. The attribution of mental states to others.

These varieties are not directly related, according to Wittgenstein. Consider self-attribution of a mental state, as in the expression, "I have a toothache". Taken literally, this expression would be interpreted like the expression, "I have a splinter", by which you mean that you have noticed that a small and unwelcome piece of wood has come along and is now embedded in your skin. Similarly, "I have a toothache" suggests that you have directed your gaze inward and noticed that a toothache has come along and set up residence in your tooth. Of course, you do not literally mean to make or defend such a claim. In Wittgenstein's view, the expression "I have a toothache" is more like a substitute for a natural expression, such as a groan. There is nothing more to say about your expression except that it sounds *as if* you are reporting a self-observation even though you are not. Similarly, if you said, "I have a mind", then you speak as if you possessed an object that was hidden from view accessible only to you. In fact, you are merely expressing in language that you are thinking or feeling.

Now consider the attribution of mental states to others, as in the expression, "Fred has a toothache". Characteristically, Wittgenstein viewed this expression as a move in a *language game*. That is, such expressions are a way that people have of using language in conducting their interactions with others. The statement is about Fred, how the speaker relates to Fred, and hints at what interactions may follow. For example, by claiming that Fred has a toothache, you may be explaining his behavior to someone else, or urging someone else to treat Fred extra-nicely. Similarly, if you said, "Fred has a mind", then you are merely saying that Fred appears to be lost in thought, or should be treated with a certain dignity, and so on.

Note that, on Wittgenstein's account, no inference about Fred is performed by the speaker. The speaker is not arguing that it must be the case that Fred is having a certain kind of experience. Nor is the speaker guessing at what sort of feeling Fred might be having. The speaker is merely applying the concept of *toothache* or *mind* as dictated by the ordinary practices of English speakers.

As with Descartes' account, Wittgenstein's solution to the other minds problem faces some difficulties. Among them are the following:

1. People in other cultures, speaking other languages, may have quite different concepts about minds. In Japanese culture, for example, mental states are held to reside, at least in part, in the public realm [18]. Gestures, facial expressions, tone of voice, etc., which are considered merely as *clues* to inner experience in Western cultures, are considered to be *continuous* with mental states in Japanese culture. In that event, it is natural to wonder whether or not the mental concepts in a given language provide adequate or even correct solutions to the other-minds problem.

2. There are people to whom ordinary linguistic concepts do not readily apply. Consider those suffering from conditions such as *totally locked-in syndrome* (TLIS). Such people experience total loss of motor functions but retain their consciousness. Because of this condition, people suffering from TLIS do not display any of the usual outward signs of mental life, so that mental concepts would not apply to them. Yet, the presence or quality of their mental life remains an open question and a topic of scientific theorizing [14].

3. Finally, there is the issue of non-humans. Normal mental concepts do not readily apply to animals, computers, or even aliens. Yet, the question of whether or not such things have minds continues to nag at us.

As different as Descartes' and Wittgenstein's accounts are, they have in common an aversion to the abductive approach to the other-minds problem. That is, both reject the solution that the existence of other minds is a hypothesis that best explains why other people are the way they are. The abductive approach was taken up by Turing in his answer to the question, "Can machines think?"

3 Thinking Machines

The difference between Turing's approach and that of Wittgenstein becomes apparent when Turing immediately rejects a solution to the other-minds problem based on concepts present in ordinary language. Indeed, Turing remarks that such an approach would be "dangerous", equivalent to settling the matter by Gallup poll [30, p. 433]. Turing does not explain this remark, but it can be understood as follows. The meanings of terms such as *think* are somewhat contingent on historical accident. In other words, their meaning may well have developed differently if history happened to be different than it is. Modern English developed in an era in which complex computers were entirely absent and hardly imaginable. Thus, the verb *think* applies only to the objects that obviously had minds at that time, namely people. To apply this concept and then conclude that machines cannot think is no more than to observe that there were no computers around in the 15th or 16th Centuries. This fact hardly provides a good reason to reject the possibility that computers could have minds.

The case is reminiscent of a remark attributed to the computer scientist Edsger Dijkstra, when accepting the Turing Award in 1972. He said that, "The question of whether a computer can think is no more interesting than the question of whether a submarine can swim" [1]. Dijkstra was expressing despair with the whole issue of machine cognition; however, the example is instructive. Most people would probably say that submarines do not swim. This view is simply a reflection of the fact that the verb *swim* developed at a time when there were no mechanical devices that propelled themselves through the water. Only fish, whales, and some birds had that ability. Had submarines been commonplace in the 15th Century, we might well have no trouble applying the verb *swim* to a submarine.

Since everyday concepts about minds reflect irrelevant historical bias, Turing rejects any solution to the other-minds problem based on them. Instead, in a move reminiscent of Descartes, Turing appeals to linguistic fluency. The ability to hold an intelligent conversation would be evidence that the speaker has a mind. So, when we encounter anyone who can converse, we would take their fluency as evidence that they have a mind. The same principle, Turing holds, should be applied to computers just the same as human beings. The resulting inference could be represented as follows:

If a computer can think, then it can converse

A computer can converse
_____

Therefore, a computer can think

This inference is clearly an abductive one, in the sense that it has the form of the fallacy of *affirming the consequent*, a paradigmatic case of abductive inference:

$$p \supset q$$

$$q$$
_____

Therefore, p

One clear difficulty with this inference is that no computer could converse when Turing wrote his article. Turing spends much effort in his article to make a case that there are no good *a priori* reasons to reject the premise that computers can converse, and that people would tend to agree with this claim if only their biases about artificial computers could be filtered out of their judgments.

Scholars continue to discuss and critique Turing's proposal and his presentation of it [10]. Here, I want to comment only on two features of Turing's solution to the other-minds problem, namely his focus on *universal computers* and his assumption that a thinking computer could be a kind of brain-in-a-vat.

4 Universal Computers

When Turing claims that a computer can think, the kind of computer that he has in mind is a *universal computer*. Put simply, a universal computer is a computer that can mimic the behavior of any other computer. Any program that can run on one such machine can be run on another one. In effect, *universal computer* defines a class of machines that can all run any software that can be run on any other computer.

To make the concept clearer, contrast it with the notion of a *special-purpose computer*. A special-purpose computer would be able to run some kinds of programs but not other kinds. For a practical example, consider an Automated Teller Machine (ATM). Such a machine can be programmed to dispense cash, update bank accounts, and so on, but probably cannot be programmed to drive a car or write a novel. There are some programs that can be run on some computers but not on an ATM. Thus, an ATM is a special-purpose computer and not a universal computer.

Turing confines his attention to the prospect that universal computers can think. He does not state why special-purpose computers should not be considered in this regard. One motivation for his preference might be that it affords simplicity for the argument he proceeds to make. The choice of universal computers directs attention towards software and away from hardware—the hardware being the physical details of the computer running the software. If a suitably programmed universal computer can think, then any other universal computer can run the same software and also think. In this situation, we would attribute the computer's success at thinking to the software that it is running, and not to the hardware that comprises it. If, on the contrary, a special-purpose computer could be programmed to think, then it would be hard to distinguish the contribution of the software and the hardware to its thinking abilities.

Another motivation for focusing on software and not hardware is that it responds to an intuition known today as *multiple realizability* [22]. Roughly speaking, multiple realizability means that mental states can be realized or implemented by different physical states. Where digital computers are concerned, multiple realizability means that a particular computational procedure can be implemented on any two computers, despite their physical differences. The Microsoft Windows operating system, for example, can run on a PC or a Macintosh, even though PCs and Macs tend to be somewhat different in their physical configuration. So, perhaps Turing confined his attention to universal computers because attributing thinking ability to software alone is in obvious agreement with the intuition of multiple realizability.

Of course, not everyone shares this intuition. Searle, for example, argues that running a particular program is not sufficient for there to exist a mental state in a computer [25]. Instead, he argues that biologically-based brains are in some wise necessary for thinking to occur.

5   A Chip in a Vat

In discussing thinking computers, Turing does not describe the surroundings in which such a computer would reside. However, he seems to have in mind a large machine sitting placidly in a room somewhere, isolated from the world except for a teletype connection through which it sends and receives short text messages. This picture would hardly be surprising since Turing helped to design just such a machine, the Mark I computer at the University of Manchester.

Such a machine bears a strong resemblance to the brain-in-a-vat of science fiction: Using advanced surgical techniques, a evil scientist removes the brain from an unwitting victim. The brain is suspended in a vat of nutrient and hooked up to a computer that simulates the inputs that the brain would receive if it were still in the head of the victim leading his normal life. The victim is deceived: it seems to him that he is still moving about the world as before, talking to others, smelling flowers or exhaust fumes, eating ice cream, and so on. However, it is all a fabrication conjured up by the evil scientist.

The brain-in-a-vat raises many interesting philosophical problems. However, the main problem that it raises for us is the fact that the brain comes into the story already programmed, as it were. All the software that the brain needs to process its inputs and produce its thoughts and outputs is already present in it. All that is left for the evil scientist is to perfect his surgical techniques and arrange his vat.

The same cannot be said for Turing. The problem that he faces is whether or not a universal computer can be programmed to think, so he cannot assume that it comes with such a program. Instead, he must indicate how the program for his chip-in-a-vat could be developed. There are two obvious ways for developing such software:

1. Software could be developed from first principles.
2. Software could be developed through some sort of learning procedure.

The first method is unpromising. It is hard to imagine how a team of programmers could create thinking software simply by dint of a conceptual analysis of the requirements of thinking. A new sorting algorithm might be generated in this fashion, but not a thinking algorithm.

Turing opts for the second method. That is, he describes how a computer might learn to think through changing its software in the light of experience. Turing gives a lengthy (and often overlooked) discussion of how a computer could teach itself to think much as a child is taught to think by its parents and educators. He imagines a process analogous to natural selection, in which the software learns through making and correcting errors.

Whatever the merits of this proposal, one of its central tenets is that there is no need for special hardware in order to complete this task, unlike the case for a human child:

> It will not be possible to apply exactly the same teaching process to the
> machine as to a normal child. It will not, for instance, be provided with legs,
> so that it could not be asked to go out and fill the coal scuttle. Possibly it
> might not have eyes. But however well these deficiencies might be overcome
> by clever engineering, one could not send the creature to school with out the
> other children making excessive fun of it. It must be given some tuition. We
> need not be too concerned about the legs, eyes, etc. The example of Miss Helen
> Keller shows that education can take place provided that communication in
> both directions between teacher and pupil can take place by some means or
> other. [30, p. 456]

In short, the computer does not need a true physical presence in the world,
as would be provided by arms, legs, eyes, and so on. Instead, like a brain in
a vat, it needs only a source of information for inputs and a destination for
outputs.

Why would Turing rule out a universal computer that has a true physical
presence in the world? It is not really because such a computer would be
ridiculed by school children. Instead, Turing is merely being consistent with
his aim of attributing thinking ability strictly to software. If the computer
that learned to think had arms, legs, eyes, etc., then critics might complain
that these special pieces of hardware were indispensable to the whole process.
Such a claim would undermine Turing's view that thinking is attributable
strictly to programming. Thus, Turing confines his remarks to a computer
that is, in effect, a chip in a vat.

Turing's view of the other-minds problem relies on at least two assumptions, that thinking ability is a matter of software, and that no special form of
hardware or physical presence is at issue. So, if we were to attribute thinking
ability to a digital computer (or human being, for that matter), we would be
making a hypothesis about the software that it is running, and not about its
physical configuration or situation.

This view was widely shared in the early days of cognitive science but
recent developments have led to their reassessment. In following sections, I
will summarize some of the developments that have tended to undermine
Turing's perspective on computers and the other-minds problem.

6   Neuroscience

The rise of connectionism in Cognitive Science was motivated, in part, by a
desire to formulate computer programs that could learn naturally and from
examples, instead of being elaborately programmed in advance. It was argued
that software that employs a brain-like means of processing information would
answer to this purpose [23]. The result was the neural network. The trend
towards increasing neurological plausibility has been accelerated partly by the
availability of brain scanning technologies such as fMRIs. Today, researchers
are contemplating detailed reconstruction of the human brain in the form of
a computer especially designed and built to mimic human brain function [12].

As far as the other-minds problem goes, one of the important outcomes of neuroscience is the development of *mind-brain identity* theory. This theory is a view of the mind-body problem in which the mind is identified with the brain: The mind is what the brain does [28]. In other words, when you are thinking, this means just that a given process is unfolding in your brain. Neurons are firing, neurotransmitters are being released and captured across synapses, and so on. This process constitutes your thinking. If it were not for this process, no thinking would occur.

It is important to note that this inference does indeed rely on claims about the brain, that is, the hardware on which the mind is operating. Physical details about the brain determine how thinking occurs. For example, consider the effect of drugs on both the brain and the mind. It is well understood that the use of drugs, such as alcohol or Ritalin, have an effect on both the functioning of the brain and the nature of thinking.

Consider Ritalin (methylphenidate). In many parts of the brain, neurons signal each other, in part, by a release of the neurotransmitter catecholamine from one neuron to its neighbor. In normal people, there is an optimal level of catecholamine present between neurons. When levels depart from this optimal state, the result is a cognitive dysfunction. For example, according to the *catecholamine hypothesis* [21], Attention-Deficit Hyperactivity Disorder is caused by a deficit of catecholamines in the synapses between neurons. Hyperactivity and an inability to concentrate are results of this dysfunction. Ritalin works by increasing the amount of catecholamines in the synapses at a given time, according to this hypothesis.

By the same token, normal people have started to use Ritalin as a cognitive enhancer. In other words, normal people can increase their ability to concentrate and think clearly by taking Ritalin. A survey conducted by the journal *Nature* [24] showed that about one in five scientists surveyed admitted to using cognitive enhancers like Ritalin in order to improve concentration or memory [20]. The effect of Ritalin on thinking is due, of course, to the particular physical details of the human brain. The unfolding of thought in the human mind could not be understood without knowledge of particular details of the brain.

It might seem, then, that the solution to the other-minds problem is to conclude that only beings with human brains have minds. Such a result would be disappointing since, as Wittgenstein pointed out, we do not normally need a description of people's brains in order to attribute minds to them. However, there are good reasons to resist such a narrow view of the mind-brain identity thesis. Note that the term *brain* applies to a wide variety of physical things. Each human being has a brain different in many details from the next human being. Human brains differ from animal brains in yet further details. We can even speak of the brains of slugs, or even call the intellectual leader of a company or research group the "brain" of the organization. My point is that the term *brain* refers not to a specific brain, nor to the brains of a particular species, but to an entity that occupies a special role within an organism.

The special role of the brain is, roughly speaking, to integrate information about the outside world and to coordinate the organism's response to that information. Biologists wanting to locate the brain of a novel and exotic organism would examine both its behavior and the layout of its physical components, looking for a mechanism that fulfills this role.

In view of the mind-brain identity thesis, the other-minds problem turns into a kind of *other-brains problem*. This problem is not a trivial one because many, diverse sorts of things could count as brains. And evidence for the presence of a brain would comprise some combination of physical and behavioral data that is hard to describe precisely in advance. For the present, it is sufficient that brain function and design is an indispensable consideration in any resolution to the other-minds problem, contrary to Turing's view.

I should point out, however, that this conclusion does not imply that digital computers cannot think. As Turing pointed out, the "executive unit" (CPU) of a digital computer acts as a sort of brain. Therefore, computers with the right sort of executive unit might qualify as having minds. Such a computer would be a special-purpose computer, one equipped with physical components specially configured for the job of thinking.

7 Situatedness

When considering how a computer could learn to think, Turing faced two alternatives. One alternative would be to fit the computer out with special sensors and limbs to allow it wander about the world and revise its programming in the light of experience. The other alternative would be to construct a simulated environment in which the learning program could learn to think. In this virtual world, as we would now call it, the computer could revise its programming in the light of simulated experience, and thus become a thinking being.

Turing rejected the first alternative because it did not gibe with his view that thinking ability is strictly a matter of software and not hardware. Besides this conviction, the first alternative poses many difficulties, such as figuring out what kind of limbs and sensory organs the computer would need, and how they should work. The history of modern robotics has certainly confirmed the difficulty of this task.

However, the second alternative is hardly more promising, as the roboticist Rodney Brooks has forcefully pointed out [2]. Brooks uses the term *situated* to describe a being that is richly connected to the real world and relies upon this connection for its normal functioning. Normal human beings are situated in the sense that they are embedded in a world of real objects and forces, and they rely upon their ability to sense and affect this world in the course of their thinking.

Compared to situated beings, beings in a simulated world face several obstacles in the path to learning to think. These obstacles come in two varieties, as Brooks points out:

1. The programmers who construct the virtual world are apt to incorporate or emphasize elements that are not crucial to the task. One example of this problem might be chess. Early research in Cognitive Science often focused on games such as chess or checkers. Turing mentions checkers as a good place to start in the project of developing thinking computers. Although chess ability does signify advanced intellectual attributes, it has not proven appropriate as a paradigm for thinking ability in general.
2. The programmers who construct the virtual world are apt to omit elements that crucial to the task. Consider object recognition. The ability to tell shoes apart from socks is not hard for even a young child but challenges even an advanced object recognition software system. Yet, the founders of Artificial Intelligence set object recognition as a summer project for an undergraduate student in 1966 [17]! Brooks has recently predicted that artificial object recognition systems should have capabilities equivalent to human 2-year olds only with another 50 years of research [3].

Because of these sorts of issues, we should be skeptical that programs developed in simulated environments could learn to think. Instead, it is more likely that a computer suitability situated in the real world could learn to do so.

It is worth adding that the real world environment in which a roboticized computer (which I will just call a *robot* from now on) will consist of not only inanimate objects like socks and shoes but also other thinking beings. Thus, a robot learning to think will have to acquire social intelligence, the ability to deal with and think about other robots like itself [7]. Human beings are intensely social on the whole, as exemplified by the current fascination with social technologies such as Facebook. In all likelihood, the same will be true of any thinking machine.

The importance of situatedness to the existence of a mind is made clear in the case of *sensory deprivation* in human beings. Sensory deprivation occurs when input from one or more senses is filtered out. It can be induced in various ways, although thorough deprivation has been induced in clinical settings by having people float in a soundproof vat of tepid water. Under profound sensory deprivation, the human mind can become unglued:

> It is clear that the stability of man's mental state is dependent on adequate perceptual contact with the outside world. Observations have shown the following common features in cases of sensory deprivation: intense desire for extrinsic sensory stimuli and bodily motion, increased suggestibility, impairment of organized thinking, oppression and depression, and, in extreme cases, hallucinations, delusions, and confusion. [26]

One of the hallmarks of having a mind is the ability to organize thoughts in a coherent and goal-directed fashion. In the absence of feedback from the external world, thinking patterns quickly lose coherence. Although there may continue to be feelings or conscious experiences, we might be reluctant to say that a mind is present in the case of someone whose cognition has become completely disordered.

The contrast between the importance of situatedness to human thinking and to Turing's notion of a thinking computer could hardly be greater. The nature or even existence of a mind in a human being suffering from extreme sensory deprivation, or indeed from Totally Locked In Syndrome, is dubious. For Turing, the existence of a mind in a computer that has never had much connection to the real world seemed quite plausible. Human experience is very much at odds with Turing's intuitions in this respect.

8 Embodiment

In human beings, the connection between brain and world is mediated by the body. The human brain appears to be highly tuned both to the kind of sensory information that human bodies are equipped to deliver, and to the kind of actions that human bodies afford. In his article, Turing implicitly denies that embodiment is important to thinking ability [13]. As we have seen, he envisions thinking ability strictly as a matter of the ability to run the right software.

Research in Cognitive Science tends to undermine this view. The fact that human beings have a certain kind of body seems profoundly to influence how they think. For example, Lakoff and Johnson [15] have shown how metaphorical language reveals a systematic connection between bodily form and conceptual understanding of the world. Consider the following piece of dialog from the movie *A fistful of dollars* [16]:

– You've gone out of your mind, Ramon!
– No, I've come to my senses, Esteban.

In this exchange, two metaphors for minds are used. In the first line, the mind is treated as if it were a location: Having a mind is like being in a special location; being insane is like leaving that location for another. In the second line, having a mind is also like being in a special location, one that connects you with your senses. This exchange suggests two things. First, it suggests that movement of the body through space provides a model for people to think about abstract events. Many similar metaphors confirm that bodily experience serves as a basis for understanding abstract ideas or experiences. Second, the lines suggest that having a mind is understood in these terms. That is, the process of thinking is conceived as if it were an orderly movement of the body through physical space. It seems natural enough for embodied beings to arrive at such a conceptual scheme. However, it is hard to see why a non-embodied intellect, such as the thinking machine Turing proposes, would do likewise.

Other studies of the human mind lead to similar conclusions. For example, on the *somatic marker hypothesis* [6], human beings rely on their bodies to help them think about the world and how to act in it. A somatic marker is a basic representation of a person's bodily state that is integrated with that person's cognitive representation of the world. The marker provides a kind of "gut feeling" that captures a persons' attitude towards a plan or a given state of affairs. In a card game, for example, a person may have a gut feeling, like a knot in the stomach, that taking another card from the deck would be a mistake. Damasio argues that a particular brain structure, the ventromedial prefrontal cortex, performs the function of associating bodily feelings with abstract cognitions. If this theory, or one like it, is true, then the linkage between brain and body is indispensable to the nature and understanding of the human mind. If so, then the human mind is again unlike the non-embodied intellect that Turing envisions in his picture of thinking machines.

9   The Extended Mind

As a final point, Cognitive Science has produced another challenge to Turing's picture of thinking machines in the form of what Clark and Chalmers call *active externalism* [4]. We have noted above that being situated in a real world of objects and peers is crucial to an understanding of human minds. Clark and Chalmers argue that human minds exploit this connection by recruiting external objects into the process of thinking itself. They discuss a thought experiment about Otto, a man with Alzheimer's disease. Due to his condition, Otto is unable to recall the location of the Museum of Modern Art (in New York City). However, Otto knows enough to look in his notebook, where he is in the habit of writing down new and useful information. He finds the address there and visits the museum. Contrast Otto's case with that of Inga, who simply remembers the address of the MOMA and finds her way there accordingly. Otto's notebook plays the same role for Otto that her own memory does for Inga. Provided that Otto's notebook is a reliable and accessible source of information for him, there seems to be no reason to deny that his notebook is part of Otto's memory.

Of course, examples of the recruitment of external objects as props in a thought process can be much more complex. Consider the use of a physical model, made of sticks, wire, cardboard, etc., by Watson and Crick as they worked on their theory of the structure of the DNA molecule [31]. To explore possible configurations of the model, the two scientists would tinker with the configuration of parts of their model. The model could then be examined to check for consistency with physical laws and available data about the molecule. In the act of tinkering with this object, Watson and Crick were thinking about the molecule's physical structure, a process known as *manipulative abduction* [19]. It became a part of the thinking process itself.

Active externalism suggests that people exploit their situated nature in order to recruit external objects into the thinking process. In doing so, the mind offloads some of the effort involved in dealing with the world onto the world itself. Since the thinking machines that Turing considers are not situated, they cannot engage in or display this activity.

10 Reflections

Turing's view of the other-minds problem rested on two assumptions:

1. Thinking ability is a matter of software and not hardware, and
2. Thinking ability could be achieved through simulation alone.

As a result, to attribute a mind to another being, whether human or artifact, is to hypothesize only that it is running a special sort of program, and says nothing about its body or the connection it has with the real world. These assumptions were shared by many researchers in the early days of Cognitive Science. However, more recent research paints a contrary picture of what it takes to have a mind. These studies suggest that details of brain structure, the connection of the brain and body, and the connection of both to the physical world, are indispensable to the nature of minds or, at least, the human mind.

On this view, which might be called an *ecological view*, to attribute a mind to another being, human or otherwise, is to hypothesize that it satisfies something like the following criteria:

1. It has a brain. The brain need not be human but should exhibit a design in which integration of sensory and motor systems is critical.
2. The brain design should facilitate sophisticated physical and social interaction with the environment.
3. The brain is coupled with the body in such a way as to support and inform its cognitive processes.
4. The brain is capable of exploiting its environment to support and inform its cognitive processes.
5. The brain can sustain a coherent pattern of thoughts, e.g., as displayed by a command of language.

The first four points summarize the fruits of recent research as discussed above. The final point is added to acknowledge the emphasis that Turing placed on fluency in his article. Fluency in language is relevant not so much in its own right but because it provides evidence that the speaker has the capability to organize thoughts in a coherent and purposeful way. As Descartes pointed out, a fluent speaker can, for example, adjust its remarks to the given situation, instead of producing random and unrelated expressions.

This ecological view is founded primarily on research into human cognition. Thus, when applying this view to the solution of the other-minds problem,

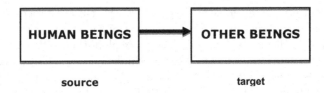

source   target

Fig. 1 An analogical abduction about other minds based on research in human beings.

we are, in effect, using normal human beings as a model against which other beings, natural or artifactual, are compared. In short, this solution to the other-minds problem is an *analogical abduction* [29]. It is an abductive inference in the sense that it supports a hypothesis about the presence of a mind, and it is analogical in the sense that it takes normal human beings as exemplary, mind-possessing creatures. This view is represented graphically as in Figure 1.

By contrast, Turing's solution was abductive but not analogical. Although his view of minds was, naturally, informed by studies of human beings, it was framed in terms of general principles about what is might take to have a mind.

11   Objections and Replies

The ecological view is not above criticism. In particular, Turing might raise the following two points against it:

1. It unfairly elevates human beings to the status of exemplary thinking beings. What evidence justifies the view that human beings are so special? Better to make judgments according to general principles than to rely on an exemplar fraught with historical accident and outright bias.
2. It stacks the deck against universal computers. If the ecological view of the other-minds problem is correct, then the claim that universal computers can have minds seems, at best, highly unlikely.

Let me address the points in reverse order. It is true that, on the ecological view, it is unlikely that a universal computer, as such, can have a mind. Perhaps the ecological view is mistaken, but it cannot be denied that we have, as yet, little evidence that universal computers can think. Sixty years of trying by some very smart and well-funded people have yet to produce a piece of software that does the trick. It may simply be the case that reality is stacked against Turing's assumption here.

The first point raises a vital issue. In making normal human beings the exemplars of thinking things, we do risk unwittingly loading our perspective with biased and arbitrary claims about what it takes for something to have a mind. Of course, it must be pointed out that to dispense with human-based research entirely would also be an enormous mistake. After all, spotty though it is, our information about human minds is the best information that we have on the subject. So, although bias is a legitimate concern, it must be admitted that humans remain the best models that we have available, even if we are still somewhat ignorant about how it is that normal human minds work.

On a more positive note, support for the ecological view is not as parochial as it might seem. Hypotheses about what how it is that humans have minds draws on research in various fields, such as biology, psychology, neuroscience, and artificial intelligence. The convergence of ideas in these fields, as well as the independence of the work in each, supports the accuracy and reliability of the ecological view.

A current, ecological view supports a solution to the other-minds problem comprising an analogical abduction to the hypothesis that some other being has a mind. This view is distinct from other views in several ways. In terms of its content, it obviously reflects advancements made in the study of human cognition. In terms of its form, it reflects the growing influence of scientific inference. Turing preferred an abductive, scientific approach because it offered an escape from various sources of bias and prejudice which people are apt to apply to the problem. Turing might argue that an *analogical* abductive approach is regressive because it exposes the inference to parochialism. This problem is one faced by anyone making use of models to support hypotheses: model selection and handling may be biased in a variety of ways, casting doubt on any conclusions drawn. Yet, the situation is not beyond repair: models are also potential sources of vital and relevant information, when handled conscientiously. The recent history of the other-minds problem is an illustration of this point.

References

1. AAAI (n.d.) Tributes. American Association for Artificial Intelligence, http://www.aaai.org/aitopics/retired/html/tributes.html (retrieved April 22, 2009)
2. Brooks, R.A.: Intelligence without reason. In: Myopoulos, R., Reiter, J. (eds.) Proceedings of the 12th International Joint Conference on Artificial Intelligence, pp. 569–595. Morgan Kaufmann, San Mateo (1991)
3. Brooks, R.A.: Rodney Brooks forecasts the future. New Scientist 2578 (2006)
4. Clark, A., Chalmers, D.: The extended mind. Analysis 58(1), 7–19 (1998)
5. Cook, J.W.: Did Wittgenstein practise what he preached? Philosophy 81(3), 445–462 (2006)
6. Damasio, A.: Descartes' Error. G.P. Putnam, New York (1994)
7. Dautenhahn, K.: Getting to know each other – Artificial social intelligence for autonomous robots. Robotics and Autonomous Systems 16(2-4), 333–356 (1995)
8. Descartes, R.: Discourse on the Method. Mclean, I. (trans.), vol. 1637, Oxford University Press, Oxford (2006)

9. Descartes, R.: Meditations on First Philosophy. Rev. ed. J. Cottingham (trans.). Cambridge University Press, Cambridge (1641/1996)
10. Epstein, R., Roberts, G., Beber, G. (eds.): Parsing the Turing test: Philosophical and Methodological Issues in the Quest for the Thinking Computer. Springer, Gronigen (2008)
11. Gabbey, A.: Reflections on the other-minds problem: Descartes and others. In: Israel, D.S. (ed.) Sceptics, Millenarians and Jews, pp. 59–69. E.J. Brill, Leiden (1990)
12. Ganapati, P.: Cognitive computing project aims to reverse-engineer the mind (Feburary 6, 2009),
    http://www.wired.com/gadgetlab/2009/02/cognitive-compu/
    (retrieved May 6, 2009)
13. Harnad, S.: Other bodies, other minds: A machine incarnation of an old philosophical problem. Minds and Machines 1(1), 43–54 (1991)
14. Kurthen, M., Moskopp, D., Linke, D.B., Reuter, B.M.: The locked-in syndrome and the behaviorist epistemology of other minds. Theoretical Medicine 12(1), 69–79 (1991)
15. Lakoff, G., Johnson, M.: Metaphors We Live By. University of Chicago Press, Chicago (1980)
16. Leone, S. (Director): A Fistful of Dollars, Motion Picture (1964)
17. Lombardi, C.: Sizing up the coming robotics revolution (May 15, 2007),
    http://news.cnet.com/Sizing-up-the-coming-robotics-revolution/
    2008-11394_3-6183596.html
    (retrieved January 21, 2010)
18. MacIntyre, A.: Individual and social morality in Japan and the United States: Rival conceptions of the self. Philosophy East and West 40(4), 489–497 (1990)
19. Magnani, L.: Abduction, Reason, and Science: Processes of Discovery and Explanation. Kluwer, New York (2001)
20. Maher, B.: Poll results: Look who's doping. Nature 452, 674–675 (2008)
21. Pliszka, S.R., McCracken, J.T., Maas, J.W.: Catecholamines in attention-deficit hyperactivity disorder: Current perspectives. Journal of the American Academy of Child and Adolescent Psychiatry 35(3), 264–272 (1996)
22. Putnam, H.: Psychological predicates. In: Capitan, W.H., Merrill, D.D. (eds.) Art, Mind, and Religion, pp. 37–48. University of Pittsburgh Press, Pittsburgh (1967)
23. Rumelhart, D.E., McClelland, J.L.: Parallel Distributed Processing, vol. 1. MIT Press, Cambridge (1986)
24. Sahakian, B., Morein-Zamir, S.: Professor's little helper. Nature 450, 1157–1159 (2007)
25. Searle, J.: Minds, brains, and programs. Behavioral and Brain Sciences 3(3), 417–457 (1980)
26. Solomon, P., Leiderman, P.H., Mendelson, J., Wexler, D.: Sensory deprivation: A review. American Journal of Psychiatry 114(4), 357–363 (1957)
27. Ter Hark, M.R.: The development of Wittgenstein's views about the other-minds problem. Synthese 87(2), 227–253 (1991)
28. Thagard, P.: Brains and the Meaning of Life. Princeton University Press, Princeton (2010)
29. Thagard, P.: Computational Philosophy of Science. MIT Press, Cambridge (1993)
30. Turing, A.: Computing machinery and intelligence. Mind 59(236), 433–460 (1950)
31. Watson, J.D.: The Double Helix. Atheneum, New York (1968)

Morality According to a Cognitive Interpretation: A Semantic Model for Moral Behavior

Sara Dellantonio and Remo Job

Abstract. In recent years researches in the field of cognitive psychology have favored an interpretation of moral behavior primarily as the product of basic, automatic and unconscious cognitive mechanisms for the processing of information, rather than of some form of principled reasoning. This paper aims at undermining this view and to sustain the old-fashioned thesis according to which moral judgments are produced by specific forms of reasoning. As critical reference our research specifically addresses the so called Rawlsian model which hinges on the idea that human beings produce their moral judgments on the basis of a moral modular faculty "that enables each individual to unconsciously and automatically evaluate a limitless variety of actions in terms of principles that dictate what is permissible, obligatory, or forbidden".[25, p. 36] In this regard we try to show that this model is not able to account for the moral behavior of different social groups and different individuals in critical situations, when their own moral judgment disagrees with the moral position of their community. Furthermore, the critical consideration of the Rawlsian model constitutes the theoretical basis for the constructive part of our argument, which consists of a proposal about how to develop a semantic, quasi-rationalistic model to describe moral reasoning. This model aims to account for both moral reasoning and the corresponding emotions on the basis of the information which morally relevant concepts consist of.

1 Introduction

Contemporary cognitive research identifies three fundamental models that describe moral behavior and in particular the processes leading to moral

Sara Dellantonio · Remo Job
Department of Cognitive and Education Sciences, University of Trento,
Rovereto, Italy
e-mail: **sara.dellantonio@unitn.it,remo.job@unitn.it**

L. Magnani et al. (Eds.): Model-Based Reasoning in Science & Technology, SCI 314, pp. 495–517.
springerlink.com

judgments [25, p. 45]. The first model is called *Humean* and is based on the idea that human beings are equipped with an innate moral sense, of an emotional kind, that drives judgments regarding right and wrong. The model foresees that an individual's perception of a morally relevant situation automatically triggers a positive or negative emotional response that leads to the moral judgment of the situation: the perceived situation will be considered as right in the case of a positive emotion and as wrong in the case of a negative emotion[1]. In fact according to this view moral pronouncements are not "judgments" in the proper sense of the word but *immediate and spontaneous intuitions that assume for the subject the appearance of evident and obvious truths.*

A second model may be defined as *rationalistic* and is based on the opposite idea that moral statements are not conceived as intuitions, but as judgments produced by conscious reasoning on explicit knowledge at the disposal of the subject[2]. This reasoning may be intended as based on utilitarian principles, but it may also be interpreted in different ways. One of the main alternative views is a *Kantian* one, according to which moral reasoning can neither be utilitarian (since morality is what allows humans to overcome particular, egoistic utility to reach an universalistic good) nor can it consist of the capacity and/or in the willingness to learn and to follow legal rules (since there is not necessarily a coincidence between what the law stated and what is right from a moral point of view)[3].

A third model – qualified as *Rawlsian* – hinges on the idea that human beings produce their moral judgments on the basis of "an evolved capacity of

---

[1] The contemporary psychological research about the Humean model became established in the 1980s mainly through Martin Hoffman [27, 28, 29] and involves authors such as Kagan [31]; Frank [15]; Schweder and Heidt [51]; Wilson [60]; Haidt [18]; Haid and Bjorklund [19].

[2] The rationalistic model originates mostly from Kohlberg's researches on morality, which are based on Piaget's developmental theory: see Kohlberg [33, 34, 35].

[3] The description we give here of this second model does not correspond entirely to Hauser's classification of the main views on morality currently discussed in the field of the cognitive debate. In fact, Hauser speaks of a Kantian model, which actually consists of a hybrid view of moral judgment according to which the perception of a morally relevant situation produces both conscious reasoning (intended mostly as utilitarian reasoning) and emotional reactions. According to this view, emotion and reasoning interact in the production of moral judgment; if their assessments diverge, the production of a judgment requires that one gets the "upper-hand" over the other. First of all, we refuse to call this rationalistic model "Kantian" because Kant's perspective on moral judgment can be connected neither with a utilitarian reasoning nor with emotional reactions. Secondly, to speak of a rational model allows us to present the positions in play first of all in their most simple versions, avoiding hybrid views where both rationality and emotions are in play. For a clarification of the difference between a purely rationalistic model and a hybrid one and for a brief overview of the positions belonging to this hybrid model see Hauser, Young, Cushman [26].

all human minds that unconsciously and automatically generates judgments of right and wrong" [25, p. 2]. More precisely: "all humans are endowed with a moral faculty – a capacity that enables each individual to unconsciously and automatically evaluate a limitless variety of actions in terms of principles that dictate what is permissible, obligatory, or forbidden" [25, p. 36]. According to this model the perception of a morally relevant situation is followed by an analysis of the situation on the basis of a moral module which decomposes it into its elements and that uses its own principles to evaluate them and to formulate moral judgments.[4] Once it has been expressed, the moral judgment causes an emotional reaction and a reasoning process aimed at justifying it *post hoc*. Even if this model is different from the Humean one in the way it describes the process that leads to the moral pronouncements, it shares with it the idea that morality is a kind of intuition based in this case not on emotive evidence, but on an *unconscious and automatic processing* of information[5].

Over recent years cognitive research has concentrated a lot of attention on the Humean and on the Rawlsian model at the expense of the traditional rationalistic model, which is mostly considered outdated[6]. These models seem indeed to be more compatible with the contemporary researches in the field of cognitive psychology, which favour an interpretation of moral behavior primarily as the product of basic, automatic and unconscious cognitive mechanisms for the processing of information rather than of some form of principled reasoning. Furthermore, the Rawlsian and some versions of the Humean model are characterized by another aspect which turns out to be extremely desirable from a cultural point of view. Indeed, if these models should turn out to be true, they would allow a tracing of a common fundament, of a biological nature, for the moral principles consisting of the specific cognitive mechanisms which drive human judgment about right and wrong. In this case, moral discourse need not be based on pure rational or sociological or even religious argumentation, but could become part of a scientific-experimental investigation of human cognition. In fact, as Hauser states, the identification of innate principles of this kind would determine "the range of both possible and stable ethical systems" [25, p. 54].

Even though this way to approach the problem of morality has become a major trend in the field of cognitive research, "mechanistic" explanations of the moral behavior often show large difficulties in explaining moral views of people in real situations outside of the experimental settings defined in the laboratory. Indeed, this paper aims at undermining the view that moral

---

[4] About the question of modularity and about the hypothesis that the central system is also organized according to (more or less rigid) modules see e.g. Samuels [50] and Carruthers [3].

[5] Among the most prominent authors that contributed to developing the so called Rawlsian Model there are Susan Dwyer [12, 13]; John Mikheil [37]; Gilbert Harman [24] and especially Mark Hauser [25].

[6] For an overview of the most important difficulties identified by this model see for example Thomas [55].

behavior can be seen as the product of a form of intuition produced by some unconscious and automatic processing of information and to sustain, on the contrary, the old-fashioned thesis according to which moral positions express judgments produced by specific forms of reasoning. This thesis does not imply that considerations of psychological or of cognitive nature no longer play any role in the understanding of moral behavior, but it does mean that moral behavior has to be explained on the basis of *conscious and non-automatic high-level cognitive mechanisms related with the thought and with the human capacity to reason and to judge.*

As critical reference our research specifically addresses the Rawlsian model[7]. We individuate two kinds of principles that many of the authors, who are in agreement with this model, believe belong (among others more controversial) to the moral faculty: the one consists of some innate moral contents, the other concerns the capacity to distinguish strictly moral principles from merely conventional norms. Furthermore, we argue that, if we assume the existence of a moral faculty based on these kinds of principles, we will not be able to account for the moral behavior of different social groups and of different individuals in critical situations, when their own moral judgment disagrees with the moral position of their community.

The critical consideration of the Rawlsian model constitutes the theoretical basis for the constructive part of our argument, which consists in a *proposal about how to develop a semantic, quasi-rationalistic model to describe moral reasoning.* This model aims to account for both the moral reasoning and the corresponding emotions on the basis of the information which morally relevant concepts consist of.

2   The Rawlsian Model and the Linguistic Faculty

The Rawlsian Model Moves from a Structural Hypothesis which is congenial to the functional architecture of the cognitive system as it is conceived by the classic cognitive science, since it is based on a faculty interpreted as a module

---

[7] The reason why we prefer to consider this model instead of the Humean one is firstly that the position of the Rawlsian model is more univocally and decidedly oriented towards a mechanistic view of moral behavior. Indeed, even though – as noticed before – both the Humean and the Rawlsian models tend to give an account of moral behavior in terms of unconscious and automatic processing of information, Humean models (which are by the way also very variegated in the specific positions they maintain) attribute in general to those mechanism a weaker role as regards the determination of the end results of the moral output. Indeed, while the Humean model just assumes that human beings are characterized by an innate affective constitution, the Rawlsian model makes a much more binding presupposition, assuming the existence of an innate moral modular faculty which works according to specific fixed principles and on specific and fixed information. A Humean model that does not imply any mechanistic view of moral judgments is for example the one proposed by Prinz [45].

and explains moral judgments in analogy with Chomsky's grammaticality judgments. Its denomination is due to the fact that it was John Rawls who first explicitly drew an analogy between language and morality and who hypothesized the existence of a moral faculty based on the linguistic faculty proposed by Chomsky.[48, p. 55]

Actually, the possibility of using the model of Universal Grammar in order to determine the functioning of a hypothetical modular moral faculty has also been put forward by Chomsky himself:

> The acquisition of a specific moral and ethical system, wide ranging and often precise in its consequences, cannot simply be the result of "shaping" and "control" by the social environment. As in the case of language, the environment is far too impoverished and indeterminate to provide this system to the child, in its full richness and applicability. [...] it certainly seems reasonable to speculate that the moral and ethical system acquired by the child owes much to some innate faculty. [7, pp. 152–153]

This idea met with big success both in psychological and in philosophical research and it has been further developed by authors like Susan Dwyer [12, 13], John Mikheil [37], Gilbert Harman [24], and especially Mark Hauser (2006) who tried to elaborate, in a concrete fashion, a modular Rawlsian model of moral competence.

As Chomsky's quotation already suggests, among the main arguments these authors appeal to in order to argue for the existence of a modular moral faculty analogous with the linguistic faculty there is the so-called "poverty of stimulus argument". Originally provided by Chomsky[8], this argument is based on the observation that children learn language early and easily, even though its rules are never taught to them and the stimuli available to them to reconstruct these rules on their own are extremely poor, fragmentary and asystematic. Since the experience the child can rely on is not – according to Chomsky – sufficiently rich to justify such a fast and easy learning process, we must assume that humans are endowed with an innate linguistic faculty that organizes and completes the available experience in a way that allows the learning of language. The authors who maintain the Rawlsian position apply the same argument to moral competence. They claim that the everyday experience children can rely on in order to learn the moral rules of their own group is partial and underspecified. Therefore, in order to account for the fact that moral rules are learned precisely and quickly in early childhood, we must assume the existence of a specific faculty that organizes and completes children's experiences to allow the learning of the moral rules of their group[9].

In order to give an idea about the way this moral faculty is supposed to work, these authors refer again to Chomsky and to his *Principles and*

---

[8] About this see in particular Chomsky [4, pp. 2–11], and Chomsky [5, pp. 5–6].

[9] Among the most important authors which appeal to the "poverty of stimulus argument" in relation to moral learning there are Mikheil [37]; Dwyer [12]; Harman [24]; Mahlmann [36]; Mikheil [38]; Nichols [42].

Parameters Theory [7, p. 62 ff.]. According to this theory (whose general structure in the field of linguistics has actually been developed in various ways) the Universal Grammar is made on the one hand of universal principles which are common to all real and possible languages, and on the other hand of universal parameters, which complement these principles for some particular aspects. The idea behind parameters is that some basic characters of the grammar of natural languages are variable, even though only a *limited and defined* set of variations is possible. Parameters have the function of defining these possible variations. One of the hypothesized parameters governs, for example, the ordering of subject, verb and object in the phrase which is supposed to vary in different languages according to a very limited number of options. According to Chomsky's theory even a minimal linguistic experience is sufficient for the child to set the parameter in the right manner depending on the language he is learning and so to organize from that moment on further stimuli according to that parameter.

The existence of parameters accounts for the fact that children learn the particular grammar of their own language in its specificity and difference to other languages. In the same vein, the hypothesis that the moral faculty is also made of parameters is introduced to explain the fact that children sharing common universal moral principles are able to learn the particular variation of these principles adopted by their own community. Once the grammar of a particular language has been learned on the basis of the principles and parameters, the children will be able to use it to produce spontaneously, and without reflection, well-formed sentences in that language. In the same way, once the moral grammar of a group has been learned, this will allow the member of the group to produce *spontaneously and without conscious reflection* moral judgments that reflect that grammar and its underlying principles.

3   The Problems with This View

As the idea of the Universal Grammar proposes that our linguistic competence is driven by innate unconscious, operating principles, the idea of a Universal Moral Grammar entails a view of moral judgments, according to which they are the automatic and immediate product of unconscious moral principles. Moral positions are no longer conceived as a form of judgment, resulting from complex, principled reasoning, as common sense used to think, but they are understood in analogy to language as the result of a creative mechanism which produces immediate moral intuitions before and independently from conscious thought.

The possibility to identify an automatic mechanism for the production of moral intuitions has, according to some, the reassuring effect to make all humans appear essentially similar and moral. A proof for that is the fact that the analogy between language and morality (often called *"linguistic analogy"*) lets us take for granted that we can speak of a "moral competence", while the opposite idea of "moral incompetence" appears paradoxical. However, it should not be disregarded that these reassuring aspects are the outcome of a naturalistic conception according to which our moral sense is part of the human nature and its judgments are entirely driven by our cognitive mechanisms. As Jesse Prinz points out:

> Recently researchers have begun to look for moral modules in the brain, and they have been increasingly tempted to speculate about the moral acquisition device and innate faculty for norm acquisition akin to the celebrated language acquisition device promulgated by Chomsky [...]. All this talk of modules and mechanism may make some shudder, especially if they recall that eugenics emerged out of an effort to find the biological sources of evil. Yet the tendency to postulate an innate moral faculty is almost irresistible. For one thing, it makes us appear nobler as a species, and for another, it offers an explanation of the fact that people in every corner of the globe seems to have moral rules. Moral nativism is, in this respect, an optimistic doctrine – one that makes our great big world seem comfortingly smaller. [46, p. 367]

These reflections show unequivocally that embracing a mechanistic view of morality has an important cultural and social impact. Still, we are not so much interested in the actual desirability of this view; rather we aim at investigating the general plausibility of the thesis that moral positions are the product of a modular moral faculty. In this regard we need to consider *what the unconscious principles this module is supposed to work with are, and whether these principles permit accounting for moral behavior as it manifests itself in the everyday life of subjects and of social groups.*

In order to address this aspect we need to distinguish two different groups of principles. The *first* group consists of principles that cannot be considered specifically moral and that are supposed to explain how the moral faculty structures the information in terms which are suitable for a moral evaluation, for example identifying in the stream of perception specific actions or events, their direct or indirect consequences or understanding whether someone caused them or is responsible for them.[10] The *second* group consists of principles that are specific to the moral competence and that drive directly the decision about what is permissible, obligatory, or forbidden. In order to carry out any further reflection about the moral faculty, we need first of all to examine this second group of principles.

As far as this second group of principles is concerned, the literature is characterized by hypotheses which are also very different from each other. Some of them recur more often and are considered more important. (A) First of all many authors agree that among the working principles of the moral faculty there are at least some *innate moral content* like the prohibition of murder, harming, stealing, cheating, lying, breaking promises, and committing adultery[11]. Since these principles are evidently subject to a lot of exceptions all

---

[10] See for example Hauser [25, pp. 8, 21, 41, 45–48, 166–182].

[11] See for example Hauser [25] who refers back to Mikheil [37].

the time, they are usually interpreted in a parametric way. This means that, even though they are considered to be universal, they are also supposed to consist of some parametric variables that assume a specific value just in the "moral community" a person happens to grow up in[12]. The most important variable in this respect is the determination of who is worthy of moral consideration and who is not. Susan Dwyer describes this problem in a very clear manner introducing the notion of "*schweeb*".

> Let us [...] define a schweeb as "creature with the highest moral status". A very basic principle of all possible [internalized] moralities might be "Schweebs are to be respected" or "Given the choice of saving the life of a schweeb or saving the life of a non-schweeb, always save the life of a schweeb." [13, p. 249]

Since human groups never identified univocally or permanently who qualifies as a "*schweeb*" (or, in less imaginative juridical words, as a "person", i.e. as a subject who is recognized as having specific rights), authors that maintain a Rawlsian view consider this notion to be a parameter in the sense that all groups decide and impose to each member who is a "*schweeb*" according to its own moral rules. In this way it becomes possible to explain why each culture prohibits, for example, the killing of some people even though it permits the killing of some other people according to criteria which resemble the logic of the ingrouping and outgrouping.

(B) A second type of principle, which is usually considered characteristic of the moral faculty, expresses a capacity. Specifically the capacity to distinguish situations which have to be considered properly moral and need to be evaluated using moral rules, from other kinds of situations where they may face a form of disrespect for some social rules but without any real moral issue being in play. This principle is considered analogous to one assumed in relation to the linguistic faculty which is supposed to account for the capacity children show to select just auditory inputs of a linguistic kind distinguishing them from other kinds of sounds. In the case of the moral faculty it is maintained that people need to distinguish on the one hand authentically moral situations and violations – which are mainly associated with harming – and on the other, situations and violations which deserve social blame (like going to the office in pyjamas) or which cause revulsion (like licking the lavatory seat) without being morally relevant[13]. In addition to Hauser, Susan Dywer also attributes to this capacity a great importance for the definition of a moral faculty. She maintains that the poverty of stimulus argument applies to it. If children can begin to distinguish properly moral situations at a very early stage of their cognitive development, without explicit instruction and even though they do not seem to have enough experience to reconstruct the

---

[12] See Dwyer [12, p. 177].

[13] The examples mentioned here have also been investigated experimentally. See below in this section for reference.

criteria commonly used to identify properly moral situations, then we need to admit that this capacity has to be innate[14].

The discussion of this issue began with the publication of Elliot Turiel's studies according to which moral norms needed to be sharply differentiated from other kinds of norms for being:

1. *objective* (in the sense that their prescriptive power is not supposed to depend on extrinsic authorities);
2. *general* (in the sense that their prescriptive power is not perceived as limited to a particular group, place or time, but it is considered to extend to any group, place or time); and
3. *important* (in the sense that their violation is perceived as something extremely serious that harms the wellbeing of the people who experience it)[15].

Furthermore these investigations go together with others which aim to show that non-moral disapproval is experienced as very different from moral disapproval even when it comes with deeply negative emotions evoked by moral violations[16].

The issue which needs to be faced in relation to this principle and to the previous one is whether they have to be considered plausible and, furthermore, whether it is plausible to assume that they are part of something like a moral faculty.

(B) Let us consider this question starting from this second principle according to which – to sum up briefly – humans are already able at a very early stage of their cognitive development to determine in a universal and transcultural manner which situations require a moral evaluation, since they are connected with harming, and which others need to be evaluated through milder norms concerning good taste and customs. If we could establish that humans do have a capacity like this, this would allow us to define shared and indisputable moral issues valid for every cultural group in any time, distinguishing them from other kinds of situations which do not deserve consideration from a moral point of view. However, as Jonathan Haidt and his colleagues showed, the identification of a distinction of this kind which binds the morality issue with harming, is only a mystification resulting from the naive Western laity of educated classes. In reality people or groups characterized by strong systems of values derived for example by religion, or even

---

[14] See for example Dwyer [12, 13]. For critics to the application of the poverty of stimulus argument to this competence see Prinz [46, pp. 392–395]. See also Dwyer's reply to Prinz in Dwyer [14].

[15] Among the most important studies which embrace the thesis of the psychological realty of the distinction between moral and conventional rules see for example: Turiel [57] 1983; Nucci [43]; Turiell; Killen, Helwig [59]; Smentana, Braeges [53]; Smentana [52]; Tisak [56]; Nucci [44]; Turiel [58].

[16] About this aspect see for example: Zahn-Waxler, Radke-Yarrow, Wagner, Chapman [61]; Nichols [40], Nichols [41, pp. 23–25].

just less educated people, do not share the same indissoluble association between morality and harming, but they do see as morally binding norms related to other kinds of facts. They also see as morally relevant norms which well-educated Westerners often consider to be "matters of custom" (like the prescriptions related to specific foods, clothing or festivals) or "matters of good taste" (for example the ones that seem repugnant or inappropriate like licking the lavatory seat or cleaning it with the national flag or masturbating with a dead chicken)[17]. Furthermore, people of low social status or coming from conservative, strongly religious backgrounds tend to also consider morally relevant lifestyle choices like having a relationship with a person of the same sex. For these reasons the existence of a universal principle that determines which actions are subject to a moral judgment and which are not, seems mostly a form of wishful thinking that does not find any correspondence in the way different cultures or even different groups belonging to a same culture think.

(A) An analogous reflection can also be carried out in relation to the supposed *innate moral contents* of the moral faculty, which according to Hauser and the pioneering study of Mikheil that he and others rely on, include things such as murder, harming, stealing, cheating, lying, breaking promises, and committing adultery[18]. As mentioned previously, the authors who introduced these principles are aware of the fact that their application in different cultures or groups is liable to many exceptions and they try to solve this problem interpreting them in a parametric way. According to this interpretation it is the cultural context which determines to whom they apply and who is an exception to them. Universal principles assume therefore the following form: "Murder/harming/stealing from/cheating/ etc. members of the group X is forbidden (morally wrong)", while X is determined contextually, depending on which creatures are "schweebs" (i.e. deserve the highest moral status) in the considered culture. Still, even though we give a parametric interpretation of these principles, at least some of them already appear at first glance unlikely and derived by a naive projection of the values of our contemporary Western society back in time, or elsewhere in space. This is decidedly the case for example of adultery, since adultery could be considered an innate principle just if we admit that the original condition of human communities is monogamy instead of polygamy, which anthropology considers to be false[19].

---

[17] All the proposed examples including the following one are taken from famous experiments of Heidt and colleagues. See for example Haidt, Koller, Diaz [23]; Heidt [18]; Heidt, Joseph [22]; Haidt, Graham, [21]. For a general report of these studies which takes Heidt's part see for example Kelly, Stich [32].

[18] See Hauser [25, p. 48]; Mikheil [37].

[19] Domestic groups have taken very different forms during the human history: before nuclear family became established, humanity went through a condition of promiscuity and subsequently of matrifocality, polygyny, and polyandry. About this see for example [49].

If adultery can be considered a borderline case which has been incautiously included in the list of the universal principles, the point is that the whole idea of the universal contents proposed by the Rawlsian account seems to be highly problematic. Let us consider this aspect discussing the case of principles which, at first glance, appear shared and transcultural like murder or harming. A good way to pose the problem is suggested by Sripada and Stich [54, p. 282], according to whom it does not make sense to appeal to universal principles as being valid in every culture like "murder is wrong", since these principles express just purely analytical sentences, i.e. sentences which are true in virtue of their meaning only. The fact that murder is wrong is, for example, already implicit in the meaning of "murder" which can be expressed in terms of "killing someone in an impermissible way" and which does not say anything about what kind of killing counts as murder and what does not (think for example about executions which are considered admissible or inadmissible depending on the political systems). In this sense "murder" is just an empty form of moral discourse which can be arbitrarily filled with all kinds of killing which the judging subject considers inadmissible. The truly moral problem does not therefore concern the question about whether all cultures agree to condemn murder, but rather the question about the criteria they use to decide which forms of killing are admissible and which are inadmissible. This way to pose the problem suggests a different "linguistic analogy" from the one embraced by the Rawlsian model, which is based on semantics instead of syntax (to which Chomsky's thesis primarily applies). It suggests that *what we really need to understand in order to face the moral problem is under which condition a subject categorizes an event as an act of murder, harming, stealing, cheating etc. rather than as an act of a different kind which can be considered as morally legitimate.*

Even though this semantic interpretation of the linguistic analogy opens up new questions related to the fact that there is not a unique recognized theory of categorization, and that therefore the reliability of the semantic model we propose depends on the reliability of the semantic theory we choose, in our opinion it still permits us to approach the moral problem in a more productive way. Firstly, it permits us to individuate more clearly the role and the weight of the cultural influence on moral judgment. Secondly, tracing back the moral problem to semantics helps us to avoid, in part, strong assumptions about the existence of automatic, innate cognitive mechanisms and restores the function of reasoning in moral choices. Affirming the idea that reasoning does play a function in moral choices allows us to account for some very relevant aspects of the moral discourse which have been mostly ignored by the Rawlsian model.

Indeed the views which conceive moral positions as intuitions – i.e. as an automatic output of a mechanism which has assimilated the cultural conventions of the culture of origins (think about the Universal Grammar as a means to learn the specific grammar of the native language) – identify somehow the moral rules with the rules imposed by a culture, while at times making a morally right choice means distancing oneself from culturally imposed rules. A well-known and often debated example in this direction is the case of Germans during the Third Reich who disagreed with race laws issued by their own government, even though they belonged to the Aryan community and believed in the nationalistic ideals of the German Right. This phenomenon, which has often been defined as "tacit dissension" is of great importance for the moral psychology since it shows the need to assume that humans are also capable of autonomous, subjective moral choices, which are separate from the "corporative moral".

In fact, dissent can hardly be explained in the theoretical framework of the Rawlsian model which stresses the automaticity of moral judgment in analogy with the automaticity of the process that lead us to produce grammatically well-formed sentences in our language. But, if in the case of grammar we can assume that sentences are well-formed because they correspond with the grammatical rules of our languages, in the case of morality it is possible that something is judged as morally wrong by some member of a group even though the group itself accept is as right. Considered more generally, the problem is how to account for the possibility to qualify an act as morally wrong or as morally right independently from the rules that a particular culture adopts at a certain time. This problem can also assume a different form which is of immediate relevance for the Rawlsian model: even though a cultural group determines who counts as a "schweeb" in that group (for example the Aryan, in the case of the Third Reich), a moral model needs to account for the fact that sometimes someone can disagree with the criteria adopted by the group and recognize that other people (for example the Jew) deserve to be regarded as "schweeb". The general point here is that the Rawlsian model does not account for phenomena like these concerning subjective autonomous moral judgments.

4   Which Theory of Concepts?

In the field of cognitive research, when categorization is addressed, people tend to confuse two different levels of the analysis. On the one hand there is the categorization in the sense of the *prelinguistic* cognitive procedure that "puts together" similar instances, forming the "conceptual units" children need in order to learn their first language; and on the other hand there is the categorization in the usual linguistic sense. In the first case the point at issue is the way in which infants or a hypothetical Robinson Crusoe who grew up alone on an island – or even animals – recognize some instances as similar to each other and group them together in "conceptual units", which do not depend on any linguistic, conventionally determined category. In the second, the matter under discussion are the forms of categorization carried out by means of the language, when adult subjects group instances together on the basis of similarities suggested by the semantics of the language they learned. The

difference between these levels of analysis can be further clarified by introducing two notions of semantics: an *Externalized Semantics* (*E-semantics*)
and an *Internalized Semantics* (*I-semantics*)[20]. The notion of E-semantics
accounts for the semantics in its conventional and public dimension, which
is characterized by rules whose aim is to assure the possibility of intersubjective communication. The notion of I-semantics addresses instead the issue
of which information people use (internally, i.e. in their mind) to carry out
categorizations and to understand the linguistic meaning (as it is codified in
the E-semantics).

These two notions of semantics can clearly not be considered completely
independent from each other, but they necessarily merge together through
linguistic learning: It is mostly for this reason that many studies think it is
not necessary to distinguish between them. To explain this aspect it is useful
to think about what happens when an infant learns his first language: in order
to learn linguistic meanings, the infant must already be able to carry out some
form of categorization which allows him to formulate a hypothesis about the
possible use of the words he hears. Nevertheless, once he has learnt his first
language, the specific concepts of *that* language (of *that* E-semantics) retroact
on his previous prelinguistic way to categorize, determining a new way for him
to see the world. I- and E- semantics are connected through a double binding:
the I-semantics set the fundamental criteria of the E-semantics, in the sense
that it would not be possible to learn E-semantics if it was not compatible
with the I-semantics; this compatibility with the I-semantics is what the E-
semantics of all existing languages must have in common. However, once it
has been learned, the E-semantics affects the I-semantics of subjects, which
at least to some extent become conformed to it. The thesis we will try to
support in the following is that a cognitive account of morality must rely on
the relation between *I-semantics* and *E-semantics*.

Let us focus first of all on the categorization at the level of the *E-semantics*.
At this level we think that a theory with great explanatory power is the so
called *theory-theory* which – on the basis of a Quinean view on language – interprets the semantic systems as complex and highly structured theories abut
the world whose elements are interlinked and determined by each other[21].
According to this view people's beliefs are articulated bodies of knowledge,
which work like theories: concepts express single elements of these theories
and are identified on the basis of the rule they play in the whole of people's
beliefs[22]. What this view suggests is that when a subject learns a language
(an E-semantics), he acquires the belief system of which this language is an

---

[20] This distinction is diffusely discussed mostly by Ray Jackendoff who develops
it on the basis of Chomsky's analogous differentiation between I-Language e E-
Language: see for example Chomsky [6], §2.2 e §2.3 and Jackendoff [30, p. 22]

[21] See for example Quine [47]. In the field of the cognitive science the so-called
theory-theory has been introduced and developed first of all by Carey [1, 2] and
Murphy, Medin [39].

[22] About this aspect see specifically Carey [2].

expression. Since concepts are the constituents of beliefs, they depend on the belief system they are connected with and vary according to it.

As already pointed out by several authors on the basis of arguments very different from each other, this semantic theory cannot account for all aspects of categorization. One argument for this has already been mentioned: if we want to account for the possibility to learn a first language, we need to admit that children are able to produce on the basis of their experience some primary internal categorization – a primary I-semantics – which they can use to make hypotheses about the possible groupings underlying the language they are learning. Once they produce some primary grouping strongly related to the perceptive experience, linguistic learning proceeds: this prelinguistic "conceptual core" can be specified in different ways making more and more precise and abstract differentiation through the adding of features carried by the linguistic meanings (the E-language) and by the beliefs about the world it expresses[23].

This way to interpret the relation between I- and E-semantics implies already a very precise theory of concept according to which the conceptual system of adult subjects (described by the I-semantics in the form it assumes after linguistic learning) will be characterized by a double structure. Concepts are supposed to consist on the one hand of a core corresponding to the primary grouping produced by the prelinguistic categorization, and on the other hand of a periphery of cultural features imported from the E-semantics which cover and specifies this core.

Differently from the periphery of the concept, the core cannot be seen as an articulated and interlinked body of knowledge, as the *theory-theory* suggests. We think that the explanation of the core needs to rely on a different semantic theory of prototypical character. We cannot enter here into details about the kind of theory of concepts we consider to be best in order to explain how the core works[24]. For the aims of this work it is sufficient to point out how – according to our view – the relationship between E- and I-semantics imposes the embracement of a form of *Dual Theory able to account for both the components of concepts, the cognitive core connected with the perceptual dimension of instances and a theoretical periphery made of more complex features added to the core by language.*

5   A Semantic Model for the Moral Judgment

The view on concepts we proposed here is just a draft of a theory of concepts. Still the elements we brought into the discussion are enough for the aim of showing how moral issues can be approached using a semantic theory. In order to introduce this aspect it is useful to go back to the conclusions we

---

[23] For an articulated critique to this view in relation to the problem of language acquisition see Dellantonio [8, Ch. IV].

[24] About this aspect see Dellantonio, Pastore [9]; Dellantonio, Pastore [10].

drew from Sripada and Stich's thesis: once an act is categorized as "murder" it has already been judged as morally wrong, while *admissible* voluntary acts which make a person die are not categorized as "murder", but otherwise (e.g. as executions, as self-defence etc.). What a theory of concepts can help to understand in this regard is first and foremost *when (under which conditions) a form of killing is categorized as "murder"*.

The concept of "murder" has indubitably at its core the idea of life and of the loss of life; this is the idea of death. However, not any death is linked to the concept of "murder"; "murder" implies the idea of that particular form of loss of one's life which is caused by an external voluntary intervention. Furthermore, "murder" differs from "killing": any living being (such as a mosquito) can be killed and the act of killing does not need to be done by a human being (for example a rock falling from a cliff can kill someone); "murder" denotes, on the contrary, only the killing of people by people. A murder can be characterized by different degrees of intentionality (and can be qualified for example as voluntary, involuntary or justifiable); however it describes an avoidable action carried out by a human being at the expense of another human being which leads to the loss of life. Since human beings perceive life as something extremely desirable and positive, whereas they perceive the loss (of anything) as something negative, the core of the concept of murder already carries a very "negative value", which is heightened even more by the fact that the loss caused by a murder is perceived as voluntary and evitable.

This observation permits the addition of another element that has been omitted until this point concerning the emotional dimension of moral judgments. If we admit that some perceptive elements of concepts are characterized by "values" which can assume both a positive or a negative sign, we may well suppose that these values comes along with corresponding emotions which match them in intensity and orientation. In respect to the example we are considering this means that if an act is categorized as "murder", *then this act is perceived as having a negative value which comes along with a corresponding negative emotion*. Furthermore, since this value and the corresponding emotion are connected with the elements of the conceptual core – i.e. with the loss of the life by means of a voluntary act – then we can suppose that all concepts characterized by this core have the same negative value coming along with the same negative emotions even though they can cover this core with a periphery made of further cultural features, which are supposed to justify or dignify this act (think about concepts like "execution" or "human sacrifice").

To clarify this idea, showing how not only the features of the core but also the features of the periphery contribute to determining the value of a concept and the emotion that comes along with it, it is useful to consider two further examples, namely the derogatory concepts of "Jew" and "nigger". The concepts of "Jew" and "nigger" have a common core, which is actually shared by all concepts defining human beings to a particular race or descent. This core is made of perceptible characteristics belonging to all human beings such as their "physical form" (legs, arms, head etc.) and their salient behavioral traits (like the movements and the reactions to the situations which are typical just of humans)[25]. This common core must not be over-interpreted, in the sense that we cannot attribute to it any complex and morally binding characteristics, since it is supposed to consist just of perceptible features. Remaining as neutral as possible, we can say that the features this core is composed of are those which allow us to recognize other people as *conspecifics*[26].

In the hypothesis we are considering the categorization of subjects as "Jews" or "niggers" is due to a specific cognitive operation which couples to itself a neutral perceptive core with a periphery of negative cultural features. Whereas in the case of "nigger" this cognitive operation is carried out setting apart an element of the core relating to a specific distinctive somatic trait (*the dark skin* of this conspecific) and associating the negative features to those specific element, in the case of "Jew" the core does not have any distinctive perceptible characteristics[27], so the negative features typically associated with this concept in its derogatory use are just linked to other non-perceptible feature of the periphery like the race, the descent or the religion.

What the discussion of these examples suggests is on the one hand the cognitive system of humans allow them to recognize spontaneously other humans as conspecifics, i.e. as similar to themselves and as belonging to the same class. Still, on the other hand, the fact that the core of a concept can be associated with any kind of cultural peripheral features rules out the conclusion that – since someone is perceived as conspecific – he must also be perceived as a "person" (i.e. as a subject with binding rights). Indeed, as history teaches us, when derogatory cultural features are added to the core, it can occur that other humans are seen as conspecifics of inferior dignity. Nevertheless, even though it is possible to denigrate specific human subjects or groups through all kinds of negative features, according to the view we are putting forward it is impossible to categorize a human being as something totally non-human. The idea that concepts are characterized by a perceptive core which, in this case, consists of human features, imposes restrictions to the possible categorizations and therefore also opposes semantic relativism according to which everything can be seen as everything, depending on the feature we attribute to it.

This remark has important consequences with regard to moral judgment. Indeed if we admit (on the basis of an independent argument which cannot be

---

[25] About this aspect see Dellantonio, Pastore [9].

[26] About this see also Dellantonio, Pastore [9]; Dellantonio, Pastore [11].

[27] The "distinctiveness" of the features is always measured in relation to the characteristics of the group who carried out the categorization.

presented here[28]) that the recognition of something as a conspecific triggers off empathic responses of some kind, then we can claim that an empathic response will be always triggered off when an instance is categorized through a concept whose core is constituted by the feature "conspecific", even in case of derogatory concepts like "Jew" or "nigger". Still, in the case of derogatory categorization, such a response is suffocated by other opposite reactions connected with the negative features of the periphery.

If we consider the example of "nigger" from this perspective we can conclude that this is characterized by a positive value of the core "conspecific" which is connected with a corresponding positive emotional reaction and by a negative value of peripheral features which is connected with a corresponding negative emotional reaction. In the case of "nigger" the negative value of the peripheral features and the negative emotions that come with it preponderate over the core and over its positive value. The fact that we categorize someone through the derogatory concept "nigger" does not imply any moral judgment by itself, but if a moral reasoning concerns someone categorized as "nigger", the negative value of the concept will influence the overall result of the reasoning.

The argument we are proposing here suggests that the concepts subjects use when they reason about morally relevant situations can be decomposed into the features they consist of and that each of these features can be evaluated as being positive or negative and associated with a corresponding emotion. Reconsider the examples proposed previously: life carries a positive value, the loss of life carries a negative value, greed (typically associated with the prejudice against Jews) carries a negative value, and lack of intelligence (typically associated with the prejudice against black people) carries a negative value. If we determine how a subject categorizes the elements of morally relevant situations and then describe the features of the concepts he uses, then we can weigh up the positive and negative values of these features and have a precise indication of the way in which he perceives the situation from a moral point of view, namely if he is more inclined to see it as morally right or morally wrong. Moreover, since the value of the features goes along with corresponding emotions, the analysis of the value of the features will also give a reliable indication of the emotions triggered off by the situation. Indeed, since reasoning is carried out on the basis of propositions, which consists of concepts, if morally relevant concepts could be described in detail in terms of their features and of their specific value obtained, by weighing up all positive and negative values of the single features these features and values could be used to develop a *semantic model of moral reasoning*. Moreover, weighing up all positive and negative values of the single features will also allow us to determine the emotional orientation of the subject toward the conceptualized objects. *Such a model could therefore be instrumental in*

---

[28] An argument of this kind could for example be developed on the basis of the connection between moral sense and perspective taking: see Gordon [17] and Goldman [16].

establishing whether a situation will be judged and perceived – cognitively and emotionally – as morally right or morally wrong and why.

For this model it is of great importance to take into consideration the distinction between a core and a periphery of concepts. Indeed, even though we do not believe that the features of the core should be weighted more than the features of the periphery, the features of the core still take in some sense priority over the periphery, since the core remains stable over time and has to be considered universal, while the periphery is culturally determined. In fact, it is because of the intersubjective and intercultural stability of the core that the semantic model we are proposing becomes capable of explaining why sometimes people can disagree with their own "corporative moral", which is produced by the conventional norms of the group, they belong to. To clarify this let us return to the examples of "nigger" or "Jew". Even though a group conveys to its members that all black people have to be categorized as "nigger" or that all people of a certain religion have to be categorized as "Jew"(meant in a derogatory sense), still the member of this group cannot avoid perceiving "niggers" and "Jews" as conspecifics. The fact that they are always identified as conspecifics can be used to overcome prejudice since it ensures the possibility to re-categorize them through different peripheral features: it ensures, for example, the possibility to consider *on the basis of a conscious principled reasoning* that – since they are conspecifics – they must have the same basic biological features white people have and must therefore be considered "persons". Once you re-categorize black people or Jews as "person", the overall reasoning about the morality of something like racial laws, extermination, slavery etc. changes completely. Furthermore, in our view the same change also affects the emotions involved in the consideration of these situations explaining how it happens that in certain cases reflection drives or changes the course of emotions[29].

We do not have an answer for the question about why only certain people but not others in certain cases produce categorizations which are different from the ones conveyed by their own culture. The processes that are going on in these cases are most probably very similar to the ones that lead to the introduction of new categorizations in scientific theories and therefore to the change of previous scientific theories. Still, the matter we are actually interested in is just the kind of reasoning involved in the re-categorization of morally relevant concepts. This reasoning is indeed not just purely abstract and based on deliberate reflections about principles or rules, but it is a much more "embodied" form of thinking in which the subject is lead to change perspective and see things differently on the basis of information he already has by decomposing it and recomposing it differently. According to this interpretation moral reasoning is not a form of reflection completely open or free from any cognitive constraint, on the contrary it describes a cognitive procedure

---

[29] These reflections are similar to the tradition of research about dehumanization and moral disengagement which becomes established in the 1970s and is still one of the strong points of social psychology.

of information processing[30]. Nevertheless, it remains a form of reasoning in the sense that it is a non-automatic and conscious processing of information.

6 Concluding Remarks

The central thread of the discussion of this paper can be traced back to the thesis widely shared in the contemporary cognitive research according to which "conscious moral reasoning often plays no role in our moral judgments, and in many cases reflects a post-hoc justification or rationalization of previously held biases or beliefs" [25, p. 25]. The model of moral behavior mostly committed to the idea of a moral mechanism of a psychological kind which – once it is filled with cultural information – automatically produces some sort of moral intuition is the Rawlsian one. We address this model first of all in order to make a critical point about the fact that – if we accept the idea of a moral mechanism – we cannot any longer provide for a plausible explanation of the concrete dynamics of moral judgments in the case of groups whose culture is very different from the one of well-educated Westerners, nor in the case of controversial situations in which people must decide whether to accept as moral the norms of their own community or whether to refuse them.

In the second part of the paper we propose an alternative cognitive theory about morality, according to which our moral positions are judgments stemming from particular forms of conscious reasoning. In the hypothesis we consider these forms of reasoning consisting of specific operations on the concepts and specifically on the features these concepts are made of. In fact, our analysis suggests the possibility to delineate a semantic model of moral reasoning based on the features of the concepts used to categorize the elements of morally relevant situations, which also includes a description of the emotions that come along with moral judgments.

Differently from the Rawlsian model, the semantic model we propose does not conceive moral positions as being entirely determined by cultural and cognitive factors. On the contrary it suggests that the same situation can be categorized in various ways and that the moral judgment of a subject can change depending on how it has been categorized. Even though members of a group are lead to see some particular categorizations as more immediate and natural than others, conscious reasoning may allow in certain conditions to reconceptualize differently specific elements of morally relevant situations by breaking down and recomposing in a new way the perceptual and cultural features of the concepts we initially used to categorize them with. This view has the advantage of supporting both the idea of a psychological procedure underlying moral thinking, and a rationalistic conception of morality

---

[30] This conclusion implies that we still need a cognitive explanation of moral behavior and that we cannot come along with a purely cultural theory of morality. About this aspect see Dellantonio, Pastore [10, pp. 139–178].

according to which moral judgments are produced by conscious and principled reasoning.

References

1. Carey, S.: Conceptual Change in Childhood. MIT Press, Cambridge (1985)
2. Carey, S.: Knowledge acquisition: Enrichment or conceptual change? In: Carley, S., Gelman, R. (eds.) The Epigenesis of Mind: Essays on Biology and Cognition, Hillsdale, pp. 257–291. Lawrence Erlbaum Associates, Mahwah (1991)
3. Carruthers, P.: Moderately massive modularity. In: O'Hear, A. (ed.) Mind and Person, pp. 67–90. Cambridge University Press, Cambridge (2003)
4. Chomsky, N.: Recent Contributions to the theory of innate ideas. Synthese 17, 2–11 (1967)
5. Chomsky, N.: Reflections on Language. Pantheon Books, New York (1975)
6. Chomsky, N.: Knowledge of Language. Its Nature, Origin and Use. Praeger, New York (1986)
7. Chomsky, N.: Language and Problems of Knowledge: The Managua Lectures. MIT Press, Cambridge (1988)
8. Dellantonio, S.: Die interne Dimension der Bedeutung. Externalismus, Internalismus und semantische Kompetenz. Peter Lang Verlag, Hamburg, New York (2007)
9. Dellantonio, S., Pastore, L.: What do concepts consist of? The role of geometric and proprioceptive information in categorization. In: Hanna, P., McEvoy, A., Voutsina, P. (eds.) An Anthology of Philosophical Studies, pp. 91–102. ATINER, Athens (2006)
10. Dellantonio, S., Pastore, L.: Teorie morali e contenuto cognitivo. Cognitivismo, postmoderno e relativismo culturale. In: Meattini, V., Pastore, L. (eds.) Identità, individuo, soggetto, pp. 139–178. Mimesis, Milano (2009a)
11. Dellantonio, S., Pastore, L.: Struttura categoriale e categorizzazione. Un'ipotesi sull'origine della rappresentazione semantica. In: Dellantonio, S., Pastore, L. (eds.) Percezione, rappresentazione e coscienza, pp. 195–230. ETS, Pisa (2009b)
12. Dwyer, S.: Moral competence. In: Murasugi, K., Stainton, R. (eds.) Philosophy and Linguistics, pp. 169–190. Westview Press, Boulder (1999)
13. Dwyer, S.: How good is the linguistic analogy? In: Carruthers, P., Laurence, S., Stich, S. (eds.) The Innate Mind. Culture and Cognition, vol. 2, pp. 237–256. Oxford University Press, Oxford (2006)
14. Dwyer, S.: How not to argue that morality isn't innate: comments on Prinz. In: Sinnott-Armstrong, W. (ed.) Moral Psychology. The Evaluation of Morality: Adaptations and Innateness, vol. 1, pp. 407–418. MIT Press, Cambridge (2008)
15. Frank, R.: Passions within Reason: The Strategic Role of Emotions. Norton, New York (1988)
16. Goldman, A.I.: Simulating Minds. The Philosophy and Psychology, and Neuroscience of Mindreading. Oxford University Press, New York (2006)
17. Gordon, R.: Sympathy, simulation, and the impartial spectator. Ethics 105, 729–742 (1995)
18. Haidt, J.: The emotional dog and its rational trail: A social intuitionist approach to moral judgment. Psychological Review 108, 814–834 (2001)

19. Haidt, J., Bjorklund, F.: Social intuitionists answer six questions about moral psychology. In: Sinnott-Armstrong, W. (ed.) Moral Psychology. The Cognitive Science Of Morality: Intuition and Diversity, vol. 2, pp. 181–217. MIT Press, Cambridge (2008)

20. Heidt, J., Joseph, C.: Intuitive ethics: How innately prepared intuitions generate culturally variable virtues. Daedalus 133(44), 55–66 (2004)

21. Haidt, J., Graham, J.: When morality opposes justice: Conservatives have moral intuitions that liberals not recognize. Social Justice Research 20, 98–116 (2007)

22. Haidt, J., Joseph, C.: The moral mind. In: Carruthers, P., Laurence, S., Stich, S. (eds.) The Innate Mind. Foundations and the Future, vol. 3, pp. 367–391. Oxford University Press, Oxford (2007)

23. Haidt, J., Koller, S., Dias, M.: Affect, culture, and morality, or is it wrong to eat your dog? Journal of Personality and Social Psychology 65, 613–628 (1993)

24. Harman, G.: Explaining Value. Oxford University Press, Oxford (1999)

25. Hauser, M.D.: Moral Minds. How Nature Designed our Universal Sense of Right and Wrong. HarperCollins Publisher, New York (2006)

26. Hauser, M.D., Young, L., Cushman, F.: Reviving Rawls's Linguistic Analogy: Operative Principles and the Causal Structure of Moral Actions. In: Sinnott-Armstrong, W. (ed.) Moral Psychology. The Cognitive Science Of Morality: Intuition and Diversity, vol. 2, pp. 107–143. MIT Press, Cambridge (2008)

27. Hoffman, M.L.: Development of prosocial motivation: Empathy and guilt. In: Eisenberg-Berg, N. (ed.) Development of Prosocial Behavior, pp. 281–313. Academic Press, New York (1982)

28. Hoffman, M.L.: Affective and cognitive processes in moral internalization: An information processing approach. In: Higgins, E.T., Ruble, D., Hartup, W. (eds.) Social Cognition and Social Development: A Socio-Cultural Perspective, pp. 236–274. Cambridge University Press, New York (1983)

29. Hoffman, M.L.: The contribution of empathy to justice and moral judgment. In: Eisenberg, N., Strayer, J. (eds.) Empathy and its development, pp. 47–80. Cambridge University Press, New York (1987)

30. Jackendoff, R.: Languages of the Mind. Essays on Mental Representation. MIT Press, Cambridge (1992)

31. Kagan, J.: The Nature of the Child. Basic Books, New York (1984)

32. Kelly, D., Stich, S.: Two theories about the cognitive architecture underlying morality. In: Carruthers, P., Laurence, S., Stich, S. (eds.) The Innate Mind. Foundations and the Future, vol. 3, pp. 348–366. Oxford University Press, Oxford (2007)

33. Kohlberg, L.: Stage and sequence: The cognitive-developmental approach to socialization. In: Goslin, D.A. (ed.) Handbook of Socialization Theory and Research, pp. 347–380. Rand McNally, Chicago (1969)

34. Kohlberg, L.: From is to ought: How to commit the naturalistic fallacy and get away with it in the study of moral development. In: Mischel, T. (ed.) Cognitive Development and Epistemology, pp. 151–235. Academic Press, New York (1971)

35. Kohlberg, L.: The Psychology of Moral Development: The Nature and Validity of Moral Stages. Harper and Row, New York (1984)

36. Mahlmann, M.: Rationalismus in der praktischen Theorie: Normentheorie und praktische Kompetenz. Nomos Verlag, Baden-Baden (1999)

37. Mikheil, J.: Rawls' Linguistic Analogy. Ph.D. Thesis, Cornell University Press (2000)

38. Mikheil, J.: Comment on Sripada. In: Sinnott-Armstrong, W. (ed.) Moral Psychology. The Evaluation of Morality: Adaptations and Innateness, vol. 1, pp. 353–359. MIT Press, Cambridge (2008)
39. Murphy, G., Medin, D.: The role of theories in conceptual coherence. Psychological Review 92(3), 289–316 (1985)
40. Nichols, S.: Norms with feeling: Toward a psychological account of moral judgment. Cognition 84, 221–236 (2002)
41. Nichols, S.: Sentimental Rules: On The Natural Foundations of Moral Judgment. Oxford University Press, Oxford (2004)
42. Nichols, S.: Innateness and moral psychology. In: Carruthers, P., Laurence, S., Stich, S. (eds.) The Innate Mind. Structure and Contents, pp. 353–370. Oxford University Press, Oxford (2005)
43. Nucci, L.: Children's conceptions of morality, social conventions and religious prescription. In: Harding, C. (ed.) Moral Dilemmas: Philosophical and Psychological Reconsiderations of the Development of Moral Reasoning, pp. 137–174. Precedent Press, Chicago (1986)
44. Nucci, L.: Education in the Moral Domain. Cambridge University Press, Cambridge (2001)
45. Prinz, J.J.: The Emotional Construction of Morals. Oxford University Press, Oxford (2007)
46. Prinz, J.J.: Is morality innate? In: Sinnott-Armstrong, W. (ed.) Moral Psychology. The Evaluation of Morality: Adaptations and Innateness, vol. 1, pp. 367–406. MIT Press, Cambridge (2008)
47. Quine, W.V.O.: Two dogmas of empiricism. In: Quine, W.V.O. (ed.) From a logical point of view, pp. 20–46. Harvard University Press, Cambridge (1961)
48. Rawls, J.: A Theory of Justice. Belknap Press, Cambridge (1971)
49. Remotti, F.: Contro natura. Laterza, Roma Bari (2008)
50. Samuels, R.: Massively modular minds: Evolutionary psychology and cognitive architecture. In: Carruthers, P., Chamberlain, A. (eds.) Evolution and the Human Mind: Modularity, Language and Meta-cognition. Cambridge University Press, Cambridge (2000)
51. Schweder, R.A., Heidt, J.: The future of moral psychology: Truth, intuition, and the pluralist way. Psychological Science 4, 360–365 (1993)
52. Smentana, J.: Understanding of social rules. In: Bennett, M. (ed.) The Development of Social Cognition: The Child as Psychologist, pp. 111–141. Guilford Press, New York (1993)
53. Smentana, J., Braeges, J.: The development of toddlers' moral and conventional judgments. Merril-Parmer Quarterly 36, 329–346 (1990)
54. Sripada, C.S., Stich, S.: A framework for the psychology of norms. In: Carruthers, P., Laurence, S., Stich, S. (eds.) The Innate Mind. Culture and Cognition, vol. 2, pp. 280–301. Oxford University Press, Oxford (2006)
55. Thomas, L.: Morality and psychological development. In: Singer, P. (ed.) A Companion to Ethics, pp. 464–475. Blackwell, Oxford (2006)
56. Tisak, M.: Domains of social reasoning and beyond. In: Vasta, R. (ed.) Annals of Child Development, vol. II, pp. 95–130. Jessica Kingsley, London (1995)
57. Turiel, E.: The Development of Social Knowledge. Cambridge University Press, Cambridge (1983)
58. Turiel, E.: The Culture of Morality: Social Development, Context, and Conflicts. Cambridge University Press, Cambridge (2002)

59. Turiel, E., Killen, M., Helwig, C.: Morality: Its Structure, functions, and vagaries. In: Kagan, J., Lamb, S. (eds.) The Emergence of Morality on Young Children, pp. 155–244. University of Chicago, Chicago (1987)
60. Wilson, J.Q.: The Moral Sense. Free Press, New York (1993)
61. Zahn-Waxler, C., Radke-Yarrow, M., Wagner, E., Chapman, M.: Development of concerns for others. Developmental Psychology 28, 126–136 (1992)

The Symbolic Model for Algebra: Functions and Mechanisms

Albrecht Heeffer

Abstract. The symbolic mode of reasoning in algebra, as it emerged during the sixteenth century, can be considered as a form of model-based reasoning. In this paper we will discuss the functions and mechanisms of this model and show how the model relates to its arithmetical basis. We will argue that the symbolic model was made possible by the epistemic justification of the basic operations of algebra as practiced within the abbaco tradition. We will also show that this form of model-based reasoning facilitated the expansion of the number concept from Renaissance interpretations of number to the full notion of algebraic numbers.

1 Symbolic Reasoning Is Model-Based Reasoning

We previously introduced the idea of considering algebraic problem solving as a model-based activity [16]. This allowed us to characterize the emergence of symbolic algebra in the sixteenth century as a transition from a geometrical model to a symbolic one. While the solution methods for algebraic problems, introduced in Europe by Latin translations from Arabic, are not by themselves geometrical, the validation for the rules of finding the roots of equations depended on a geometrical model. Geometrical proofs from the Arabic and abbaco tradition may have been derived from the practice of geometrical algebra which goes back to Old-babylonian algebra. Jens Høyrup has convincingly demonstrated how Old-Babylonian scribes did not solve equations as proposed by Neugebauer, but depended on a naive cut-and-paste geometrical model [18]. Jörgen Friberg has further shown that the geometrical algebra in book II of Euclid's *Elements* "appears instead to have been a

Albrecht Heeffer
Post-doctoral fellow of the Research Foundation Flanders (FWO Vlaanderen),
Ghent University, Belgium
e-mail: **albrecht.heeffer@ugent.be**

L. Magnani et al. (Eds.): Model-Based Reasoning in Science & Technology, SCI 314, pp. 519–532.
springerlink.com © Springer-Verlag Berlin Heidelberg 2010

direct translation into non-metric and non-numerical 'geometric algebra' of key results from Babylonian metric algebra" [12]. Greek geometric algebra can thus be considered a generalization of Babylonian metric algebra using the same geometric model.

While the geometrical model continued to provide an epistemic justification for the rules of algebra in the Arabic and abbaco tradition, it also had its limitations. Geometrical models lost their intuitive appeal once problems went beyond the three dimensions. Also notions such as negative quantities or negative surfaces are impossible or very difficult to represent geometrically. Within the abbaco tradition geometrical models were actually rarely used. We only find them in the treatises by Maestro Dardi di Pisa (w. 1344) Antonio de' Mazzinghi (c.1353–c.1383) , Maestro Bendetto da Firenze (1429–1479) and Piero della Francesca (1416–1492). During the abbaco period which is to be situated between 1300 and 1500, algebraic practice slowly moved towards a symbolic model. This is not immediately evident from the treatises they have left behind as there is little use of symbolism in these texts. However, we argue that symbolism was introduced into mathematics as a consequence of this process towards symbolic reasoning and not as a precondition. Algebraic symbolism developed into its present form during the sixteenth and early seventeenth century. The conditions for the transition towards a symbolic model were prepared by the practice of abbaco masters. The main condition was the epistemic justification of the basic operations of arithmetic. Once there was a strong belief that current mathematical practices had a general validity, it became possible to apply these operations in an abstract way, without accounting for the values they were dealing with. After several centuries of abbaco practice, this belief in the validity of the operations became so strong that it allowed for the acceptance of anomalous results, such as negative and imaginary quantities. The main mechanism of the symbolic model is that the practices of arithmetical operations were adopted within a model in which one makes abstraction of the actual values. This mechanism can be explained by the principle of permanence of equivalent forms.

2 The Principle of the Permanence of Equivalent Forms

George Peacock was together with George Boole, August De Morgan and Duncan Gregory one of the founders of Cambridge's Analytical Society. This group laid the foundations of what they called algebra of logic, and would later become symbolic logic. This symbolic logic in turn would lay the new foundations for formalism in mathematics with Frege and Hilbert. Peacock was the first to coin the term "symbolic algebra". In 1830, he published his *Treatise on Algebra*, in two books. The first book is on *Arithmetical algebra*, the second on *Symbolic algebra*. Peculiarly, both works use symbols, but in arithmetical algebra: "we consider symbols as representing numbers, and the operations to which they are submitted as included in the same definitions" [23, p. ix]. What this means is that Peacock formulates (arbitrary) restrictions on the operations of algebra so that the results always remain natural numbers. A quadratic equation is therefore not allowed in arithmetical algebra as it can lead to negative, irrational or imaginary roots. Symbolic algebra is then seen as a generalization of arithmetical algebra in which all its truths are preserved. For this property he coined the term "the principle of the permanence of equivalent forms". Forms which are equivalent within arithmetic (for any choice of natural numbers) therefore remain equivalent in symbolic algebra. These forms implicitly define the laws of associativity and commutativity for addition and multiplication and the law of distribution. Such an approach would later lead to the axiomatization of arithmetic and other branches of mathematics. Now, from the point of model-based reasoning, not only does Peacock's symbolic algebra use a symbolic model, so does his arithmetical algebra. The operations allowed in his arithmetical algebra preserve closure for the natural numbers, while his symbolic algebra allows for all the operations valid for the arithmetic of natural numbers, integers, irrational numbers and complex numbers. If we use the principle of permanence of equivalent forms to this last class of numbers, we can characterize the history of symbolic algebra until the advent of quaternions, introduced in 1843 by Hamilton. Quaternions do not preserve the commutative law for multiplication and lead to the idea of multiple possible algebras. We would like to demonstrate that Peacock's principle of the permanence of equivalent forms is a fruitful framework for studying changes in the history of the number concept. We will first show that the arbitrary limitations put on Peacock's arithmetical algebra also appear in the *Arithmetica* by Diophantus. Further we will demonstrate that such limitations on operations were gradually lifted from algebraic practice and that because of a process of epistemic justification of basic operations a symbolization of algebra became possible. We will show by two examples that some important developments and changes of the number concept can be explained as a form of model-based reasoning within this framework.

3 The Arithmetical Algebra of Diophantus

3.1 The Myth of Syncopated Algebra

The *Arithmetica* by Diophantus has often been considered a transition point between rhetorical and symbolic algebra. In his study on Greek algebra [21], the German scholar Georg Heinrich Ferdinand Nesselmann coined the term 'syncopated algebra' for such intermediate phase. His tripartite distinction has become such a common-place depiction of the history of algebraic symbolism that modern-day authors even fail to mention their source. The repeated use of Nesselmann's distinction in three *Entwickelungstufen* (steps in the development) on the stairs to perfection is odd because it should be considered a highly normative view which cannot be sustained within our current assessment of the history of algebra. Its use in present-day textbooks can only be explained by an embarrassing absence of any alternative models. We have pointed out three serious problems with Nesselmann's approach [17] which we here summarize.

3.1.1 A Problem of Chronology

Firstly, if seen as steps within a historical development, as is most certainly the view by many who have used the distinction, the three phases suffer from some serious chronological problems. Nesselmann places Iamblichus, Arabic algebra, Italian abbacus algebra and Regiomontanus under rhetorical algebra ("Die erste und niedrigste Stufe") thereby covering the period from 250 to 1470. The second phase, called syncopated algebra, spans from Diophantus's *Arithmetica* to European algebra until the middle of the seventeenth century, including Viète, Descartes and van Schooten. The third phase is purely symbolic and constitutes modern algebra with the symbolism we still use today. Though little is known for certain about Diophantus, most scholars situate the Arithmetica in the third century which is about the same period as Iamblichus (c. 245–325). So, syncopated algebra overlaps with rhetorical algebra for most of its history. This raises serious objections and questions such as "Did these two systems influence each other?" With the discovery of the Arabic translations of the *Arithmetica* [26, 25] we now know that Diophantus was translated and discussed in the Arab world ever since Qustā ibn Lūqā's book (c. 860). So if the syncopated algebra of Diophantus was known by the Arabs why did it not affect their rhetorical algebra? If the Greek manuscripts used for the Arab translation of the *Arithmetica* contained symbols, we would expect to find some traces of it in the Arab version.

3.1.2 The Role of Scribes

The earliest extant Greek manuscript, once in the hands of Planudes and used by Tannery, is the thirteenth-century Codex Matritensis 4678 (ff. 58–135). The extant Arabic translation published independently by Jacques Sesiano and Roshdi Rashed was completed in 1198. So no copies of the *Arithmetica* before the twelfth century are extant. The ten centuries separating the original text from the earliest Greek copy is a huge distance. Two important revolutionary changes took place around the ninth century: the transition of papyrus to paper and the replacement of the Greek uncial or majuscule script by a new minuscule one. The transition to the new script was very uniform and drastic to a degree which puzzles today's scholars. From about 850 every scribe copying a manuscript would almost certainly adopt the minuscule script. Transcribing an old text into the new text was a laborious and difficult task, certainly not an undertaking to be repeated when a copy in the new script was already somewhere available. It is therefore very likely that all extant manuscript copies are derived from one Byzantine archetype copy in Greek minuscule. Although contractions where also used in uncial texts, the new minuscule much facilitated the use of ligatures. This practice of combining letters, when performed with some consequence, saved considerable time and therefore money. Imagine the time savings by consistently replacing ἀριθμτὸς, which appears many times for every problem, by ς in the whole of the *Arithmetica*. The role of professional scribes should therefore not be underestimated. Although we find some occurrences of shorthand notations in papyri, the paleographic evidence we now have on a consistent use of ligatures and abbreviations for mathematical words points to a process initiated by mediæval scribes rather than to an invention by classic Greek authors. Whatever syncopated nature we can attribute to the *Arithmetica* it is mostly an unintended achievement of the scribes.

3.1.3   Symbols or Ligatures?

A third problem concerns the interpretation of the qualifications "rhetorical" and "syncopated". Many authors of the twentieth century attribute a highly symbolic nature to the *Arithmetica*. Let us take Cajori as the most quoted reference on the history of mathematical notations. Typical for Cajori's approach is the methodological mistake of starting from modern mathematical concepts and operations and looking for corresponding historical ones. He finds in Diophantus no symbol for multiplication, and addition is expressed by juxtaposition. For subtraction the symbol is ⋔. As an example he writes the polynomial $x^3 + 13x^2 + 5x + 2$ as $\kappa^v\,\overline{\alpha}\,\varsigma\,\overline{\eta}\,⋔\,\delta^v\,\overline{\iota\gamma}\,\overset{\circ}{\mu}\,\overline{\beta}$ where κ^v, δ^v, ἀριθμτὸς are the third, second and first power of the unknown and $\overset{\circ}{\mu}$ represents the units. Higher order powers of the unknown are used by Diophantus as additive combination of the first to third powers.

Cajori makes no distinction between symbols, notations or abbreviations. In fact, his contribution to the history of mathematics is titled *A History of Mathematical Notations*. In order to investigate the specific nature of mathematical symbolism one has to make the distinction between symbolic and non-symbolic mathematics. This was, after all, the purpose of Nesselmann's threefold phases. We take the position together with Thomas Heath, Paul Ver Eecke and Jacob Klein, that the letter abbreviations in the *Arithmetica* should be understood purely as ligatures [19, p. 146]:

> We must not forget that all the signs which Diophantus uses are merely word abbreviations. This is true, in particular for the sign of "lacking", ⋔, and for the sign of the unknown number, ς, which (as Heath has convincingly shown) represents nothing but a ligature for ἀριθμτὸς.

Even Nesselmann acknowledges that the "symbols" in the *Arithmetica* are just word abbreviations ("sie bedient sich für gewisse oft wiederkehrende Begriffe und Operationen constanter Abbreviaturen statt der vollen Worte"). In his excellent French literal translation of Diophantus, Ver Eecke consequently omits all abbreviations and provides a fully rhetorical rendering of the text as in "Partager un carré proposé en deux carrés" (II.8, "Divide a given square into two squares"), which makes it probably the most faithful interpretation of the original text.

This objection marks our most important critique on the threefold distinction: symbols are not just abbreviations or practical short-hand notations. Algebraic symbolism is a sort of representation which allows abstractions and new kinds of operations. This symbolic way of thinking can use words, ligatures or symbols. The distinction between words, word abbreviations and symbols is in some way irrelevant with regards to the symbolic nature of algebra.

We will now show that the solution method of Diophantus often reflects the characteristics of Peacock's arithmetical algebra in the way solutions are guided by arbitrary limitations on the possible solutions.

3.2 Diophantus's Number Concept

The definition of number states that "all numbers are made up of some multitude of units, so that it is manifest that their formation is subject to no limit" [14, p. 7]. Thus zero and one were not considered numbers, only natural numbers higher than one are multitudes. Negative numbers were considered "absurd". Irrational solutions do not appear at all since they were not considered numbers. Fractions are acceptable as they can be brought to the same denominator and thus become multitudes. The value of the *arithmos* can only be a number which satisfies this concept of numbers. Therefore zero can never be a solution. Where quadratic problems lead to a positive and a negative root, Diophantus always takes the positive solution. In case of two positive roots, the smaller one is used. Other types of solutions are not allowed in Diophantus's arithmetical algebra.

3.3 The Restrictions of Arithmetical Algebra

We will now demonstrate by means of two examples from book IV that Diophantus's algebra resembles Peacock's arithmetical algebra in putting arbitrary restrictions on the operations to avoid irrational and negative solutions. Not only is Diophantus avoiding such solutions, the process of resolving the indeterminacy of many of his problems precisely depends on these restrictions.

3.3.1 Avoiding Non-rational Solutions

The first problem IV.10 asks to find two cubes the sum of which is equal to the sum of their sides [14, p. 172]. The solution starts with the choice of two and three *arithmoi* for the sides of the two cubes. If we use x for *arithmos* then we arrive at the identity $5x = 35x^3$, with an irrational result for x. In Arabic algebra or abbaco algebra this would pose no problem at all. Now the next step is interesting. Diophantus remarks that the solution would become rational if we find "two cubes the sum of which has to the sum of their sides as the ratio of a square to square". The cubes 5^3 and 8^3 satisfy this condition and this choice leads to the rational solution $(\frac{125}{343}, \frac{125}{343})$. Heath adds a long footnote to this problem that a general solution can be obtained by dividing the equation $x^3 + y^3 = x + y$ by $(x + y)$. This observation is true of course, but is a typical approach of symbolic algebra which would not be endeavored by Diophantus. The diophantine approach is to first 'probe' the problem with the most simple choice of $2x$ and $3x$ for the sides of the cubes. He then notices that this leads to a non-rational solution and uses the problematic expression to find the condition which guarantees a rational solution. It is precisely the restriction of "the ratio of a square to square" which guarantees a rational solution that resolves the indeterminacy of the problem. This way of reasoning is not coincidental but systematic to many problems of the *Arithmetica*. Problems 9, 10, 11, 12, 14, 18, 24, 28, 31 and 32 of book IV explicitly state conditions to make the result rational.

3.3.2 Avoiding Negative Solutions

Problem 27 of book IV shows how Diophantus adds conditions to avoid a negative even when the final solution would be positive. The problems asks for two numbers such that their product minus either gives a cube [14], 168. He takes for the first number $8x$ and for the second $x^2 + 1$ so that the product minus the first $8x^3 + 8x - 8x$ is a cube. Then he notices that the product minus the first $(8x^3 + 8x - x^2 - 1)$ becomes a problem as it should be equated to $(2x - 1)^3$ to get rid of the cube term. He therefore calls this "impossible". Heath remarks that the expression can be equated to either $(2x - \frac{1}{12})^3$ or $(\frac{1}{12}x - 1)^3$ with a positive rational solution for both. However, this is not the point. The salient point is that Diophantus chooses new initial conditions in order to guarantee a positive result with $8x + 1$ for the first and x^2 for the second. Now $(8x^3 + 8x - x^2 - 1)$ can be equated with $(2x - 1)^3$ and the result will be positive. Here again the choice of conditions to resolve the indeterminacy is guided by his limitations on the conception of number.

3.4 Expanding the Number Concept

Jacob Klein, a student of Heidegger and interpreter of Plato, wrote a long treatise in 1936 on the number concept starting with Plato and the development of algebra from Diophantus to Viète [19]. It became very influential for the history of mathematics after its translation into English in 1968. For Klein it is not the evolution of solution methods for solving equations which follows some logical path but the ontological transformation of the underlying

concepts within an ideal Platonic realm. He restricts all other possible understandings of the emergence of symbolic algebra by formulating his research
question as follows: "What transformation did a concept like that of *arithmos* have to undergo in order that a 'symbolic' calculating technique might
grow out of the Diophantine tradition?" [19, p. 147]. According to Klein it is
ultimately Viète who "by means of the introduction of a general mathematical symbolism actually realizes the fundamental transformation of conceptual
foundations" [19, p. 149]. Klein places the historical move towards the use of
symbols with Viète and thus ignores important contributions by the abbaco
masters, by Michael Stifel [28, 29], Girolamo Cardano [7, 8] and the French
algebraists Jacques Peletier [24], Johannes Buteo [6] and Guillaume Gosselin
[13]. The new environment of symbolic representation provides the opportunity to "the ancient concept of arithmos" to "transfer into a new conceptual
dimension" [19, p. 185]. As soon as this happens, symbolic algebra is born:
"As soon as 'general number' is conceived and represented in the medium of
species as an 'object' in itself, that is, symbolically, the modern concept of
'number' is born" [19, p. 175].

Of course, Klein is right that the expansion of the number concept is crucial
to the emergence of symbolic algebra but we do not endorse a philosophy
where concepts realize themselves with the purpose to advance mathematics.
The line of influence is in the opposite direction. The algebraic practices of
abbaco masters facilitated the expansion of the number concept. But before
we will demonstrate this for negative and imaginary numbers we first have
to understand the primary conditions for such a process. The acceptance of a
new kind of solutions to algebraic problems becomes possible only when there
is a strong belief in the validity of accepted practices. We will now discuss
how these practices were epistemically justified.

4 Epistemic Justification of Basic Operations

4.1 Example of Abstraction: Multiplying Binomials

A good example of the process of abstraction as a necessary condition for the
transition to a symbolic mathematics is found in the multiplication procedure
for two binomials. The procedure of crosswise multiplication, "multiplicare
in croce" is a recurring topic in almost all abbaco treatises. The interesting
aspect is that the method applies to a wide variety of 'numbers' and still
follows the same procedure. The procedure is often accompanied by a diagram
showing the terms in a crosswise fashion. In our interpretation the diagram
functions as a validation for the procedure rather than being an essential
element in the application of the procedure. Let us look at an example in the
abbaco treatise by Paolo Gherardi where he multiplies two rational numbers
[4, p. 16]:

Se noi avessimo a multipricare numero sano e rocto contra numero sano e rocto, sì dovemo multipricare l'uno numero sano contra l'altro e possa li rocti in croce. Asempro a la decta regola. $12\frac{1}{2}$ via $15\frac{1}{4}$ quanto fa? Però diremo: 12 via 15 fa 180. Or diremo: 12 via $\frac{1}{4}$ fa [3], echo 183. Or prendi il $\frac{1}{2}$ di $15\frac{1}{4}$ ch'è $7\frac{5}{8}$, agiustalo sopra 183 e sono $190\frac{5}{8}$ e tanto fa $12\frac{1}{2}$ via $15\frac{1}{4}$. Ed è facta.

Here the multiplication of two fractional numbers $12\frac{1}{2}$ and $15\frac{1}{4}$ is surprisingly treated as the product of two binomials $(12+\frac{1}{2})(15+\frac{1}{4})$ instead of the product of two fractions $\frac{25}{2}$ and $\frac{61}{4}$. In the pseudo-Paolo dell'abbaco treatise, the author explicitly refers to the two methods, one by multiplication of binomials and the other as a multiplication of two fractions [1, p. 28]. We therefore understand the method of multiplying binomials as a general procedure for multiplying all kinds of entities that can be expressed as binomials. It suffices to identify the elements of the two binomials in the crosswise diagram to justify the method. We thus find it applied to the multiplication of surds and the multiplication of polynomials. As shown in the Figure 1, Maestro Dardi uses the crosswise multiplication for calculating the square of $(\sqrt{5} + \sqrt{7})$ as $12 + \sqrt{140}$. Many abbaco treatises also treat the multiplication of two algebraic binomials in the same way, such as $(x - 2)(x - 3)$.

Fig. 1 Maestro Dardi's scheme for crosswise multiplication of surd binomials (from Chigi M.VIII.170 $f.7^r$).

4.2 Epistemic Justification of the Rules of Signs

The justification for the rules of signs build further on the justification schemes for the multiplication of binomials. These rules define the result of arithmetical operations on combinations of positives and negatives. These rules were common in cultures that recognized and calculated with negative quantities such as China and India. The Brāhmaspuṭhasiddhānta of c. 628 includes all the rules of sign for addition, subtraction, multiplication and division [10]. They also appear in Arabic works from the eleventh century. In Europe there was no recognition of negative quantities and therefore a formal treatment of the rules of signs appeared much later. These rules were known implicitly and were applied within the abbacus tradition, for example in the multiplication of irrational binomials in Fibonacci (1202; [5, p. 370], [27, p. 510]).

Its epistemic validation stems from correctly applying the rules for multiplying binomials by cross-wise multiplication in which you add all the subproducts. The first of such proofs in European mathematics appeared in a

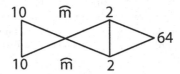

Fig. 2 Maestro Dardi's scheme for the justification of the rules of signs by crosswise multiplication (from Chigi M.VIII.170 $f.4^v$).

treatise of c.1343 by Maestro Dardi titled *Aliabraa argibra* ($f.4^v$, [11, p. 44]). It explains why a negative multiplied by a negative makes a positive. It is repeated in various other manuscripts dealing with algebra during the fifteenth century. The reasoning goes as follows: we know that 8 times 8 makes 64. Therefore $(10 - 2)$ times $(10 - 2)$ should also result in 64. You multiply 10 by 10, this makes 100, then 10 times -2 which is -20 and again 10 times -2 or -20 leaves us with 60. The last product is $(-2)(-2)$ but as we have to arrive at 64, this must necessarily be $+4$. Therefore a negative multiplied by a negative always makes a positive. The strong belief by abbaco masters in the correctness of the operations made it possible to extend algebra from the domain of natural numbers to a domain which includes negative numbers. Using the principle of permanence of equivalent forms, the "proof" is the only possible justification for the rules of signs which preserves the distributive law and the law of identity for multiplication (1 times any number n equals n). Reasoning within a symbolic model which was epistemologically justified and led to the creation of new objects on the object level: negative numbers in arithmetic. Luca Pacioli lists the rules of signs for all basic operations in his *Summa* of 1494 [22]. Interestingly, he does not refer to whole numbers, fractions, surds or cossic numbers. The rules are formulated abstractly, as in "a negative divided by a negative makes a positive", which we would expect within a symbolic context.

5 Expansion of the Number Concept

If our first example were negative numbers, our second example comes as no surprise. Cardano was the first to perform a calculation with - what is now known as - imaginary numbers, in chapter 37 of the *Ars Magna* named de *regula falsum ponendis*, or the "Rule of Postulating a Negative" [9, III, p. 287], [8, English p. 219].

Here the context is slightly different. In the early sixteenth century quadratic problems in algebra were reduced to a standard form for which a canonical rule could be applied. Depending on the sign of the coefficients, three different rules were applied, already known from the first Arabic work on algebra. Furthermore, a geometrical proof was known for these rules, so there was a justified belief in their correctness. Now Cardano was confronted with a quadratic problem which leads to imaginary roots. However, he does not reduce the problem to an equation but tries to reason geometrically. The problem is as follows:

> The second species of negative assumption involves the square root of a negative. I will give an example: If it should be said, divide 10 into two parts the product of which is 30 or 40, it is clear that this case is impossible. Nevertheless, we will work thus: We divide 10 into two equal parts, making each 5. These we square, making 25. Subtract 40, if you will, from the 25 thus produced, as I showed you in the chapter on operations in the sixth book, leaving a remainder of -15, the square root of which added to or subtracted from 5 gives parts the product of which is 40. These will be $(5 + \sqrt{-15})$ and $(5 + -\sqrt{-15})$.

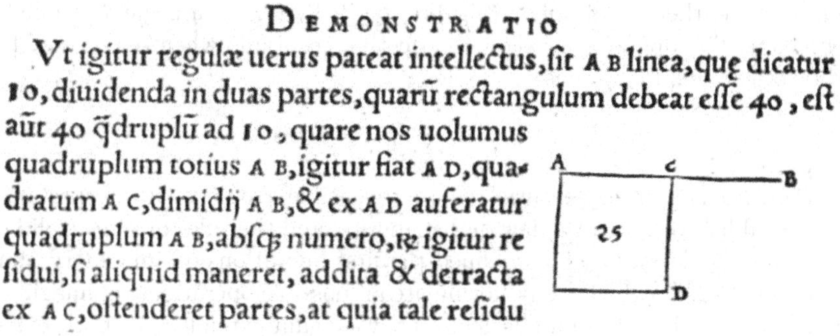

D E M O N S T R A T I O

Vt igitur regulæ uerus pateat intellectus,fit A B linea,quę dicatur **1 0**, diuidenda in duas partes,quarū rectangulum debeat effe **40** , eft aūt **40** q̃druplū ad **1 0** , quare nos uolumus quadruplum totius A B,igitur fiat A D,qua*
dratum A C,dimidij A B,& ex A D auferatur
quadruplum A B,abfcꝗ numero,ꝛ igitur re
fidui,fi aliquid maneret, addita & detracta
ex A C,oftenderet partes,at quia tale refidu

Fig. 3 Cardano picturing a negative surface (from the *Ars Magna*, 1545, *f.*66ʳ).

By using a geometrical demonstration, he tries to get a grasp on the new concept (Figure 3). He proceeds as with the standard demonstration of the rule for solving quadratic problems. Let AB be the 10 to be divided. Divide the line at C into two equal parts. Square AC to AD. Since 40 is four times 10 this corresponds with the rectangle $4AB$. Now $\sqrt{-15}$ corresponds with the subtraction of AD by the larger $4AB$. Thus, Cardano finds that this strange new object is a negative or a missing surface. This makes no sense to him and he therefore writes that the problem is impossible. Cardano still struggled with the interpretation of $\sqrt{-15}$, but was "putting aside the mental tortures involved" and performed the multiplication of the surd binomials

$$(5 + \sqrt{-15})(5 + \sqrt{-15})$$

correctly to arrive at $25 + 15$ or 40. Multiplying the two binomials produces four terms. The first is evidently 25. The second and third $(5(\sqrt{-15}) + 5 + (-\sqrt{-15})$ are canceled out by their signs, whatever their value is. The innovation lies in the fourth:

$$(\sqrt{-15})(-\sqrt{-15}) = -(-15)$$

Cardano was well aware of the "proofs" of the rules of signs, as the one by Maestro Dardi we discussed, and proceeds in a similar way. Here the product of the two terms must be 15 to arrive at the sum of 40. The multiplication of a positive root with a negative must lead to something negative. However, the result must be +15, therefore the product of two roots of minus 15 must be minus 15. Again, the reasoning takes place within the symbolic model. No reasonable interpretation could be given to the root of minus 15. Cardano attempted a geometrical interpretation, but a negative surface makes no sense either. Actually, Cardano is here using an abductive reasoning step which is explained by the chapter heading "Posing a negative". The occurrence of the root of a negative is an anomaly to the Renaissance conception of number. To make the reasoning acceptable he opts the most convenient hypothesis which fits into the rhetoric of abbaco algebra: begin by posing a negative value for the cosa. The operations become perfectly acceptable when one poses that the cosa stands for -15 as:

$$(\sqrt{x})(-\sqrt{x}) = -(x)$$

It took another three centuries before a sensible geometrical interpretation was established. Using the law of distribution and the rules of signs, valid on the symbolic level, Cardano defined the first operations on imaginary numbers. In 1572, Bombelli would formulate all possible operations on imaginary numbers by which the number concept was again extended. In both examples new mathematical objects were created by reasoning within the symbolic model. Through the epistemological justification of correctly performing the operations, these new objects, negative numbers and imaginary numbers, became accepted.

References

1. Arrighi, G. (ed.): Paolo Dell'Abaco, Trattato d'aritmetica. Domus Galilaeana, Pisa (1964)
2. Arrighi, G. (ed.): Antonio de' Mazzinghi. Trattato di Fioretti, secondo la lezione del codice L.IV.21 (sec. XV) della Biblioteca degli Intronati di Siena, Domus Galilaeana, Pisa (1967)
3. Arrighi, G. (ed.): Libro d'abaco, Dal Codice 1754 (sec. XIV) della Biblioteca Statale di Lucca, Cassa di Risparmio di Lucca, Lucca (1973)
4. Arrighi, G. (ed.): Paolo Gherardi, Opera mathematica: Libro di ragioni - Liber habaci. Codici Magliabechiani Classe XI, nn. 87 e 88 (sec. XIV) della Biblioteca Nazionale di Firenze. Pacini-Fazzi, Lucca (1987)
5. Boncompagni, B.: Scritti di Leonardo Pisano, matematico del secolo decimoterzo (2 vols.). vol. I.: Liber Abbaci di Leonardo Pisano, vol II.: Leonardi Pisani Practica geometriae ed opuscoli, Tipografia delle Scienze Matematiche e Fisiche, Rome (1857/1862)
6. Buteo, J.: Logistica. Gulielmum Rovillium, Lyon (1559)

7. Cardano, G.: Practica arithmetice, & mensurandi singularis. In: qua que preter alias cōtinentur, versa pagina demnstrabit, Bernardini Calusci, Milan (1539)

8. Cardano, G.: Ars Magnaæ, sive, De regulis algebraicis lib. unus: qui & totius operis de arithmetica, quod opus perfectum inscripsit, est in ordine decimus, Johann Petreius, Nürnberg (1545) English translation by Witmer R.T. Ars Magna or the Rules of Algebra. M.I.T. Press, Cambridge, Mass (1968). Reprinted by Dover Publications, New York (1993)

9. Cardano, G.: Opera omnia (10 vols.), Jean Antoine Huguetan and Marc Antione Ravaud, Lyon (1663)

10. Colebrooke, H.T.: Algebra with Arithmetic of Brahmagupta and Bhaskara. London (1817)

11. Franci, R. (ed.): Maestro Dardi (sec. XIV) Aliabraa argibra. Dal manoscritto I.VII.17 della Biblioteca Comunale di Siena. Quaderni del Centro Studi della Matematica Medioevale 26, Università di Siena, Siena (2001)

12. Friberg, J.: Amazing Traces of a Babylonian Origin in Greek Mathematics. World Scientific Publishing, Singapore (2007)

13. Gosselin, G.: Gvlielmi Gosselini Cadomensis Bellocassii. De arte magna, seu de occulta parte numerorum, quae & algebra, & almucabala vulgo dicitur, libri qvatvor. Libri Qvatvor. In: quibus explicantur aequationes Diophanti, regulae quantitatis simplicis, [et] quantitatis surdae. Aegidium Beys, Paris (1577)

14. Heath, T.L. (ed.): Diophantus of Alexandria: A Study in the History of Greek Algebra. Cambridge (second edition, 1910, With a supplement containing an account of Fermat's Theorems and Problems connected with Diophantine analysis and some solutions of Diophantine problems by Euler) Dover, New York (1964) (1885)

15. Heeffer, A.: Abduction as a strategy for concept formation in mathematics: Cardano postulating a negative. In: Pombo, O., Gerner, A. (eds.) Abduction and the Process of Scientific Discovery, pp. 179–194. Colecçáo Documenta, Centro de Filosofia das Ciências da Universidade de Lisboa, Lisboa (2007)

16. Heeffer, A.: The emergence of symbolic algebra as a shift in predominant models. Foundations of Science 13(2), 149–161 (2008)

17. Heeffer, A.: On the nature and origin of algebraic symbolism. In: Van Kerkhove, B. (ed.) New Perspectives on Mathematical Practices. Essays in Philosophy and History of Mathematics, pp. 1–27. World Scientific Publishing, Singapore (2009)

18. Høyrup, J.: Lengths, Widths, Surfaces: A Portrait of Old Babylonian Algebra and its Kin. Springer, Heidelberg (2002)

19. Klein, J.: Greek mathematical thought and the origin of algebra. MIT Press, Cambridge (1968) (Transl. E. Brann of Die griechische Logistik und die Entstehung der Algebra, Berlin (1934-6))

20. Neugebauer, O., Sachs, A.: Mathematical Cuneiform Texts. American Oriental Society (1945)

21. Nesselmann, G.H.F.: Versuch einer kritischen Geschichte der Algebra, vol. 1. Die Algebra der Griechen, Berlin (1842) Reprinted, Frankfurt (1969)

22. Pacioli, L.: Summa de arithmetica geometria proportioni: et proportionalita. Continetia de tutta lopera. Paganino de Paganini, Venice (1494)

23. Peacock, G.: A Treatise on Algebra. J. Smith, Cambridge (1830)

24. Peletier, J.: L'algèbre de Jaques Peletier du Mans. Départie en 2 livres, J. de Tournes, Lyon (1554)

25. Rashed, R.: Diophante. Les Arithmétiques. Tome III (livre IV) et tome IV (livres V, VI, VI). Texte établit et traduit par Roshdi Rashed, Les Belles Lettres, Paris (1984)
26. Sesiano, J.: Books IV to VII of Diophantus' Arithmetica in the Arabic translation attributed to Qustā ibn Lūqā. Springer, Heidelberg (1982)
27. Sigler, L.: Fibonacci's Liber Abaci. In: A Translation into Modern English of Leonardo Pisano's Book of Calculation, Springer, Heidelberg (2001)
28. Stifel, M.: Arithmetica Integra. Petreius, Nürnberg (1545)
29. Stifel, M.: Die Coss Christoffe Ludolffs mit schönen Exempeln der Coss. Gedrückt durch Alexandrum Lutomyslensem. Zu Königsperg in Preussen (1553)
30. Viète, F.: In artem analyticam isagoge. Seorsim excussa ab Opere restituae mathematicae analyseos, seu algebra nova. J. Mettayer, Tournon (1591)

The Theoretician's Gambits: Scientific Representations, Their Formats and Content

Marion Vorms

Abstract. It is quite widely acknowledged, in the field of cognitive science, that the format in which a set of data is displayed (lists, graphs, arrays, etc.) matters to the agents' performances in achieving various cognitive tasks, such as problem-solving or decision-making. This paper intends to show that formats also matter in the case of theoretical representations, namely general representations expressing hypotheses, and not only in the case of data displays. Indeed, scientists have limited cognitive abilities, and representations in different formats have different inferential affordances for them. Moreover, this paper shows that, once agents and their limited cognitive abilities get into the picture, one has to take into account both the way content is formatted and the cognitive abilities and epistemic peculiarities of agents. This paves the way to a dynamic and pragmatic picture of theorizing, as a cognitive activity consisting in creating new inferential pathways between representations.

1 Introduction

Philosophers of science have traditionally approached theoretical representations (i.e. theories, models, concepts) from an abstract point of view, by idealizing away both from the actual means of representation used in scientific practice, and from the actual reasoning of scientists who use these representations. From such a perspective, contents are therefore considered as independent both from the form in which they are expressed and from the cognitive abilities and epistemic peculiarities of the agents. In consequence, two logically equivalent representations (e.g. equations of motion in

Marion Vorms
Institut d'Histoire et de Philosophie des Sciences et des Techniques (CNRS),
Paris, France
e-mail: `marion.vorms@ens.fr`

L. Magnani et al. (Eds.): Model-Based Reasoning in Science & Technology, SCI 314, pp. 533–558.
springerlink.com © Springer-Verlag Berlin Heidelberg 2010

polar and cartesian coordinates) are generally considered as descriptions of the same model (e.g. the harmonic oscillator): philosophers of science usually assume that, despite their differences, two such representations have exactly the same content.

In this paper, I adopt a different perspective, by considering scientific representations as tools for theorizing; by "theorizing", I refer to a certain class of cognitive activities implying the construction, use, and development of theoretical hypotheses. Assuming that scientists do not reason *in abstracto* by contemplating abstract logical or mathematical structures, but rather by manipulating concrete representing devices, I shall focus on the external representations they construct and use in their day-to-day practice. I assume that the main function of such representing devices is to enable scientists to draw inferences concerning the systems they stand for. The overall purpose of this paper is to analyze a crucial – though oft-neglected – feature of the functioning of scientific (external) representations as inferential tools, namely the importance of what I call their *format*.

Some studies in cognitive science and Artificial Intelligence on the use of external representations in problem-solving [28, 35, 36] and decision-making [21] show that the very way in which data are displayed (e.g. a list of numerals as opposed to its corresponding graph) has important consequences on the agents' performances. Indeed, two representations coding the same information can nevertheless convey it in different ways, thus facilitating different cognitive processes and making such or such piece of information more or less easy to access. Such differences I shall call differences in *format*. However, studies emphasizing the importance of formats are almost always concerned with tasks involving *data* manipulation and processing by agents in order to achieve a particular task, and few analyzes (if any) have been given of the importance of such phenomena for the use and manipulation of *theoretical* representations. Theoretical representations, as opposed to mere presentations of data, are representations expressing *hypotheses* about a certain domain of phenomena. If one considers theorizing as a kind of cognitive activity, which consists in reasoning with theoretical hypotheses and exploring their consequences, it becomes legitimate to inquire into the consequences of a change in format for theoretical representations as well.

In this paper, my aim is twofold. My main goal is to show that formats, whose importance is quite widely acknowledged in the case of data display, have notable consequences on theorizing as well. In order to assess such consequences and to evaluate their bearing on a philosophical understanding of the content of scientific representations, further analysis is needed of the fact that representations in different formats have different inferential affordances for agents. Giving such an analysis is the second, subordinate aim of this paper.

Firstly (section 2), I shall give a few examples showing the importance of formats for both data manipulation and theorizing. In order to show that a change in representation sometimes induces a change in the agents' reasoning processes, I shall restrict to examples of representations in different formats, which are nevertheless logically – or informationally – equivalent. In section 3, I propose to clarify the very notion of format, as it is used in describing the cases in section 2. It will appear that the intuitions underlying the use of this notion are not fully captured by an account of the syntactic and semantic rules according to which information is coded within a representation (its "symbol system", in Goodman's sense). Indeed, the most relevant feature of the format of a representation, in my analysis, is that it determines the *inferential affordances* or *potential* of this representation *for agents with limited cognitive abilities*. As we will see, the inferential affordances of a representation depend both on the way information is displayed *and* on the cognitive abilities of this representation's users. I will therefore argue that, as soon as one acknowledges the importance of the format under which a certain informational content is displayed for its users's performances, then one has to take into account these two parameters (information display and cognitive abilities of agents). I will finally draw a few consequences of this analysis for a study of theorizing, conceived as the exploration of the content of theoretical representations (section 4).

2 The "Representational Effect": Data Displays and Theoretical Models

In this section, I wish to show that what Zhang [36] has coined "the representational effect" has important consequences for theorizing as well, and not only for tasks involving data manipulation. "Representational effect" refers to the fact that various representations displaying the same information in different ways do not facilitate the same cognitive behavior. After having briefly recalled what it consists in in the case of data display (subsection 2.1), I shall take two examples (the equations of Classical Mechanics and Feynman's diagrams) highlighting the importance of the representational effect for the use of theoretical representations (subsection 2.2).

2.1 External Representations and the "Representational Effect"

Nobody would deny that external representations – as opposed to internal or mental ones – sometimes prove practically indispensable to perform various cognitive tasks, such as problem-solving or decision-making: laying out a mathematical operation in order to solve it, drawing a graph from a set of data in order to see easily the relation between two variables, or constructing a diagram in order to solve a geometrical problem are common practices. For instance, although it is in principle possible to divide 346 by 7 by mere mental computation, such an operation is quite difficult and costly – and doomed to error – for an average agent. One would rather use paper and pencil to keep track of the various steps of the computational process. Moreover, humans have invented special procedures for displaying numerals[1], which turn the solution of a division into a simple manipulation. All pupils have learned how to lay out a division and how to reach its solution by following simple transformation rules of this kind of device:

$$
\begin{array}{c|c}
3\ 4\ 6 & 7 \\
\hline
6\ 6 & 4\ 9 \\
3 &
\end{array}
$$

In virtue of its particular spatial display and of its "internal dynamics", this external device, so to speak, "computes" the solution on behalf of the agent. Manipulating it exempts one from drawing various inferences that would otherwise be indispensable.

Note that the cognitive advantage of this procedure does not rely merely on its being externalized: on the one hand, one could imagine that a trained agent be able to "lay out" the division in his/her mind's eye; on the other hand, using paper and pencil to write down this division problem by following a different procedure could prove much more costly[2]. The very advantage of this procedure rather relies on what I propose to call the "format" of the device shown above, namely the way data are displayed[3], which determines the processes agents have to follow in order to extract information. This

---

[1] See [24] for a review of the artifacts and procedures that were invented, throughout history, to serve as "cognition amplifiers", and which can be thought of as the ancestors of our modern computers.

[2] My focus on external representations is rather based on expediency (it makes the study of formats easier) than on a commitment to any particular thesis concerning the relation between external and mental representations. One can acknowledge that most cognitive processes involve, indeed, both external and mental representations – and should therefore be studied as distributed processes, as suggested by the advocates of distributed cognition [17, 15, 35] – without committing oneself to the metaphysical version of the extended mind thesis, as advocated by Andy Clark and David Chalmers [3, 2]. Note, however, that an interesting empirical question would be: are the processes actually performed on external representations a mere externalization of internal processes that would be performed mentally on the "same representations", if merely imagined? In other words, are these two kinds of processes similar in some relevant sense, the internal ones being only too complex and requiring too much memory skills to be performed without any external aid, but being in principle describable by the same algorithms? Another, related, question is whether agents using external devices have to construct an internal model of the problem to be solved, the external device serving as an aid in this construction process (see [35]). In this paper, I shall not tackle these issues, which have no bearing on my argument.

[3] Zhang [35] speaks of the "form" of the "graphic display".

becomes clear when one considers several representations displaying the same data in different ways.

Consider, for example, a list of numerical data corresponding to the temperature in Paris over a year (the bracketed numerals represent the months, and the numerals in bold characters represent the values of the corresponding temperature in degrees Celsius): [1]**2.5** ; [2]**3**; [3]**9**; [4]**15**; [5]**18**; [6]**23**; [7]**25**; [8]**23**; [9]**22**; [10]**13**; [11]**6**; [12]**3**. From such a list, one can draw a graph, as in figure 1 below.

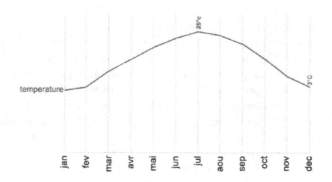

Fig. 1 Annual evolution of temperature in Paris.

The data displayed in these two representations are identical. In other words, the list and the graph contain exactly the same information – since the graph was drawn from the list, and the numerals shown in the list can be retrieved by properly reading the graph. However, as Larkin and Simon [28] would put it, these two representations are *informationally equivalent*, but *computationally different*: though containing the same information, they do not require nor do they facilitate the same cognitive operations. Consequently, the graph and the list do not make the various pieces of information they contain equally accessible to the agents.

Consider, for instance, the task of assessing the global evolution of temperature from January to June. In virtue of the spatial relationships between the points of the graph, one does not need to memorize and then compare the numerals standing for the values of temperature at different times, in order to finally infer the global evolution of temperature over the year; the graph displays in an immediately accessible[4] form the temporal evolution of temperature. The spatial display of the graph, again, computes this information

---

[4] For a definition of the notion of accessibility, see [32], where I rely on John Kulvicki's notions of "extractability", "syntactic salience" and "semantic salience" [23].

on behalf of the agent. On the other hand, if one wants to know the precise value of the temperature in June, one would rather use the list, since this value is explicitly[5] displayed in the list, whereas it is not so in the graph.

To sum up, differences in the representational format imply differences in the *cost* and in the *type* of the cognitive processes required to access the various pieces of information contained in the two representations. Some pieces of information are easier and quicker – less costly – to access within the graph, while others are so within the list. Entering data into the graph as well as computing information within it do not consist in the same type of processes as entering data and computing them within the list. Certainly, as Larkin and Simon acknowledge [28, p. 67], "ease" and "quickness" are not precise concepts, and it seems therefore difficult to give a measure for the cost of a cognitive task. Similarly, our present knowledge of cognitive processes is too poor to enable us to assess precisely the difference between two *types* of cognitive processes (e.g. those involved in the reading of a list of numerals as opposed to those involved in the reading of a graph). However, without knowing what "happens in the head" of an agent, and without being able to precisely model these processes, we have an intuitive grasp of what a type of cognitive operation is, by analogy with the notion of algorithms; similarly, it seems *prima facie* possible to assess the quickness of a cognitive process by measuring the amount of time involved in performing this task, or by counting the number of (at least conscious) steps involved in it. For an average user with normal cognitive abilities, it seems quite uncontroversial that the process of assessing the general evolution of temperature between June and December (stating whether it increases or decreases) is much less costly by using the graph than the list[6].

Cognitive scientists and AI researchers nowadays pay a growing attention to the role of external representations in tasks involving complex information-processing (see [36] for a review[7]). Some have underlined the importance of what Zhang called the "representational effect" – namely the consequences of the format on the agents' performances in various cognitive tasks, such as

---

[5] The notion of explicitness needs a further analysis as well. Here, I take it in the intuitive sense – which is also the sense Larkin and Simon [28] seem to rely on – corresponding to the idea that an information is explicitly represented when no inference is needed to access it. For a more refined analysis of the implicit/explicit distinction, see [20].

[6] For a more detailed analysis, see [32].

[7] See also Jiajie Zhang's online bibliography on external representations (thanks to Alex Kirlik for indicating me this link):
http://acad88.sahs.uth.tmc.edu/resources/ExtRep_Bib.htm

problem-solving[8] and decision-making[9]. However, although some have suggested that some kinds of representations are particularly well suited to the expression of some kinds of information[10], no clear account of what I have proposed to call "format" has been given[11]. I shall come back to this in section 3. Let me first turn to a few examples revealing the existence of a representational effect in the use of theoretical representations as well.

2.2 The Representational Effect and Theoretical Models

Till now, I have been considering external devices displaying data to be processed by agents in order to achieve simple cognitive tasks. As such, the contents of these representations are sets of data, which were collected by empirical inquiry and entered into the representing device by following rather simple rules. However, theoretical models, such as, for instance, the equation of the simple pendulum, are not mere displays of data. They are rather representations expressing *hypotheses* about a wide range of phenomena and systems' behavior, thus enabling scientists to explain and predict these phenomena. Their *content*, as such, is much richer and more complex to define than the content of the representations considered above.

Indeed, analyzing the content of scientific representations and accounting for their explanatory and predictive power is one of the central problems in the philosophy of science. Philosophers have generally addressed this problem by giving a logical reconstruction of the relation between theoretical representations and the phenomena they stand for. Therefore, logical equivalence has long been taken as a criterion of identity of content for scientific representations: two representations are considered scientifically equivalent if they are inter-deducible, thus having the same set of empirical consequences. On

---

[8] Zhang [35, 36, 37] shows that different representations of a common abstract structure can generate dramatically different representational efficiencies, task complexities, and behavioral outcomes. He moreover suggests [34] that all graphs could be systematically studied under a representation taxonomy based on the properties of external representations.

[9] Kleinmutz and Schkade [21] showed that different representations (graphs, tables, lists) of the same information can dramatically change decision-making strategies.

[10] Larkin and Simon [28] suggest that different kinds of representations typically display, in an explicit form, different kinds of information: diagrams preserve topological relations, outlines preserve hierarchical relations, and languages are well fitted to display logical or temporal relations. For an analysis of the types of reasoning associated with the use of graphs and diagrams, see the works by Tufte [30, 31]; for diagrammatic logic, see [29, 27].

[11] Note, incidentally, that differences in format can happen between different types of representations (e.g. linguistic *versus* diagrammatic) as well as between representations belonging to the same broad type (e.g. graphs in different coordinate systems, arabic *versus* roman numerals).

such a view, one could feel reluctant to attribute any importance to a mere change in the presentation of this content.

However, if one considers scientific representations as tools for theorizing, namely for drawing inferences enabling agents to explore the content of these representations – and therefore to gain knowledge concerning the systems they stand for, and concerning their link to other representations –, it becomes legitimate to pay attention to the very format of these representations, since it might have consequences for their inferential role for agents, *in practice* – despite their equivalence *in principle*. Let me briefly give two examples showing that formats matter for the use of theoretical representations as well.

2.2.1 The Equations of Classical Mechanics

Consider, first, the equations of Classical Mechanics (CM). Solving a problem in CM, say, predicting and explaining the dynamical evolution of a system, typically consists in finding the functions that describe the temporal evolution of the position and velocity of the system under study. First, one writes down the differential equations governing the dynamics of the system with the help of the information one has about it; and then, one solves these equations.

The equations of CM can be formulated in different ways, according to the kind of coordinate system used to describe the motion of a physical system. One distinguishes generally between the Newtonian and the analytical (Lagrangian and Hamiltonian) formulations. According to the problem at hand, using one or the other formulation can dramatically facilitate both the processes of writing the equations and of solving them. The Newtonian formulation relies on a description of the configuration of systems by means of Cartesian (and sometimes polar) coordinates, which represent the position and velocity of each point of the system at some instant t. Newtonian equations of motion, which govern the dynamical evolution of a system so represented, have the form of Newton's Second Law ($\mathbf{F} = m\mathbf{a}$, where $\mathbf{F}$, the force, and $\mathbf{a}$, the acceleration, are vectorial quantities). The first step in solving a problem in this framework consists in specifying the forces exerted on the various points of the system in order to write down the corresponding equations. In other words, the Newtonian *format* requires that one enters data concerning the forces, since the value of forces is explicitly displayed in the Newtonian equations.

In the case of constrained systems, namely systems with internal forces maintaining constraints between different points (thus preventing them from moving independently from each other) the identification and specification of each force is practically impossible. In such cases, the Lagrangian formulation is more appropriate: it relies on a description of the configuration of systems by means of so-called "generalized" coordinates, which correspond to the degrees of freedom of the system. Transforming the description of the system from Cartesian coordinates into generalized ones (q_i) enables one to write down the Lagrangian equations of the system, without needing to know the forces maintaining the constraints. The Lagrangian equations have the following form:

$$\frac{d}{dt}\left(\frac{\partial L}{\partial \dot{q}_i}\right) - \frac{\partial L}{\partial q_i} = 0.$$

Forces do not appear in them; the dynamical evolution of the system is entirely governed by a scalar quantity, the Lagrangian L, which typically corresponds to the difference between the kinetic and potential energies of the system. These equations implicitly contain the forces maintaining the constraints, since they are expressed in coordinates taking these constraints into account. Therefore, it is possible to retrieve information concerning the forces; nevertheless, the constraints need not appear explicitly. What may be called the Lagrangian *format* consists in presenting, in an explicit form, information concerning the energy of the system, rather than information concerning the forces. Despite their equivalence to the Newtonian equations, the Lagrangian ones are therefore much more appropriate to cases where some forces are unknown.

In some cases, though, Lagrangian equations are only partially integrable; it is thus impossible to achieve the second step of the problem-solving process, namely finding the analytical solutions of the equations. The Hamiltonian formulation, which uses a different kind of generalized coordinates[12], enables one to change an intractable Lagrangian equation into two corresponding first order equations, by means of mathematical transformations called "Legendre transformations". Such first order equations ($\frac{\partial H}{\partial p_i} = \dot{q}_i$ and $\frac{\partial H}{\partial q_i} = -\dot{p}_i$) are integrable. The Hamiltonian H typically equals the total energy of the system.

One can easily show the inter-deducibility of these three kinds of equations. Nevertheless, as we have seen, changing from one to the other can considerably enhances our problem-solving capacities: changing from Cartesian to generalized coordinates sometimes facilitates the process of writing the equations, which is otherwise practically impossible; changing from a Lagrangian to a Hamiltonian representation transforms one intractable equation into two tractable ones. As Paul Humphreys [16] would put it, despite their equivalence *in principle*, these various equations are not equally usable *in practice*. Their formats neither require nor facilitate the same processes.

Moreover, this variety of formulations has consequences on the development of the theory itself. The Lagrangian formalism is suitable for the expression of the theory of relativity, and the Hamiltonian formalism is used in Quantum Mechanics. As Feynman said concerning the various ways of expressing the law of gravitation (*via* Newton's law, field theory, or minimum principles), these formulations are "equivalent scientifically. [...] But psychologically, they are very different" [9, p. 53].

---

[12] Lagrangian generalized coordinates have the dimension of positions and of velocities, whereas Hamiltonian ones have the dimension of positions and momenta.

2.2.2 Feynman's Diagrams

Consider now the case of Feynman's famous diagrams (see Figure 2). Feynman first introduced them in 1948, as a mean to help physicists get rid of the infinities of quantum electrodynamics (QED) which prevented them from giving predictions about complex interactions of atomic particles[13]. At that time, he presented them as "mnemonic devices" [18, p. 52] to complete complex higher order calculations without confusing or omitting terms, that task being practically impossible by means only of the mathematical formulae.

A year later, Dyson [5, 4] demonstrated the equivalence of Feynman's diagrams with the mathematical derivations given at the same time by Schwinger [25, 26] as a workable calculational scheme for QED. Moreover, Dyson made Feynman's methods "available to the public"[14], by codifying the rules for constructing the diagrams, stating the one-to-one correspondence of features of the diagrams to particular mathematical expressions.

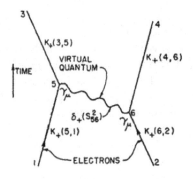

Fig. 2 Diagram corresponding to the following equation: $K^{(1)}(3,4;1,2) = -ie^2 \int \int K_{+a}(3,5)K_{+b}(4,6)\gamma_{a\mu}\gamma_{b\mu} \times \delta_+(s_{562}^2)K_{+a}(5,1)K_{+b}(6,2)d\tau_5 d\tau_6$ (drawn from [8, p. 772]).

As is well known, this was the beginning of an amazingly successful career for these diagrams, which, as Kaiser [19] describes in detail, were eventually used in almost every field of theoretical physics. The diagrams were initially intended to relieve physicists' memory and help them in performing difficult calculations; they finally became genuine theoretical tools, which went beyond the theoretical frame within which they had first been designed. They indeed played a crucial role in research in high energy post-war physics, and are still taught and used today. Beyond the equivalence of mathematical formulae and diagrams in principle, the latter acquire a genuine independence. As Kaiser [19, p. 75] suggests – thus echoing Feynman's quote about the

---

[13] For historical and technical details, I refer the reader to the works of David Kaiser [18, 19], from which I drew all my material concerning this case.

[14] Dyson, Letter to his parents, 4 Dec. 1948, quoted by Kaiser [19, p. 77].

law of gravitation –, Dyson demonstrated "the mathematical [...] equivalence between Schwinger's and Feynman's formalisms", but "by no means" their "conceptual equivalence"[15].

In both cases (equations of CM and Feynman's diagrams), representations whose equivalence can be mathematically proven happen to have different consequences in problem-solving and theory-development. According to the intuitive understanding of the notion of "format" suggested by the toy examples given in subsection 2.1, it makes sense to claim that all these are differences in format. Newtonian, Lagrangian, and Hamiltonian equations have different formats; so do Schwinger's formulae and Feynman's diagrams. Although they contain, at least partially, the same information, these representations do not convey this information the same way. Constructing and using them do not consist in the same cognitive operations: identifying and specifying the forces exerted on a system by manipulating vectors does not amount to the same process as identifying the Lagrangian of the system; drawing a diagram enabling one to visualize the different quantities to be remembered in a calculation does not consist in the same operation as performing this calculation by means of a mathematical formula. Since theorizing often consists in drawing inferences by means of theoretical representations in order to explore their content – be it in order to draw predictions concerning particular phenomena or in order to inquire into the logical relationships between various theoretical hypotheses – one can therefore say that formats do matter for theorizing.

In section 4, I shall come back to these two examples, and draw a few consequences for our understanding of both theorizing and the content of theoretical representations. Beforehand, in the next section, I will analyze further the very notion of format; in particular, I shall examine the idea that the format of a representation determines the inferential affordances of this representation for its users.

3 Formats and Inferential Affordances

As we have seen above, representations in different formats can be informationally equivalent though computationally different. This means that such representations can contain the same information – have the same *informational content* – without making it equally accessible to agents with limited cognitive abilities. In other terms, although the informational content of a

---

[15] So-called "conceptual role semantics" or "inferential role semantics" (see, e.g., [12]), which states that the content of a representation consists in (or depends on) its role in the inferential processes of agents could help us give a precise meaning to Kaiser's suggestive remark: diagrams and equations neither facilitate nor require the same inferential processes. Defending such a view would enable us to state that what Feynman calls a "psychological difference" sometimes counts as a genuine conceptual difference.

representation seems independent from the way this representation is formatted, its format (partially) determines the *inferential procedures* agents have to follow in order to access the various pieces of this content. Representations in different formats do not have the same *inferential affordances* for agents; they neither facilitate, nor require the same inferential procedures.

Before elaborating on such a view, a clarification of the very notion of "informational content" is needed. Although the informational content of a representation depends neither on the way it is coded within it, nor on the inferential procedures agents have to follow in order to access it, it is certainly determined by some features of the representation, namely the syntactic and semantic features in virtue of which this representation has such content. In section 3.1, I propose a definition of the notion of informational content, by appealing to the Goodmanian analysis of symbol systems.

In section 3.2, I shall come back to the notion of format. The intuitive use of this notion obviously relies on an analogy with computer science: a format serves to specify a procedure for both encoding data and retrieving them. *Prima facie*, the format of a representation seems to be one and the same thing as its symbol system: indeed, the symbol system of a representation is the set of syntactic and semantic rules according to which information is encoded within it. However, I shall argue that such rules do not correspond to the actual inferential procedures agents have to follow when they intend to retrieve pieces of information from the representation. Therefore, appealing to the notion of symbol system is insufficient to account for the very idea that representations in different formats have different inferential affordances for agents. I shall conclude that, if one wants to pay attention to the inferential procedures agents have to follow in order to access the content of a representation (rather than concentrate on its mere informational content), then one has to take into account *both* the way information is encoded *and* the cognitive abilities of agents.

3.1 Symbol Systems and the Informational Content of Representations

There are various theories of information[16], but it is unnecessary for my purpose to enter into any detail: let me simply state that a piece of information is a proposition that can be object of belief. One can start by defining the informational content of a representation as *the set of all the pieces of information that an agent mastering its symbol system could in principle extract from it.*

Here, "symbol system" is understood along the lines of Nelson Goodman's definition [11], namely, by analogy with a language, as a set of syntactic and semantic rules. Its syntax defines the set of relevant perceptual properties and their rules of arrangement and transformation. The semantics governs the

---

[16] See [10] for a recent account.

way these perceptual properties so arranged denote different elements of the domain of reference of the system. According to Goodman, a particular set of marks (visual or auditory) is a representation of some feature of the world (its target)[17] in virtue of a symbol system. Following the initial definition of the informational content given above, one can say that the symbol system under which a certain set of marks has to be interpreted determines its informational content. This definition, though, requires several refinements.

As I said in the introduction, one of the main functions of scientific representations is to enable agents to draw inferences concerning the systems they stand for. This can be generalized to non-scientific representations as well, if one concentrates on their epistemic use (as opposed, for instance, to their aesthetic use), namely their use in a knowledge-seeking enterprise. However, the very phrase "drawing inferences concerning a system" needs to be refined. Indeed, the epistemic enterprise of using a representation to gain knowledge concerning its target consists in *two* inferential steps, which are often simultaneous, but need to be distinguished.

For such an enterprise to be successful, one has to be able to interpret the information carried by the graph as information *about* some particular target. One needs to be aware of the source and the precision of the data one can extract from the representation. For instance, one has to know what approximations and idealizations were made in collecting the data. Consider the graph in figure 1. If one believes that the values shown in it represent the average temperature over one month, whereas measurements were made at 8am every 10th of the month, one's epistemic enterprise fails. Likewise if one takes the graph to be a representation of the variations of temperature in Paris, whereas the measurements were in fact made in Madrid.

But, before inferring – soundly or not – from the features of the graph to the features of its target, one has to *know how to read the graph itself* in order to extract information from it. This is what I call *mastering the system* under which the graph functions. For a given graph, the symbol system that defines it determines which of its perceptual features are syntactically relevant, and how they are to be interpreted, *within* the graph. Let me insist: reading off the information from the graph – even before interpreting it as about some target – requires the knowledge of the system's semantics (although not necessarily of its actual referent), and not just of its syntax. When a teacher draws a graph on a blackboard in order to teach how to read such a representation, he does not intend the graph to represent the evolution of any real quantity. Nevertheless, there is a sense in which the graph "tells" its readers that the intended quantity increases or decreases over time. It does contain such information, whether or not it is true of any real place. In the following, I shall concentrate on this second – in fact, logically first – step of the reading of a representation: the very extraction of the information it contains within

---

[17] The target can be a material object, properties of an object, the evolution of the value of some quantity, the relation between various quantities, an event, a pattern, etc.

it, independently of the true or approximately true statements its user may formulate about its intended target[18].

Of course, this definition of the informational content of a representation involves some further difficulties. First, some pieces of information are presented in an explicit[19] way, while other pieces require some inferential process to be extracted. Performing such inferences requires the mastering of the rules of the system. But it is far from clear, for a given representation, what rules are to be included in the system's syntax and semantics. For instance, the graph in figure 1 "says" in an explicit way that the temperature in July is 25 degrees Celsius, and that the temperature in December is 3 degrees Celsius. One would also like to say that it contains – though implicitly – the information that the temperature in December is 22 degrees less than in July. Therefore, it seems reasonable to consider that the basic rules of arithmetic, which enable one to make the subtraction, are part of the system's rules – or at least are supposed to be mastered by any user of such representation. Now, should we consider that the transformation rules from Celsius to Fahrenheit are part of the system's rules? If they are, then the graph also says – implicitly – that the temperature in July is 77 degrees Fahrenheit. If not, then the graph does not contain such information. Here, one has to acknowledge that whether these transformation rules are part of the system depends on the context in which the graph is used.

Moreover, the informational content of a representation seems to be context-dependent for (at least) one more reason. As Haugeland [13] suggests, one should distinguish between what he calls the "bare-boned" content of a representation and its "fleshed-out" content. The bare-boned content of a representation needs no further assumption to get extracted. On the other hand, the fleshed-out content is obtained via a deduction which implies some background knowledge. As Haugeland notes, this distinction does not correspond to the implicit/explicit distinction. Some implicit information can be extracted from the graph without any further *factual* knowledge: inferring that the temperature in December is 22 degrees less than in July only requires mastering the rules of arithmetic. However, inferring, for instance, that the temperature in Paris in July is 15 degrees less than in Madrid implies possessing the knowledge that the temperature in Madrid in July is 40 degrees Celsius. In some contexts – and particularly scientific ones –, a considerable amount of background knowledge is indispensable to derive important

---

[18] Whether one can genuinely speak of representation and informational content when there is no actual referent is a difficult issue, one I shall not tackle here. Since I will not consider issues concerning successful representations or misrepresentations, I shall not use the term "information" as a success term; rather, I use it to refer to any propositional content that an agent who knows how to read the graph – who masters its symbol system – can extract from it, whether or not this agent is mistaken concerning the target, and whether or not there is any such target.

[19] See footnote 5.

information from a representation. Here again, what knowledge is supposed to be possessed by the user of a representation depends on the context. Therefore, the informational content of a representation is context-dependent.

However, acknowledging this fact does not dangerously challenge the characterization of the informational content of a representation in terms of its symbol system. In any case, the very system in which a representation has to be read is settled by the context: to borrow one of Goodman's examples [11, pp. 229-230], whether a "black wiggly line" has to be read as an electrocardiogram or as a drawing of Mt Fujiyama obviously depends on the context (for instance, on the caption). One can therefore state that, for a given set of marks, once what is part of the system under which it functions is settled by the context, this system fully determines the informational content of the set of marks. Although this set of marks' having a content certainly depends on the existence of some agent, since nothing is a representation unless someone uses it as such, this agent is an ideal one, who perfectly masters the system, and whose cognitive abilities are not limited.

So far, nothing has been said about the *actual cognitive processes that agents with limited abilities have to run* when they seek information within a representation, nor about the practical possibility for them to access different pieces of its informational content. Symbol systems are sets of objective syntactic and semantic rules, which determine the informational content of representations. The inferential procedures agents have to follow when they want to extract pieces of this content certainly depend on the symbol system of this representation: the differences in the procedures required in order to read the list and the graph in figure 1 are obviously due to objective differences in the way data are structured within each of them; the graph and the list are not constructed along the same syntactic and semantic rules. In other words, the inferential affordances of a representation obviously depend on its symbol system. However, as we will see in the next section, merely referring to its symbol system is not sufficient to account for the inferential affordances of a representation, and therefore to fully capture the intuition underlying the use of the notion of format.

3.2 The Inferential Affordances of a Representation Are Agent-Relative

As suggested above, the intuitive use of the notion of format relies on an analogy with computer science. Following this analogy, the format of a representation might be said to determine a set of procedures for constructing, transforming, and interpreting this representation. *Prima facie*, there is no reason why one should not describe these procedures by referring to the symbol system of the representation, namely the rules governing the way data are structured within it.

However, as a further analysis of the computer science analogy reveals, these procedures do not only depend on the way data are structured (on the symbol system), but also on the "processor" which is operating on them[20], namely on the agents themselves, and on their cognitive peculiarities. Indeed, a difference in symbol system may induce a difference in the inferential affordances of the representation for one and the same agent, but the same set of marks functioning under the same symbol system can also have different inferential affordances for different agents. Laplace's demon – whose cognitive abilities are unlimited –, a computer, a trained agent and a child do not process data the same way. For these different "agents", the very same representation (the same set of data structured following the same syntactic and semantic rules) may have different inferential affordances.

To be clear, let me take a simplistic example. Imagine two representations, A and B, which have the same informational content, consisting of two pieces of information x and y. Suppose that, in virtue of the way data are structured within A, and of my own cognitive abilities, I need to extract x first if I want to access y. On the other hand, B does not enable me access x unless I extract y first. Obviously, this difference in the inferential procedures needed to access x and y is grounded in a difference in symbol system (though it does not affect the informational content)[21]. However, suppose now that there exist cognitive agents different from me (let's say, Martians) for whom the situation is the other way around. Accessing x and y within A requires the same inferential procedures for me as accessing x and y in B would require for Martians. The operations these fictional agents have to perform in order to access respectively x and y with B are the same as the operations I have to perform if I use A. The symbol systems of A and B are the same for Martians and for me (A and B encode data along the same objective rules). However, their inferential affordances are not the same for Martians and for me.

Without even appealing to fictional agents such as Martians, there exist many inter-individual differences in virtue of which the same representation does not have the same inferential affordances for two different agents. A trained agent may be able to see immediately the form of the solutions of an

---

[20] As Larkin and Simon [28, p. 67] note, "when we compare two representations for computational equivalence [as opposed to informational equivalence], we need to compare both data and operators. The respective value of sentences and diagrams depends on how these are organized into data structures and on the nature of the processes that operate on them". Larkin and Simon refer to Anderson [1] who argues "that the distinction between representations is not rooted in the notations used to write them, but in the operations used on them" [28, p. 68].

[21] Indeed, the change in symbol system is accompanied by a change in the perceptual properties, therefore canceling the effects on the informational content: consider for example a map of temperature where reddish colors would stand for warm areas, and blueish colors for cold areas. If I both change the system – the rules according to which red stands for warm and blue for cold – and the colors, the content remains unmodified.

equation, while a beginner might need to perform various operations (some-times with the help of paper and pencil) in order to reach the same conclusion. In some cases, a change in the perceptual properties of a representation (e.g. the addition of colors on a black and white diagram, facilitating the extraction of some information) counts as a change in its inferential affordances for some agents (agents with normal perceptual abilities) though not for others (color-blind persons).

More generally, one can state that the objective rules in virtue of which a certain set of marks has a certain content *underdetermine* the procedures one would actually follow in searching information. Consider again the graph in figure 1, and suppose one wants to calculate the difference of temperature between October and February. The system's rules in virtue of which the graph contains this information allow for a great variety of procedures in order to find it: one can add a graduation to the graph (based on the two numerical values which are given), draw lines from the points of the graph corresponding to February and October to the graduated scale, and then make the subtraction; one can also measure the distance between the height of the graph in february and in october and compare it to the distance between July and December (which we know corresponds to $25 - 3$), etc. Therefore, one cannot say that the syntactic and semantic rules of a representation determines a set of fixed and objective procedures, independently from a particular user in a particular situation. These rules do not correspond to the actual procedures agents have to follow. The procedures one will perform in order to access such or such piece of information also depend on one's cognitive abilities, skills, habits, preferences, background knowledge, etc.

As a consequence, in order to fully capture the intuition underlying the use of the notion of format, the right unit of analysis is not the symbol system of a representation, but the formatted representation insofar as it is used by agents with limited cognitive abilities – the rules according to which data are structured *insofar as they are processed by cognitive agents.* In other words, one should focus on the *cognitive interactions of agents with formatted representations.*

To sum up, the format of a representation determines its inferential affordances (or potential) for *a particular agent.* The inferential affordances of a representation for a particular agent might be defined as the set of procedures this agent, given his/her cognitive abilities and epistemic peculiarities, should follow in order to access the various pieces of the informational content of this representation. In the following section, I propose to draw some consequences of such a view for an analysis of theorizing. Meanwhile, it will appear that such a definition of the inferential affordances of a representation is untenable: not only is it impossible, in practice, to state the explicit algorithms agents have to implement in order to use a representation, but in fact the very idea of a set of procedures to be implemented is a rough idealization, which does not do justice to the complexity of the cognitive interactions of agents with external representations.

4 Theorizing as "Work in Progress"

In section 2, we have seen that formats matter to the use of theoretical representations as well. The analysis of section 3 led me to conclude that the actual processes one has to follow in order to extract information from a representation depend both on the way data are structured and on one's cognitive abilities. In other words, the inferential affordances of a representation are agent-relative. How relevant is such an agent-relative notion to an analysis of the role of theoretical representations in scientific practice?

In subsection 4.1, I argue that the recognition of the agent-relativity of the inferential affordances of a representation might shed light on our understanding of expertise and learning. In subsection 4.2, I propose to go one step further, by reassessing the normative dimension of my analysis of formats – as determining procedures agents *have to* follow in order to access the content of a representation. Finally, in subsection 4.3., I draw a few consequences of my view for an analysis of the content of theoretical representations.

4.1 Agent-Relativity of Inferential Affordances: Consequences on Expertise and Learning

Let me come back to the examples of section 2, which reveal the importance of formats for theoretical representations as well (as opposed to mere data display). As we have seen, the different forms of equations of CM have different inferential affordances for agents. However, one can assess these differences without referring to any particular agent. Newtonian equations are practically useless in describing the motion of a constrained system for *all agents*. Certainly, the differences in inferential potential between the equations of CM would not apply to Laplace's demon, since differences in format do not affect the informational content of a representation. But, in order to assess the inferential differences between the equations of CM, it seems legitimate to assume a *standard agent* without considering inter-individual differences. Unlike idealized agents for whom differences of formats would not matter, standard agents have limited cognitive abilities. Such a standard agent would be the typical user of this kind of representation: in the case of the graph, a human adult with normal cognitive abilities; in the case of the equations of CM, someone mastering the rules of the calculus and having a fair training in physics. For a layman without such training, it makes little sense to compare the Lagrangian and the Hamiltonian formats. If one does not even know how to solve a differential equation, whether integrable or not, one would not gain anything if provided with the Legendre transformations, in addition to a non-integrable Lagrangian equation. Therefore, the agent-relativity of the inferential affordances of representations does not seem, *prima facie* a relevant feature for an analysis of theoretical representations.

In some cases, though, it proves useful to pay attention to inter-individual differences, and therefore to acknowledge that the same representation can be used in different ways by different agents (and thus have different inferential affordances for them). As I will now suggest, this might shed light on expertise and learning (the process of becoming an expert), conceived of as the deepening and sharpening of one's understanding of a theory.

Beginners and experts, trivially, do not use differential equations in the same way. However, that does not mean that beginners always use them in a faulty way: even in conforming to the rules of the calculus, beginners may follow longer and less efficient inferential paths than experts. One could therefore think of learning as a process consisting in progressively modifying the procedures of use of some representations. By acquiring the skills enabling one to use a type of representation in an efficient way, one reduces, so to speak, the inferential path leading to a problem's solution. Learning how to use a certain type of equations, and becoming more and more skillful in it, consists in modifying their inferential affordances by learning new transformation rules – new *inferential paths*[22].

Accordingly, expertise could be thought of as the ability to use certain representations in an optimally efficient way. As suggested by Andrea Woody [33], becoming an expert consists in acquiring an "articulated awareness" of the representations used in this field. The more expert you are, the more easily you draw inferences with these representations. Moreover, in addition to solving problems more quickly, the expert has a deeper understanding of the very content of theories, namely of the deductive relationships between the various hypotheses this theory consists in. Deepening one's understanding of a theory therefore consists in progressively modifying the inferential architecture of its various principles and hypotheses, by developing new inferential paths between them. Consider again the equations of CM: whereas the beginner might find it difficult to understand why Newtonian and Lagrangian equations are equivalent, the trained physicist can "see immediately", so to speak, their equivalence. Indeed, he is able to transform the ones into the others very quickly.

Now, whether one considers the inferential affordances of a representation for a beginner or for an expert, the view of formats I have proposed still has a normative dimension. Indeed, according to the above analysis, the format of a representation for a particular agent determines the set of procedures this agent (given his/her limitations, skills, background knowledge, etc.) *has to* follow in order to access the various pieces of information contained in it. In other words, given a particular agent, there seems to be something as *the right way* of using a representation. Moreover, by suggesting that expertise could be assessed by referring to the efficiency of the way one uses a certain type of representations, I have assumed that there exists an optimal way of using scientific representations in order to access their content. In the next

---

[22] As noted by Kuhn [22], learning consists in acquiring skills (know-how) rather than learning explicit rules.

subsection, I shall argue that the very idea of an agent's optimally mastering the rules of a certain type of representations, and of the existence of a set of procedures *to be followed*, relies on an illegitimate idealization.

4.2 Who Is the Expert?

Consider again the case of Feynman's diagrams. The interest of this case is not exhausted in the comparison between diagrams and mathematical formulae (which can be made without referring to inter-individual differences). As their letters and personal papers show, Feynman and Dyson explicitly disagreed on the legitimate use and status of the diagrams within QED. Dyson conceives of them as secondary, psychological aids to the performance of mathematical calculations. On various occasions, he claimed that their use would be illegitimate if they had not been proven rigorously derivable from mathematical formulae: "until the rules were codified and made mathematically precise, I could not call [Feynman's method] a theory." [6, p. 127]. For him, diagrams were means to "visuali[ze] the formulae which [he derived] rigorously from field theory" [6, pp. 129-130][23]; they had a meaning only within QED, to which they added nothing except cognitive tractability.

On the other hand, Kaiser reports that Feynman never felt the need to show how to derive diagrams from mathematical expressions, and expressed clearly on various occasions his theoretical preference for diagrams over mathematical formulae: "All the mathematical proofs were later discoveries that I don't thoroughly understand but the physical ideas I think are very simple." (Feynman, Letter to Ted Welton, 16 Nov. 1949, quoted in [19, p. 178])[24]. Hence, unlike Dyson, he thought of diagrams as primary and more important than any mathematical derivation that might be given. In addition to being mnemonic devices, they provided an intuitive dimension to the theory, and Feynman took them as "intuitive pictures" [19, p. 176]. As Dyson notes, Feynman "regard[ed] the graph as a picture of an actual process which is occurring physically in space-time" [6, p. 127][25]. Rather than visualizations of the formulae, they were primary visualizations of the physical processes themselves. Despite their agreement on the in-principle equivalence of the diagrams and the formulae, Feynman and Dyson did not construct and use diagrams in the same way, and in the final analysis did not even "see" the same thing in them.

As Kaiser suggests, this difference in use by the two physicists can be explained by referring to their own theoretical commitments and preferences. Unlike Dyson, who demonstrated how to cast both Feynman's diagrams and

---

[23] Quoted in [19, p. 190].

[24] Feynman also spoke of the "physical plausibility" of the diagrammatic approach (quoted in [19, p. 177]).

[25] Quoted in [19, p. 190].

Schwinger's equations within a consistent field-theoretic framework[26], Feynman's renormalization approach, from which the diagrammatic method arose, was based on particles, rather than on fields. More generally, as Kaiser notes, Feynman had a preference for a semi-classical approach, and worked almost entirely in terms of particles, trying to remove fields from theoretical descriptions altogether. Such theoretical commitments and interests, together with individual preferences for some kinds of reasoning (Feynman expressed on various occasions his favoring "visualization" over abstract calculation) must have contributed to giving the diagrams different inferential affordances for Feynman and Dyson. Dyson deduces them from mathematical formulae, whereas Feynman draws them intuitively: each one relates them in a different way to other representations, and, finally, to the physical world.

In addition to Feynman and Dyson's using the diagrams according to different rules, Feynman himself, as well as other physicists, continuously modified their rules of use. Kaiser [19] gives an impressive analysis of the "plasticity" of diagrams throughout their "spreading" in theoretical practices in modern physics. He studies their varying uses and interpretations in different contexts and "schools" (Oxford and Cambridge, Japan, Soviet Union). Despite Dyson's efforts, the rules of construction and interpretation of the diagrams have not been strictly followed. Moreover, this is the reason why they were so successful: rather than mere calculation tools, they were genuine discovery tools that contributed important theoretical developments to modern physics, by being used and applied in new fields.

Who is the "standard user" of these diagrams? Feynman? Dyson? Others? Who is the expert who follows the procedures corresponding to the inferential affordances of the diagrams *for experts*? The inferential affordances of diagrams are obviously different for Feynman and for Dyson (since they do not use them the same way). Moreover, these affordances constantly changed for Feynman himself. Stating that the format of the diagrams determines a set of procedures their user must conform to would amount to missing some essential aspects of their role in theorizing.

I suggest that, far from being an exception, this is an exemplary case of the way representations are used in theorizing. Let me come back to the example of CM. In the case of the Newtonian, Lagrangian, and Hamiltonian equations, there seem to exist fixed sets of procedures that any expert masters. Let's suppose that this is so, and that every (trained) physicist today uses them by following the same processes. Reducing them to representations whose inferential affordances are strictly fixed would nevertheless prevent us from noticing essential aspects of scientific invention and discovery. Consider Hamilton's use of the Legendre transformations: this innovation does not rely on any empirical novelty, but rather consists in the introduction of new transformation rules within mechanics, which results in a modification of the

---

[26] Dyson [7, p. 23] claims that he contributed to allow "people like Pauli who believed in field theory to draw Feynman diagrams without abandoning their principles".

inferential affordances of the various equations. I suggest that theory development often consists in such a process of modifying the procedures followed in using the representations. As a chess player who, knowing the rules, invents a new gambit, the theoretician modifies the inferential processes he/she performs in order to solve problems and sometimes develops novel connexions between different equations. In the case of CM, as well as in the case of Feynman's diagrams, stating that the format of the different equations determines a set of rules agents have to follow in order to extract some pieces of information relies on an illegitimate idealization: *the* right algorithm does not exist.

These considerations imply that we should reassess the normative dimension of my previous analysis of formats and inferential affordances. The consequences of what I intuitively meant by "format" are too context and agent-dependent to be settled in terms of a fixed set of procedures *to be applied*. An agent's use of a representation depends on the particular situation in which he/she is involved and on his/her particular goals. As the case of Feynman's diagrams show, the inferential affordances of a representation for the same user – as well as for the community – change over time: the inferential affordances of diagrams are not the same for Feynman in 1948 and ten years later. Likewise, the inferential affordances of the Lagrangian equations were modified by Hamilton's adding to mechanics new rules of transformation, which changed the role of the various equations of the theory in the scientists' reasoning processes. Therefore, the inferential affordances of a representation are fundamentally dynamical in character; they should to be defined (beside the perceptual properties of the representation and a minimal set of construction and interpretation rules) in reference to a particular situation, involving a particular agent, with particular skills, theoretical commitments, preferences, reasoning habits, as well as interests and intentions in the particular inquiry in which he/she is involved.

By this, I do not mean that a philosophical analysis of the use of representations in theorizing and the importance of formats has to be strictly descriptive. Once the highly agent-dependent and situation-dependent nature of the use of representations has been acknowledged, it is certainly worth trying to find the theoretically interesting regularities in the use of representations in different contexts, and there is certainly room for normative claims. For instance, acknowledging that the inferential affordances of a representation are relative to its users' background knowledge and skills could help us analyze the role and virtue of various kinds of representation in scientific teaching and popularization. In these activities, as well as in theory development by experts, there are definitely successful as well as failing strategies. In analyzing those cases, a certain degree of idealization is required, and it is worth ignoring some inter-individual differences and assuming *types* of agents (therefore speaking of the inferential affordances of a representation for beginners, for experts, etc). According to the kind of question to be studied, different levels of idealization might be justified: if one is interested in popularization, one should pay attention to various inter-individual differences and to the cognitive abilities of the laypersons – although one can assume *types* of laypersons; on the other hand, if one is interested in the inferential differences between Lagrangian and Hamiltonian equations, one should definitely assume a type of agents – experts – who use differential equations the same way. Note, however, that the case of Feynman's diagrams shows that inter-individual differences between experts (or groups of experts[27]) are sometimes also worth taking into account.

4.3 Formats, Theorizing, and the Content of Theoretical Representations

Finally, the view I have been defending enlightens some common features of various theoretical activities, in particular learning and theory development, which are usually not treated under the same heading. Indeed, as suggested above, both might be thought of as processes consisting in the modification of the inferential affordances of some representations – i.e. of the way one uses them. Contrary to what the idea of formats as determining a fixed set of procedures to be followed suggests, experts, as well as students, continue to deepen their understanding of the theoretical hypotheses they develop and use, by inventing new inferential paths between representations. This is what physics students do. This is what Hamilton did, with this (important) difference: he did it first and made his modification publicly available.

Let me clearly state that my point does not amount to saying that scientists' reasoning does not obey any rule and that there exists no difference between a sound inference and a wrong one. Of course, solving a differential equation implies that one conforms oneself to a whole set of calculation rules; if one obeys them, one cannot deduce contradictory results from the different types of equations of mechanics. There is a sense in which these equations are equivalent; one cannot draw just anything from them. Just like a chess game, the inferential affordances of a representation partially depend on a set of rules which are objective, in the sense that these rules do not depend on the users and on the situation. My point is to claim that these rules are highly insufficient to determine the way representations are used; reducing our analysis of theoretical representations to an idealized – positivist-like – image of a

---

[27] Such a view enables us to characterize scientific communities by referring to their sharing types of representations and using them the same way, as already suggested by Kuhn [22]. As Kuhn emphasized, different communities can use the same "symbolic generalizations" – e.g. Schrödinger's equations – in consistent (i.e. conforming to objective mathematical rules) but *different* ways (applying them to different cases and giving them different interpretations). These equations do not have the same inferential affordances for these different practitioners. Kaiser's study [19] of the uses and interpretations of the diagrams by different schools is an example of the fruitfulness of this view.

fixed set of rules results in a narrow conception of theorizing, which does not enable us to capture the complex processes which are at play in theorizing, and particularly the creative dimension of theorizing[28]. Theoretical representations such as the equations of mechanics or Feynman's diagrams are highly sophisticated tools of calculation and inquiry; as the analogy with the chess-player's inventing new gambits suggests, the process of drawing new bridges between them and inventing new ways to connect them to the phenomena – and therefore of modifying the procedures of use of these representations – are potentially infinite. Theorizing partially consists in developing these rules in order to *explore the content* of representations.

Acknowledging that such exploration is potentially infinite might finally offer a new standpoint to think of the content of theoretical representations, and of their predictive and explanatory power. Indeed, in virtue of the objective rules' underdetermining the inferential procedures one may follow in using them, it is always possible to draw new consequences and "discover" new relations between them and the phenomena. In other words, the famous "theoretician's dilemma" formulated by Hempel [14] as a consequence of the reductionist demand on theoretical terms does not arise: theoretical representations are not theoretical because they seem to refer to some unobservable entities, but rather because they allow theoreticians to create novel connections between them by manipulating them and modifying their inferential affordances.

5 Conclusion

Philosophers of science generally consider theories and models as abstract entities, whose representational relationships with the phenomena have to be elucidated by formal reconstruction. I hope to have shown that one cannot understand the explanatory and predictive fruitfulness of scientific representations without taking into account the particular form of what Humphreys calls the "concrete pieces of syntax" [16], which are used in theory learning, application, and development, and whose rules of construction and interpretation are not fixed. As he suggests, one should give up the "no-ownership perspective" [16] characteristic of most philosophy of science, and pay attention to the *computational* dimension of theorizing. Moreover, I have shown that, once agents and their limited cognitive abilities get into the picture, it becomes impossible to draw a clear-cut frontier between epistemic differences, which would count for all humans, and purely psychological differences.

---

[28] Note, incidentally, that Kuhn, although he strongly criticized the positivist image of theorizing conceived of as mere application of rules, also missed this creative dimension. He indeed considered the mathematical development of a theory such as CM as a "purely formal" work, as opposed to conceptual innovation. My analysis of theorizing aims at showing that conceptual novelty sometimes arises from formal invention.

This is particularly true in the case of scientific representations, which are complex and sophisticated tools, and not mere displaying of data. Taking into account particular agents, with particular skills, involved in particular situations, finally enables us to enlighten some essential aspects of theorizing, which are often neglected. From this perspective, theorizing partially consists in constructing and manipulating representations, whose role in the agents' reasoning processes – whose inferential affordances – change(s) with the agents' abilities and interests.

Acknowledgements. I would like to thank Anouk Barberousse, Paul Humphreys, Alex Kirlik, Pierre Jacob, and two anonymous referees for comments on earlier drafts and/or for helpful discussions.

References

1. Anderson, J.: Representational types: A tricode proposal. Tech. Rep. 82.1, Office of Naval Research, Washington, D.C. (1984)
2. Clark, A.: Supersizing the Mind. Embodiement, Action, and Cognitive Extension. Oxford University Press, Oxford (2008)
3. Clark, A., Chalmers, D.J.: The extended mind. Analysis 58(1), 7–19 (1998)
4. Dyson, F.J.: The S matrix in quantum electrodynamics. Physical Review 75, 1736–1755 (1949)
5. Dyson, F.J.: The radiation theories of Tomonaga, Schwinger, and Feynman. Physical Review 75, 486–502 (1949)
6. Dyson, F.J.: Advanced quantum mechanics, Mimeographed notes from lectures delivered at Cornell (1951)
7. Dyson, F.J.: Old and new fashions in field theory. Physics Today 18, 21–24 (1965)
8. Feynman, R.: Space-time approach to quantum electrodynamics. Physical Review 76(6), 769–789 (1949)
9. Feynman, R.: The Character of Physical Law. MIT Press, Cambridge (1965)
10. Floridi, L.: Information. In: Floridi, L. (ed.) The Blackwell Guide to the Philosophy of Computing and Information, pp. 40–61 (2004)
11. Goodman, N.: Languages of Art. An Approach to a Theory of Symbols, 2nd edn. Hackett Publishing Company (1976) (1968/1976)
12. Greenberg, M., Harman, G.: Conceptual role semantics. In: Lepore, E., Smith, B.S. (eds.) The Oxford Handbook of Philosophy of Language, pp. 295–322. Clarendon Press, Oxford (2006)
13. Haugeland, J.: Representational genera. In: Ramsey, W., Stich, S., Rumelhart, D. (eds.) Philosophy and Connectionist Theory, Lawrence Erlbaum, Hillsdale (1991); Reprint in Haugeland, J.: Having Thought, pp. 171–206. Harvard University Press, Cambridge (1998)
14. Hempel, C.G.: The theoretician's dilemma. In: Feigl, H., Scriven, M., Maxwell, G. (eds.) Concepts, Theories, and the Mind-Body Problem. Minnesota Studies in the Philosophy of Science, vol. 2. University of Minnesota Press, Minneapolis (1958)
15. Hollan, J., Hutchins, E., Kirsh, D.: Distributed cognition: Toward a new foundation for human-computer interaction research. ACM Transactions on Computer-Human Interaction 7(2), 174–196 (2000)

16. Humphreys, P.W.: Extending Ourselves. In: Computational Science, Empiricism, and Scientific Method. Oxford University Press, Oxford (2004)
17. Hutchins, E.: Cognition in the Wild. MIT Press, Cambridge (1995)
18. Kaiser, D.: Stick-figure realism: Conventions, reification, and the persistence of Feynman diagrams, 1948-1964. Representations 70, 49–86 (2000)
19. Kaiser, D.: Drawing Theories Apart: The Dispersion of Feynman Diagrams in Postwar Physics. University of Chicago Press, Chicago (2005)
20. Kirsh, D.: When is information explicitly represented? In: Hanson, P. (ed.) Information, Language, and Cognition, University of British Columbia, Vancouver (1991)
21. Kleinmuntz, D., Schkade, D.: Information displays and decision processes. Psychological Science 4(4), 221–227 (1993)
22. Kuhn, T.S.: Second thoughts on paradigms. In: Suppe, F. (ed.) The Structure of Scientific Theories, pp. 459–482. University of Illinois Press, Champaign (1974/1977)
23. Kulvicki, J.: Knowing with images: Medium and message. Philosophy of Science 77(2), 295–313 (2010)
24. Nickerson, R.S.: Technology and cognition amplification. In: Sternberg, R.J., Preiss, D.D. (eds.) Intelligence and Technology: The Impact of Tools on the Nature and Dvelopment of Human Abilities. Lawrence Erlbaum, Mahwah (2005)
25. Schwinger, J.: On quantum-electrodynamics and the magnetic moment of the electron. Physical Review 73, 416–417 (1948)
26. Schwinger, J.: Quantum electrodynamics. Physical Review 74, 439–461 (1948)
27. Shin, S.: The Logical Status of Diagrams. Cambridge University Press, Cambridge (1994)
28. Simon, H.A., Larkin, J.H.: Why a diagram is (sometimes) worth ten thousand words. Cognitive Science 11, 65–99 (1987)
29. Stenning, K., Oberlander, J.: A cognitive theory of graphical and linguistic reasoning: Logic and implementation. Cognitive Science 19(1), 97–140 (1995)
30. Tufte, E.: The Visual Display of Quantitative Information. Graphics press, Cheshire (1983)
31. Tufte, E.: Envisioning Information. Graphics press, Cheshire (1990)
32. Vorms, M.: Formats of representation in scientific theorizing. In: Humphreys Paul, W., Imbert, C. (eds.) Representations, Models, and Simulations, Routledge (forthcoming)
33. Woody, A.: More telltale signs: What attention to representation reveals about scientific explanation. In: Philosophy of Science Proceedings of the 2002 Biennial Meetings of the Philosophy of Science Association, vol. 71(5), pp. 780–793 (2004)
34. Zhang, J.: A representational analysis of relational information displays. International Journal of Human-Computer Studies 45, 59–74 (1996)
35. Zhang, J.: The nature of external representations in problem solving. Cognitive Science 21(2), 179–217 (1997)
36. Zhang, J.: External representations in complex information processing tasks. Encyclopedia of library and information science 68(31), 164–180 (2000)
37. Zhang, J., Norman, D.A.: Representations in distributed cognitive tasks. Cognitive Science 18, 87–122 (1994)

Modeling the Epistemological Multipolarity of Semiotic Objects

Zdzisław Wąsik

Abstract. For practitioners of semiotics the most controversial questions constitute the status and nature of the semiotic object equalized with the sign separated from its object(s) of reference or encompassing its object(s) reference, i.e., whether the sign is a unilateral entity or a plurilateral unit comprised of interrelated constituents, or a relation (a network of relations) between those constituents. Further questions refer to the manifestations of signs, namely, whether they appear in material or spiritual (corporeal or intelligible, physical or mental), concrete or abstract, real or ideal forms of being, being examined subjectively or objectively in their extraorganismic or intraorganismic manifestations. Accordingly, signs are approached either extra- or introspectively, through individual tokens or general types, occurring in the realm of man only; in the realm of all living systems, or in the universe of creatures, extraterrestrial and divine in nature. These varieties of sign conceptions exhibit not only differences in terminology but also in the formation of their visual presentations. Bearing in mind the need for their analysis and comparison, the practitioner of semiotic disciplines has to find a parameter or a matrix that would contain features and components characteristic for particular approaches to their forms of being and manifestation. Within the framework of this article, the adept readers will be provided with a theory-and-method related outlooks on the token and type relationships between the mental and concrete existence modes of semiotic objects and their objects of reference. Having reviewed all hitherto known sign conceptions, it will be demonstrated how their two main components, the *signans* and *signatum*, may be modeled with their collective and individual properties as oscillating between the possible four epistemological positions: logical positivism, rational empiricism, empirical rationalism, and absolute rationalism.

Zdzisław Wąsik

Department of Linguistic Semiotics and Communicology, Philological School of Higher Education in Wrocław, Kolegium Karkonoskie in Jelenia Góra and Department of Linguistic Semiotics, Adam Mickiewicz University in Poznań, Poland

e-mail: zdzis.wasik@gmail.com

L. Magnani et al. (Eds.): Model-Based Reasoning in Science & Technology, SCI 314, pp. 559–569.

springerlink.com

1 Asking for the *Genus Proximum* of the Sign among Other Semiotic Objects

Primary attention in semiotics is given to the manifestation forms of signs and their objects of reference. To distinguish the sign from a non-sign, one has to ask for its *genus proximum*. In this context, the Scholastic formula *aliquid stat pro aliquo*, "something that stands for something else", applied by Karl Bühler (cf. [2]) who defined the sign as "the sensorially perceivable phenomenon" ["das sinnlich wahrnehmbare Phänomen"] (quoted in [15, p. 96]), does not appear to be advantageous for the goals of linguistics, because it requires elaborating an extensive set of *differentia specifica*. In order not to oscillate between various conceptions, one needs a parameter for all notional scopes of the term *sign*. And such a proposal may be deduced from the classification of semiotic *phenomena*, specified as something that is recognized as it appears or as it is experienced by senses, or sensible *stimuli*, understood as something that excites the body and activates the mind.

The definitional properties of the sign as a cognizable *phenomenon*, as exposed in Figure 1., may be determined by the four positively marked levels of (i) implicative vs. non-implicative, (ii) artificial vs. non-artificial, (iii) non-semantic vs. semantic or (iv) arbitrary vs. non-arbitrary phenomena.

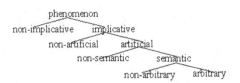

Fig. 1 A classificatory approach to cognizable and apprehendible phenomena.

An alternative explanation can be also made in terms of *stimuli* preferred by mind-and-body-centered practitioners of semiotics. Correspondingly, as illustrated in Figure 2., one may utilize also the division between: (i) non-associated vs. associated, (ii) non-intentional vs. intentional, (iii) non-inferred vs. inferred, (iv) conventional vs. non-conventional stimuli. To be more precise, n the last instance, following the usage according to which the term conventional is often reduced, especially in the linguistic works, to the meaning of the term arbitrary, i.e., free and non-motivated.

Taking into consideration the definition of language as a system of signs, one has to be aware that the sign, with respect to its notional scope, should occupy as such the lowest place in the hierarchy of *phenomena* or *stimuli* (Cf. [15, pp. 96–97]).

That's why to reach the nature of language one should go through the characteristics of semiotic objects, as presented in Figure 3, which play the role of: *index* as an implicative phenomenon (or a non-associated stimulus), *symptom* as an implicative non-artificial phenomenon (or an associated non-intentional

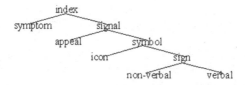

Fig. 2 A classificatory approach to sensorially perceivable and mentally apperceivable stimuli.

Fig. 3 Verbal and non verbal signs among other types of semiotic objects.

stimulus), *signal* as an implicative artificial phenomenon (or an associated intentional stimulus), *appeal* as an implicative artificial non-semantic phenomenon (or an associated intentional non-inferred phenomenon), *symbol* as an implicative artificial semantic phenomenon (or an associated intentional inferred stimulus) , *icon* as an implicative artificial semantic non-arbitrary phenomenon (or an associated intentional inferred non-conventional stimulus), and *sign* as an implicative artificial semantic arbitrary phenomenon (or an associated intentional inferred conventional stimulus).

In view of such a hierarchy of semiotic objects, the linguistic sign may be specified as an: (1) implicative vs. non-implicative phenomenon or associated vs. non-associated stimulus, i.e., index vs. non-index (2) artificial vs. non-artificial (natural) phenomenon or intentional vs. non-intentional stimulus, i.e., indexical symptom vs. signal (3) inferred vs. non-inferred phenomenon or semantic vs. non-semantic stimulus, signaling appeal vs. symbol (4) arbitrary vs. non-arbitrary stimulus or conventional vs. non-conventional phenomenon, i.e., iconic symbol vs. signifying symbol.

2   On the Typology of Sign Conceptions, Their Manifestation Forms and Ontological Status

An overview of semiotic thinking has shown that the manifestation forms of the sign are expressed in: (I) the unilateral sign concept in which the *sign-vehicle* and the *referent* are treated as separate entities, (II) the bilateral sign concept where its *signifier* and its *signified* constitute a twofold mental unity, (III) the concept of semantic triangle in which the *sign-vehicle*, the *meaning* (thought or notion), and the *referent* form separate parts, or (III') the trilateral sign where the *sign-vehicle*, the meaning as *interpretant* generating

/an/other sign/s/, and the *object of reference* constitute a threefold unity (for detailed references see inter alia [13, 14, 15, 16]).

These varieties of sign conceptions exhibit not only differences in the terminology but also in the formation of their visual representations. Bearing in mind the need for their comparison, one has to elaborate a matrix that contains all features and constituents specific for particular approaches to their ontological status, as presented in Figure 4.

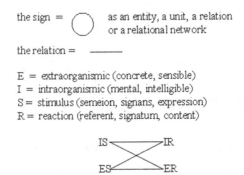

Fig. 4 A matrix for the typology of signs with respect to their ontological status.

Through the application of a unified scheme, presented in Figure 4, one can expose at least seven types of signs according to their ontological status (discussed from a historical perspective in [13, pp. 209–217]; cf. Figure 5, adapted from [15, p. 100]).

Some conceptions are distributed in semiotic writings, e.g., IA (Karl Bühler, cf. [2]), IB (St. Augustine, cf. [1]), IC (Wiesław Łukaszewski, cf. [8]), ID (Louis Trolle Hjelmslev, cf. [5]), II (Ferdinand de Saussure, cf. [12]), II' (Sydney Macdonald Lamb, cf. [6]), IIIα (William of Ockham, cf. [9]), IIIβ (Charles Kay Ogden, Ivor Armstrong Richards, cf. [10]), IIIγ (John Lyons, cf. [7]), IIIδ, & IV (Pierre Guiraud, cf. [4]). Some others are potential only or occur as parts of more complex schemes, as, for example, ID within II, and IB within IIIb, IC within IIIg. However, do not fit within the proposed framework, but may be explained in terms of accepted primitives, the conceptions of (III') the trilateral sign (Gottlob Frege, cf. [3], Charles Sanders Peirce, cf. [11]), and (III") the sign as a triadic relation (Charles Sanders Peirce, cf. [11]).

So far, the sign as a quadrilateral unity or as a fourfold (tetradic) relation (i.e., a relational network between four "sign-arguments") has been not postulated. Anyway, types (IV') and (IV") might be admissible (assuming that the bilateral or trilateral sign or the sign as a dyadic or triadic relation are also tenable).

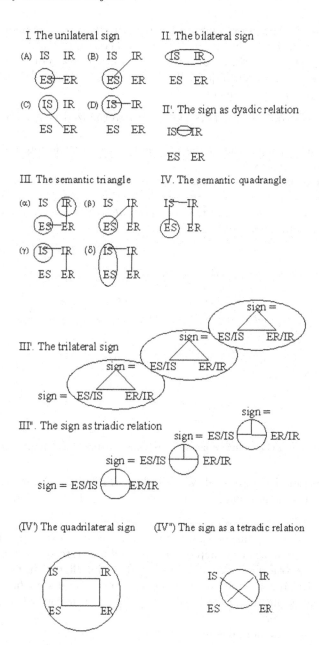

Fig. 5 Manifestation forms of the sign.

Bringing the explanatory and illustrative primitives to a common denominator, as it is illustrated in Figures 4 and 5 and consequently also 8 (adapted from [15, pp. 99 and 102], cf. also [16, p. 128]), one may point out that all concepts of signs and their objects of reference embrace four elements of (IV) a semantic quadrangle, namely:

- an externalized *repraesentans* (i.e., the externalized sign as a concrete *signans*),
- an internalized reflection of the *repraesentans* (i.e., the internalized sign as a mental *signans*),
- an externalized *repraesentatum* (i.e., the externalized referent as a concrete *signatum*), and
- an internalized reflection of the *repraesentatum* (i.e., the internalized referent as mental *signatum*).

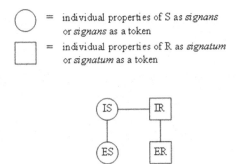

Fig. 6 *Signans* and its *signatum* as separate constituents of a semantic quadrangle.

Having in mind the possibility of mutual relationships between the constituents of a semantic quadrangle, as illustrated in Figure 7 (adapted from [16, p. 129]), namely:

- a sensible extraorganismic sign (ES) and its sensible extraorganismic referent (ER) rendered as concrete *signans* and concrete *signatum*,
- a sensible extraorganismic sign (ES) and its intelligible intraorganismic referent (IR) rendered as concrete *signans* and mental *signatum*,
- an intelligible intraorganismic sign (IS) and its sensible extraorganismic referent (ER), and
- an intelligible intraorganismic sign (IS) and its intelligible intraorganismic referent (IR), one may notice the existence of four kinds of ontological implications in the observed and inferred reality (A, B, C, D).

In the context of the notion of signs, which function as meaning-bearers/carriers in the domain of other semiotic objects, as heterogeneous presents itself the notion of *meaning*, which should be, therefore, discussed separately.

(A) Concrete *signans* stands for concrete *signatum*, i.e., ES implies ER

(B) Concrete *signans* evokes mental *signatum*, i.e., ES implies IR

(C) Mental *signans* is referred to concrete *signatum*, i.e., IS implies ER

(D) Mental *signans* is associated with mental *signatum*, i.e., IS implies IR

Fig. 7 Multipolarity of *signans* and *signatum*.

3 The Relationship between *Signans* and Its *Signatum* as a Token and/or a Type

Another kind of distinction that can have an impact upon a number of multipolar relationships between the constituents of sign conceptions depends on the answer to the question whether the sign is to be regarded as a token or a type (in the sense of a specimen or a class, an item or a kind). This distinction, however, between tokens and types concerns not only the manifestation forms of signs but also of the objects they stand for, refer to or signify, represent, evoke or indicate, namely to those objects, which are named *signata*.

In a much-generalized way, it is initially assumed that the main task attributed to signs consists in their capacity of representation. In order to state what the representation of a certain sign is, one has to determine the status of the *repraesentans* to which it corresponds, i.e., whether it is localized at the level of indication, signalization, symbolization, or signification. Specifying the concept of a verbal sign (a word, a name, a locution or a text element, and the like) for the tasks of linguistics, one should rather opt for its narrow understanding at the level of signification, i.e., as an arbitrary semantic intentional associated stimulus.

It is a matter of epistemological preferences as to what kinds of names are ascribed to the constituents or the entities of the domain of signification. They may be specified as the *signifier* and the *signified* or *signans* and *signatum*, *repraesentans* and *repraesentatum*, *significans* and *significatum*. They may be also treated separately as *sign* and *designate*, *sign* and *significate* or *name* and *designate* (*nominatum*, *signum* and *significatum*, *signum* and *signatum*, *designator* and *designatum*, and the like).

To avoid any adherence either to a psychological or to a logical frame of reference, it is proposed (see [15, pp. 101–102]) to use the term *signatum* for the *repraesentatum* of the sign as the most neutral. In fact, the term *designate* entails the counterpart of a *name* considered under the aspect of truth, and the term *significate* connotes rather the meaning of an abstract correlate constituting the *raison d'être* of the sign.

Taking for granted (with reference to terminological distinctions adapted from [15, *passim*]) that *signans* and its *signatum* – either real or fictitious, corporeal or intelligible, observed or concluded – constitute objects in the

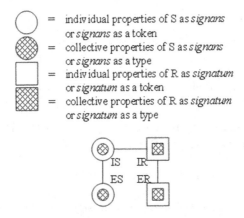

Fig. 8 *Signans* and *signatum* as a token and a type in a unified scheme of the semantic quadrangle.

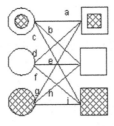

Fig. 9 The relationship between *signans* and its *signatum* as a token and a type, inclusively and exclusively.

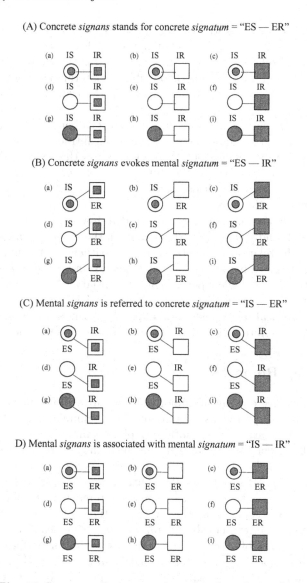

Fig. 10 *Signans* and *signatum* in multipolar relationships.

philosophical sense manifesting themselves as tokens and/or types, one can explain them within the framework of a unilateral sign conception, in accordance with Bühler's view (cf. [2], discussed in [15, pp. 94 and 96]), as constituents of a semantic quadrangle. (Cf. respectively the terms in Figure 6 and 8.)

However, the relationship between *signans* as a token and type, as a singular token, or as an abstract type and its *signatum* as a token and type, as singular token, or as an abstract type, appears to be more complicated when they appear in 3 times 3 kinds of combinations (cf. Figure 9).

The potential relations: a, b, c, d, e, f, g, h, i (as shown in Figure 9) may be multiplied by the four kinds of possible unilateral sign conceptions: A, B, C, D (cf. Figure 10) and interpreted within the framework of a semantic quadrangle.

Thus, one can expose in reality 36 combinations between *signans* and its *signatum*, as an extraorganismic objective or an intraorganismic subjective token and/or a type, inclusively or exclusively, on the basis of distinctions provided in Figure 7 (cf. [16, p. 129] and 2b (cf. [15], pp. 102–103) which illustrate the epistemological multipolarity of semiotic objects and their meaning related objects of reference.

As one may figure out from Figures 7 and 10 in comparison with Figure 5 (cf. [14], p. 52 f. and [15], p. 103), depending on the choice of an epistemological position, the relationships between the constituents of the unilateral sign concept, selected as a matrix, namely the *signans* and its /de/*signatum* may oscillate between: (A) logical positivism, referential antipsychologism = concretism, (B) rational empiricism, psychological logicism = moderate psychologism, (C) empirical rationalism, logical psychologism = moderate psychologism, and (D) absolute rationalism, extreme psychologism = mentalism.

Concluding Remarks

In the investigative domain of semiotics, there are various meaning bearers, produced and interpreted as meaningful by communicating subjects. The practitioner of a particular semiotic discipline cannot neglect the fact that there are also other theories of sign and meaning, borrowed from cultural sciences, in which the verbal sign is considered as a type of cultural object having significance for instrumental and utilitarian purposes of human agents as the subjects of culture. In such human-centered models (for details see [16]), the emphasis is placed on the semiotic activity of the subject of culture who subsumes and interprets all cultural objects as significant, i.e., as relevant from a certain point of view (a) when they are useful for performing certain functions with respect to his purposes, goals or aims in view, (b) because they possess certain values for the satisfaction of his needs, wants or expectations. Included into the realm of semiotic objects, verbal expressions that bear certain meanings might be also modeled from the perspective of their users, producers or receivers as correlates of certain functions, values or contents deduced from the domain of their references.

References

1. Augustine, St. [Augustinus, Aurelius (Hipponensis)], On Christian Doctrine. In: Robertson, Jr. D.W. (ed. & trans.) De doctrina Christiana, Liber secundus, I. 1, Liberal Arts Press, New York, [397] (1958)

2. Bühler, K.: Theory of Language. The Representational Function of Language. In: D.F. Goodwin (trans.) Sprachtheorie. Die Darstellungsfunktion der Sprache. Jena: Gustav Fischer. John Benjamins, Amsterdam, Philadelphia (1990 [1934])

3. Frege, G.: Über Sinn und Bedeutung. Zeitschrift für Philosophie und philosophische Kritik 100, 25–50 (1892)

4. Guiraud, P.: Semiology. Gross, G. (trans.) Routledge & Kegan Paul, London, Boston. La sémiologie. Paris: Presses universitaires de France (1975 [1971])

5. Hjelmslev, L.: Prolegomena to a Theory of Language. F.J. Whitfield (trans.). Waverly Press, Baltomore. Omkring sprogte-oriens grundlaeggelse. Festskrift udgivet af Københavns. Universitet. København (1953 [1943])

6. Lamb, S.M.: Language and Illusion. Toward a Cognitive Theory of Language. Rice University, Houston TX (1991)

7. Lyons, J.: Introduction to Theoretical Linguistics. Cambridge University Press, Cambridge (1968)

8. Łukaszewski, W.: Pojęcia i systemy pojęciowe z psychologicznego punktu widzenia (Concepts and conceptual systems from a psychological viewpoint). In: Acta Universittis Wratislaviensis, vol. 197. Studia Linguistica I, pp. 73–95 (1974)

9. Ockham, W.: Summa logicae. In: Boehner, P. (ed.) Franciscan Institute, St. Bonaventure, NY. St. Bonaventure University, [ca 1323]) (Ockham's "Theory of Terms": Part I of the Summa Logicae, Loux, M. J. trans. & introd. Notre Dame, IN: University of Notre Dame Press (1951)

10. Ogden, C.K., Richards, I.A.: The Meaning of Meaning. A Study of the Influence of a Language Upon Thought and of the Science of Symbolism. Kegan Paul, Trench, Trubner & Co., London (1923)

11. Peirce, C.S.: Division of signs. In: Hartshorne, C., Weiss, P. (eds.) Collected Papers of Charles Sanders Peirce. Elements of Logic, 2.28, vol. 2. Harvard University Press, Cambridge (1932 [1897])

12. de Saussure, F.: Course in General Linguistics. W. Baskin (trans.). Philosophical Library, New York Cours de linguistique générale. Ch. Bally, A. Sechehaye (in coop. with) A. Riedlinger (eds.). Payot, Lausanne & Paris (1959 [1916])

13. Wąsik, Z.: The consequences of epistemological positions for the foundations of Linguistic sign theories. In: Kardela, H., Persson, G. (eds.) New Trends in Semantics and Lexicography. Proceedings of the International Conference at Kazimierz, December 13-15, 1995, pp. 209–217. Swedish Science Press, Umeå (1995)

14. Wąsik, Z.: An Outline for Lectures on the Epistemology of Semiotics. Wydawnictwo Uniwersytetu Opolskiego, Opole (1998)

15. Wąsik, Z.: Epistemological Perspectives on Linguistic Semiotics. Peter Lang, Frankfurt am Main, et al. (2003)

16. Wąsik, Z.: Between praxeological and axiological semiotics in the cultural sphere of intersubjective meanings. Chinese Semiotic Studies 2, 124–134 (2009)

Imagination in Thought Experimentation: Sketching a Cognitive Approach to Thought Experiments

Margherita Arcangeli

> To be told that thought experiments are a certain kind of imaginary fantasy might be informative, but [...] why should an *imaginary fantasy* of *any* sort alter our fundamental beliefs about reality?
>
> Michael Bishop [3, p. 26]

> The less simple exercises of imagination can exploit and inherit the contents of our beliefs and perceptions to such an extent that, within a sufficiently complicated game, discoveries are made possible.
>
> Kevin Mulligan [33, p. 61]

Abstract. We attribute the capability of imagination to the madman as to the scientist, to the novelist as to the metaphysician, and last but not least to ourselves. The same, apparently, holds for thought experimentation. Ernst Mach was the first to draw an explicit link between these two mental acts; moreover – in his perspective – imagination plays a pivotal role in thought experimentation. Nonetheless, it is not clear what kind of imagination emerges from Mach's writings. Indeed, heated debates among cognitive scientists and philosophers turn on the key distinction between sensory and cognitive imagination. Generally speaking, we can say that sensory imagination shares some processes with perception, cognitive imagination with the formation of belief. Both the vocabulary used in the literature on thought experiments and what I refer to as "Machian tradition" indicate imagination as a notion of central importance in the reasoning involved in thought experiments. However, most authors have really focused on *sensory* (in particular, visual) imagination, but have neglected the second kind. Moreover, some authors attribute to Mach the idea that it is *visual* imagery that is primarily at work in thought experiments. I claim another interpretation is possible, according to which Mach can be said to deal with cognitive imagination. The main aim of this paper is to retrace Mach's original arguments and establish

Margherita Arcangeli
Institut Jean Nicod, CNRS-EHESS-ENS, Paris, France
e-mail: `margheritarcangeli@gmail.com`

L. Magnani et al. (Eds.): Model-Based Reasoning in Science & Technology, SCI 314, pp. 571–587.
springerlink.com   © Springer-Verlag Berlin Heidelberg 2010

a connection with the cognitive literature on imagination. I will argue that imagination *tout court* could play a role in thought experimentation. Once imagination is seen as the key to the "cognitive black-box" of the thought experiment, we will have moved a step closer to a simulative imagining-based account of thought experimentation.

1   A Cognitive Approach to Thought Experimentation

Over the past twenty years the expression "thought experiment" (TE) has become part of the philosophical vocabulary. Although the term Gedanken-experiment was first introduced by Ernst Mach at the end of the XIXth century, appeal to thought experimentation has a much older history – in philosophy, as well as in natural and social sciences. Famous examples range from Plato's ring of Gyges [40], to Burge's arthritis [5], and include Locke's inverted spectrum [29], Galileo's TE on free-fall [13], Kant's on handedness [24], Einstein's lift [12], Heisenberg's γ-ray microscope [20], Putnam's brain in the vat [42], and so on.

However, what is a TE? Although there is no completely unanimous answer to this question, almost all authors involved in the debate on TEs agree that to perform a TE is "to reason about an imaginary scenario with the aim of confirming or disconfirming some hypotheses or theory" [15, p. 388].

This definition may motivate two approaches to thought experimentation. One might take an epistemic approach and regard TEs as sources of knowledge; but then, as noted by Kuhn [26], this raises a crucial question: how can a pure thought experiment yield new empirical knowledge, without the input of new data?

Alternatively, one might adopt a cognitive route, and thus recast the problem in terms of reasoning and imagination; which in turn leads to the question: what goes on in the head of someone performing a TE?

Most of the literature on TEs focuses on the former epistemic approach, and more precisely on the proper function of TEs. Differences aside, the literature is unanimous in indicating the function of TEs as that of producing knowledge to be exploited in a theoretical choice. But how does a TE fulfil its function? In the debate this is the most controversial and the least developed question[1].

---

[1] The debate over thought experimentation has focused on three key questions: what is a TE? What is its function? How does a TE fulfil its function? A variety of possible answers has emerged and on some points there is strong disagreement among philosophers: do TEs belong to the theoretical realm or to the experimental one? What kind of knowledge do they really produce (e.g. new/old, a priori/a posteriori)? To what extent are TEs a reliable source of information (i.e. the problem of scientific and philosophical TEs)? What role do TEs play in processes of rational theory choice? What kind of reasoning is involved in thought experimentation (e.g. hypothetical, counterfactual, deductive, inductive, imaginative, simulative model-based)? See [45] and [32] for a brief overview.

Nonetheless, closer inspection provides a lead: TEs seem to produce knowledge mediated by a reasoning process involving the use of imagination.

In talking about reasoning and imagination the subject comes into view: a TE is an object that presupposes a subject who carries it out in her head. Nevertheless, in the debate more attention has been devoted to the object TE, rather than to the subject who performs it. For instance, trying to answer the question about *how* a TE fulfils its function, many attempts have been made to describe the structure of a TE, without really taking into account what goes on in the head of the subject. We can say that TEs, viewed as an item to be disassembled, have been studied from an objectual point of view. On the contrary, in pursuing a cognitive approach one shifts the attention to the subject who performs the TE, thus taking a subjective perspective to thought experimentation.

Although thought experimentation can be approached from either an epistemic or a cognitive perspective, the debate has privileged the epistemic over the cognitive, and sometimes confused the two. Undoubtedly both approaches are dealing with the same cognitive process, but it seems more useful to differentiate these two aspects to have a clearer view of how they will interact. Hence it is useful to keep in mind that there is no competition between the two approaches; rather a cognitive approach may provide new guidance for the epistemic one. Indeed, even though it is beyond the scope of this paper to go into the details of the epistemic debate, it should be noted that on some points there is strong disagreement among philosophers. The contribution of a cognitive approach might smooth down these divergences, for example by shedding light on the typology of the knowledge gained, or by providing new criteria for distinguishing TEs from other types of reasoning. Moreover, in this respect a collaboration between the two approaches paves the way to a new frame for studying thought experimentation, which might give rise to rules for successful thought experimentation and guidelines for a taxonomy of TEs.

Opening the "cognitive black-box" of TEs would be useful not only *per se*, thus, I call for a cognitive approach to thought experimentation. Moreover, to open the box we need a key: my hypothesis is that imagination is such a key.

In the next section of the paper I will provide a framework for clarifying how imagination can be defined. Relying on the cognitive literature on imagination, I will define the notion of *recreative* imagination and its varieties (i.e. sensory and cognitive imagination).

In the third section I will come back to the debate on TEs, in search for a starting point for a cognitive approach. In this respect I claim that the literature offers two converging leads: the vocabulary, and Mach and his tradition. The upshot is that imagination emerges as a notion of central importance relative to TEs. Therefore, I will show how the cognitive literature on imagination may help to reveal that both the vocabulary and the Machian tradition have really focused on a sensory (in particular, visual) kind of imagination, whereas Mach himself would have gone beyond this narrow meaning of imagination.

I maintain that Mach was on the right track in allowing for imagination *tout court* to play a role in thought experimentation. Moreover, if we consider the simulative account of imagination provided by the cognitive literature, imagination itself could be seen not only as the key to the "cognitive black-box" of TEs, but also as one step closer to a simulative imagining-based account of thought experimentation, or at least as a useful starting point for future research about the kinds of reasoning involved in this activity.

2   The Garden of Imagination

"Imagination" is a complex notion, in which different meanings have become stratified. One only needs to think that we attribute the exercise of imagination to the madman as the scientist, to the novelist as the metaphysician, and not least to ourselves. Semantic history aside, imagination is defined by the cognitive literature as a *recreative* faculty [8]. The widespread view is that imagination re-creates, i.e. simulates, other kinds of mental states.

The cognitive literature defines different kinds of non-imaginative mental states; beliefs, desires, percepts and emotions are the most important among them. What mental states could the imagination recreate? Some authors [33, 17] claim that imagination is able to recreate a wide range of mental states. Nevertheless, while doubts have been raised about the existence of imaginative desires and emotions [35, 8], most authors concur in thinking that there are imaginative beliefs and imaginative percepts.

Hence, there are at least two kinds of imagination: cognitive (I_C) and sensory (I_S) imagination. I_C has belief as its counterpart mental state, in the sense that it produces mental states which simulate belief states (i.e. belief-like imaginings); whereas perception is the counterpart of I_S, which produces mental states which simulate perceptual states (i.e. percept-like imaginings). The I_C/I_S distinction has been variously labeled in the literature; however, it is not our purpose here to analyze the interplay of these alternative dimensions, sets and taxonomies. Let us simply note that there is a common trend that has I_S belonging to the non-propositional realm, and I_C – *stricto sensu* – to the propositional one.

An example may help clarify the distinction between the two kinds of imagination. I am walking in my grandmother's garden and I am looking at these beautiful birches which surround me. Suddenly I imagine a beech. This could mean that I simulate perceiving (seeing, touching, etc.) a beech, in other words I imagine$_S$ a beech. But it could also be that I simulate believing that there is a beech in the garden – I imagine$_C$ that there is a beech.

In the first case the content of my imagining is non-propositional (e.g. "I visually imagine *a beech*"), as would the content of its counterpart ("I visually perceive *a beech*"), were it occurrent. Analogously, in the second case the content of my imagining is propositional (e.g. "I imagine that *there is a beech*"), as would the content of the recreated mental state ("I believe that *there is a beech*")[2].

The dichotomy between the propositional and non-propositional in turn leads to the one between non-visual and visual. The distinction between I_C and I_S could also be drawn on the basis of pictoriality: in the I_S case, and not in the I_C one, I will form an image of a beech in my mind. Unsurprisingly, in the literature I_S is commonly tied to imagery, and I_C is linked with supposition. It is debatable whether these pairs of notions can in fact be legitimately conflated; for present purposes, however, this is an acceptable working hypothesis[3].

Considerations about the content of imaginings have lead Gregory Currie and Ian Ravenscroft to an important insight. They advocate the view that neither perceiving nor believing is part of the content of the imaginative state. If I imagine$_S$ a beech, it does not mean that I imagine *seeing a beech*, but I imagine seeing *a beech*. Similarly if I imagine$_C$ that there is a beech, I am not imagining *believing that there is a beech*, but I am imagining believing that *there is a beech*.

Imagination is the capacity to have "states that are not perceptions or beliefs [...], but which are in various ways like those states – like them in ways that enable the states possessed through imagination to mimic and to substitute" for perceptions or beliefs [8, p. 11]. Nowadays experimental data support the claim that (visual) I_S and (visual) perception share neural mechanisms, and plausibly the same holds for I_C and belief. Hence I_S and I_C preserve some features of their counterparts, and for that reason we can consider imagination as recreative[4].

However, two points ought to be kept in mind. Firstly, that the resemblance between the imagining and its counterpart is cognitive, not objectual [1]. In other words, imagination simulates mental states, more than the real events they have as their objects. This view does not purport to deny our capacity to mentally model physical processes; rather, it aims to relieve imagination of an obligation to depict faithfully reality, and calls for a more rigorous use

---

[2] The propositional/non-propositional dichotomy actually appears less than ideal for the purpose of carving I_C and I_S at the joints, indeed some objections might be raised. However, it is beyond the scope of this paper to go into the details of this issue. For further discussion see [33] and [10].

[3] Some further clarifications may be needed. Firstly, some authors have pointed out that the boundary line between the propositional and the visual is not so clear ([43]). Secondly, explanations in cognitive science appeal to many different kinds of mental representation (i.e. imagery). The question of their format has been discussed at length, and has received answers that go from pictorialist views to descriptivist ones. Thirdly, even though according to some authors supposition is nothing more than I_C [46, 36], others class it outside imagination [38, 14].

[4] The literature does not specify what exactly is preserved by the imagining, although I consider this to be an interesting open question. See [33] for further discussion.

of the term "imagination". So perhaps pictoriality is more closely related to I_S than to I_C, although it is not necessary to I_S. In my grandmother's garden, an image of a beech in my mind's eye might trigger or improve my percept-like imagining, in a way that my belief-like imagining would not. But the image of the beech is neither essential to my imaginative act nor is it an imaginative act itself – in Currie's and Ravenscroft's sense.

Secondly, recreation does not mean perfect reproduction. Imagination preserves some features of its counterpart, but this similarity is broken up by other characteristics proper to imagination itself[5]. Indeed, in imagination we are able "to project ourselves into another situation and to see, or think about, the world from another perspective" [8, p. 1].

3 Thought Experimentation and Imagination

3.1 The Vocabulary

A closer look at the expressions used in the literature to describe thought experimentation suggests that TEs should be counted as imaginary experiments (IEs). We encounter the expressions "imaginary illustration" [9], "expérience fictive" [11], "imaginary experiment" [41, 25, 37], fictitious/imaginary example [21], imagined experiment/observation [21, 28]. These phrases are actually ambiguous. Indeed, they could be interpreted to mean that a TE is just an experiment carried out in the mind, or an experiment that figures in the content of some imagining, or an act of imagining the performance of an experiment. Nevertheless, other expressions indicate that imaginative ability plays an important role in thought experimentation [3, 32]. Thus, we may at least say that TEs are experiments carried out in the mind, with the aid of the subject's imagination.

At the same time, given the previous considerations on the complexity of the notion of imagination, one is led to ask the more specific question – which kind of imagination is involved? Nowhere do the authors engaged in the debate over thought experimentation explicitly specify an answer to this question. Nevertheless, one can remark that the assimilation between TE and IE depends on the notion of observability, besides the fact that other expressions point to the perceptual realm.

James Brown [4] speaks of "seeing" the laws of nature and he claims that the pictorial and sensory aspect is essential to thought experimentation, and on that point Richard Arthur [2] agrees with him. Nersessian [34] stresses that when we perform a TE, we perceive ourselves as observers. Martin Cohen takes the second rule of good thought experimentation to be assumed that

---

[5] According to the literature, what characterizes imagination is the relationship with the will: I_S and I_C are will-dependent, whereas their counterparts are will-independent. It follows that the imaginative state has a higher degree of freedom than its counterpart (both I_S and I_C are belief and truth-independent). See [33].

the TE must be imaginable, i.e. "the clearer the picture, the stronger the image, the better the experiment" [6, p. 106]. David Gooding [18] claims that visualization is a necessary and sufficient condition for a TE. John Norton [37] admits that TEs involve visualization, but denies its epistemic role. Jeanne Peijnenburg and David Atkinson [39]) claim that a TE should give a sudden and exhilarating insight (i.e. a seeing in, like the Latin term "intuition", derived from the compound verb *intueri*). Thus, thought experimentation seems to involve a sensory kind of imagination – specifically, visual.

Moreover, if we take into account the expressions referring to pictoriality, it seems that in thought experimentation imagination can recreate not only mental states, but also states of affairs. For instance, in re-performing the TE of Galilei one would mentally re-produce the tower of Pisa, the two weights, the rope that binds them, etc.

According to Currie and Ravenscroft neither I_C nor I_S would be involved in thought experimentation, since the role of the latter is just to mirror the situation described in the experiment. When speaking of the mental process underlying TEs, we should refer to it by a different terminology (e.g. "icastic emulation"), and avoid the term "imagination" - or at least qualify it by an adjective (e.g. "*icastic* imagination"), in order to differentiate it from *recreative* imagination. I agree with Currie and Ravenscroft that discharging imagination of the obligation to resemble faithfully reality seems to take us closer to understanding TEs. Still, we are in my opinion unwarranted to conclude that recreative imagination has no role to play in thought experimentation[6].

Regarding the assimilation between TE and IE, although to consider TEs imaginary experiments is not inaccurate, I maintain that it might be misleading: TEs are complex processes involving imagination and not mere "images of experiment". If we do not keep this in mind, we run the risk not only of making thought experimentation a "handmaiden" of the real one, but also of putting the two kinds of experimentation in competition with each other. Unfortunately, this is what has mostly happened in the debate over thought experimentation, and it is not a coincidence that the adjective "imaginary" has often been used to give the noun "experiment" a negative connotation.

As Roy Sorensen has pointed out, "imaginary" is "a fairly clear case of a 'falsidical' adjective" [44, pp. 218-219]. Although the imaginary unit and imaginary numbers are "real entities" for mathematicians, as is the Imaginary for sociologists and psychologists, by contrast imaginary friends, imaginary worlds, imaginary fears and beliefs are commonly understood to be fictitious entities. The emphasis is put on the negative aspects, on what they lack in

---

[6] We might actually use the term "recreative imagination" in both cases, but in two different senses. One can recreate a mental state, by putting oneself in a simulated state, but without representing it. Alternatively, one recreates states of affairs when one represents them in the context of an imaginary scenario; in this case, we don't need the concept of a mental state but only that of a state of affairs, to recreate it in the imagination.

order to be *true* friends, worlds, fears and beliefs. The same holds for TEs when they are called "expériences fictives", or "fictitious/imaginary examples". To be more specific, the vocabulary conveys the idea that a TE should be regarded as an IE, inasmuch as it is not an experiment, but rather a visualization of an experiment "in the mind's eye".

3.2 Mach and His Tradition

3.2.1 The Machian Tradition

The issue of the importance of I_S and pictoriality arises also on closer examination of what I call the Machian tradition. I refer by this expression to those authors who have tried to pursue Mach's approach in analyzing TEs on the basis of a psychological theory. To the Machian tradition, thus defined, belong therefore both Tamar Szabó Gendler and the model-based approach – developed by Nenad Miščević, Nersessian and Rachel Cooper.

Gendler is among those philosophers who have integrated the empirical psychological data in their methodological analysis of TEs. Gendler [16] establishes a link between the research of cognitive scientists and philosophers such as Kosslyn, Damasio, Reisberg, Shepard and Cooper, and the analysis of Stevin's TE on the inclined plan, and finds that in some TEs, imagery plays a key epistemic role. In such cases, similar to Stevin's TE, knowledge is achieved not by inductive or deductive inferences, but in a "quasi-observational" way.

Stevin was dealing with the force needed to keep a weight on an inclined plane from sliding down, and he concluded that the force required is inversely proportional to the length of the plane. In order to explain his result he proposed a thought experiment, which involves imagining the manipulation of a chain of fourteen balls, draped over a triangular prism with frictionless sides of unequal slope (see Fig. 1).

Following Gendler, certain quasi-sensory intuitions are evoked by contemplating this imaginary set-up. On the basis of these intuitions, Stevin formed a new belief about the natural world. Analogously, for us the presence of a mental image plays a pivotal cognitive role for grasping Stevin's law of the inclined plane (i.e. the force to the weight is equal to the ratio of the height to the length of the plane). "This is not to say that *all* scientific experiment involve such imagistic reasoning [...]. There will, no doubt, be many cases where the role of the imagery is simply heuristic. But there will also be cases where the role of the imagery is [...] epistemically crucial" [16, p. 1161].

What about the kind of imagination involved in thought experimentation from Gendler's point of view? Although she deals with imagery, as we have seen, this term is commonly intrinsically tied to I_S. Indeed, the term imagery is used by Gendler in connection with words which refer to a sensory kind of imagination. Gendler thus seems to take I_S to be the kind of imagination playing the key epistemic role in thought experimentation.

Fig. 1 Stevin's chain.

Nevertheless, Gendler does not speak about imagination in terms of mental simulation, and instead points to pictoriality[7]. What is epistemically crucial is the imaginary set-up, that is, the recreation of states of affairs and not of mental states. Hence she seems to consider "icastic imagination", more than "recreative imagination", to be the epistemically relevant imagination in (some scientific) TEs.

A question arises: is it really necessary to mentally represent the imaginary scenario of Stevin's TE (i.e. the prism with the balls) in order to perform his TE successfully? The most plausible supposition is that it is a matter of choice. Some TEers might perform better than others by using the prism with the balls as a (quasi) picture in the mind's eye. Others, on the other hand, might need external support, such as pen and paper to draw the image. It is not by chance that pictures invariably form an integral part of the standard narrative for Stevin's TE.

The image of the prism with the chain, represented by way of a mental image or a physical picture, seems to be an aid to trigger the subject's projection into an imaginary scenario, where she should simulate to perceive or believe that the plane is frictionless, that the rope is totally flexible, that since a perpetual motion is absurd the weight of four balls offsets the weight of two balls, that a force is required in order to keep an object from sliding down on an inclined plane, which varies inversely with the length of the plane.

I do not want to deny icastic imagination a role in thought experimentation. However, I claim that recreative imagination can also be epistemically

---

[7] Abell and Currie have formulated a simulative account of imagery. They remarked that: "the mistake is to take pictures as the basic category and then to try to assimilate mental imagery to that [...] there is a category more basic than either: mental simulation" [1, p. 431].

fruitful and may even be more important than icastic imagination. Moreover, both I_S and I_C seem to have a role to play.

The model-based approach also belongs to the Machian tradition. It calls on the literature of model-based reasoning in cognitive science to explain how we gain new knowledge in TEs. Among the authors advocating this approach (e.g. Gooding, Bishop), three are to be considered as its developers, namely Miščević, Nersessian and Cooper. However, these authors have advanced different theses. They agree with one another on maintaining that in thought experimentation we gain knowledge through manipulating a model, but they rely on different notions of model.

Contrary to Cooper, Nersessian and Miščević appeal to the cognitive literature concerning "mental modeling", and more specifically to the notion of mental model proposed by Johnson-Laird. A mental model is a structure stored in short or long term memory and it is defined by cognitive scientists as a third type of mental representation, half way between propositional and pictorial ones[8]. Indeed, mental models are structurally analogous to that which they represent, but not all such models can be visualized.

What is the link between mental models and imagination? Imagination too builds mental models, much like perception or linguistic comprehension. And mental models can capture spatial configurations not only of the real world, but also of imaginary worlds [22]. Among philosophers of mind dealing with fiction, the role played by recreative imagination is well-known, although establishing which kind (i.e. I_S or I_C) is the main one is more problematic. "A typical narrative text or film is a *prop* that induces one to adopt the factual-reader or factual-observer perspective" [17, p. 287]. Hence it is not by chance that mental models are involved in understanding fictional stories, as well as real ones. Building a mental model of the fictional world may aid our reasoning within this world, i.e. to project ourselves into the fictional situation and to simulate perceiving or thinking from this perspective.

TEs are presented in narrative form, so it is plausible that the reasoning underlying thought experimentation is closely related to the one employed in fiction. Both Miščević and Nersessian seem aware of this. Even though they tend to emphasize the role of mental modeling, rather than of imagination, references to imagination are included in their analysis.

In Miščević's account we can retrieve the focus on pictoriality. On his account, imagination emerges as synonymous of "picturing in front of one's inner eye" [31, p. 217]. A TEer uses imagination to imagine things and to rearrange these imagined objects, more than to simulate mental states. In addition Miščević claims that the mental model is a "quasi-spatial picture" and has a "concrete and quasi-spatial character" [31, p. 220]. This analysis,

---

[8] In his famous book on mental models, Johnson-Laird argues on the one hand that the semantics of the mental language maps propositional representations onto mental models of real or imaginary worlds; on the other hand that mental images are to be considered as views of mental models [22, p. 156].

like Gendler's, focuses thus on icastic imagination, rather than on recreative imagination.

Nersessian argues that the mental model manipulated in thought experimentation is neither a picture in the head nor a linguistic representation; following Johnson-Laird, she holds it is rather a structural analogue of the situation depicted in the TEal narrative [34, p. 297]. Nevertheless, she highlights the role of non-propositional (i.e. visualizable) representations much more than Miščević: the reasoning proper to TEs is entirely rather than partially non-propositional. For that reason, Nersessian maintains that deductive and inductive inferences do not have a part in thought experimentation. Moreover, even though she does not explicitly refer to recreative imagination, we can say that she points to I_S as the pivotal kind of imagination that is engaged in thought experimentation.

Cooper disagrees with both Miščević and Nersessian. She argues on the one hand that "both accounts are based on contestable empirical data", and on the other hand that "whether the thought experimenter reasons through the situation via manipulating a set of propositions, or a mental picture, or even plasticine characters makes no difference" [7, p. 341][9]. For Cooper, a TEer manipulates a model, but not necessarily a mental one, and she can carry out deductive or inductive inferences, as well as diagrammatical ones. Imaginings play a role in Cooper's account of TEs, but again imagination is assimilated to visualization (i.e. pictoriality): performing a TE is to imagine "the situation unfolding in our mind's eye" [7, p. 332].

I agree with Cooper that pictoriality is not necessary to thought experimentation. As I have said, icastic imagination may aid to trigger or to improve recreative imagination; still, it is open to a TEer to use either diagrams or concrete objects in order to obtain the same imaginative projection. Recreative imagination appears essential to thought experimentation, *as much* in its propositional (I_C), as its non-propositional form (I_S).

In my opinion, Cooper is also right in underlining that the hypothesis of mental models is a moot one. Despite not being unanimously accepted by cognitive scientists, it has nevertheless given rise to a promising research program [23][10]. I would thus not be inclined to deny that mental models have a specific role in thought experimentation; yet it remains problematic to identify those that do and specify how they are linked to imagination. What I do want to suggest is that imagination has a primary function in thought experimentation.

An important lesson to be drawn from the Machian tradition is that TEs are species of simulative-based reasoning. This idea is implicit in Gendler's analysis, but only in the model-based approach is it made fully explicit and

---

[9] Another point of disagreement with Nersessian and Miščević is that for them "models are restricted to simulating the way in which phenomena would unfold in the real world". Cooper replies that "the thought experimenter may model a world in which some laws of nature are suspended or altered" [7, p. 341].

[10] On the topic, see also Paul Thagard's contribution in these proceedings.

linked to the notion of (mental) model. Nevertheless, if we consider the simulative account of imagination provided by the cognitive literature, imagination itself could be seen as a step closer to a simulative account of thought experimentation.

To sum up, our analyzes of the vocabulary and the Machian tradition lead us to conclude that pictorial or sensory imagination is the main kind of imagination at work in thought experimentation. We now take a closer look at what Mach himself had to say on the matter.

3.2.2 Ernst Mach

Mach was not only a physicist: he also had a great passion for philosophy. Despite refusing to identify himself as a philosopher, he works on a topic in which philosophy and science seemed to meet: the *Gedankenexperiment*. Moreover, Mach was the first to link thought experimentation to a psychological theory[11].

In spite of his empiricism, he saw TEs as an important scientific tool with which ideas, data and theoretical hypotheses can be put to work. TEs are peculiar epistemic devices: they are grounded in previous experiences, which are accumulated by our innate propensity to experiment. More precisely, Mach introduced the term *Gedankenexperiment* in order to define a technique of scientific discovery, which bridges the gap between the domain of ideas and that of real experiments.

On Mach's view the process of scientific discovery can be characterized as our taking notice of what is overlooked in perception, yet gets fortunately stored in our memory, and then recovered by imagination. Hence we can say that according to Mach imagination plays a pivotal role in thought experimentation. Indeed he argued that performing a TE is to "combine circumstances" in imagination [30, p. 452][12]. So what kind of imagination emerges from Mach's writings?

According to some authors, such as Sorensen and Gendler, visual imagery already appears as an essential feature of TEs in Mach's analysis. Indeed, Mach seems to conceive of imagination as visualization; furthermore, on his view the power of imagination rests on its ability to resemble reality. Indeed,

---

[11] Nersessian's words summarize the Machian point of view perfectly: "while thought experimenting is a truly creative part of scientific practice, the basic ability to construct and execute a thought experiment is not exceptional. The practice is highly refined extension of a common form of reasoning [...] by which we grasp alternatives, make predictions, and draw conclusions about potential real-world situations" [34, p. 292]. Moreover, she remarked that the naturalistic approach provided by Mach is still up-to-date, even though the sensationalist psychology and the biological theories on which he relied on are now outmoded.

[12] Quotations from Mach 1896 refer to the 1973 English translation.

he wrote: "the possibility of thought experiments rests upon our ideas as being the more or less exact copy of the facts" (*ibid.*)[13].

Other authors, such as Kevin Mulligan, would argue instead that Mach is concerned primarily with cognitive imagination, insofar as he considers the variation of circumstances and suppositions to be essential to the imaginative processes in thought experimentation. Analogously, the cognitive literature on imagination seems to point to supposition, that is to I_C, as the basic kind of imagination involved in thought experimentation. For instance, Mulligan remarks that very often TEs begin with a "suppose that" and he claims that "all philosophers [...] who introduce thought experiments formulate suppositions" [33, p. 55]. Kendall Walton distinguishes TEs from fiction on the basis of the kind of imagination engaged. He maintains that thought experimentation calls for supposition, whereas fiction for "a more substantial sense of imagining" [47, pp. 43-44]. But even in the literature on thought experiment Sorensen, breaking away from the mainstream, has pointed out that it "is one thing to imagine physical things being manipulated and another to vary what one supposes" [44, p. 221]. In the first case we are doing an IE, in the second one a TE[14].

Nevertheless, a closer analysis of the expressions used by Mach shows that he did not look for a particular kind of imagination. In his writing on TEs it is possible to uncover a tension between an imagination conceived only in "icastic" terms, and a liberation of imagination from a duty to resemble faithfully reality. This kind of more "liberal" imagination, which could represent without resembling, can also lead to conjectures of an idealized form. Clearly, Mach did not distinguish between "icastic" and "recreative" imagination, nor between cognitive and sensory imagination, but it is likely that he would have considered suppositional elements as constitutive of an imaginative project. Thus, I claim that Mach was on the right track in allowing for "recreative" imagination, in both its forms (I_S and I_C), to play a role in thought experimentation.

The Machian perspective seems to be promising for at least two reasons. Firstly, it allows to clarify the reason why a TE should be considered as an

---

[13] However, this is not the case for TEers such as "the dreamer, the builder of castles in the air, the poet of social or technological utopias". They "combine circumstances in their imagination not encountered in reality, or deem these circumstances to be accompanied by results not bound to reality" [30, p. 451]. Nevertheless, following an insight provided by Sorensen, we could say that Mach deals with 'familiarity' rather than with 'resemblance' [44, p. 65]. It should be noted that Sorensen, contrary to Gendler, thinks that imagery is never essential to thought experimentation [44, p. 68].

[14] According to Sorensen we are prone to consider TEs as IEs (i.e. visualization as a necessary condition to thought experimentation), because sometimes a TE contains an imaginary experiment, which is only one essential module of a TE, but not the TE itself. I consider this to be an interesting hypothesis, which however is beyond the scope of this paper.

experiment, rather than a mere visualization of an experiment "in the mind's eye". As we saw in the analysis of Stevin's TE, it is not necessary for TEers to have internal pictures of the situation depicted in the TEal narrative. In order to grasp Stevin's intuition about the force that is needed to keep a weight from running down on an inclined plane, a TEer should consider herself as an observer, engaged in a "bodily feel" [19, p. 305] that puts her in the position to "perceive" and "believe" – just as in a real experiment. Recreative imagination seems to be the cognitive mechanism that is conducive to this kind of experience. Hence, the Machian perspective sheds light on the cognitive mechanism underlying thought experimentation.

A more thorny question is whether adopting a simulative imagining-based account of thought experimentation might help clarify why TEs can be regarded as sources of knowledge; that is, how a TEer switches from percept-like and belief-like mental states to knowledge. The Machian perspective lacks an explicit answer to this question, but paves the way for it. Indeed, one could even say that for Mach TEs ought to be seen as an example of how imagination can recreate complex mental states, with both experiential and non-experiential aspects, which are conducive to knowledge. Only by regarding thought experimentation to encompass other kinds of imagination besides the visual, can we see TEs as examples of those complex activities which make imagination "a guide to *knowability*" [10, p. 117].

To sum up, although Mach's work on TEs represents an important source for all the subsequent debate, few scholars have developed his psychological perspective on TEs. In fact, Mach's analysis of TEs contains *in nuce* a first sketch of a cognitive approach to thought experimentation, in which imagination in *all* its forms has a pivotal role.

4   All the Imagination in Thought Experimentation

Thought experimentation is a widespread activity in different domains; we have at our disposal two approaches to its analysis. Focusing one's attention to the object TE as an epistemic tool will lead to take an epistemic perspective. If instead one tries to uncover the cognitive processes underlying the performance of TEs, one will pursue a cognitive approach. The latter is less developed in the literature on thought experimentation, but I have argued that imagination emerges as a good starting point.

On a naïve approach, imagination is tied to mental model physical processes but, as we have seen, imagination is defined by the cognitive literature as a recreative faculty dealing with cognitive rather than objectual resemblance. The mental states acquired via imaginative activity simulate other mental states, such as percepts (i.e. sensory imagination) and beliefs (i.e. cognitive imagination), more than the real events that they have as their objects.

The vocabulary used in the debate over thought experimentation points to a visual kind of imagination as the main one involved in TEs. Moreover, great importance is attached to pictoriality, that is the recreation of physical processes; however, this leads to consider TEs as a mere visualization "in the mind's eye" of an experiment.

The Machian tradition also privileges a sensory – specifically, visual – kind of imagination. However, TEs have emerged, from the model-based approach in particular, as a species of simulative-based reasoning, which does not necessarily involve pictoriality. Indeed, the upshot of our analysis of Mach was that imagination is no longer constrained to depict faithfully reality; corollaries to this are that sensory imagination can be seen in terms that are not only pictorial, and the rehabilitation of cognitive imagination. This is in line with the hypothesis suggested by some cognitive philosophers, who have pointed out that cognitive imagination is also engaged in thought experimentation.

Once a cognitive approach to TEs is developed on the basis of the notion of imagination as defined by the cognitive literature, we shall have moved closer to a simulative imagining-based account of thought experimentation. TEs are not simple "images of experiment", but complex processes involving imagination *tout court.*

Acknowledgements. For comments and discussions, I am grateful to Jérôme Dokic, Daniela Tagliafico, Valeria Giardino, Giulia Terzian and Alessandro Arcangeli. I would also like to thank the organizers, the participants of the conference MBR'09 and the anonymous referees for helpful comments and suggestions.

References

1. Abell, C., Currie, G.: Internal and external pictures. Philosophical Psychology 12(4), 429–445 (1999)
2. Arthur, R.: On thought experiments as a priori science. International Studies in the Philosophy of Science 13(3), 215–229 (1999)
3. Bishop, M.: An epistemological role for thought experiments. In: Shanks, N. (ed.) Idealization in Contemporary Physics, Poznan Studies in the Philosophy of the Sciences and the Humanities, Rodopi, Amsterdam/Atlanta, GA, vol. 63, pp. 19–33 (1998)
4. Brown, J.R.: The Laboratory of the Mind: Thought Experiments in the Natural Sciences. Routledge, London (1991)
5. Burge, T.: Individualism and the mental. Midwest Studies in Philosophy 4, 73–121 (1979)
6. Cohen, M.: Wittgenstein's Beetle and Other Classic Thought Experiments. Blackwell, Oxford (2005)
7. Cooper, R.: Thought experiments. Metaphilosophy 36(3), 328–347 (2005)
8. Currie, G., Ravenscroft, I.: Recreative Minds. Clarendon Press, Oxford (2002)
9. Darwin, C.: The Origin of Species by Means of Natural Selection, or the Preservation of Favoured Races in the Struggle for Life. John Murray, London (1859), http://darwin-online.org.uk/

10. Dokic, J.: Epistemic perspectives on imagination. Revue internationale de philosophie 243, 99–118 (2008)
11. Duhem, P.: La Théorie physique: son objet, sa structure. Vrin, Paris (1914)
12. Einstein, A., Infeld, L.: The Evolution of Physics. The Growth of Ideas from Early Concepts to Relativity and Quanta. Simon & Schuster, New York (1938)
13. Galilei, G.: Discorsi e dimostrazioni matematiche intorno a due nuove scienze. Louis Elsevier, Leida (1638)
14. Gendler Szabó, T.: The puzzle of imaginative resistance. Journal of Philosophy 97(2), 55–81 (2000)
15. Gendler Szabó, T.: Thought Experiment, pp. 388–394. Encyclopedia of Cognitive Science. Nature/Routledge, New York/London (2002)
16. Gendler Szabó, T.: Thought experiments rethought – and reperceived. Philosophy of Science 71, 1152–1164 (2004)
17. Goldman, A.: Simulating Minds: The Philosophy, Psychology, and Neuroscience of Mindreading. Oxford University Press, New York (2006)
18. Gooding, D.: What is experimental about thought experiments? In: Hull, D., Forbes, M., Okruhlik, K. (eds.) PSA 1992, vol. 2, pp. 280–290. Philosophy of Science Association, East Lansing (1993)
19. Hacking, I.: Do thought experiments have a life of their own? Comments on James Brown, Nancy Nersessian and David Gooding. In: Hull, D., Forbes, M., Okruhlik, K. (eds.) PSA 1992, vol. 2, pp. 302–308. Philosophy of Science Association, East Lansing (1993)
20. Heisenberg, W.: Physikalische Prinzipien der Quantentheorie. Hirzel, Leipzig (1930)
21. Hull, D.: A function for actual examples in philosophy of science. In: Ruse, M. (ed.) What the Philosophy of Biology Is: Essays Dedicated to David Hull, pp. 309–321. Kluwer, Dordrecht (1989)
22. Johnson-Laird, P.N.: Mental Models: Toward a Cognitive Science of Language, Inference and Consciousness. Harvard University Press, Cambridge (1983)
23. Johnson-Laird, P.N.: The history of mental models. In: Manktelow, K., Chung, M. (eds.) Psychology of Reasoning: Theoretical and Historical Perspectives, pp. 179–212. Psychology Press, New York (2004)
24. Kant, I.: Von dem ersten Grunde des Unterschiedes der Gegenden im Raume. In: Buchenau, A. (ed.) Vorkritische Schriften, 1912th edn., vol. II, pp. 375–383. Bruno Cassirer, Berlin (1768)
25. Koyré, A.: Galileo's Treatise De Motu Gravium: The Use and the Abuse of Imaginary Experiment. Revue d'Histoire des Sciences 13, 197–245 (1960)
26. Kuhn, T.: A function for thought experiments. In: L'aventure de la science, Mélanges Alexandre Koyré, Hermann, Paris, vol. 2, pp. 307–343 (1964); Reprinted in Kuhn [27]
27. Kuhn, T.: The Essential Tension. University of Chicago, Chicago (1977)
28. Lennox, J.G.: Darwinian thought experiments: A function for just-so stories. In: Horowitz, T., Massey, G. (eds.) Thought Experiments in Science and Philosophy, pp. 223–245. Rowman and Littlefield (1991)
29. Locke, J.: An Essay Concerning Human Understanding. Thomas Dring e Samuel Manship, London (1690/1694)
30. Mach, E.: Über Gedankenexperimente. Zeitschrift für den physikalischen und chemischen Unterricht 10, 1–5 (1896); Translated by Price Price, W.O., Krimsky, S.: On thought experiments. Philosophical Forum 4/3, 446–457 (1973)

31. Miščević, N.: Mental models and tought experiments. International Studies in the Philosophy of Science 6, 215–226 (1992)
32. Moue, A.S., Masavetas, K.A., Karayianni, H.: Tracing the development of thought experiments in the philosophy of natural sciences. Journal for General Philosophy of Science 37, 61–75 (2006)
33. Mulligan, K.: La varietà e l'unità dell'immaginazione. Rivista di estetica 11(2), 53–67 (1999)
34. Nersessian, N.J.: In the theoretician's laboratory: Thought experimenting as Mental Modelling. In: Hull, D., Forbes, M., Okruhlik, K. (eds.) PSA 1992, vol. 2, pp. 291–301. Philosophy of Science Association, East Lansing (1993)
35. Nichols, S.: Review: Recreative Minds. Mind 113, 450 (2004)
36. Nichols, S., Stich, S.: Mindreading. Oxford University Press, Oxford (2003)
37. Norton, J.: Why thought experiments do not transcend empiricism. In: Hitchcock, C. (ed.) Contemporary Debates in the Philosophy of Science, pp. 44–66. Blackwell, Oxford (2004)
38. Peacocke, C.: Imagination, experience and possibility: A berkeleian view defended. In: Foster, J., Robinson, H. (eds.) Essays on Berkeley, pp. 19–35. Clarendon Press, Oxford (1985)
39. Peijnenburg, J., Atkinson, D.: When are thought experiments poor ones? Journal for General Philosophy of Science 34, 305–322 (2003)
40. Plato: The Republic. Cosimo, New York (2008) Translated by Benjamin Jowett
41. Popper, K.: On the use and misuse of imaginary experiments, especially in quantum theory. In: The Logic of Scientific Discovery, Hutchinson, London, pp. 442–456 (1959)
42. Putnam, H.: Brains in a vat. In: DeRose, K., Warfield, T. (eds.) Skepticism: A Contemporary Reader, pp. 27–42. Oxford University Press, Oxford (1992)
43. Shimojima, A.: The graphic-linguistic distinction. Artificial Intelligence Review 15, 5–27 (2001)
44. Sorensen, R.: Thought Experiments. Oxford University Press, Oxford (1992)
45. Stöltzner, M.: The dynamics of thought experiments – comment to Atkinson. In: Galavotti, M. (ed.) Observation and Experiment in the Natural and Social Sciences, pp. 243–258. Kluwer, Dordrecht (2003)
46. Vendler, Z.: The Matter of Minds. Clarendon Press, Oxford (1984)
47. Walton, K.: Morals in fiction and fictional morality. In: Marvelous Images: On Values and the Arts, pp. 27–45. Oxford University Press, Oxford (2008)

Representations of Contemporaneous Events of a Story for Novice Readers

Barbara Arfé, Tania Di Mascio, and Rosella Gennari

Abstract. We are working on a story comprehension tool for novice readers, among whom are 6–8 olds in Italy. The tool also asks them to reason on the temporal dimension of stories. In the design of the tool, we stumbled on the following question: how we can render qualitative temporal relations of a story with a visual representation that is conceptually adequate to novice readers. The question triggered the trans-disciplinary work reported on in this paper, written by a cognitive psychologist, an engineer and a logician. The work primarily consists in an experimental study with 6–8 old novice readers, first and second graders of an Italian primary school. We read them a story, and then asked them to visually represent certain contemporaneous relations of the story. The results of the experiment shed light on the variety of strategies that such children employ. The results also triggered two novel experimental studies that are reported on in the conclusion to this paper.

1 Introduction

Temporal reasoning is one of the many cognitive skills that children must develop in order to integrate well in our society [17, p. 3]: "of the many cognitive skills which children must master in order to become proficient members of their cultures, the acquisition of commonsense time concepts is among the

Barbara Arfé
UniPD, via Venezia 8, 35131 Padova
e-mail: `barbara.arfe@unipd.it`

Tania Di Mascio
UnivAQ, Monteluco di Roio, 67100 l'Aquila
e-mail: `tania.dimascio@univaq.it`

Rosella Gennari
FUB, Piazza Domenicani 3, 37100 Bolzano
e-mail: `gennari@inf.unibz.it`

L. Magnani et al. (Eds.): Model-Based Reasoning in Science & Technology, SCI 314, pp. 589–605.
springerlink.com   © Springer-Verlag Berlin Heidelberg 2010

most essential". The study of [4] supports the relevance of temporal features of texts as viable cues for facilitating the coherent interpretation of the texts.

Reasoning coherently with time concepts is acquired indirectly through narration, and evolves with age and experience. Language specific factors also affect the age at which temporal connectives are comprehended and mastered. In general, after the age of 5, normally developing children become able to make deductions with temporal relations, reasoning on sequences of events with "before" and "after" [13]. This ability seems to develop further from the age of 7 to that of 9, when children seem to be able to master the "while" temporal connective, e.g., see [5].

We are working on an e-story comprehension web tool. The tool's users include Italian primary-school children, who are novice readers [14]. It originates from LODE, a logic-based web system for the literacy of deaf readers, e.g., see [6] and [12]. The tool invites readers to reason on the temporal dimension of an e-story by using qualitative temporal relations, between pairs of events of the e-story, that can be expressed with "before", "while" or "after". An example is "Mammy hen is worried about her little Gino, while Gino is telling stories to the wolf".

Based on [15], numerous studies already showed significant comprehension gains when people can visualize while reading, as reported in [9]. Our web tool aims to be visual as for: (1) the interface; (2) the story's main events; (3) the e-story's qualitative temporal relations.

A non-trivial challenge comes forward in the creation of such a visual tool, namely, how we can render qualitative temporal relations of a story with a visual representation that is conceptually adequate to primary-school children, where the term "conceptual adequacy" is used in the sense of [10].

We took over the challenge and conducted an exploratory evaluation with fifty-six 6–8 olds in order to assess how they would visually represent "while" relations between events of a story.

This paper reports on the results of our experimental work with novice readers, and paves the way for the development of our visual tool. More precisely, Section 2 overviews related work on the visualization of qualitative temporal relations, mainly in the AI and HCI literatures. The overview lays the groundwork for Section 3, which gives the rationale and goals of our experimental work.

Then the paper delves into the details of the experiment. As we expect that this paper can appeal to a heterogenous class of readers, ranging from logicians working in the field of knowledge representation to cognitive psychologists and educators, we try to be as detailed as possible in the description of the experiment, without (we hope) sacrificing readability. So, Section 4 explains the experiment modus operandi. Section 5 is concerned with the user analysis that the experimenters conducted. Section 6 explains the experiment design. Section 7 details the user teaching. Section 8 reports on the experiment execution.

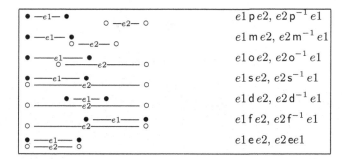

Fig. 1 The atomic Allen relations.

After the details of the experiment are so explained, we bite the crucial bit and analyze the results of the experiment in Section 9. We discuss about our results in Section 10. Section 11 reports on ongoing-work with two categories of progressively more skilled text comprehenders: ten old children (10-15 olds), and ten adults (16-30 olds). The goal of this experimental work is to start investigating whether the old children and adults would produce other visual representations. Section 12 concludes this paper.

2 Related Work

Stories' qualitative temporal relations can be expressed in terms of the so-called *atomic relations of Allen*. Figure 1 gives their standard representation over the real line. For instance, consider the following story excerpt:

Mammy hen is worried about her little Gino, while Gino is telling stories to the wolf.

In terms of the Allen relations, the above story excerpt states that the relation d is feasible between the event "Mammy hen is worried about her little Gino" and the event "Gino is telling stories to the wolf".

According to the experimental studies of [10], the visual representation of the atomic Allen relations in Figure 1 seems to be cognitively adequate for adults. However the visualization of disjunctions of atomic Allen relations is challenging. As for this, [8] has an interesting proposal, capable of expressing certain disjunctions such as "before or meets", see Figure 2. More recently, [2] proposed three alternative visual metaphors, see Figure 3. Their metaphors are based on concrete objects and phenomena from the physical word: elastic bands, springs and paint strips.

Notice that all such visual representations of qualitative temporal relations are diagrammatic and meant for adults. Moreover, they are all based on linear spatial orderings.

Fig. 2 The visualization of [8].

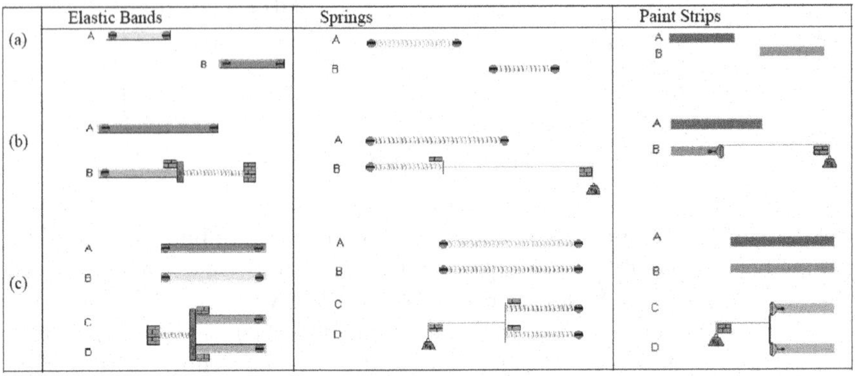

	Elastic Bands	Springs	Paint Strips
(a)			
(b)			
(c)			

Fig. 3 Visualisations of [2].

Such a situation naturally triggers the following challenge: would novice readers spontaneously produce similar visual representations of the temporal information contained in a story, in general, and of qualitative temporal relations, in particular?

To the best of our knowledge, there are very few studies that assess such or similar questions with novice readers. For instance, the experimental study of [11] evaluates the preschool children's ability in mapping a sequence of three temporal events onto spatial relations on a panel. The authors tested whether the children would spontaneously produce and comprehend a spatial linear ordering. According to their results, children's "mapping appears to be influenced by cultural conventions (such as the left-to-right direction of reading and writing), but space is used to represent non-spatial relations spontaneously before these cultural conventions are learned" (ibidem, p. 394).

However, the spontaneous visual representations of contemporaneous temporal relations, like "while", seems by far less explored than that of sequential temporal relations, like "before" or "after". Our evaluation does not consider sequential temporal relations, which are more studied, and instead concentrates on "while" sentences of a story.

3   Rationale and Goals of the Evaluation

We asked 6–8 olds of an Italian primary school to represent "while" relations of an entire story by drawing, and by arranging the illustrations of the correlated events on an A4 paper.

We opted for a story in our evaluation instead of, say, brief isolated sentences like in [11]. First of all, a multiple-sense connective is particularly sensitive to the context. This is the case of "while" in English as well Italian. For instance, it can correspond to "during" as well as to "during or finishes". Secondly, stories give children a meaningful context, and in general children are more successful at interpreting texts where the context reinforces the interpretation of the temporal connective, as reported in [18].

We chose the first short story, of circa 250 words, from the picture-book [7] for children, older than 5. The story characters and settings are well defined and clearly depicted, there are "while" relations between significant temporal events, and the language is suitable for 6–8 olds alike.

We also decided to read the story, instead of letting the children read the story on their own. The decoding skills of some of the children involved in the evaluation might still have been rather immature, and this might have affected their story comprehension. Story telling, instead, is an experience familiar to all, from their preschool year.

The evaluation that we report on in this paper aims at assessing the following questions:

(G1) the type and quality of their visualization strategies, e.g., the dimension of events in drawing;
(G2) if class, grammar comprehension, and working memory capacity affect their visualization strategies.

In the following, we first give the necessary details about the experiment, and then report on our analyzes.

4 Method

The experimental evaluation we conducted is based on a classical HCI user based schema, e.g., see [3]. It consists of the following steps.

- User Analysis, that is, the description of the involved users and definition of users' categories.
- Experiment Design, that is, the definition and description of study models, metrics and tasks.
- User Teaching, that is, exhaustive explanation for people involved in the evaluation about the modus operandi in the experiment sessions.
- Experiment Execution, that is, the description of experiment sessions, a.k.a., experiment diary.
- Result Analysis, that is, the collection of data and description of significant results.

In the following, we describe the experimental evaluations, using the aforementioned steps.

5 User Analysis

Our experiment participants were fifty-six children at end of their school year. They were subdivided in classes as reported in Table 1.

Table 1 Class composition.

Class acronym	Class description
(1a)	First-year class, with seventeen 6–7 olds
(2a)	Second-year class, with nineteen 7–8 olds
(2b)	Second-year class, with twenty 7–8 olds

As for the (1a) and (2a) classes, their working memory was assessed by means of the digit span test (DS). Their grammar comprehension was assessed by means of an Italian standard receptive grammar test (briefly G). See [16]. Each child of (1a) and (2a) thus obtained a DS score (min 0, max 28) and a G score (min 0, max 16). This was not possible with (2b) due to time restraints. On (1a) and (2a), we can then compare:

- the children's DS scores to the DS median value (DSm), that is equal to 9 for (1a) and 12 for (2a);
- the children's G scores to the G median value (Gm), that is equal to 10 for (1a) and 12 for (2a).

The comparison allows us to group children of (1a) and (2a) in the DS/G categories of Table 2. Note that a child of the X class has *high DS score* (*low DS score*) if the child has DS score higher than or equal to DSm (has DS score lower than DSm).

Similarly, a child of the X class has *high G score* (*low G score*) if the child has G score higher than or equal to Gm (has G score lower than Gm).

Such categories and terminology are then used in the result analysis and discussion.

Table 2 DS/G categories, according to the DS (digit span) or G (grammar) scores.

Measured skill	DS/G categories	DS/G scores
Working memory	high DS score	$DS \geq DSm$
	low DS score	$DS < DSm$
Grammar comprehension	high G score	$G \geq Gm$
	low G score	$G < Gm$

6 Experiment Design

We divided our experiment into three phases:

1. the *Pre-test Phase* is the assessment of working memory and grammar comprehension;
2. the *Transparency Phase* is the arrangement of transparent illustrations on a white paper;
3. the *Drawing Phase* is the spontaneous drawing.

The Pre-test Phase served to classify children according to their working memory and grammar comprehension levels, as explained in Section 5.

The other two phases served to study the capability of visually representing the "while" relations. Whereas the children's preferred visual patterns of representation may emerge in the Transparency Phase, common drawing strategies of children can emerge in the Drawing Phase. We cannot assume that all the experiment's children equally comprehended the "while" relations, and we performed Pearson chi-square tests for these two phases.

The tasks for children in the three phases are summarized in Table 3. Note that all the sentences assigned to children narrate significant episodes of the story. According to the story, the "while" between the first and the second event introduced in the TW1 and TW2 tasks can be mapped into the "during" Allen relation. Notice that DW2 uses the same sentence as TW2 does. The "while" of DW3 is vaguer, and can be mapped to "during or finishes".

Table 3 Task description.

Phase	Task acronym	Task description
Pre-test	DS	Repeat aloud the sequence of numbers read by the evaluator.
	G	An examiner reads a sentence. The child is asked to point to one out of four pictures, which he or she thinks to portray the sentence best.
Transparency	TW1	Represent the following with transparencies: "The wolf reaches Gino while Gino is picking up strawberries in the woods".
	TW2	Represent the following with transparencies: "Mammy hen and the other animals go to the wolf's house while Gino is telling stories to the wolf".
Drawing	DW2	Draw "Mammy hen and the other animals go to the wolf's house while Gino is telling stories to the wolf".
	DW3	Draw "While mammy hen is worried about her little Gino, Gino is telling stories to the wolf".

7 User Teaching

Before performing the experiment, on June 1 2009, the two experimenters met
the classes' teachers and school dean. During the meeting, they discussed the
organization of the experiment (e.g., meeting time, sequence of tasks), and
their respective roles in the experiment. More precisely, teachers were asked
to support the experimenters and children in all the phases of the experiment.
Teachers were asked to provide considerable support in the Pre-test Phase.
This phase is a sort of vis--vis interview with each child, and as such it
requires a considerable amount of time.

Then the experimenters explained the experiment modus operandi, detailed as follows.

- There should be a relaxed and playful atmosphere during the experiment,
  e.g., the story is told imitating the animals' voices.
- The absolute respect of privacy must be clear to children. A personal
  number identifies each child, unequivocally but anonymously.
- The usage of transparent illustrations is explained.
- One side of the A4 sheet is dedicated to spontaneous drawing (for DW2
  and DW3). The other side is used for transparencies (TW1 and TW2).

Finally, experimenters and teachers fix the amount of time that children can
spend on each task, especially DW2 and DW3.

8 Experiment Execution

The experiment took place at school, a familiar environment for children.
At 8:00, the two experimenters met the teachers at their school. Then the
experiment is divided in three consecutive sessions, one session per class.

Session I is carried on with (1a), a first-year class. The experiment phases
   are in the following order: Pre-test; Transparency; Drawing. The correlated
   tasks are assigned in the following order: DS, G; TW1; DW2.

Session II is carried on with (2a), a second-year class. The experiment
   phases are in the following order: Pre-test; Drawing; Transparency. The
   correlated asks are assigned in the following order: DS, G; DW3; TW2.

Session III is carried on with (2b), a second-year class. There is only the
   Drawing Phase with the DW2 task.

In Sessions I and II, drawing and placing transparencies are executed in
different order so as to augment the independency between the DS/G scores,
and the child's capability of visually representing "while" relations. In Session
III, time restraints compelled us to choose either transparencies or drawings.
We opted for the latter as it gives more freedom to children.

The remainder of this sections explains the three sessions in details,
whereas Table 4 gives only the essential information.

Session 1, date: 05 June 2009

S1.1 08:3008:45. The evaluators meet the (1a) children in the classroom. They explain children the assigned tasks, e.g., "...children, please, help us! We need to explain 'while Gino is telling stories, mammy hen is worried' to younger children...". Then the evaluators give each child his or her personal number.

S1.2 08:4509:15. First, one of the evaluators demonstrates how the DS and G tasks must be performed with the aid of one child. Then the evaluators and the teacher record the DS and G data.

S1.3 09:159:30. First, one of the evaluators read the story while children are listening carefully. Then the evaluators distribute a transparency set, and an A4 white sheet per child. Finally, the evaluators ask the children to write their number on their paper sheet.

S1.4 09:3009:50. First, one of the evaluators reads the sentence of the TW1 task. Then the children arrange transparencies on the paper sheet. Finally, the evaluators trace the pattern created by the children on the white paper.

S1.5 09:5010:30. First, one of the evaluators reads the sentence of the DW2 task. Then the children draw on the blank side of the paper sheet.

S1.6 10:3010:40. The experimenters and teacher collect paper sheets, leaving the transparencies as presents.

Session 2, date: 05 June 2009

S2.1 10:4511:00. As (S1.1) of Session 1.
S2.2 11:0011:15. As (S1.2) of Session 1.
S2.3 11:1511:30. As (S1.3) of Session 1.
S2.4 11:3012:00. Like (S1.5) of Session 1 with DW3.
S2.5 12:0012:20. Like (S1.4) of Session 1 with TW2.
S2.6 12:2012:30. As (S1.6) of Session 1.

Session 3, date: 05 June 2009

S3.1 12:4513:00. As (S1.1) of Session 1.
S3.2 13:0013:15. Like (S1.3) of Session 1.
S3.3 13:1513:30. Like (S1.5) of Session 1.
S3.4 13:3014:00. Like (S1.6) of Session 1.

Table 4 Sessions, classes and assigned tasks.

Session	Class	Phase order	Task order
I	(1a)	Pre-test, Transparency, Drawing	DS, G, TW1, DW2
II	(2a)	Pre-test, Drawing, Transparency	DS, G, DW3, TW2
III	(2b)	Drawing	DW2

9 Results Analysis

This section only gives the most significant results.

9.1 *Pre-test Phase for Measuring the Working Memory and Grammar Comprehension*

The Pre-test Phase is concerned with the grammar (G) and working memory tests (DS) of the (1a) and (2a) classes. In our analysis, children get grouped using the DS/G categories of Table 2 above. The analysis results are shown in Table 5 below.

Table 5 Results of the Pre-test Phase.

School class	DS/G category	Number of children
(1a), 17 children	high DS score	10
	high G score	11
(2a), 19 children	high DS score	10
	high G score	11

9.2 *Transparency Phase*

The Transparency Phase is concerned with the TW1 and TW2 tasks of arranging transparencies. TW1 is administered to the (1a) first-year class, whereas TW2 is conducted with the (2a) second-year class.

 In order to represent each event, children had to arrange the transparencies of the involved actors and background elements on the paper sheet. For instance in case of TW1, one out of the two events is "Gino is picking up strawberries in the wood". The correlated transparencies are the strawberries in the wood, and Gino.

9.2.1 The TW1 Transparency Task with the (1a) Class

Let us recall the sentence that children were asked to represent: "The wolf reaches Gino while Gino is picking up strawberries in the woods". Children from (1a) employed three major strategies in arranging transparencies:

Strategy 1: the transparencies of each event are horizontally arranged; one event is aligned above the other;

Strategy 2: the transparencies of one event are vertically aligned, those of the other event are horizontally aligned; the two events are horizontally aligned;

Strategy 3: the transparent actors of the two events are horizontally aligned, and the correlated background elements are aligned beneath.

The top image of Figure 4 shows an example of Strategy 3. Circa 11,8% children opted for Strategy 1, 23,5% went for Strategy 2, and 23,5% of them chose Strategy 3.

Only the relation between the grammar comprehension and the adoption of Strategy 3 is close to significance: $\chi^2(1) = 3,66$, $p = .056$.

9.2.2   The TW2 Transparency Task with the (2a) Class

Let us recall the sentence that children were asked to represent: "Mammy hen and the other animals go to the wolf's house while Gino is telling stories to the wolf". Children from (2a) employed three major strategies in arranging transparencies:

Strategy 1: the transparencies of each event are horizontally arranged; one event is aligned above the other;

Strategy 2: only one event is represented and its transparencies are horizontally aligned;

Strategy 3: no clear arrangement emerges.

The bottom image of Figure 4 shows an example of Strategy 1. The horizontal linear arrangement of both events (Strategy 1) is the most frequent for the TW2 task with (2a): 50% of them chose it. Only 17% of the (2a) class represented only one event, adopting Strategy 2. 33% of (2a) gave no clear order between events, or used only the transparency of one character without representing any event (Strategy 3).

The (2a) children with higher G scores employed Strategy 1, aligning the events horizontally, significantly more frequently than the children with lower G scores: $\chi^2(1) = 5,84$, $p < .01$.

9.3   Drawing Phase

The Drawing Phase is concerned with the DW2 and DW3 tasks of drawing. DW2 is conducted with the (1a) and (2b) classes, whereas DW3 is with the (2a) class. Strategies for representing the "while" relations can be grouped as follows:

(A) spatial arrangement of the events,
(B) other drawing strategies.

As for the spatial arrangement, three major linear arrangements emerged:

(A1) horizontal, that is, two separate events along a horizontal line,
(A2) vertical, that is, two separate events on a vertical line,
(A3) diagonal, that is, two separate events on a diagonal.

Representations consisting of one single event or no clear distribution of events are reported as "other".

Fig. 4 The top image shows the TW1 task by a (1a) child, placing actors horizontally at the top and background elements beneath. The bottom image shows the TW2 task by a (2a) child, placing one event below the other.

Then we found out three other drawing strategies, explained as follows.

(B1) Children draw a background common to both events. That is, the child represents the two events within the same background frame (e.g. sky, ground).

(B2) Children blend both events in a unique scene or keep them in two separate scenes. The child represents the two events as part of a single scene (e.g., mammy hen and the other animals are close to Gino, while Gino is telling stories).

(B3) Children represent two separate events and join them with a link (e.g., a path), albeit they happen in different locations in the story.

Let us see such strategies in the context of the DW2 and DW3 tasks.

9.3.1   The DW2 Drawing Task with the (1a) and (2b) Classes

Let us recall the sentence that children were asked to draw: "Mammy hen and the other animals go to the wolf's house while Gino is telling stories to the wolf". Table 6 recaps the most frequent linear arrangements, adopted by (1a) and (2b) children. Table 7 shows how many of the (1a) and (2a) children drew a common background; see also B1 in Subsection 9.3. Table 8 shows how many of the (1a) and (2a) children represented the two events in two separate scenes; see also B3 in Subsection 9.3.

Table 6 Spatial (linear) arrangements in DW2.

Class	Horizontal	Vertical	Diagonal	Other
(1a)	77,00%	12,00%	0,00%	11,00%
(2b)	67,00%	9,50%	0,00%	23,50%

Table 7 Background strategy in DW2.

Class	Common background	No background	Other
(1a)	59,00%	12,00%	29,00%
(2b)	19,00%	62,00%	19,00%

Table 8 Scenario strategy in DW2.

Class	Single scene	Two scenes	Other
(1a)	65,00%	12,00%	23,00%
(2b)	48,00%	29,00%	23,00%

According to Table 6, the predominant representation is horizontal. Moreover a horizontal representation with the core event ("Gino was telling stories to the wolf") on the right and the other on the left was predominant for all the experiment participants. A representation with the core event on the left and the secondary on the right was rare, but significantly more frequent among the younger children, that is, (1a) children: $\chi^2(1) = 3,75$, $p < .05$. For an example, see the two drawings in Figure 5.

Note also that the younger children adopted the common background strategy (59%), which is instead the least employed by (2b) children (19%), see Table 7.

As Table 8 shows, the majority of children employed the single scene strategy for representing the "while" of DW2 ("Mammy hen and the other animals go to the wolf's house *while* Gino is telling stories to the wolf").

Fig. 5 The top drawing for task DW2 is by a (2b) child, and the bottom one for the same task is by a (1a) child.

9.3.2 The DW3 Drawing Task with the (2a) Class

Let us recall the sentence that children were asked to draw: "While mammy hen is worried about her little Gino, Gino is telling stories to the wolf". The (2a) children tended to mostly represent the two DW3 events diagonally (33%). The majority of them (61%) represented the two events in two separate scenes (see B2 in Subsection 9.3), depicted as physically separate, and distant on the paper sheet. Circa 50% of them also tended to include a common background. Circa 44% of them correlated the two events with a link (see B2 in Subsection 9.3).

Such results seem to contrast with those of their (2b) peers on DW2. Moreover, (2a) children, with low G scores, tended to adopt the common background strategy more than their peers with high G scores, $\chi^2(1) = 3,53$, $p = .06$.

10 Discussion

Let us revisit the two main goals of our evaluation, as stated in Section 3:

(G1) the type and quality of 6–8 olds' visualization strategies;
(G2) if class, grammar comprehension, and working memory capacity affect their visualization strategies.

As our result analysis shows, linear representations are frequently used by first and second graders alike, across tasks.

In the drawing tasks, other interesting strategies emerged and that we can use in the design of our visual tool, like the predominant adoption of a common background by first graders, the drawing of a concrete link (e.g., a path) between the correlated events when these are placed in two different scenarios. See also Subsection 9.3.

Our results also show that class and grammar comprehension skills (DS and G scores) can have an impact on the child's visualization strategies. Contrary to our expectations, verbal working memory did not affect the children's performances at any level in our study.

Given the same task, older children (in second-year classes) tended to represent the "while" relation more abstractly, e.g., the younger children adopted the common background strategy (59%), which is instead the least employed by (2b) children (19%) in the DW2 task. A representation with the core event on the left and the secondary on the right was rare, but significantly more frequent among the younger children. More generally, the visual patterns of the younger children seemed more conventional, with a dominant horizontal orientation.

The visual representation of the "while" between events also seems to depend on the type of events, and the type of Allen relation that the "while" corresponds to, as the DW3 drawing task suggests (it is vague, in that it corresponds to the Allen relation "during or finishes", and it happens in two different locations). In fact, when the "while" is as in the DW3 task, a number of second graders opt for more concrete drawing strategies, that is, the introduction of a common background (50%) and a link between the two events (44%). However, other studies are needed to assess such hypotheses.

Noticeably, grammar skills seem to play a role in discriminating those children that can be in trouble in representing the temporal relations of a story. For instance, children with low G scores in our study tended to represent a single event, when asked to depict relations between two events. This result is confirmed across the different types of tasks.

A final remark on the evaluation methodology is in order: according to our preliminary findings, drawings offer a richer set of information than transparencies, not only about the patterns of visual representations that are most common among children, but also on the strategies that children spontaneously adopt to render complex temporal relations.

11   Ongoing and Future Work

The adoption of linear strategies across tasks and types of children (according to their grammar comprehension, working memory, and class) is a relevant information for the design of our visual story-comprehension tool. The choice of a spatial linear representation is also supported by the preliminary results of [1], which seem to indicate that 7–8 olds can comprehend certain qualitative temporal relations of a story better with a spatial linear representation, based on the one in Figure 1, than with a choice-box textual visualization.

Moreover, an assumption consistent with the results of this paper is that the level of abstraction of the produced visual representations depends on the age and the type of "while" (see Section 10).

Currently, we are investigating whether old children and adults would indeed produce other or more abstract visual representations than young children did. In order to evaluate this, we conducted a preliminary experimental study with old children and adults, divided in two categories according to their experience in reading: ten old children (10–15 olds), and ten adults (16–30 olds). All underwent the Pre-test Phase, and their results with the digit span and grammar receptivity tests were high (see also Section 5). The experimenters let them read the experiment story, and handed out the transparencies of the Transparency Phase (see Table 3). Therein, a remarkably different strategy emerged: circa 70% of the experiment groups overlapped transparencies in order to represent the "while" relation, a strategy that 6–8 olds never employed.

In order to completely assess our assumption, a forthcoming study will extend the exploratory evaluation reported in this paper to a richer variety of "while" temporal relations, and thus provide us with a solid insight on the level of abstraction that novice readers may employ in such diverse cases.

12   Conclusions

This paper mainly discusses our experimental study with 6–8 old novice readers, first and second graders of an Italian primary school. Our analyzes reveal interesting common strategies that children employ for visually representing "while" temporal relations of a story. In particular, literacy maturity seems to play a relevant role in discriminating the sophistication and abstraction of the children's visual representations. The results of this work have triggered the novel evaluation that we are currently analyzing, and that involves old children and young adults, and a forthcoming one with a richer variety of temporal contemporaneous relations. In the conclusion to this paper, we also briefly reported on them.

Acknowledgements. The third author was partially supported by a CARITRO grant. We thank the school children and staff for their participation in the experiment. Our thanks are also due to Dario, Lorenzo and Vittoria for granting the authors first-hand daily experience on the world of young children.

References

1. Arfé, B., Gennari, R., Mich, O.: Evaluations of the LODE Temporal Reasoning Tool with Hearing and Deaf Children. Tech. rep., TR of the MCES 2009 AAAI symposium (2009)
2. Chittaro, L., Combi, C.: Representation of Temporal Intervals and Relations: Information Visualization Aspects and their Evaluation. In: IEEE (ed.) Proc. of TIME (2001)
3. Di Mascio, T., Catarci, T., Santucci, G., Dongilli, P., Franconi, E., Tessaris, S.: Usability Evaluation in the SEWASIE project. In: L.E.A. (ed.) Proc. of HCI 2005 (2005)
4. Duran, N.D., McCarthy, P.M., Graesser, A.C., McNamara, D.S.: Using Temporal Cohesion to Predict Temporal Coherence in Narrative and Expository Texts. Behavior Research Methods (2007)
5. Ge, F., Xuehong, T.: Temporal Reasoning on Daily Events in Primary School Pupils. Acta Psychological Sinica 34, 604–610 (2002)
6. Gennari, R., Mich, O.: Constraint-based Temporal Reasoning for E-learning with LODE. In: Proc. of the Thirteenth International Conference on Principles and Practice of Constraint Programming (2007)
7. Gunthorp, K., Cassinelli, A.: Gino il pulcino e altre storie. Giunti (2002)
8. Hibino, S., Rundensteiner, E.A.: User Interface Evaluation of a Direct Manipulation Temporal Visual Query Language. In: Proc. of the ACM Multimedia Conference (1997)
9. Johnson-Glenberg, M.C.: Web-based Training of Metacognitive Strategies for Text Comprehension: Focus on Poor Comprehenders. Reading and Writing 18 (2007)
10. Knauff, M.: The cognitive adequacy of allen's interval calculus for qualitative spatial representation and reasoning. Spatial Cognition and Computation 1(3), 261–290 (1999), http://dx.doi.org/10.1023/A:1010097601575
11. Koerber, S., Sodian, B.: Preschool children's ability to visually represent relations. Developmental Science 11(3), 390–395 (2008)
12. Lode, the demonstration (2009), http://lodedemo.fbk.eu (retrieved June 11, 2009)
13. McColgan, K., McCormack, T.: Searching and planning: Young children's reasoning about past and future event sequences. Child Developmental Science 11 (2008)
14. Orsolini, M., Fanari, R., Cerracchio, S., Famiglietti, L.: Phonological and Lexical Reading in Italian Children with Dyslexia. Reading and Writing 22, 933–954 (2009)
15. Paivio, A.: Dual-coding Theory: Retrospect and Current Status. Canadian Journal of Psychology 45, 255–287 (1991)
16. Rustioni, D., Lanscaster, M.: Prove di valutazione della comprensione linguistica. In: Organizzazioni Speciali (1994)
17. Scott, C.L.: The Development of Some Working Time Concepts in Pre-school Children. Tech. Rep. ED407155, ERIC (1997)
18. Winskel, H.: The acquisition of temporal reference cross-linguistically using two acting-out comprehension tasks. Journal of Psycholinguistic Research 33 (2004)

Understanding and Augmenting Human Morality: An Introduction to the ACTWith Model of Conscience

Jeffrey White

Abstract. Recent developments, both in the cognitive sciences and in world events, bring special emphasis to the study of morality. The cognitive sciences, spanning neurology, psychology, and computational intelligence, offer substantial advances in understanding the origins and purposes of morality. Meanwhile, world events urge the timely synthesis of these insights with traditional accounts that can be easily assimilated and practically employed to augment moral judgment, both to solve current problems and to direct future action. The object of the following paper is to present such a synthesis in the form of a model of moral cognition, the ACTWith model of conscience. The purpose of the model is twofold. One, the ACTWith model is intended to shed light on personal moral dispositions, and to provide a tool for actual human moral agents in the refinement of their moral lives. As such, it relies on the power of personal introspection, bolstered by the careful study of moral exemplars available to all persons in all cultures in the form of literary or religious figures, if not in the form of contemporary peers and especially leadership. Two, the ACTWith model is intended as a minimum architecture for fully functional artificial morality. As such, it is essentially amodal, implementation non-specific and is developed in the form of an information processing control system. There are given as few hard points in this system as necessary for moral function, and these are themselves taken from review of actual human cognitive processes, thereby intentionally capturing as closely as possible what is expected of moral action and reaction by human beings. Only in satisfying these untutored intuitions should an artificial agent ever be properly regarded as moral, at least in the general population of existing moral agents. Thus, the ACTWith model is intended as a guide both for individual moral development and for the development of artificial moral agents as future technology permits.

Jeffrey White
KAIST, South Korea
e-mail: `jbenjaminwhite@mail.com`

L. Magnani et al. (Eds.): Model-Based Reasoning in Science & Technology, SCI 314, pp. 607–621.
springerlink.com

1

The ultimate goal of A.I., generally, is the construction of a fully embodied and fully autonomous artificial agent. This task poses special challenges, of course, especially the reconciliation of neural research with traditional thinking on intelligence and autonomy [7]. In the development of autonomous moral agents, some authors have contended that the starting point is in the selection of a suitable moral framework for implementation into moral machines [22][1]. However, I disagree with this tact. Though the necessary and sufficient physical mechanisms cannot yet be articulated, either for artificial or natural moral agents (humans), the approach that the following work takes is to first specify the necessary architecture, and then to see what moral framework arises from the proper function of that architecture[2].

The architecture at issue is the ACTWith model of moral cognition, or the ACTWith model of conscience. The scope of the present work forbids exhaustive review of pertinent research from diverse fields all touching on the issues of conscience in practice and theory, artificial morality, and neurological mechanisms at work in moral cognition. However, a brief review is necessary in order to indicate important points of reference. The ACTWith model is at root a bottom-up hybrid architecture, originally informed by Ron Sun's CLARION architecture.[21] However, as it is developed here, it is intentionally task and implementation non-specific, being essentially a model of control of information processing[3]. The model builds from two key insights into moral cognition from neurology, disgust and mirroring[4]. It is essentially a model of situated cognition, and although developed independently, it is consistent with work from situationist psychology[3], and represents a strong form of embodiment[8].

The scope of this paper forbids an exhaustive inquiry into the nature of conscience[5]. However, in this section, I will provide some disambiguating

---

[1] This seems to mirror the method in which moral theory is often pursued, as well.

[2] That neurology, especially, has not already delivered the final word on human morality is a common misconception amongst many. Though one might presume the case closed on moral theory, that we must only wait for the neurologist to tell us what the brain tells us is right and wrong, this is an overly hasty position. Even if the neurosciences level some incontrovertible facts, there remains the issue of interpretation of these facts and the integration of such into existing practices. For discussion, see [9, 15].

[3] For relative advantages to this approach, see [10].

[4] See [11, 23, 27] for initial discussion. The view put forward here is not to be confused with that of popular "mindreading" theorists. I have trouble with this program for reasons too detailed to develop here, but one issue involves the disputes within the body of researchers themselves over what mindreading actually amounts to. See [12] and [13] for examples.

[5] I take on this task in my current book manuscript Conscience: the mechanism of morality, forthcoming with publication expected 2010. See [1, 4, 5, 6, 26, 14, 18, 24, 25, 28] for an introduction to some basic issues in conscience, especially concerning its naturalization and psychological interpretation.

remarks, first in regards to conscience itself, and later in regards to conscience and consciousness.

Conscience is an old term for a family of phenomena, ranging from voices that warn of impending wrong action to providing the fundamental basis for international humanitarian law. It is an extremely complex concept, often confused with consciousness, and more often burdened with seemingly contradictory tasks as it has traditionally been associated with such things as self-preservation on the one hand and altruistic selflessness on the other. Even the seemingly simple and most familiar characterization as a warning voice carries deep implications that demand some specification. For instance, conscience as that universally recognized voice which rises against acting towards morally repulsive ends cannot be merely a simple voice[6]. After all, for it to fulfill even this seemingly simple function, the operations of conscience must extend through all levels of end selection. In order to reject some ends while endorsing others, conscience must act as the steering mechanism of the entire embodied complex that is the moral agent. And that is a very complex concept, indeed. In simple terms and for purposes of introduction, conscience can initially be understood as naming the extended homeostatic function of body to sustain personal integrity in the face of a changing environment, presented in the basic ACTWith model as a generic mechanism which regulates the opening and closing to environmental input, a process which leads to the accumulation of experience[7] which is used to guide future operations of the same mechanism.

Conscience is historically, and linguistically, related to consciousness[8]. In fact, the term conscience precedes that of consciousness by some 300 years, and it is from conscience that the term consciousness originally derives[9]. However, the historical use of these terms is beside the point, now, as consciousness receives a great deal more attention than does conscience, and either clearly represent two very distinct aspects of the human condition, however less than clear their namesakes remain.

We may gain clarity on both terms by exploiting their structural similarities. Both consciousness and conscience consist of conjunctions between a prefix "con-" and a root, "sciousness" and "science", either of which carry individual connotation. "Con-" means "together", or "with". It is a prefix that indicates synthesis. "Sciousness" was proposed by William James in the

---

[6] And to say that it is raises further questions about the nature of verbal language and the origins of symbols, themselves.

[7] Initially understood as memory, see [20], but eventuating in embodied adaptations due to peripheral attunements, i.e. hormones and general metabolism, over time.

[8] Consciousness, as well, has been understood as an extension of homeostatic mechanisms. See [19]

[9] See for example
http://www.etymonline.com/index.php?search=conscience\&searchmode=none.
Last accessed February 15, 2010.

10th chapter of his landmark text, Principles of Psychology, to be a foundation for consciousness. He employed introspection, the only psychological tool available to him at the time, to inquire into the nature of consciousness and found a rolling stream of sensation that receded from his introspective projections just outside his conscious reach. "Con-sciousness", thus, can be taken to mean the synthesis of merely felt moments into discretely realized phenomena[10]. Accordingly, sciousness can be understood as the felt ground of all discrete thought, consisting of clear and distinct ideas in the classical Cartesian sense of self-awareness [2, 16, 17].

"Science", the relative root of the term "conscience" conveys a strikingly different sense, at least on initial inspection. Typically, "science" implies a specific field of knowledge and inquiry, constituted by certain systematic principles of relation between a specific and select body of objects. Examples such as Chemistry, consisting of chemists working in the field of entities related by chemical laws and constitutive of chemical theories over a specific set of chemical objects, make this use of the word "science" clear enough.

Yet, there is something universal about the use of the word "science" that ties all of the seemingly discrete fields of inquiry together, and it is from this universal implication that the term "con-science" should be construed. This universal nature is that "science" as the root of "con-science" represents what it is to be in *any* field of *any* set of objects, however non-specific, which are bound by any principles however non-systematic. In effect, "science" can be taken to name the field in which each each person is individually (and persons are collectively) embedded, and in terms of which he or she seeks successful action and even truth. It can be understood as the "scene" from within which one sees and understands the world, and from within which and in terms of which one acts, experiences, further understands (learns), or fails. "Science", in this sense, is reducible to "situation" in a very strong sense, being the irreducible complex of agent and environment, understood from the perspective of the experiencing agent, or subject. "Con-science" can be understood, then, as the synthesis of embodied situations, the "what it feels like" to be in a place at a time, and in such processing produces information on the differences – both merely felt and otherwise cognized – between the relative value of one situation with any other.

In this way, conscience, understood fully as an embodied mechanism, serves as a motivational and self-preserving extension of basal homeostatic mechanisms common to all sufficiently complex organisms[11]. An organism that is able to evaluate the relative values of situations will seek those situations that feel good, and avoid those that feel bad, as these situations are effectively environments in terms of which that organism must subsequently reach

---

[10] This is effectively the operation employed through the use of mathematical algorithms in hybrid models. For discussion on James and sciousness on this point, see [21].

[11] Again, these issues are more adequately developed in *Conscience: the mechanism of morality*.

homeostatic equilibrium. Conscience, thus, and morality by further extension, operate according to this logic, but present themselves in recognizable forms only in organisms of necessary complexity, such as human beings.

The scope of the present paper does not permit a thorough explication of the relationship between conscience and consciousness, or of the place of conscience as part of an organism's homeostatic mechanisms. But, the preceding brief account does specify the guiding role of conscience in the motivation of any autonomous moral agent, artificial or otherwise, and opens the window to develop in simpler terms a generic mechanism from which we might conceive a moral framework emerging, that being a framework from which moral action proceeds and in terms of which moral judgment can be based. In the next section, I will detail the basic ACTWIth model in a more easily appropriated form derived from hybrid neural net models.

2

The ACTWIth model is a four-step cycle, with two belonging to a top (rational) level and two to a bottom (affective) level, with each step a related mode of information processing. This structure is captured in the name, "ACTWIth"[12]. "ACTWith" stands for "As-if" "Coming to Terms With". "As-if" involves feeling a situation out, while "Coming to Terms With" involves defining the situation in terms of the things originally felt. The model consists in 4 modes:

As-if (closed) coming to terms with (closed)

As-if (open) coming to terms with (closed)

As-if (closed) coming to terms with (open)

As-if (open) coming to terms with (open)

These modes are intended to represent the bare minimum for the eventual emergence of morality. The closed modes are derived from the mechanism of disgust, while the open modes are from mirroring mechanisms, both affective and action oriented. While the systems in which these mechanisms must operate are not here specified, at the level of implementation, inspiration might be drawn from primate or from human brains, as some researchers are doing in non-moral realms presently, or they can be taken purely from computational intelligences, which are also common in the study of human learning, cooperation, and motivation. In any event, it is not the purpose of the present work to detail potential applications.

Altogether, the four modes can be visualized as follows (see Figure 1).

---

[12] ACTWith, either in name or function, bears no deliberate relationship with the famous ACT-R model.

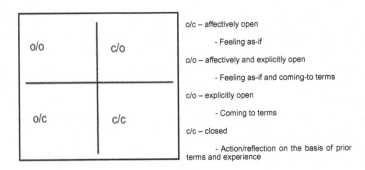

Fig. 1 Basic ACTWith model consisting of four static modes.

In order to illustrate the individual modes, it is useful to imagine that each represent a certain personality type which might arise through the habitual application of one of the four modes at the exclusion of the others. For instance, consider the mode o/c. This personality is open to other situations affectively, but not at the level of discrete reason. If a person were to habitually engage in this mode when dealing with others, he or she would present genuine sympathy for the situations in which these others were finding themselves, but would only be capable of understanding the significance of those situations in light of his or her own prior understanding. Contrast this mode with that of c/o. This personality is closed, affectively, but open at the level of discrete reason. If a person were to habitually engage in this mode, he or she would not be able to feel what it is like to be in another's situation, but would be interested in having an explanation for why that person is in that situation, how he or she plans to get out of it, and etcetera. The first may seem warm, but "flaky", while the second may seem cold, and calculating. Ether represent personality types that are common, enough, to be easily recognized as archetypes.

The o/o and the c/c modes are the most interesting, and the most recognizable. The o/o mode, when habitually employed, represents the genuine saint. This personality is both affectively open to another's situation as well as genuinely interested in understanding what it is like to be in that situation at least insofar as that other understands it. In practice, this sort of person is exceptionally rare, while the habits that lead to its realization are the object of many if not most religions. Buddhist practitioners (of some strains) stand out as exemplifying this mode as habitually employed. Meanwhile, the c/c mode is the opposite of the o/o mode. Persons habitually employing this mode are selfish, arrogant sorts who come off both as cold and calculating. This personality is perhaps most recognizable, as it represents a being who is both unable to feel what it is like to be in another's situation, as well as being disinterested in understanding why he or she is in that situation and how or why he or she would plan to leave it. This is the mode of the psychopath.

Different personality types can be rendered more finely by recognizing that these modes may be habitually employed only in certain types of situations. As "another situation" equally means one's own or another's situation, one may be completely open to one's own different situations (o/o) while being indifferent to those of other persons (c/c) or interested in them solely insofar as understanding those situations fortifies his or her own understanding of his or her own place in the world (c/o). Over the long course of personal development, it is easy enough to see how the habitual employment of one or another of these four modes of information processing can lead to a wide diversity of personality types.

To articulate these four modes in static terms, in terms of habitual employment at the exclusion of one another, is useful for illustrative purposes. However, any realistic model of agency must be dynamic. The ACTWith model is, fully developed, a cycle of information processing. It can be represented thusly (Figure 2):

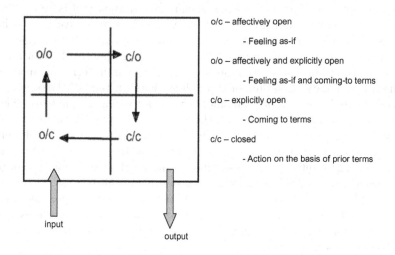

o/c – affectively open

- Feeling as-if

o/o – affectively and explicitly open

- Feeling as-if and coming-to terms

c/o – explicitly open

- Coming to terms

c/c – closed

- Action on the basis of prior terms

Fig. 2 The Beating Heat of Conscience.

This model is "the beating heart of conscience", recognizing the fact that the conscience has traditionally been associated with the beating of a heart, and capitalizing on the input/output life-preserving dynamic common to both the human heart and to less complex organisms, such as the common bivalve. However, where the bivalve is effectively a slave to it external environment, being as it is rooted to a sea floor and capable only of feeding from what the tides bring, more complex organisms are able to seek out and to avoid situations that are either beneficial or contrary to integrity (physical or otherwise) and survival.

In order to illustrate the effect of this cycle, it may serve to demonstrate two modes, these being with a conscience, "conscientious", or with a heart, and being "without a conscience"[13]. Consider the following scene. A cold and lonely agent is making his way down an icy city street when he stumbles upon a man, dirty and disheveled and obviously very cold, sitting over a steaming man-hole cover. The man is wet from the steam, dressed in rags, and in the bitter wind, the stinking vapor - his only source of heat - turns to ice in his ratty beard. At first glance, the man is ill, with spots of pus dried from broken sores upon his windburned lips, and his feet are bloody through the ragged boots that hang over the side of the manhole cover into the dirty slush that rings it.

At the instant that the agent comes onto the scene, he has a chance to either open to the plight of the poor man, or to close to it. Let's consider the open mode, first. In opening to the poor man, the agent will perhaps at first have to overcome disgust in order to mirror both the feeling that is expressed by the man as well as mirror potential action paths, as opening means feeling as if the agent were that other man, however momentarily[14]. First, the agent will appreciate the situation from the agent's own prior experience (o/c). Then, as the agent opens to the other in genuine compassion, the agent is amenable to coming to an understanding of the situation from the perspective of the other. This is the mode of concern, again with "con-" playing its typical role, and "-cern" meaning the being of a cognitive agent altogether, in thought and in feeling, as the agent comes to appreciate the situation in terms of the other, perhaps through conversation, or through the careful study of the other's actions and expressions, whether momentarily or for a longer time (o/o). Then, as the situation sinks in, the agent in the words of Adam Smith "makes himself at home" in the situation, (c/o). Thusly, the agent is able to feel the difference between his own situation and that of the other, as the terms to which he has come are backfed into his own prior understanding. The feeling of being literally moved in compassion for another is he product of this process. Finally, the agent will be able to reflect on his new experience, and either open once again to the situation, searching for greater understanding[15], or act – perhaps by offering the poor man some charity – and move on to other situations, enriched for the new experience (c/c).

---

[13] Strictly speaking, as the model suggests, one is never literally "without a conscience", one merely fails to employ certain modes of cognition at morally appropriate times, thereby demonstrating immoral or amoral behavior, while at once – through routine – become an immoral or amoral person by habit if not by reputation.

[14] Strictly speaking, as the model suggests, one is never literally "without a conscience", one merely fails to employ certain modes of cognition at morally appropriate times, thereby demonstrating immoral or amoral behavior, while at once – through routine – become an immoral or amoral person by habit if not by reputation.

[15] This is a much deeper process, one of trading situations in a strong sense, than that represented by mindreading theorists.

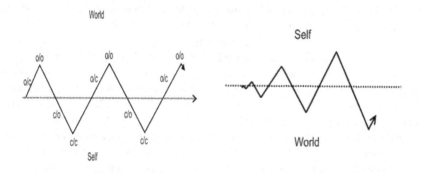

Fig. 3 Stitching one's self into the world.

The closed case is effectively much easier to demonstrate. The agent, upon the sight of the man, closes to him in disgust, and during this cycle of processing opens instead to the agent's own future or past situations, perhaps reliving a trip to Disney World or imagining what it will be like to eat with a mistress. The agent simply walks by, and though the cycle of cognition that is the beating heart of conscience proceeds uninterrupted, the agent "without a conscience" has a heart only for its own self.

In many ways, it is easy to see how the closed agent has certain advantages over the open agent. Especially in a world whose customs, largely shaped by latter-day corporate capitalism, favor those who act selfishly and without regard for the situations that others are left in due to one's own selfish actions, the closed mode has the advantage of delivering its habitual employer to positions of relative success and material wealth. The habitual employer of the open mode, on the other hand, suffers and is in fact increasingly burdened as more and more persons fall to desperate situations in the wake of the selfish stampede for success.

In either case, the lesson is that agents shape their environments through their actions[16]. The agent shapes the world through action, thereby setting out the terms to which it must come in future iterations, and so on. Self and world, what one knows and does, are not only inseparable but are increasingly related on this picture. As the agent opens to the world, the agent takes up the understanding of this situation, and carries it into the next situation, and so on. Thus, in opening and in closing to the world, the agent becomes the product of the terms generated. This process is illustrated in Figure 3.

In the diagram on the left, the process of opening and closing to the world is given in ACTWith processing terms. In the diagram on the right, there is illustrated the potential for personal growth that is the promise of the

---

[16] The student of Philosophy will recall that this is the essential message and the primary motivation behind J.S. Mill's Utilitarianism, and may also recall the central role of conscience therein.

habitually open mode, which leads to what existentialist have called the "beautiful soul" and that phenomenologists have called "authenticity".

At this point, the role of conscience in personal freedom, freewill, can be clarified. As is shown in the preceding figure, and as is alluded to in the preceding discussion, the role of conscience in freedom is that it serves as the mechanism which makes the freedom of self-determination a real possibility. Conscience is not the seat of something more, some radical freedom, that permits an agent to perform any action willy-nilly without regard to past or prior constraints imposed by very real facts about the agents embodiment and its capacity to adapt to new and changing situations. Instead, conscience is a steering wheel of sorts, a gentle handle on personal self-transformation and, perhaps, even personal transcendence, though any further discussion on these issues is beyond the immediate scope of this paper[17].

The question we are left with as we turn to consider what sort of moral framework emerges from the model developed thus far has less to do with what one will do in a given situation, and more to do with who one wishes to become as the product of his or her experience from either opening or closing to situations as they change. This leads us to the final section, and returns us to recent considerations on the possibility of autonomous moral agents.

3

The job of conceiving of autonomous moral agents is difficult enough, but the task becomes more difficult for the fact that human moral agency is not so well understood. Even in cases where one would think the matter settled, such as an application of traditional moral theory to the conception of moral agency, there is the additional problem of misinterpretation of moral theory that must be dealt with before one can attend to the issue of agency design.

Consider Kant's moral theory in this light. According to some commentators, the role of conscience in Kant's moral theory is merely that of the traditional voice of conscience, warning against immoral action. Conscience is simply recast as the representative voice of the categorical imperative. On this account, conscience rises to awareness when one is considering an action which would violate the categorical imperative[24]. Still other appropriations of Kant's moral theory, specifically into discussions of on the possibility of autonomous moral agents, fail to consider conscience at all [22].

However, such accounts are not consistent with Kant's greater moral theory. In Kant's moral theory, fully explored, conscience plays a central role not merely in deliberation over action, but in the process of becoming a moral person and, in fact, in a process which, mirroring Stevan Harnad's famous "symbol grounding" problem, grounds moral action through Kant's infamous "goodwill". Issues of space forbid a full exposition of these claims. So, in this section, I will simply lay out a cursory interpretation of Kant's moral theory

---

[17] These issues are fully explored in *Conscience, the mechanism of morality*.

along these lines, illustrate how it might arise through the proper functioning of an ACTWith endowed agent per the discussion in the previous two sections, and finally redraw Kant's categorical imperative in terms consistent with both.

Let us first consider what Kant means by morality, not forgetting the example of the poor man from the last section. In Section 35 of *The Metaphysics of Ethics*, Kant tells us that "[...] although it is no direct duty to take a part in the joy or grief of others, yet to take an active part in their lot is [...]" and that we ought not "avoid the receptacles of the poor, in order to save ourselves an unpleasant feeling, but rather to seek them out". As well, we ought not "[...] desert the chambers of the sick nor the cells of the debtor, in order to escape the painful sympathy we might be unable to repress, this emotion being a spring implanted in us by nature, prompting to the discharge of duties, which the naked representations of reason might be unable to accomplish". In these lines, Kant paints a picture of an affectively motivated moral agent, compelled as a "spring" overcoming reason in discharging of what he terms one's moral duty. Just what this moral duty is will become clear in a moment. And, what is this "spring"? It is conscience, not misunderstood as mere warning light for the categorical imperative, itself understood as a purely rational directive, but instead understood as the spring that motivates a person according to the logic of moral affect.

The affect central to Kant's moral theory is goodwill. What is good will? Earlier, in the first section of The Metaphysics of Ethics, Kant tells us that the good will is "to be considered, not the only and whole good, but as the highest good, and the condition limiting every other good, even happiness [...]" And, later, in the second section "That, we now know, is a good will whose maxim, if made law universal, would not be repugnant to itself". Thus, it is good will both that one aspires to (insofar as one wishes to be moral) and that guides action along the way. Here, it is important to note that repugnance is another word for disgust, both of which are not concepts belonging to reason, where typical misinterpretations on Kantian ethics place the locus of moral motivation (in rationality), instead.

How does good will work to motivate to moral ends via moral actions? By Kant's account, goodwill alone is not enough. One must also have in mind some exemplar, some other embodied agent, whether real or ideal, in light of which one may, at least initially, model ones actions, and thus eventually one's self. The emotion that signifies the importance of these examples is reverence, and in fact the object of reverence serves as the measuring stick for one's own moral worth. Kant tells us, in the notes to chapter 1, that "What is called a moral interest, is based solely on this emotion". And what is reverence? Without prying any further detail directly from Kant's own writings in support of the claim, it can be understood, in contemporary terms, to involve the employment of mirroring capacities of the human body to emulate, and so train, one's self to adopt and thus become like another human being, whether that being be, on Kant's account, real or ideal. Moral interest, thus, is fundamentally to become the best person one can become.

In these two concepts, reverence and goodwill, the opening and closing functions of the ACTWith model are plotted onto Kant's moral theory. So, where is conscience in all of this? Conscience is the binder of the two. In a section of the Metaphysics of Ethics interestingly entitled "Prerequisites towards constituting man a moral agent", Kant affirms that one's understanding is the limit whereby he or she can determine right or wrong, writing that "obligement can extend only to the illuminating his understanding as to what things are duty, what not". And this returns us to the notion of moral duty, and to the question what is this "spring implanted in us by nature" that motivates a person to seek to fulfill this duty. Both the duty that is attached to action, and the spring that motivates to one's highest potential as a person, to become worthy of reverence through the exhibition of goodwill, are the subjects of conscience. To this end, Kant writes:

> The only duty there is here room for, is to cultivate one's conscience, and to quicken the attention due to the voice of a man's inward monitor, and to strain every exertion (i.e., indirectly a duty) to procure obedience to what he says.

In other words, one's highest potential is to be conscientious, and one's primary duty in action is to maximize this potential through conscientiousness, the habitual act thereof maximizing one's understanding, and so expanding one's potential to recognize his or her obligation to others in the fulfillment of moral duty. It is a cycle. And, it is easy to see that this process leads directly to the "beautiful soul"[18].

Finally, shortly after the preceding statement, Kant spells out this duty for conscientious moral agents when passively serving as models and guides for others, according to the same logic of disgust and mirroring:

> The compunction a man feels from the stings of conscience is, although of ethical origin, yet physical in its results, just like grief, fear, and every other sickly habitude of mind. To take heed, that no one fall under his own contempt, cannot indeed be my duty, for that exclusively in his concern. However, I ought to do nothing which I know may, from the constitution of our nature, become a temptation, seducing others to deeds which conscience may afterwards condemn them for.

Altogether, we have a portrait of Kantian moral theory which can be understood as a direct extension of the mechanisms at work in the ACTWith model. Accordingly, it serves to reconsider the categorical imperative in light of these results. Arguably, the most famous form of the categorical imperative is the following, and the one which Kant himself prefers as he restates

---

[18] It is also worth noting that Kant equates one's giving oneself over to these emotions with freewill, in short because such opens the potential for one's becoming the best person one can become, and such a result is, on his understanding, the universal aim of every person.

it in chapter 2 of *The Metaphysics of Ethics*: "Act according to that maxim which thou couldst at the same time will an universal law". In light of the present results, especially in view of the role of conscience in the preceding appropriation of Kant's moral theory, this imperative can be rewritten in the following forms:

1. Do not become through action (or inaction) an object of self-disgust.
2. And, conversely: Do become through action (or inaction) an object of reverence.
3. And, most simply: Do not put another into a situation that you would not seek for your own[19].

4   Conclusion

This paper has put forward a model of moral cognition consistent both with neurological insights into human motivation to moral action and to becoming a moral person. What are the implications of this proposal? Ideally, it serves in two ways. One, it may redirect focus in the development of autonomous moral agents away from the post-hoc introduction of ethical systems or principles, either as strictures or as measures of moral performance, and toward the development of morally productive architectures from the ground up. As technology develops, limitations to applications increasingly derive from the conceptions which drive and inspire these applications rather from the technology, itself. In terms of moral agents, thus, it is up to the moral philosopher to prefigure these potential applications by providing frameworks of the broadest possible scope with the greatest possible explanatory power. The future of the development of autonomous moral agents, in my mind, depends on this. The ACTWith model proposed here is intended to serve as a starting point in exactly this way.

Two, it may open the way for computational, control, and systems theories of moral agency to be employed increasingly as tools in the analysis and augmentation of human moral conduct. The flow of information from man to machine is bi-directional. It goes both ways. As these models are developed, they require testing and evaluation, and the only method available is against direct human experience. Further, in the testing, we human beings stand to learn something about ourselves that may have lain hidden without the mediation of the models under review.

Finally, it is my hope that the ACTWith model serves as an introspective guide for the moral practice of actual, living people whose interests rest alongside that put forward by Immanuel Kant and so many other moral philosophers before and since: to become, through reflection, and perhaps through the use of what may be called "moral mediators" in the spirit of Lorenzo Magnani's "epistemic mediators", the best people that they can possibly become.

---

[19] Which might invite a violation of either 1 or 2.

References

1. Ames van, M.: Conscience and calculation. International Journal of Ethics 47, 180–192 (1937)
2. Bailey, A.R.: The strange attraction of sciousness: William james on consciousness. Transactions of the Charles S. Peirce Society 34, 414–434 (1998)
3. Barsalou, L.W.: Perceptual symbol systems. Behavioral and Brain Sciences 22, 577–660 (1999)
4. Beiswanger, G.: The logic of conscience. The Journal of Philosophy 47, 225–237 (1950)
5. Boutroux, E.: The individual conscience and the law. International Journal of Ethics 27, 317–333 (1917)
6. Boutroux, E.: Liberty of conscience. International Journal of Ethics 28, 59–69 (1917)
7. Brooks, R., Stein, L.: Building brains for bodies. Autonomous Robots 1, 7–25 (1994)
8. Clark, A.: Embodiment and the philosophy of mind. Current Issues in Philosophy of Mind 43, 35–52 (1998)
9. Dean, R.: Does neuroscience undermine deontological theory (2010), doi:10.1007/s12152-009-9052-x
10. Eliasmith, C.: How we ought to describe computation in the brain (2010), http://www.arts.uwaterloo.ca/~celiasmi/cv.html (last accessed February 15, 2010)
11. Gallese, V., Keysers, C., Rizzolatti, G.: A unifying view of the basis of social cognition. Trends in Cognitive Sciences 8, 396–403 (2004)
12. Goldman, A.: Hurley on simulation. Philosophy and Phenomenological Research 77, 775–788 (2008)
13. Hurley, S.: Understanding simulation. Philosophy and Phenomenological Research 77, 755–774 (2008)
14. Klein, D.B.: The psychology of conscience. International Journal of Ethics 40, 246–262 (1930)
15. Lavazza, A., De Caro, M.: Not so fast: On some bold claims concerning human agency (2010), doi:10.1007/s12152-009-9053-9
16. Natsoulas, T.: The sciousness hypothesis - part i. The Journal of Mind and Behavior 17, 45–66 (1996)
17. Natsoulas, T.: The sciousness hypothesis - part ii. The Journal of Mind and Behavior 17, 185–206 (1996)
18. Olson, R.G.A.: Naturalistic theory of conscience. Philosophy and Phenomenological Research 19, 306–322 (1959)
19. Ramachandran, V.: A Brief Tour of Human Consciousness. Pearson Education, New York (2002)
20. Reid, M.D.: Memory as initial experiencing of the past. Philosophical Psychology 18, 671–698 (2005)
21. Sun, R.: The Duality of Mind: A Bottom-Up Approach to Cognition. L. Erlbaum and Associates, New Jersey (2002)
22. Tonkens, R.: A challenge for machine ethics. Minds & Machines 19, 421–438 (2009)
23. Umilta, M., Kohler, E., Gallese, V., Forgassi, L., Fadiga, L., Keysers, C., Rizzolatti, G.: I know what you are doing: A neurophysiological approach. Neuron. 31, 155–165 (2001)

24. Velleman, J.D.: The voice of conscience. Proceedings of the Aristotelian Society 99, 57–76 (1999)
25. Ward, B.: The content and function of conscience. The Journal of Philosophy 58, 765–772 (1961)
26. William, W.: Some paradoxes of private conscience as a political guide. Ethics 80, 306–312 (1970)
27. Wilson, E.: Consilience: The Unity of Knowledge. Random House, New York (1998)
28. Wright, W.K.: Conscience as reason and emotion. Philosophy Review 25, 676–691 (1916)

Analog Modeling of Human Cognitive Functions with Tripartite Synapses

Alfredo Pereira Jr. and Fábio Augusto Furlan

Abstract. Searching for an understanding of how the brain supports conscious processes, cognitive scientists have proposed two main classes of theory: Global Workspace and Information Integration theories. These theories seem to be complementary, but both still lack grounding in terms of brain mechanisms responsible for the production of coherent and unitary conscious states. Here we propose – following James Robertson's "Astrocentric Hypothesis" – that conscious processing is based on analog computing in astrocytes. The "hardware" for these computations is calcium waves mediated by adenosine triphosphate signaling. Besides presenting our version of this hypothesis, we also review recent findings on astrocyte morphology that lend support to their functioning as Local Hubs (composed of protoplasmic astrocytes) that integrate synaptic activity, and as a Master Hub (composed, in the human brain, by a combination of interlaminar, fibrous, polarized and varicose projection astrocytes) that integrates whole-brain activity.

1 Introduction

Recent research focusing on the participation of astrocytes in glutamatergic synapses has revealed their role in several cognitive functions: learning, perception, conscious processing and memory formation/retrieval. The discovery of the participation of astrocytes as active elements in these processes has led to the construction of broader models, composed by functional units of two neurons and one astrocyte, the tripartite synapses.

Alfredo Pereira Jr.
Institute of Biosciences, State University of São Paulo, Botucatu, Brazil
e-mail: apj@ibb.unesp.br

Fábio Augusto Furlan
School of Medicine, University of Marília, Marília, Brazil
e-mail: fabioaugustofurlan@yahoo.com.br

L. Magnani et al. (Eds.): Model-Based Reasoning in Science & Technology, SCI 314, pp. 623–635.
springerlink.com

Astrocyte terminations wrap the synaptic cleft (in some brain regions, each astrocyte can contact up to 140,000 synapses) and respond to presynaptic input by means of calcium waves and release of gliotransmitters that modulate neural activity. Neighboring astrocytes are coupled by gap junctions forming a functional syncytium. In a series of publications [1, 2, 3] we have described how human cognitive functions, including conscious processing, can be modeled by an ensemble of tripartite synapses connected by the astrocytic syncytium.

We have argued [3] that the dynamical process that boosts neuro-astroglial communication is the synchronization of neuronal graded and action potentials. Synchronization of large populations of neurons, in several medium to high frequencies (from theta to gamma), increases glutamate release from neurons to astrocytes. Glutamatergic activation of the inositol triphosphate (IP$_3$) pathway in astrocytes, beyond a given threshold, elicits coherent, amplitude modulated calcium waves with the potential of integrating local information.

In this paper we propose that calcium standing waves in astrocytic microdomains activate adenosine triphospate (ATP) signaling to adjacent domains. ATP signals cross gap junctions and regenerate the calcium wave in the adjacent domain. Combining such waves and ATP signaling, the result is a "domino effect", by which an ensemble of calcium standing waves are coupled by ATP signaling through astroglial gap junctions. In a situation of global brain synchronization, this communication and processing system possibly contributes to integrate sensory patterns from distinct neuronal populations into a conscious episode. After formulating this hypothesis, we gather recent evidence about astrocyte morphology to argue that the astrocytic network can function as a biophysical "Global Workspace".

2   The Glutamatergic Tripartite Synapse

In Neurobiology, associative learning and memory formation are classically illustrated at the synaptic level by means of a model composed of two (the pre- and postsynaptic) connected neurons, and their respective inter and intracellular signaling pathways (Figure 1). Glutamatergic heterosynaptic converging input to a neocortical or hippocampal neuron activates alpha-amino-3-hydroxy-5-methyl-4-isoxazolepropionic acid receptors (AMPAR) and the resulting depolarization opens N-methyl-D-aspartic acid receptors (NMDAR) of the NR2A subtype, promoting calcium ion entry that cause membrane potentiation related to associative learning (mostly by means of a signaling cascade and gene expression that leads to an increase in AMPA-dependent response).

Considering the participation of astrocytes, we present a diagram of molecular mechanisms present in tripartite synapses (Figure 2), which contributes to explain physiological bases of cognitive functions.

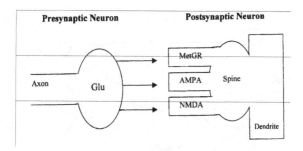

Fig. 1 The Glutamatergic Synapse: Glutamate (Glu) released from the presynaptic neuron's axon terminal is spread in synaptic space and bind to three different kinds of receptors (AMPA, NMDA and Metabotropic Glu Receptors – MetGR) located at the postsynaptic neuron membrane. The three kinds of receptors activate signal-transduction pathways that converge into the dendritic spine.

Glutamate released from astrocytes to postsynaptic neurons in tripartite synapses binds to extrasynaptic NMDA receptors of the NR2B subtype, which drives slow inward calcium currents (SIC), causing a delayed depolarization and an increase of calmodulin-dependent protein kinase subtype II (CaMKII) phosphorylation and AMPA excitability (a process we called "meta-potentiation"), or, alternatively, triggering a process of long term depression (LTP). By means of this feedback, astrocyte cognitive processing can have an effect on learning, memory and behavior.

3 Cognitive and Conscious Processing: From Neurons to Astrocytes

The building-blocks of conscious experiences - the prototypical contents – can be related to local field electromagnetic (EM) signatures. These patterns are activated over baseline through a matching with the spatio-temporal structure of incoming spike trains. The matching function has been modeled in neural networks using e.g. Adaptive Resonance Theory, a theoretical tool that is useful for the understanding of neural mechanisms underlying conscious perception and learning (see [4]).

Human consciousness is composed of episodes that dynamically combine a large number of prototypical patterns. As local fields are too weak to allow direct magnetic interaction of all regions of the brain, biological evolution led to the development of several forms of communication, able to produce integrated whole-brain spatio-temporal waves that correlate with occurrence of conscious episodes.

There are several mechanisms of integration in the brain, some of them related to *attention* processes. Their function is both to *select* local fields that participate in the composition of the conscious episode, and to *bind* the

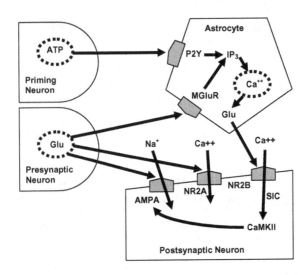

Fig. 2 The Tripartite Synapse. Astrocyte calcium waves are primed by purinergic transmission mediated by metabotropic (P2Y) receptor, and also by GABAergic and cholinergic neurons (not shown in the picture). Glutamate (Glu) released by the presynaptic neuron binds with both astroglial (MGluR; possibly also astroglial NMDA receptors, not shown in the picture) and postsynaptic neuronal (AMPA and NMDA containing the NR2A subunit) receptors. Synergic action of MGluR and other astroglial receptors activate the inositol triphosphate (IP3) pathway, inducing the release of calcium ions from internal stores (mitochondria and endoplasmatic reticulum) to prompt Glu release (and induce calcium waves in adjacent cells by means of adenosine triphosphate signaling). Astroglial Glu binds mostly with neuronal NMDA receptors containing the NR2B subunit (NR2B), causing slow calcium ion entry [slow inward currents (SIC)] and binding to calmodulin-dependent protein kinase subtype II (CaMKII), then sustaining the excitatory activity of the neuron by means of AMPA phosphorylation.

fields' informational content forming the focus of attention (including the figure-background distinction proposed by Gestalt psychologists).

Axonal firing is the main form of long-range communication in the brain, but has limitations. It is not able to directly transmit the waveform signature of each neuron, because the firing threshold is determined by *constant parameters* [5]. In other words, the amplitude modulation that characterizes each EM signature pattern is lost in axonal transmission; since all spikes have approximately the same amplitude. Axonal transmission operates with a system of discrete pulses, usually described as a binary code in the artificial neural network literature. The coding of the axonal message is based only on the frequency and phase of the pulses.

Axonal firing is also limited by the point-to-point architecture of axon-dendrite connections. This architecture does not support the integration of distributed information, since there is not a brain center where all circuits converge. There are limited convergence zones [6] that integrate brain activity regionally. Supplementary mechanisms, such as electrical synapses, astrocytic calcium waves, nitric oxide spread and hormonal signaling are likely to help to integrate local fields into a whole-brain spatio-temporal EM wave.

Brain cognitive mechanisms leading to the formation of conscious episodes can be described in three steps, which hopefully cover the main steps in the generation of *perceptual consciousness*. Other modalities of human consciousness (abstract thinking, planning the future, aesthetic and moral judgment, self-consciousness) will not be discussed here, but they can be accounted by the consideration of higher-order relations between brain systems, specially the "executive system", involving large brain networks connecting the hippocampus and frontal cortex with the parietal and temporal associative areas.

3.1 Amplification Mechanisms

The actualization of a potential conscious pattern begins with the stimulation of the respective specialized neuronal assembly by means of an afferent spike train or an endogenous brain signal that matches with the assembly's EM signature and excites it beyond baseline. Departing from this initial excitation, *recurrent circuits*, widely present in the brain, promote the amplification of the assembly's EM signature. In recurrent circuits, *excitatory loops* are formed: initial excitatory postsynaptic potentials (EPSPs) generate spike trains that activate EPSPs in other neurons that generate spike trains that reinforce the initial EPSPs.

In the awake state, inter-neuronal communication in thalamocortical networks is boosted by tonic (i.e., pulsed) spiking provided by cholinergic activation. The dependence of amplification on cholinergic mechanism makes acetylcholine one of the major transmitters involved in conscious processing (see [7]).

Excitatory loops are counterbalanced by the activation of inhibitory interneurons, which release neurotransmitter gamma-amino-butyric acid (GABA) to other neurons. This transmitter and its membrane receptors contribute to inhibit membrane activity by controlling the flux of cloride ions.

The amplification of prototypical contents by recurrent circuits involves the following operations:

a) neuronal assembly A, which generates EPSPs corresponding to an prototypical content, sends spike trains to other assemblies B:
b) reentrant signaling (see [6, 8]) – i.e., reentrant spike trains – from B to A releases transmitters that bind to membrane receptors to sustain the original EPSPs.

Therefore, a recurrent circuit neural population amplifies its EM pattern signature using recurrent circuits that activate, in each neuron belonging to the circuit, the same mechanisms involved in generating the original pattern:

a) the action of metabotropic receptors on the ionotropic ones; the effect of this activation is membrane depolarization, which also controls the opening of voltage-dependent channels (VDCC);

b) the action of hormones/neuropeptides and other molecular effectors sustaining the feedback pathways;

c) Ca entering on NMDA and VDCC, activating cellular short-term potentiation that feedback on the membrane.

3.2   Broadcasting Mechanisms

Broadcasting involves the formation of *wave packets* composed of carrier waves modulated in amplitude and phase [9]. The carrier wave - one that is not really a continuous wave as in radio transmission - is composed of a series of neuronal EPSPs distributed along brain networks. Each excited local assembly's field signature determines the spatio-temporal structure of spike trains that modulate the target EPSPs of the assemblies with which they are connected. This pattern is then propagated serially to the next connected assemblies, generating the wave packet.

Broadcasting uses neuronal communication by means of spiking activity and other complementary mechanisms that modulate the carrier fields. Understanding large-scale integration modes in the brain requires aggregated neuron-cables typical of the output and input between systems, such as basal ganglia or the hippocampus, to and from polymodal cortical areas. The resulting phase portrait embodies interference patterns of EM signatures of all the neuronal assemblies involved in the wave packet.

Two of the most important mechanisms for the broadcasting of local field signatures are neuronal *oscillations* and *synchrony*. High-frequency oscillations are based on electrical transmission through gap junctions [10, 11]. Gap junctions are regions of contact of neuron membranes that contrast with chemical synapses, where membranes do not have contact. Synchrony is dependent on inhibitory neurons in the thalamocortical system [12]. Both mechanisms are coupled, since GABA-releasing inhibitory interneurons communicate by electrical synapses [13].

Both oscillations and synchrony are important broadcasting mechanisms that possibly contribute to the binding of distributed neuronal activity. Singer [14] showed the relation between oscillatory synchrony and the physiology of the NMDA receptor, one of the major players in the determination of local field signatures. Engel et al. [15] suggested that "oscillatory signals may be well suited as carrier signals for a temporal code". Recently, the relation of synchrony and oscillatory activity in the propagation of local field signatures in the visual system was clarified by Samonds and Bonds: "the reliable synchrony at response onset could be driven by spatial and temporal correlation of the stimulus that is preserved through the earlier stages of the visual system. Oscillation then contributes to maintenance of the synchrony to enhance reliable transmission of the information for higher cognitive processing" [16]. Another important result is that gamma oscillations can be generated from an endogenous source [17].

3.3 Selection Mechanisms

Selective mechanisms operate on the amplified and broadcasted EM patterns, to determine the participants in the spatio-temporal wave that correlates with a conscious episode. Selection processes occur in successive hierarchical steps, beginning at primary sensory areas, and progressing to associative areas, interplay of frontal and posterior areas and finally inter-hemispheric rivalry.

These hierarchical mechanisms provide "top-down" control from higher to lower hierarchical levels, producing a relative stability of the final conscious processes, in spite of changes at the lower levels. Patterns that do not reach consciousness in a given moment may remain activated at the lower processing levels and have a future impact by priming the system.

The selection of patterns involves the balance of excitation and inhibition (see [18]) providing signal-to-noise ratio. This balance is involved in the homeostasis of brain activity, habituation and shifting the focus of attention.

Opponent-processing circuits is a mechanism present in all modalities [19, 20, 21]. One sub-group of neurons in an opponent process circuit is tuned via potentiation due to past learning. An afferent or an endogenous stimulation disinhibits the opposite sub-group by parameter-matching. The homeostasis is disturbed and changes from its baseline setting, and returns to baseline when the external afferent is inactive.

Opponent-processing is based on the interplay of neurotransmitters and modulators. Glutamate-GABA is the major opponent pair for the fundamental balance of excitation and inhibition. Cholinergic modulation of the network shapes the dominant focus and maintains open relays from peripheral signals. Neuromodulators dopamine and serotonin are higher-order controllers of large neuronal networks, influencing the balance of excitation and inhibition related to the selection of patterns. Neuropeptides such as orexin A can control sustainment via action at multiple loci in thalamocortical, corticothalamic and corticocortical circuits.

The selective processes involve large recurrent circuits such as the thalamocortical (see [22]) and the striatum-thalamo-cortical (see [23]) ones; e.g., the basal ganglia pathway to the thalamic reticular nucleus and the hippocampal pathway to cortical layer 1 are involved in gating tonic recurrent loop circuits.

3.4 Calcium Waves and Conscious Processing

Astrocytes receive digital-like signals from neurons and convert them in wavelike patterns, having calcium ions and ATP signaling as the vehicle for

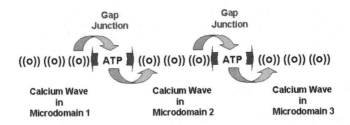

Fig. 3 The "Domino Effect": Neuronal input on astrocyte glutamate metabotropic receptors prompt the release of calcium ions from internal stores (mitochondria and endoplasmatic reticulum) and induce calcium waves in astrocyte microdomain 1; this wave prompts adenosine triphosphate (ATP) signaling through gap junction hemichannels, which induces vibrational states of calcium waves in microdomain 2, and so on.

information processing. Contrary to neurons, one astrocyte can communicate these wave-like patterns to other astrocytes, allowing large-scale wavelike computing.

According to a model developed by De Pittà et al. [24], the dynamics of astrocyte calcium waves may encode information about external stimuli in amplitude and/or frequency modulation. When neuronal excitation reaches a threshold, astrocytic amplitude-modulated calcium waves are produced, allowing the integration of vibrational patterns along a population of cells.

The dynamics of such waves is "saltatory" according to Roth et al. [25], In our modeling, the amplitude-modulated, locally generated wave propagates in the astrocytic syncytium by means of a "domino effect" (Figure 3), interfering with other waves and then promoting an integration of the information embodied in the population of neurons connected to astrocytes.

We have made an approximation of this model with the prospect of a large-scale ion-trap quantum computer proposed by Kielpinski et al. [26] and with the "Astrocentric Hypothesis" advanced by Robertson [27]. According to the last author, conscious perception of a stimulus occurs when astocytic calcium waves integrate neuronal distributed information patterns.

In a theoretical perspective, the astrocytic syncytium can be viewed as a "Global Workspace" (according to the model presented by Baars [28]) that integrates patterns broadcasted from local neuronal assemblies to a brain-wide network, where it is made accessible to other local assemblies, such as motor and emotional systems (Figure 4).

We further suggest that conscious processing mediated by astrocytic calcium waves has a role in the determination of which patterns are more likely to form new memories that can be retrieved later. When a cognitive pattern is reinforced by astrocytic glutamatergic output to NMDA receptors, the chance to form long-term memories and be retrievable in the future increases. Correspondingly, the chance decreases if the pattern is "vetoed" by means of membrane depression. Postsynaptic neuronal

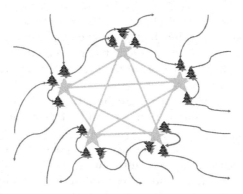

Fig. 4 A Biophysical "Global Workspace" Composed of a Network of Astrocytes Participating in Tripartite Synapses. Blue stars represent astrocytes and red trees represent neurons. Each astrocyte participates in a tripartite synapse, being activated by the presynaptic neuron and contributing to sustain or reduce the activity of the postsynaptic neuron. When neuronal potentials oscillations synchronize, each astrocyte produce a calcium waveform, mediated by ATP signaling. The radial architecture of interconnected astrocytes promotes the emergence of wave interference patterns, supporting the integration of information patterns received from neurons.

membrane potentiation or depression are thus conceived as possibilities of conscious processing having an effect on memory and behavior.

4 Discussing Astrocyte Morphology and Function

Astrocytes in the brain extensively establish a cross-talk with neurons at the synaptic level and form with their neighbors a syncytium that permeates virtually all the brain. One promising alternative to explain how unitary conscious processes are generated from distributed processing in several brain regions and circuits is the participation of astrocytes, operating as local and global integrating units (that we call Local and Master Hub, respectively) parallel to neural connections. At a local level, according to Oberheim et al. [29], "the ability of astrocytes to sense neuronal activity and in turn release gliotransmitters creates a new dimension of communication that might participate in processing of local activity independent of synaptic transmission".

In a large-scale perspective, recent findings about astrocyte morphology in the human brain also support the possibility of astrocytes mediating *global* brain activity, and therefore operating as a biological Global Workspace. The "star-like" name of astrocytes derives from the shape of some of these cells – protoplasmic astrocytes – that function as Local Hubs. Oberheim et al. [29, 30] have also identified other classes of astrocytes that connect the Local ones, composing a large network functioning as a Master Hub:

a) Interlaminar Astrocytes: "In layer 1 of the primate cortex, an area devoid of neuronal cells bodies but highly enriched with synapses, there are interlaminar astrocytes (that) extend striking long, frequently unbranched processes throughout the layers of the cortex, terminating in either layer 3 or 4. The cell bodies of these astrocytes... extend two types of processes: three to six fibers that contribute to the astrocytic network near the pial surface, and another one or two that penetrate deeper layers of the cortex";

b) Polarized Astrocytes: "These essentially unipolar cells reside in the deep layers of the cortex, near the white matter (and) extend one or two long (up to 1 mm in length)... processes away from the white matter";

c) Fibrous Astrocytes: Present in white matter, they "represent a fourth major class of astrocytes that are arguably the least distinguished between primates and non-primate mammals... In contrast to the non-overlapping domain structure of protoplasmic astrocytes, and the respect exhibited by their processes for those of their neighbors, the processes of adjacent fibrous astrocytes intermingle and overlap".

While in their 2006 paper Oberheim et al. [29] remarked that "it is unclear whether a higher level of organization among astrocytic domains exists within functionally defined regions, such as within the barrel cortex", new findings (including a fifth type of human astrocyte) by a larger team (including the same first and last authors, Oberheim and Nedergaard), reported in a 2009 paper [30], suggest that large-scale coordination do exist.

Oberheim et al. [30] report the existence of Varicose Projection Astrocytes (VPA), apparently exclusive to the human brain, presenting "more spiny processes than exhibited by typical protoplasmic astrocytes and typically extended one to five essentially unbranched, millimeter long fibers within the deep layers of the cortex. The processes of the varicose projection astrocytes did not respect the domain organization, because they traveled in all directions, piercing and traversing the domains of neighboring protoplasmic astrocytes. Their process morphology was also intriguing; the evenly spaced varicosities suggest specialized structures or compartmentalization of cellular elements along the great distance of the fibers... We hypothesize based on their distinct morphology that these cells are specialized for long-distance communication across cortical layers or even between gray and white matter." [30]

They also make a suggestion about the function of interlaminar astrocytes: "In light of the length of interlaminar fibers and the large numbers of cells that each fiber may contact in its cortical path, their capacity to respond to both purinergic and glutaminergic stimulation with calcium elevation, and thence to propagate calcium waves, is of potential importance. Of note, the calcium increases in interlaminar cells could be triggered independently in both the cell bodies and fibers. The functions of the interlaminar fibers are unknown, but these traits suggest that they may provide a network for the long-distance coordination of intracortical communication thresholds".

5 Concluding Remarks

If conscious processing requires a Global Workspace and/or a mechanism for large-scale information integration, today there is morphological evidence that evolution provided brains (specially the human one) with a network of cells that might be executing exactly this function. At this moment, this conclusion is inductive, but new findings may contribute to show that the activity of the astrocytic Master Hub correlates with conscious processing (e.g., it is deactivated during dreamless slow-wave sleep, disturbed during loss of consciousness in general anesthesia and generalized epileptic seizures, etc.).

In the context of a brain-based theory of consciousness, there are important questions that might arise, such as: "(a) if humans have more astrocytes, does that have any bearing necessarily on the question of consciousness, and (b) are there more or different astrocytes in brain regions not associated with consciousness, like the cerebellum or spinal cord, compared to those that are, like the ventral stream of the cortex?" (Bernard Baars, personal communication). Fortunately, the findings of Oberheim et al. [29, 30] give good clues about these issues.

The answer to (a) is that humans not only have more astrocytes, but have new kinds of morphologically distinct astrocytes specialized for long-range communication. As long as conscious processing requires this kind of network connectivity, these cells are good candidates to support consciousness. It is too early to conclude that they do or that they do not; empirical research will tell us what they really do.

The answer to (b) is that protoplasmic astrocytes operating as Local Hubs exist in all brain regions, but the astrocytic network that we hypothesize to operate as a Master Hub is restricted to thalamocortical and limbic areas well correlated with conscious processing (for a description of these areas, please see [31]). An exciting research program for the next future would be to check if the activity of the proposed Master Hub overlaps with the correlates of conscious processing proposed by these authors [31].

Acknowledgements. CNPQ for a research grant; Bernard Baars and Gene Johnson for suggestions and criticisms.

References

1. Pereira Jr., A., Furlan, F.A.: Biomolecular information, brain activity and cognitive functions. Annual Review of Biomedical Sciences 9, 12–51 (2007)
2. Pereira Jr., A., Furlan, F.A.: Meta-potentiation: Neuro-astroglial interactions supporting perceptual consciousness. Available from Nature Precedings (2007), http://hdl.handle.net/10101/npre.2007.760.13
3. Pereira Jr., A., Furlan, F.A.: On the role of synchrony for neuron-astrocyte interactions and perceptual conscious processing. Journal of Biological Physics 35, 465–481 (2009)

4. Grossberg, S.: The link between learning, attention and consciousness. Consciousness and Cognition 8, 1–44 (1999)
5. Edwards, J.C.W.: Is Consciousness only a property of individual cells? Journal of Consciousness Studies 12, 60–76 (2005)
6. Damasio, A.R.: Time-locked multiregional retroactivation: A systems-level proposal for the neural substrates of recall and recognition. In: Eimas, P.D., Galaburda, A.M. (eds.) Neurobiology of Cognition, pp. 25–62. The MIT Press, Cambridge (1990)
7. Perry, E., Walker, M., Grace, J., Perry, R.: Acetylcholine in mind: A neurotransmitter correlate of consciousness? Trends in Neuroscience 22, 273–280 (1999)
8. Edelman, G.M.: The Remembered Present: A Biological Theory of Consciousness. Basic Books, New York (1989)
9. Freeman, W.J.: The wave packet: An action potential for the 21st century. Journal of Integrative Neuroscience 2, 3–30 (2003)
10. LeBeau, F.E., Traub, R.D., Monver, H., Whittington, M.A., Buhl, E.H.: The role of electrical signaling via gap junctions in the generation of fast network oscillations. Brain Research Bulletin 62, 3–13 (2003)
11. Bennett, M.V., Zukin, R.S.: Electrical coupling and neuronal synchronization in the mammalian brain. Neuron 41, 495–511 (2004)
12. Steriade, M.: Sleep, epilepsy and thalamic reticular inhibitory neurons. Trends in Neuroscience 28, 317–324 (2005)
13. Galaretta, M., Hestrin, S.: Electrical synapses between GABA-releasing interneurons. Nature Reviews Neuroscience 2, 425–433 (2001)
14. Singer, W.: Search for coherence: A basic principle of cortical self-organization. Concepts in Neuroscience 1, 1–26 (1990)
15. Engel, A.K., Konig, P., Schillen, T.B., Singer, W.: Temporal coding in the visual cortex: New vistas on integration in the nervous system. Trends in Neuroscience 15, 218–226 (1992)
16. Samonds, J.M., Bonds, A.B.: Gamma oscillation maintains stimulus structure-dependent synchronization in cat visual cortex. Journal of Neurophysiology 93, 223–236 (2005)
17. Hermann, C.S., Lenz, D., Junge, S., Busch, N.A., Maess, B.: Memory-matches evoke human gamma-responses. BMC Neuroscience 5(13) (2004)
18. Marino, J., Schummers, J., Lyon, D.C., Schwabe, L., Beck, O., Wiesing, P., Obermeyer, K., Sur, M.: Invariant computations in local cortical networks with balanced excitation and inhibition. Nature Neuroscience 8, 194–201 (2005)
19. Schluppeck, D., Engel, S.A.: Color opponent neurons in V1: A review and model reconciling results from imaging and single-unit recording. Journal of Vision 2, 480–492 (2002)
20. Stecker, G.C., Harrington, I.A., Middlebrooks, J.C.: Location coding by opponent neural populations in the auditory cortex. PLoS Biology 3(3), e78 (2005)
21. Seymour, B., O'Doherty, J.P., Koltzemburg, M., Wiech, K., Frackowiak, R., Friston, K., Dolan, R.: Opponent appetitive-aversive neural processes underlie predictive learning of pain relief. Nature Neuroscience 8, 1234–1240 (2005)
22. Jones, E.G.: The thalamic matrix and thalamocortical synchrony. Trends in Neuroscience 24, 595–601 (2001)
23. Gilbert, P.F.C.: An outline of brain function. Cognitive Brain Research 12, 61–74 (2001)

24. De Pittá, M., Volman, V., Levine, H., Pioggia, G., De Rossi, D., Ben-Jacob, E.: Coexistence of amplitude and frequency modulations in intracellular calcium dynamics. Physical Review E 77, 030903-R (2008)

25. Roth, B.J., Yagodin, S.V., Holtzclaw, L., Russell, J.T.: A mathematical model of agonist-induced propagation of calcium waves in astrocytes. Cell Calcium 17, 53–64 (1995)

26. Kielpinski, D., Monroe, C., Wineland, D.J.: Architecture for a large-scale ion-trap quantum computer. Nature 417, 709–711 (2002)

27. Robertson, J.M.: The Astrocentric Hypothesis: Proposed role of astrocytes in consciousness and memory formation. Journal of Physiology 96, 251–255 (2002)

28. Baars, B.: In the Theater of Consciousness: The Workspace of the Mind. Oxford, New York (1997)

29. Oberheim, N.A., Wang, X., Goldman, S.A., Nedergaard, M.: Astrocytic complexity distinguishes the human brain. Trends in Neuroscience 29, 547–553 (2006)

30. Oberheim, N.A., Takano, T., Han, X., He, W., Lin, J.H.C., Wang, F., Xu, Q., Wyatt, J.D., Pilcher, W., Ojemann, J., Ransom, B.R., Goldman, S.A., Nedergaard, M.: Uniquely hominid features of adult human astrocytes. Journal of Neuroscience 29, 3276–3287 (2009)

31. He, B.J., Raichle, M.E.: The fMRI signal, slow cortical potential and consciousness. Trends in Cognitive Science 13, 302–309 (2009)

The Leyden Jar in Luigi Galvani's thought: A Case of Analogical Visual Modeling

Nora Alejandrina Schwartz

Abstract. In De viribus electricitatis in motu muscolari. Commentarius, Luigi Galvani offers an "analogical modeling" case where he "retrieves" the perceptual structure of the representation of the Leyden experiment pertaining to the electricity domain. In this way Galvani's suspicion and surprise about the existence of an "animal electricity" were strengthened. Using "model based reasoning", Galvani infers that what yields nervous fluid in the frog is the putting of the conductive arc of electricity on it, which also is a source of electricity.

In this paper I will analyze a historical case of scientific research that leads to establishing the "animal electricity" hypothesis. In particular, I will focus on the role that the Leyden Jar and the electrical circuits of which it was part played in Luigi Galvani's thinking. The purpose is to examine to what extent the visualization of those dispositives took part in the abduction of a scientific hypothesis pertaining to Animal Neurophysiology. I am going to show that the images of the Leyden Jar and the electrical circuits mentioned before worked out as visual models. As such, they were useful tools for the analogical solving of problems related to the electrical discharge production in animals. More specifically they were useful tools for the analogical abduction of the explanatory hypothesis about (a) a particular nervous circuit in frogs and (b) the more general fact of the existence of animal electricity.

Cognitive artifacts are representations that can modify the class of computation that a human agent uses to reason about problems [1]. *Models* are a

Nora Alejandrina Schwartz
Facultad de Ciencias Económicas, Universidad de Buenos Aires, Buenos Aires, Argentina, and Facultad de Psicología, Universidad Nacional de La Plata, La Plata, Argentina
e-mail: nora_schwartz@yahoo.com.ar

L. Magnani et al. (Eds.): Model-Based Reasoning in Science & Technology, SCI 314, pp. 637–642.
springerlink.com

case of cognitive artifacts. A very relevant function of models is to represent the world. Similarity is a factor that allows a model to be used for representing something in the relevant aspects and grades. Some models are internal representations, others are external ones [2, 3].

Many scientific practices make use of *"model based reasoning"*. These are problem solving processes that consist in doing inferences from and through models building or models recuperation and manipulation. There are different sorts of model based reasoning. One of them is *"analogical modeling"*. In this one, from the modes of representation of the source domain, relational structures and problem solutions are abstracted and are fitted to the constraints of the target domain of the new problem (4). In the case where the analogical problem solutions have as their subgoal yielding hypothesis, they may be considered *"analogical abductions"*. *"Analogical abductions based on visual models"* are cases of them [5][6].

In *De viribus electricitatis in motu muscolari. Commentarius* from 1791 [7], Luigi Galvani offers an "analogical modeling" case where he "retrieves" the perceptual structure or "image scheme" of the representation of the Leyden experiment pertaining to the electricity domain. He extracts a solution from that perceptual structure and transfers it – fitting it – to the animal physiology domain.

Galvani reports that after a long series of experiments with frogs he came to the conclusion that the muscular contractions phenomenon of the frog limbs must be attributed to electricity (Part I). He notes that contractions do not result directly from external sources of electricity: neither from artificial electricity -represented by the sparks that an electrical machine generates- nor from natural one, i.e., the atmospheric electricity capable to work over animals (Parts I and II). All this makes him wonder what the source of the electrical current is in this particular case. And, in the same work, he remarks:

> But when I brought the animal into a closed room, placed it on an iron plate, and began to press the hook which was fastened in the spinal cord against the plate, behold!, the same contractions and movements occurred as before. I immediately repeated the experiment in different pieces with different metals and at different hours of the day. The results were the same except that the contractions varied with the metals used [...]. These results surprised us greatly and led us to suspect that the electricity was inherent in the animal itself. (Part III, 18)

So, according to this fragment, Galvani makes experiments of electricity discharge using as a conductor arc a metal hook and a plate in contact with a frog and claims that *the frog itself is the source of electricity*. Immediately he becomes interested in the way the nervous fluid follows through the frog and records:

> An observation that a kind of circuit of a delicate nerve fluid is made from the nerves to the muscles when the phenomenon of contractions is produced, similar to the electric circuit which is completed in a Leyden jar, strengthened this suspicion and our surprise. (Part III, 18)

The flow of the nervous fluid of the frogs with which Galvani experiments follows the pattern of a closed circuit. The perceptual structure of the representation of the Leyden experiment has that very same property.

If the target domain has a constraint pattern which describes properties implicit in the perceptual structure of a source domain representation, it is likely that this representation will be retrieved and mapped easily. From a cognitive approach this explains Galvani's evocation of "the electrical circuit that is completed in a Leyden Jar".

How is this circuit? What is a Leyden Jar? Pietr van Musschenbroek from Holand, a famous experimentalist physicist of the Leiden University, was considered the creator of the first electrical condenser named "Leyden Jar". J.A. Nollet spread the innovation publishing an extract of a letter sent to him by van Musschenbroek in the *Memoires* of the French Academy at the beginning of 1746. In that letter van Musschenbroek reports to Nollet an experiment performed with the condenser, which became to be known as the "Leiden experiment" or the "Musschenbroek experiment". For the first time, in this experiment the path of the electrical discharge as a closed circuit becomes perceivable.

> I am going to tell you about a new but terrible experiment which I advise you not to try for yourself . . . I was making some investigations on the force of electricity. For this purpose I had suspended by two threads of blue silk, a gun barrel, which received by communication the electricity of a glass globe that was turned rapidly on its axis while it was rubbed by the hands placed against it. From the other end of the gun barrel there hung freely a brass wire, the end of which passed into a glass flask, partly filled with water. This flask I held in my right hand, while with my left I attempted to draw sparks from the gun barrel. Suddenly my right hand was struck so violently that all my body was affected as if it had been struck by lightning. . . If the flask is placed on a metal support on a wooden table, then the one who touches this metal even with the end of his finger and draws the spark with his other hand receives a great shock. [8]

The "retrieval" of the Leyden experiment representation and the "mapping" between the electrical circuit in it and the electrical circuit that goes through the frog when the muscles of its foot become contracted, strengthened Galvani's suspicion and surprise about the existence of an "animal electricity". According to the notion of "analogical modeling" this is due to the fact that once a model is available and correspondences with the phenomena of the target domain are established, the solution of the domain of the model is extracted and is transferred to the domain of the new problem. Specifically, Galvani knew that putting the conductive arc of electricity on the Jar is the mechanism that yields the electricity flow in the Leyden experiment. Once the Jar is electrified it works as a source of electricity. Making needed adjustments, this solution is translated to the biological domain. Using "model based reasoning", Galvani infers that what yields nervous fluid in the frog is putting the conductive arc of electricity on it, *which also is a source of electricity.*

Galvani transfers to the biological domain not only the explanatory hypothesis of the conductive arc but also its interpretation in terms of the Franklin's theory of a unique electrical fluid.

> One can clearly see how conveniently and neatly this phenomenon can be repeated with a plate that functions as a kind of arc, producing the afore-mentioned circuit [...]". "[...] we do not want to pass over the following information which is particularly apt in revealing the arc's significance, and I might say, capacity for bringing about such muscular contractions; [...]. (Part III, 20)

> (...)From the discovery of a circuit of nerve fluid, (an electric fire, as it were,) it naturally seemed to follow that a two-fold and a dissimilar, or rather an opposite, electricity produces this phenomenon in the same way that the electricity in the Leyden jar or the magic square is two-fold, whereby it releases in these bodies its electric fluid in a circuit. For, as the natural philosophers have shown, a flow of electricity in a circuit can take place only in a restoration of equilibrium and occurs chiefly between two opposite charges. (Part III, 21)

The Leyden experiment was understood within the frame of Franklin's theory of the electrical fluid. The notion of "electrical fluid" or "electrical fire" refers to an electrical substance into the electrified object that conveys or flows to a non electrified object easily.

Franklin held that there is only one sort of electrical fluid and not two (vitrious and resinous) like Dufay had claimed. He thought it existed in all bodies. Bodies appearing void of electrical activity have a normal or in balance amount of fluid which produces no observable effects. The process of electrification consists in taking some of the electrical fluid from a body and bringing it to another. When a neutral object gets electrical fluid, it reaches a positive state, and when a body loses some of its natural amount, it is left in a negative state.

The process of electrification can be produced by conduction or by influence. The former happens when two conductive bodies are touching each other or when they are close enough for a spark to pass between them through the air. Electrification by influence happens in an object which is only near an electrified object.

In the case of the Leyden experiment the inner layer of the Jar, the water, is a conductor of electricity. This layer is positively electrified by contact with an electrical machine. The outer layer of the Jar, for example, a metal support of the glass Jar or a hand, is also a conductor of electricity. This layer is separated from the water by an isolator, the glass, so it cannot be electrified by contact, but it can be electrified by influence: the "plus" of electrical fluid in the inner layer – the water – produces a repelling "influence" or strength on the electrical fluid which is naturally in the outer layer -the metal support;

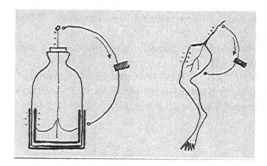

Fig. 1 The frog behaves like a Leyden Jar.
Cf. http://ppp.unipv.it/VoltaGalvani/Media/RanaBott.jpg.

and, if this layer in connected to earth, the fluid is repelled into the earth, leaving the outer layer deficient in fluid, or negatively electrified.

It is possible to restore the balance in the bottle, i.e., the inner and the outer layers of the Jar can be made to have the same quantity of electrical fluid. In order to achieve this, the inner and outer layers of the bottle must be put in communication out of it by a conductor. According to Galvani, in the frogs case, the balance of the animal's own electricity can also be restored by a conductor arc, producing the circuit of nerve fluid.

The "animal electricity" discovery shows that Galvani thought with visual models –material models or "image schemes" from these material models. Galvani made an "analogical abduction based on visual models" that "retrieves" the physical disposition of the Leyden experiment; "translates" the conductor arc explaining mechanism to the animal physiology domain and adjusts this solution to discover "animal electricity". The fact that Galvani made use of perceptual "model based reasoning" allowed him to make these discoveries.

References

1. Giere, R.: Models, metaphysics, and methodology. In: Hartmann, S., Hoefer, C., Bovens, L. (eds.) Nancy Cartwright's Philosophy of Science, pp. 123–126. Routledge, New York (2008)
2. Giere, R.: Scientific cognition as distributed cognition. In: Carruthers, P., Stitch, S., Siegal, M. (eds.) Cognitive Bases of Science, pp. 15–17. Cambridge University Press, Cambridge (2002)
3. Craig, D., Nersessian, N.Y., Catrambone, R.: Perceptual simulation in analogical problem solving. In: Magnani, L., Nersessian, N.J. (eds.) Model-Based Reasoning. Science, Technology, Values, pp. 167–189. Kluwer Academic/Plenum Publishers, New York (2002)
4. Thagard, P.: Computational Philosophy of Science, pp. 60–63. MIT Press, Cambridge (1988)

5. Shelley, C.: Visual abductive reasoning in archaeology. Philosophy of Science 2, 278–301 (1996)
6. Galvani, L.: De viribus electricitatis in motu muscolari. Commentarius. Burndy Library, Norwalk (1953)
7. Roller, D.Y., Roller, D.H.D.: The development of the concept of electric charge. In: Conant, J.B. (ed.) Harvard Case Histories in Experimental Science, vol. 2, pp. 594–595. Harvard University Press, Cambridge (1964)

Modeling the Causal Structure of the History of Science

Osvaldo Pessoa Jr.

Abstract. This paper is an overview of an approach in the philosophy of science of constructing causal models of the history of science. Units of scientific knowledge, called "advances", are taken to be related by causal connections, which are modeled in computers by probability distribution functions. Advances are taken to have varying "causal strengths" through time. The approach suggests that it would be interesting to develop a causal model for scientific reasoning. A discussion of counterfactual histories of science is made, with a classification of three types of counterfactual analyses: (i) in economic and technologic history, (ii) in the history of science and mathematics, and (iii) in social history and evolutionary biology.

1 The Model: Advances Connected by Causal Relations

This paper is part of a project of developing a computational model that describes the history of science. Such a representation stays close to the narrative of the historian of science, who writes about ideas, discoveries, instruments, theories, etc., each of which exerts influences, in differing degrees, on the appearance and confirmation of other scientific advances. These units of scientific knowledge, which are explicitly or tacitly passed among scientists, will be called *advances* (even though they might not be a positive contribution to the progress of science). The prototype of an advance is an idea, but there are other types of theoretical advances, such as explanations, laws, problems, theory development, as well as experimental advances, such as data, experiments, and instruments. Other advances include the comparison

Osvaldo Pessoa Jr.
Department of Philosophy, FFLCH, University of São Paulo, São Paulo, Brazil
e-mail: opessoa@usp.br

L. Magnani et al. (Eds.): Model-Based Reasoning in Science & Technology, SCI 314, pp. 643–654.
springerlink.com
© Springer-Verlag Berlin Heidelberg 2010

between theory and experiment, methodological theses, metaphysical assertions, projects, tacit knowledge, etc.

Advances are connected in certain ways: they influence the *appearance* of other advances, and they also affect the *degree of acceptance* of other advances. In the present approach, such a connection is taken to be a *causal* relation, not a logical one. For example, the construction of the thermopile by Nobili & Melloni in 1830 was essential for the discovery of polarization of radiant heat by James Forbes in 1836: without the thermopile, Forbes would not have discovered polarization at that moment. The thermopile may therefore be considered a "cause" of the proposal that "radiant heat may be polarized", in the sense expressed by the so-called counterfactual definition of causality: if the cause had not occurred, then the effect would not have existed (in the case of a necessary connection), or the probability of its occurrence would have been different.

Causal relations in social systems are always complicated, and one can rarely single out a necessary and sufficient condition. A cause is better represented as an "INUS condition" [9], which amounts to saying, in the above example, that many other causes acted together with the thermopile to lead Forbes to his discovery, and that probably another sufficient set of conditions (not including the thermopile, but perhaps a more sensitive mercury thermometer) could have led to his discovery.

The use of computation to model causation in the history of science has also been explored by Gerd Graßhoff & Michael May, from a different perspective. They investigate how a scientist deploys causal reasoning to construct a causal model of their subject matter, such as done by Krebs & Henseleit for the biochemical pathway underlying the urea cycle [5].

2 Probabilistic Causal Relations Express Possible Histories

Another weakening of these causal relations is that a set of conditions can at best increase the *probability* that a scientist will arrive at a certain advance in a certain interval of time. The great number of causal influences that act haphazardly on a scientist, but cannot be accounted for by the model, are considered as "noise" or random fluctuations, the dispersion of which is encompassed by probability distribution functions.

Figure 1 is an example of how a causal connection may be modeled by a probability distribution function. Advance A1 is Newton's famous experiments with sunlight and prisms, which he reported in 1672. Such investigation was a necessary condition for the discovery of advance A2, that the solar spectrum has dark lines, discovered independently by William Wollaston (1802) and Joseph Fraunhofer (1814). To express the conditional probability of A2, given A1, one may use a gamma distribution, with mean value given by 1808 (the mean between 1802 and 1814) and the standard deviation (the

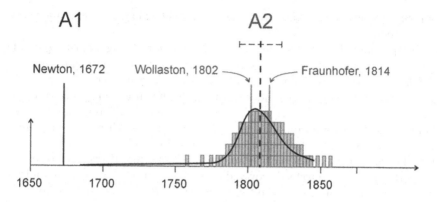

Fig. 1 Probability distribution for the appearance of advance A2, given A1.

half-width of the curve) determined by the spread of the dates of the independent discoveries.

One way to interpret such a distribution function is to think in terms of possible histories of science. Imagine one hundred worlds, created (say in 1673) from the actual world, but with small random changes (for a more detailed recipe for doing this, see [12]). In each of these possible worlds, in which A1 is given, how much time would it take for advance A2 to arise? It is natural to suppose that the time intervals would not be exactly the same, but would be distributed according to a certain curve. In Fig. 1, each possible scenario is represented by a small rectangle, placed in the year in which A2 would appear. The resulting histogram is supposed to be an approximation to the associated distribution function.

3 Causal Structure of Episodes in the History of Science

The present approach to modeling the history of science is being implemented in the LISP-like SCHEME programming language. Computer programs don't provide actual thinking and intuition, but they allow the storage of detailed information concerning the relations between advances and their causal strengths, and allow simulations to be run, which we hope might help to test different metatheoretical theses about the development of science.

Most of our historical studies has focused on the fields of optical spectroscopy and thermal radiation in the 19th century. The ultimate aim is to represent in detail the beginnings of quantum physics. In a preliminary study of the possible paths leading to the birth of the old quantum theory [10], it was suggested that there would be four main paths, the most probable not being the actual one (in the field of thermal radiation), but in the field of optical effects. A simple causal model helped to organize the study, but the conclusion was reached "intuitively", and should be qualified and refined with a more detailed causal model.

Independent discoveries offer interesting material for comparing possible histories of science. We have examined the origins of the science of magnetism in China and in Europe, up to around the 5th century, and constructed a single causal model which accounts for why the rudimentary magnetic compass was developed in China but not in Europe, based on different initial probabilities for the existence of divination techniques in both regions [11]. This followed the account of the historian of science Joseph Needham, for whom it was the widespread use of such techniques in China that allowed the directive property of lodestone to be discovered in the East. In this model, probabilities were assigned by identifying the "empirical time span" between two advances (involving the actual years in which the advances arose) with the mean of the associated distribution function $f(t)$ (such as the one in Fig. 1).

Another example illustrating the causal structure of an episode in the history of science is given in Fig. 2, which represents the actual paths leading to the independent discovery of the principle of spectral reversal, in 19th century spectroscopy (the associated probabilities are not represented in the figure). This principle states that a medium which absorbs well certain spectral lines will also emit well these lines. For example, when sodium gas is excited by an electric arc, it emits several spectral lines, especially the yellow D double lines, which appear strongly on the lower right corner of Fig. 2, in spectrum (a). On the other hand, when sodium gas intercepts light coming from another source, such as the sun, it absorbs strongly the D lines. The principle that good emitters are good absorbers, for each wavelength of light, was discovered independently by Foucault (1848), Ångström (1853), and Kirchhoff (1859), while the latter expressed such a principle as a mathematical law. Foucault's and Kirchhoff's path to discovery involved the curious observation that the sun's dark D lines get even darker as they pass through sodium gas. This is related to the phenomenon of self-reversal, depicted in spectrum (c) of Fig. 2, when the increasing intensity of the bright D line emission leads to a paradoxical darkening of the line, which is explained by the absorption of this line by the surrounding cooler sodium gas.

The same principle of spectral reversion was also discovered by Balfour Stewart (1858) in the field of infrared radiation, which at the time was called "radiant heat", and both he and Kirchhoff generalized the principle to both visible light and infrared radiation, after it became well accepted that both are essentially the same form of radiation, differing only in their wavelengths. A study of how the rate of development of these two fields was influenced by different sets of technological advances is presented in [13].

Fig. 2 illustrates how different pathways of actual discovery can be represented by causal models, involving conjunctions and disjunctions of paths. Further complications must be introduced to represent the causal strengths of advances, which varies with time (see section 5). The model does not include counterfactual scenarios.

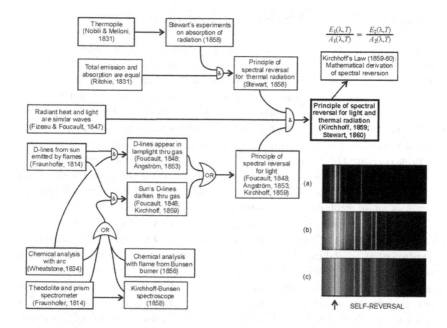

Fig. 2 Simplified causal model for the history of the discovery of the principle of spectral reversal.

There is a clear similarity between the possible histories presented here and the "investigative pathways" explored by David Gooding [3], Frederic Holmes [6], Andrea Loettgers [8], and other historians involved in the model-based reasoning approach to science. The difference involves the scale adopted by each approach.

One may classify the different scales of study in the history of science into at least five groups. (5) Global theses about the scientific institution, involving spans of hundreds of years, all of the world, and all of the scientific fields. (4) General views about the scientific change in a certain field (as done by Kuhn, Lakatos, and Laudan), usually involving decades and a whole civilization. (3) Study of a historical case, such as Darwin's ideas or science in Scotland, etc., involving months or years, and a few institutions. (2) Focus on the procedures adopted by scientists to bring about an advance; this is the scale of ethnomethodological studies, involving hours: "Faraday did this, and then that, etc." (1) Microcognition: cognitive details in the mind of the scientist, involving seconds: studies in this scale are still incipient.

The present approach to causal models in the history of science focuses mainly scale 3, while the investigative pathways are closer to scale 4. The next section will address scale 1.

4 Causal Model of Scientific Reasoning

When a scientist derives a new theoretical result, such a result is usually presented as a logical inference based on other advances. Although the connection between these advances is presented as a logical relation, a consideration of the actual circumstances of the derivation will point out which of the advances are the causes (being previously known), and which one is the effect (the new result). When a scientist justifies a result in deductive form, there are at least two possibilities for the causal history of the result: either the premisses are the actual causes of the conclusion (so the scientist actually discovered the conclusion by deductive inference from the premisses), or the conclusion was previously accepted by the scientist and led him to formulate a premiss as an explanatory hypothesis, in an abductive inference.

The present approach sees a scientist as a very complex cognitive machine that receives a large number of advances (with changing degrees of acceptance) as causal inputs and generates new advances, which will causally affect himself and other scientists. Although the present approach should impose no requirements on how human beings think, it would be interesting for the completeness of the programme if the human mind could be modeled in strict causal terms. This would satisfy a certain "causal closure" of the world, but this expression should allow for the possibility that truly stochastic, non-deterministic events could occur in nature (the existence of such events is an open question in the philosophy of physics). The present author would love to give at least a sketch of the project of describing scientific reasoning (especially abduction) in causal mechanical terms, but he has yet no clue of how this could be done, although many contributions to the "model-based reasoning" community seem to be relevant for this Hobbesian dream.

5 Causal Strength of an Advance

One must also take into account the "strength" of the causal relation. The time interval between the appearance of the first advance (the cause A1) and of the second (the effect A2) is an indication of this strength: the shorter the time, the stronger the cause. Another aspect of this concept of causal strength is that it is a measure of the degree of acceptance of the advance, and it varies with time, as scientists discuss its merits. If the advance is an idea, this discussion might involve debating its degree of confirmation, which affects the degree of acceptance of the idea. If the advance is a new instrument, different scientists must investigate its performance, which then affects how trustworthy are its measurements. If the advance is a problem, then its strength reflects how many scientists are concerned with it.

The *causal strength* of an advance may thus be defined as the potentiality that it may influence the appearance of other advances, or that it may affect the causal strength of other advances (mediated, of course, by the brains and hands of scientists, and by their social and institutional interactions) [14]. When working with causal models in the history of science, an advance should always be considered together with an estimate of its causal strength.

Fig. 3 shows graphically how the causal strength (represented by the thickness of the "strip") of a few advances in 19th-century research on optical spectroscopy and radiant heat developed through time. Advances that are successful usually start out with little support and gradually become widely accepted, such as depicted in strip (d). Some don't have a monotonic growth, such as the thesis that the dark lines in the solar spectrum originate in the solar atmosphere (strip c). This view was suggested by David Brewster and others in the early 1830's. However, during the solar eclipse of 1836, James Forbes concluded, from his observation of the spectra arising from the solar corona, that the dark lines in the solar spectrum do not arise in the sun's atmosphere. Brewster & Gladstone repeated this negative point on the eve of Kirchhoff's discoveries, who in 1859 argued convincingly that the dark lines of the solar spectrum are not caused by the earth's atmosphere, but originate from the presence of those substances in the glowing solar atmosphere.

Many advances have their causal strength going to zero, such as the thesis that radiant heat is of a different nature from visible light, after 1872 (strip b in Fig. 3).

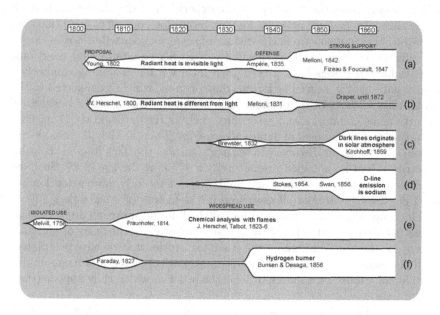

Fig. 3 Change in casual strength through time, for a few advances in 19th-century research on optical spectroscopy and radiant heat.

6   Counterfactual Histories of Science

We have extracted from the historian's discourse a model of science in which advances are connected by probabilistic causal relations. By defining such causation in counterfactual terms, we have automatically introduced the controversial notion of "counterfactual" or "virtual history". A counterfactual situation is a *possible* situation that did not happen. Is it really necessary to introduce counterfactual possibilities in a causal description of history? One can always choose to avoid counterfactual statements, but it can be argued that every causal statement implicitly implies counterfactual scenarios. For example, if one asserts that the main cause of the decline of science in France around the 1830's was its centralized organizational structure, then one is implicitly asserting that *had* such a structure *been* transformed into a more decentralized structure, as in the German countries, then French science *would have* thrived.

Counterfactual scenarios in history are always speculative, but so is the postulation of causes. In the hard sciences, a causal statement may be tested by exploring the different outcomes of an experiment for each value of the parameters controlled by the scientist. These experiments map out the possibilities of outcomes of the experimental situation, so it might be said that the possible histories (describing the measurement outcomes of the experiment) are all *actual*, since the "history" of the experimental situation repeats itself many times. In the case of social history, repetitions of sets of conditions are quite rare (but not so rare in the history of science), so attributions of causes are difficult to test, remaining speculative, just like counterfactual assertions.

Still, human beings have a very good intuition for imagining counterfactual situations [16], as well as for imagining causes, which is connected to the evolutionary fitness value for predicting the future. Historians of science frequently make counterfactual assertions, such as Harry Woolf's ([17],p. 628) comment that the pioneer of flame spectroscopy, Thomas Melvill, "was clearly on the road to major discovery in science" (such as the dark lines in the solar spectrum), had he not died prematurely in 1753, at the age of 27. Such assertions are usually made in a marginal way, but recently more attention has been given to counterfactual assertions in the history of science (see [15]).

The counterfactual histories to be sketched in our approach are very close to factual history, and much of the research investigates the delay or anticipation of an advance. Throughout the possible histories that we have postulated, each advance maintains its identity (i.e., we neglect changes of meaning due to different contexts); what changes is the order in which they appear (their causal path).

A counterfactual scenario is a possible situation that did not actually happen. But what is a "possible" situation? For our purposes, we will not be concerned with logical possibilities, as is common in the metaphysics of possible worlds, but with what has be called "temporal possibilities" (or "causal possibilities"). We start by considering that our future is "open", and the different future possibilities are partially dependent on our choices and on random events in the physical world. (If the universe were strictly deterministic, then there would be only one temporally possible scenario for the future, and only one possible history of science.)

Granted this, we can define a possible scenario as *a future possibility at some instant t_0 of the past*. According to this definition, a counterfactual history must be defined in relation to a branching time t_0 in the past (the time when the counterfactual situation "branched off" from the actual history of science). The probability attributed to a counterfactual state of affairs usually changes according to the branching time being considered.

One might ask whether it would be causally possible that bacteria were discovered on Earth without the use of optical microscopes. Suppose that there were no way of producing glass on Earth; then it is plausible to speculate that bacteria would have been discovered by some other path, not involving optical microscopes. However, there is no instant t_0 in the past from which a possible world without glass could branch off (unless, maybe, if we go back to a time close to the Big Bang). Therefore, such a scenario is not causally possible, although it is physically possible (in the sense that it doesn't violate any law of physics) and logically possible.

The notion of a "tree of possible histories" is useful in philosophy of science for clarifying different conceptions of scientific progress, such as the more traditional one of convergence to the truth (Popper, etc.) and the more relativist conception of selection of the fittest theory (Kuhn) (see [12]).

7   Counterfactual Scenarios in Different Fields

There are at least three different types of counterfactual analyses that may be done in the historical sciences. The most fruitful one comes from the field of economic history, starting with the work of Robert William Fogel [2] on railroads and the economic growth of the United States in the 19th century. There was a traditional conception that the railroads were indispensable to the American progress in the 19th century, i.e., they were a necessary cause for this progress. Fogel examined this thesis, and calculated in detail the costs and the efficiency of other alternatives, and concluded that if railroad technology were not available at the time, there was an equally efficient alternative which was transportation in waterways. According to his calculations, the gross national product that the United States in fact attained in January 1, 1890, would have been reached without railroads (but with waterways) only three months later! The option for waterways would make use of the navigable rivers and lakes, the canals already built, and also many new canals. The industrial regions that would develop would be partially different from the ones that have in fact developed in our actual world.

What allows economic calculations of plausible counterfactual scenarios is the possibility of making reasonably accurate quantitative predictions about the future. For example, the government may open a bid for a contract on an alternative form of energy, so different engineering projects may be presented, each with a possible scenario for the future. After one of them is chosen and implemented, the non-realized projects will have become counterfactual histories (since they were future possibilities at a time in the past). These counterfactual scenarios will be more accurate than the original projects, since hindsight includes information about how the circumstances actually evolved. These two elements, *predictability* and *hindsight*, make counterfactual assessments quite plausible in economic and technologic history.

A second type of counterfactual analysis is done in the history of science and mathematics. Here, the postulation of counterfactual scenarios is less accurate than in economic history, since there is no way of predicting the future of science, contrary to what happens, to a certain extent, in engineering, technology, and economics. One may predict situations related to science policy, but one cannot predict what new discoveries will be made.

However, there is distinguishing feature in the development of science and mathematics which is its *objectivity*. To put it in simple terms, natural science is an attempt to mirror reality, so this reality (which is invariant across the possible worlds) constrains the appearance of scientific advances. In more general terms, without such a commitment to scientific realism (but only to objectivity), there are "attractors" in science, mathematics, and technology (be it reality, consistency, subjective categories, material determinations, or whatever) which constrain the formulation of these disciplines. In almost all causally possible worlds, branching say after the year 1800, scientists would have discovered that the molecule responsible for inheritance has the structure of a double helix, so in this sense there is a common attractor acting on these possible histories of science.

With the advantage of *hindsight* and of the present knowledge of the field, we now know (to a large extent) what the scientists of the past were close to discovering. This allows us to imagine to what consequences slight modifications in the circumstances and choices surrounding the scientists could lead. We may conjecture what could be the different possible paths leading to a discovery, such as the quantization of energy [10]. We can investigate what consequences would have arisen if an advance appeared before or after the time it actually appeared.

But what would be the use of postulating counterfactual histories of science, of generating them with the help of a computer? Without postulating counterfactual scenarios, a lot could be done with detailed causal models, such as testing different metatheoretical theses. But if we were able to generate counterfactual scenarios that are plausible to the historian's intuitions, that would indicate that the theory of science behind these models is well constructed, and that is the ultimate aim of the present project: to contribute to a testable theory of science.

A third type of counterfactual analysis occurs in social, political, and cultural history, in the approach known as "virtual history". Here, however, the constraints are much weaker than in the two previous types: one does not have an economical rationality which allows to predict with some detail the collective choices of the agents, and neither a strong attractor as in science, mathematics, and technology. For example, what would have happened if the shot that killed John F. Kennedy had missed him? Our knowledge of human behavior tells us, for sure, that he would have immediately taken cover, and then left Houston, but what next? The number of possible scenarios increases immensely. A few events, such as the presidential election of 1964, would seem predictable: in this counterfactual scenario, Kennedy would have a high probability of being reelected. But after that, would the United States remain at war against Vietnam? Many have given their opinion, but there is no consensus (see [7]). The best one could do would be to attribute a probability around 1/2 for each alternative, but that would lead nowhere, since subsequent events would also be unpredictable.

Much more could be said about virtual history, but let us consider a final case of counterfactual reasoning, which arises in biological evolution. Biologists such as Stephen Jay Gould [4, ch. 5], Stuart Kauffman and Richard Dawkins [1, pp. 482–93] have examined the question of how biological evolution on Earth would take place if the "tape of evolution" were run back to a moment of the past, and if random variation made living beings evolve in different directions. The consensus is that the species that would appear on earth would be quite different from the present ones, and what we define as the human species would not appear for branching times earlier than a few million years ago. The paleontologist Dale Russell and the geologist S. Conway Morris have speculated on what could have happened if a great meteor had not fallen on Earth 65 million years ago, extinguishing the dinosaurs. Maybe a descendent of the troodont would have become as intelligent as we are, and be doing philosophy of science by now. Notice, however, that in spite of the great divergence in variation (although there are constraints to this), it is reasonable to suppose that intelligent beings would eventually inhabit the Earth, which is an example of convergent evolution. One may say that environmental niches act as attractors to the development of biological structures, or "ecological types". The postulation of counterfactual evolutionary histories would depend on knowledge of what variations are possible and on how selective pressures act (a knowledge that has apparently already been achieved). However, the number of possible branches would be huge, contrary to the case of appearance of advances in the history of science, and to the rational possibilities in economic history, but similarly to virtual history and to the outcome of a sports game.

References

1. Dawkins, R.: The Ancestor's Tale. Weidenfeld & Nicolson, London (2004)
2. Fogel, R.W.: Railroads and American Economic Growth. Johns Hopkins Press, Baltimore (1964)

3. Gooding, D.: Mapping experiment as a learning process: How the first electro-magnetic motor was invented. Science, Technology and Human Values 15(2), 165–201 (1990)

4. Gould, S.J.: Wonderful Life. W.W. Norton, New York (1989)

5. Graßhoff, G., May, M.: Hans Krebs' and Kurt Henseleit's laboratory notebooks and their discovery of the urea cycle – Reconstructed with computer models. In: Holmes, F.L., Renn, J., Rheinberger, H.-J. (eds.) Reworking the Bench, pp. 269–294. Kluwer, Dordrecht (2003)

6. Holmes, F.L.: Laboratory notebooks and investigative pathways. In: Holmes, F.L., Renn, J., Rheinberger, H.-J. (eds.) Reworking the Bench, pp. 295–308. Kluwer, Dordrecht (2003)

7. Kunz, D.: Camelot continued: What if John F. Kennedy had lived? In: Ferguson, N. (ed.) Virtual History, pp. 368–391. Picador, London (1997)

8. Loettgers, A.: Exploring concepts and boundaries of experimental practice in laboratory notebooks: Samuel Pierpont Langley and the mapping of the infra-red region of the solar spectrum. In: Holmes, F.L., Renn, J., Rheinberger, H.-J. (eds.) Reworking the Bench, pp. 159–182. Kluwer, Dordrecht (2003)

9. Mackie, J.L.: Causes and conditions. American Philosophical Quarterly 2, 245–264 (1965)

10. Pessoa Jr., O.: Counterfactual histories: The beginning of quantum physics. Philosophy of Science (Proceedings) 68, S519–S530 (2001)

11. Pessoa Jr., O.: Computation of probabilities in causal models of history of science. Principia (Florianópolis, Brazil) 10(2), 109–124 (2006)

12. Pessoa Jr., O.: Scientific progress as expressed by tree diagrams of possible histories. In: Mortari, C.A., Dutra, L.H.A. (eds.) Anais do V Simpósio Interna-cional Principia, pp. 114–122. Núcleo de Estudos da Linguagem - Universidade Federal de Santa Catarina, Florianópolis (2009a)

13. Pessoa Jr., O.: Independent discoveries following different paths: The case of the law of spectral reversion (1848-59). In: Cattani, M.S.D., Crispino, L.C.B., Gomes, M.O.C., Santoro, A.F.S. (eds.) Trends in Physics, pp. 259–282. Livraria da Física, São Paulo (2009b)

14. Pessoa Jr., O.: The causal strength of scientific advances. Forthcoming. In: Videira, A.A.P., Krause, D. (eds.) Brazilian Studies in the History and Philos-ophy of Science. Boston Studies in the Philosophy of Science. Springer, New York (2010)

15. Radick, G.: Introduction: Why what if? Isis 99, 547–551 (2008)

16. Roese, N.J., Olson, J.M. (eds.): What Might Have Been: The Social Psychology of Counterfactual Thinking. Erlbaum, Mahwah (1995)

17. Woolf, H.: The beginnings of astronomical spectroscopy. In: Cohen, I.B., Taton, R. (eds.) L'Aventure de la Science, Mélanges Alexandre Koyré, Hermann, Paris, vol. 1, pp. 619–634 (1964)